Design and Control Advances in Robotics

Mohamed Arezk Mellal
M'Hamed Bougara University, Algeria

A volume in the Advances in Computational Intelligence and Robotics (ACIR) Book Series

IGI Global
PUBLISHER of TIMELY KNOWLEDGE

Published in the United States of America by
IGI Global
Engineering Science Reference (an imprint of IGI Global)
701 E. Chocolate Avenue
Hershey PA, USA 17033
Tel: 717-533-8845
Fax: 717-533-8661
E-mail: cust@igi-global.com
Web site: http://www.igi-global.com

Library of Congress Cataloging-in-Publication Data

Names: Mellal, Mohamed Arezki, editor.
Title: Design and control advances in robotics / Mohamed Arezki Mellal, editor.
Description: Hershey PA : Engineering Science Reference, [2023] | Includes bibliographical references and index. | Summary: "The goal of this book is to present the most recent applications and design advances in robotics, highlighting the latest developments and applications to further the research in the field to assist many domains such as industry and medicine"-- Provided by publisher.
Identifiers: LCCN 2022023418 (print) | LCCN 2022023419 (ebook) | ISBN 9781668453810 (hardcover) | ISBN 9781668453827 (paperback) | ISBN 9781668453834 (ebook)
Subjects: LCSH: Robotics. | Robots--Control systems.
Classification: LCC TJ211 .D46 2023 (print) | LCC TJ211 (ebook) | DDC 629.8/92--dc23/eng/20220812
LC record available at https://lccn.loc.gov/2022023418
LC ebook record available at https://lccn.loc.gov/2022023419

This book is published in the IGI Global book series Advances in Computational Intelligence and Robotics (ACIR) (ISSN: 2327-0411; eISSN: 2327-042X)

British Cataloguing in Publication Data
A Cataloguing in Publication record for this book is available from the British Library.

For electronic access to this publication, please contact: eresources@igi-global.com.

Advances in Computational Intelligence and Robotics (ACIR) Book Series

Ivan Giannoccaro
University of Salento, Italy

ISSN:2327-0411
EISSN:2327-042X

Mission

While intelligence is traditionally a term applied to humans and human cognition, technology has progressed in such a way to allow for the development of intelligent systems able to simulate many human traits. With this new era of simulated and artificial intelligence, much research is needed in order to continue to advance the field and also to evaluate the ethical and societal concerns of the existence of artificial life and machine learning.

The **Advances in Computational Intelligence and Robotics (ACIR) Book Series** encourages scholarly discourse on all topics pertaining to evolutionary computing, artificial life, computational intelligence, machine learning, and robotics. ACIR presents the latest research being conducted on diverse topics in intelligence technologies with the goal of advancing knowledge and applications in this rapidly evolving field.

Coverage

- Robotics
- Intelligent Control
- Machine Learning
- Pattern Recognition
- Computational Logic
- Adaptive and Complex Systems
- Automated Reasoning
- Brain Simulation
- Agent technologies
- Heuristics

IGI Global is currently accepting manuscripts for publication within this series. To submit a proposal for a volume in this series, please contact our Acquisition Editors at Acquisitions@igi-global.com or visit: http://www.igi-global.com/publish/.

The Advances in Computational Intelligence and Robotics (ACIR) Book Series (ISSN 2327-0411) is published by IGI Global, 701 E. Chocolate Avenue, Hershey, PA 17033-1240, USA, www.igi-global.com. This series is composed of titles available for purchase individually; each title is edited to be contextually exclusive from any other title within the series. For pricing and ordering information please visit http://www.igi-global.com/book-series/advances-computational-intelligence-robotics/73674. Postmaster: Send all address changes to above address.

List of Reviewers

Doğukan Akgöl, *Dokuz Eylül University, Turkey*
Sameer Al-Dahidi, *German Jordanian University, Jordan*
Ranjit Barua, *IIEST-SHIBPUR, India*
Gökçe Biricik, *Dokuz Eylul University, Turkey*
Ana María Castillo-Rivera, *Jaén University, Spain*
Salvador Castillo-Rivera, *University of Cantabria, Spain*
Arghya Dey, *Engineering and Architecture College, India*
Ahona Ghosh, *Brainware University, India*
Mehmet Mert İlman, *Celal Bayar University, Turkey*
Davut Izci, *Batman University, Turkey*
Nikolay Karabutov, *Mirea Russian Technological University, Russia*
Showkat Ahmad Lone, *Saudi Electronic University, Saudi Arabia*
Md. Masud Rana, *Memorial University, Canada*
Debanik Roy, *Homi Bhabha National Institute, India*
Usama Saqib, *Aalborg University, Denmark*
Hamza Taş, *Manisa Celal Bayar University, Turkey*
Pelin Yildirim Taşer, *İzmir Bakırçay University, Turkey*
Edward J. Williams, *University of Michigan, Dearborn, USA*
Şahin Yavuz, *Dokuz Eylül University, Turkey*

Table of Contents

Detailed Table of Contents

Robot perception is the ability of a robotic platform to perceive its environment by the means of sensor inputs, e.g., laser, IMU, motor encoders, and so on. Much like humans, robots are not limited to perceiving their environment through vision-based sensors, e.g., cameras. Robot perception, through the scope of this chapter, encompasses acoustic signal processing techniques to locate the presence of a sound source, e.g., human speaker, within an environment for human-robot interaction (HRI), that has gained great interest within scientific community. This chapter will serve as an introduction to acoustic signal processing within robotics, starting with passive acoustic localization and building up to contemporary active sensing methods, such as the usage of neural networks and spatial map generation. The origins of active acoustic localization, which finds its roots in biomimetics, are also discussed.

Soft robots have gained superiority in outdoor applications compared to traditional robots today. This advantage is clearly due to bio-inspiration and evolving material technology. The objective of this research is to use nanotechnology to improve material qualities. For this, silicone named DragonSkin 20 (DS20), which can be employed in soft robot applications, was selected as the matrix material, while functionalized multi-walled-carbon-nanotube (MWCNT) was utilized as an additive. One of the parameters that determine the mechanical properties is the change of curing behavior. The choice of mixing technique, on the other hand, is very crucial since it affects the curing behavior. For this reason, the effects of not only the additive but also the various mixing techniques on the material behavior and curing time were reported as a result of the experiments. The results showed that the mixing methodologies plays an important role on the mechanical properties and curing time of neat and MWCNT reinforced silicone.

In this study, active vibration control of a single-link flexible manipulator has been performed. Residual vibrations affect the end-point positioning accuracy of the manipulator. To control residual vibration amplitudes, closed loop control scheme has been effectuated. Velocity feedback of the desired endpoint has been carried out and traditional proportional-integral (PI) Control has been applied. The mathematical model of the flexible manipulator has been established by the finite element method (FEM). The FEM results in modal analysis have been verified by the experimental results and then, the transient analysis has been realized by the Newmark method. Different velocity profiles have been applied to the manipulator to demonstrate the effect of PI controller on vibration amplitudes. The root mean square values and steady-state time have been calculated presented for different cases. It is observed from the results that the residual vibration amplitudes of the manipulators are successfully suppressed by the proportional-integral control gains.

Currently, numerous robotically controlled steerable needle-based methods have been offered, which can precisely maintain needle tip deflection to monitor premeditated curved paths inside soft tissue. Percutaneous interventions via MIS methods can deliver patients with improved consequences and earlier recovery times than traditional or open surgeries. Precise needle insertions are important for effective procedures, and dynamically piloted needles can intensify system exactness. Current robotic technology can overcome individuals' limitations and upturn the attainment of minimally invasive percutaneous treatment. Needle insertion can also be applied as a "satellite access" area, and parallel calculus can be pushed or broken into the renal pelvis. The main access is applied for the removal of stones, in that way avoiding the formation of various channels and efficiently caring for kidney function. Here, the authors will present the present state-of-the-art insertion needles for MIS, concentrating on the limitations and challenges still open for their application in the medical area.

Modern science and technology development enables humans to control devices using their brains. brain computer interface (BCI) system using EEG (Electroencephalogram) is a non-invasive method that uses brain signals to control robots. Brain control robots can assist people with disabilities to improve their livelihood. In our proposed work to control wheelchair movement for rehabilitation using EEG, electrodes are first placed on the subject's scalp to acquire brain signal activity. After that, the signals got filtered using fast fourier transformation (FFT) method, and then features got extracted using power spectral density (PSD). Random forest is used for the classification of wheelchair movement. For this purpose, a publicly available dataset from Kaggle is used, and an average accuracy of 96.79 is achieved. The proposed architecture has outperformed all the existing ones in its concerned domain; thus, it is suitable, cost-effective, and flexible for the users, which also helps maintain user privacy.

In this work, a fractional order internal model controller (FOIMC) is developed for the level control of a coupled tank system. The parameters of the FOIMC are optimized by the particle swarm optimization (PSO) algorithm. The controller parameters obtained by the PSO algorithm, are compared with the gray wolf optimizer (GWO) and the whale optimization algorithms (WOA), and the PSO algorithm gave better results. The performance of the proposed FOIMC is compared with the state of the art. The experimental results obtained from the laboratory setup of the cylindrical tank system using the proposed FOIMC are also presented. The step responses from the simulation and the experimental results, indicated that the proposed FOIMC has good time domain performance than the state of the art. The proposed controller showed better robustness to a change in the operating point. The novelty of the present work is, for the first time, a comparative study of latest swarm algorithms is done for the tuning of FOIMC parameters.

Lyapunov exponents (LE) are one of the most effective tools for analyzing the quality of systems in robotics. They are used to optimize the robot work, evaluate the quality of their work, and solve various tasks in control systems with robots. Calculating the full spectrum of LE is a complex problem. It relates to the identifiability recoverability and detectability of LE. The identifiability recoverability and detectability issues of Lyapunov exponents were not considered. This problem is relevant. The authors propose an approach to verify these characteristics for the linear dynamical system. It bases on the analysis of geometric frameworks (GF) that depends on the structural properties coefficient (SPC). The SPC reflects the change in Lyapunov exponents, and GF guarantees decision-making on the LE type. They obtain (1) conditions of fully detectable LE (these conditions correspond to the determination of an indicators complete set in the robot control system) and (2) s-detectability conditions with level v-no recoverability if the system contains no recoverable lineals.

The number of vehicles on the roads worldwide grows annually. With the increase in the population of on-road vehicles comes the increase in traffic-related problems such as emissions, traffic accidents, traffic jams, and increased fuel consumption, to mention but a few. The field of robotics presents the autonomous robotic platooning concept, which presents a very promising future for the transportation industry in general and intelligent transportation industry, in particular. With autonomous platooning, efficient road usage, reduced fuel consumption and emission are possible. This promising field, however, faces issues and challenges barring the deployment of autonomous platoons. In this chapter, the authors introduce the user to the concept of autonomous platooning, present the current state-of-the-art of autonomous robotic platoons and issues that they still face, and ultimately, discuss their progress and future trends.

This investigation shows the implementation of parametric accelerated life testing (ALT) as established. It involves (1) a product BX lifetime that X% of a system population is unsuccessful with ALT plan, (2) fatigue design, (3) ALTs with modifications, and (4) discernment as to whether the design(s) fulfills the objective BX lifetime. As a case study, pneumatic cylinder was investigated. The pneumatic cylinder was unsuccessful in a manufacturing line. To reproduce it, parametric ALT was carried out. At the first ALT, the metal seal made of nickel-iron alloy partially cracked and chipped with metal sound. As an action plan, material of the seal was altered from metal to silicone rubber. At the second ALT, due to seal hardening and wear, the piston seal leaked gas. The failure of the silicone seals in the laboratory tests were alike to those returned from the marketplace. The seal material was changed from silicone rubber to (thermoset) polyurethane. For the third ALT, there were no issues. Finally, the lifetime of the pneumatic cylinder was manifested to have a B1 life of 10 years.

The effectiveness of the welding process depends on some attributes, such as costs, the production time, and the quality of welding joints. This matter is more vital when dealing with the thick plates of steel for special structures in industry. To provide the quality when joining thick plates, especially with narrow gap and small chamfer angle, it is necessary to design the progressive process focusing on the main parameters, including the arc power, the translational velocity, and the rotational velocity of the tip. To evaluate the stability and the reliability of this technique, it must also perform the microstructure analysis in different zones of welding joint. In this chapter, previous publications related to this topic were gathered and analyzed. Particularly, several findings in recent studies for improving this progressive technology in the manufacturing area were discussed in detail.

Artificial intelligence in the healthcare field has gotten the attention of professionals and researchers because the automated analysis of medical images can enhance diagnosis and therapy considerably. Deep learning (DL) puts forwards tools to deal with this kind of problem. Data science is presumable to address significant changes in the study of cardiovascular imaging allowing change appreciably in the next years encouraged by this learning. For medical professionals, it is fundamental to keep track of the development to make sure that it can have a significant impact on clinical practice. DL in medical image analysis is an interdisciplinary area that requires the support of medical experience and computer techniques. Nevertheless, the optimal clinical application of DL needs tricky formulation of problems, selection of the most proper algorithms and data, and balanced interpretation of outcomes. A current challenge comes from echocardiography. The need for human interpretation has restricted echocardiography's capability for precision medicine.

Robotic-assisted surgery is a branch of engineering that develops robotic machines that allow the operator to perform surgery by maneuvering, from a distance, a robot that is not completely autonomous but capable of performing controlled maneuvers. It is a technique that has recently come into use, even in selected centers, and represents a further step in the field of minimally invasive surgery (MIS). It has basically

Preface

Robotics plays a pivotal role in many domains such as industry and medicine. Robots allow for increased safety, production rates, accuracy, and quality; however, robots must be well designed and controlled to achieve the required performance. The design and control of robotics involve many varying disciplines, such as mechanical engineering, electronics, and automation, and must be further studied to ensure the technology is utilized appropriately.

Design and Control Advances in Robotics considers the most recent applications and design advances in robotics and highlights the latest developments and applications within the field of robotics. Covering key topics such as deep learning, machine learning, programming, automation, and control advances, this reference work is ideal for engineers, computer scientists, industry professionals, academicians, practitioners, scholars, researchers, instructors, and students. The book includes 15 chapters.

Robots presently occupy a large place in diverse activities. The robots currently in service are manipulator-type robots. They are often used in modern manufacturing processes to increase the production volume as well as to improve the quality of the product. This type of robot consists of a base, carrier composed of segments, and mobile joints expressed in the degree of freedom (rotoid and prismatic), actuator, sensor, and terminal organ. It is known that these systems have defects, such as insufficient accuracy, very slow reaction time, and instability. Furthermore, to adjust them, control systems are used: classic, dynamic, and adaptive. These commands need geometric, kinematic, and dynamic modeling of the studied systems. Most often, non-linear systems are transformed into linear systems using linearization methods. Chapter 1 presents a short overview of robot manipulator control.

Due to their adaptability, flexibility, and deformability, soft robots have been widely studied in many areas. On the other hand, soft robots have some challenges in modeling, design, and control when compared to rigid robots, since the inherent features of soft materials may create complicated behaviors owing to non-linearity and hysteresis. To address these constraints, recent research has utilized different machine learning algorithms and meta-heuristic optimization techniques. First and foremost, Chapter 2 looks at current breakthroughs and applications in the field of soft robots. Studies in the field are grouped under main headings such as modelling, design and control. Fundamental issues and developed solutions were analyzed in this manner. Machine learning and meta-heuristic optimization-oriented methods created for various applications are highlighted in particular. At the same time, it is emphasized how the problems in each of the modeling, design, and control areas impact each other.

Robot perception is the ability of a robotic platform to perceive its environment by the means of sensor inputs, e.g., laser, IMU, motor encoders and so on. Much like humans, robots are not limited to perceiving their environment through vision-based sensors, e.g., cameras. Robot perception, the scope of Chapter 3, encompasses acoustic signal processing techniques to locate the presence of a sound source, e.g., human speaker, within an environment for human-robot interaction (HRI), that has gained great interest within scientific community. This chapter will serve as an introduction to acoustic signal processing within robotics, starting with passive acoustic localization and building up to contemporary active sensing methods, such as the usage of neural networks and spatial map generation. The origins of active acoustic localization, which finds its roots in biomimetics, are also discussed.

Soft robots have gained superiority in outdoor applications compared to traditional robots today. This advantage is clearly due to bio-inspiration and evolving material technology. The objective of this research is to use nanotechnology to improve material qualities. For this, silicone named DragonSkin 20 (DS20), which can be employed in soft robot applications, was selected as the matrix material, while functionalized multi-walled-carbon-nanotube (MWCNT) was utilized as an additive. One of the parameters that determine the mechanical properties is the change of curing behavior. The choice of mixing technique, on the other hand, is very crucial since it affects the curing behavior. For this reason, the effects of not only the additive but also the various mixing techniques on the material behavior and curing time were reported as a result of the experiments. The results in Chapter 4 showed that the mixing methodologies plays an important role on the mechanical properties and curing time of neat and MWCNT reinforced silicone.

In Chapter 5, active vibration control of a single-link flexible manipulator has been performed. Residual vibrations affect the end-point positioning accuracy of the manipulator. To control residual vibration amplitudes, closed loop control scheme has been effectuated. Velocity feedback of the desired endpoint has been carried out and traditional Proportional-Integral (PI) Control has been applied. The mathematical model of the flexible manipulator has been established by the finite element method (FEM). The FEM results in modal analysis have been verified by the experimental results and then, the transient analysis has been realized by the Newmark method. Different velocity profiles have been applied to the manipulator to demonstrate the effect of PI controller on vibration amplitudes. The root mean square values and steady-state time have been calculated presented for different cases. It is observed from the results that the residual vibration amplitudes of the manipulators are successfully suppressed by the Proportional-Integral control gains.

Currently, numerous robotically controlled steerable needle-based methods have been offered, which can precisely maintain needle tip deflection to monitor premeditated curved paths inside soft tissue. Percutaneous interventions via MIS methods can deliver patients with improved consequences and earlier recovery times than traditional or open surgeries. Precise needle insertions are important for effective procedures, and dynamically piloted needles can intensify system exactness. Current robotic technology can overcome individuals' limitations and upturn the attainment of minimally invasive percutaneous treatment. Needle insertion can also be applied as a "satellite access" area, and parallel calculus can be pushed or broken into the renal pelvis. The main access is applied for the removal of stones, in that way avoiding the formation of various channels and efficiently caring for kidney function. In Chapter 6, the authors will present the present state-of-the-art insertion needles for MIS, concentrating on the limitations and challenges still open for their application in the medical area.

Chapter 7 covers how modern science and technology development enables humans to control devices using their brains. Brain Computer Interface (BCI) system using EEG (Electroencephalogram) is a non-invasive method that uses brain signals to control robots. Brain control robots can assist people with disabilities to improve their livelihood. In our proposed work to control wheelchair movement for rehabilitation using EEG, electrodes are first placed on the subject's scalp to acquire brain signal activity. After that, the signals got filtered using Fast Fourier Transformation (FFT) method, and then features got extracted using Power Spectral Density (PSD). Random forest is used for the classification of wheelchair movement. For this purpose, a publicly available dataset from Kaggle is used, and an average accuracy of 96.79 is achieved. The proposed architecture has outperformed all the existing ones in its concerned domain; thus, it is suitable, cost-effective, and flexible for the users, which also helps maintain user privacy.

In Chapter 8, a fractional order internal model controller (FOIMC) is developed for the level control of a coupled tank system. The parameters of the FOIMC are optimized by the particle swarm optimization (PSO) algorithm. The controller parameters obtained by the PSO algorithm, are compared with the gray wolf optimizer (GWO) and the whale optimization algorithms (WOA), and the PSO algorithm gave better results. The performance of the proposed FOIMC is compared with the state of the art. The experimental results obtained from the laboratory setup of the cylindrical tank system using the proposed FOIMC are also presented. The step responses from the simulation and the experimental results, indicated that the proposed FOIMC has good time domain performance than the state of the art. The proposed controller showed better robustness to a change in the operating point. The novelty of the present work is, for the first time a comparative study of latest swarm algorithms is done for the tuning of FOIMC parameters.

Lyapunov exponents (LE) are one of the most effective tools for analyzing the quality of systems in robotics. They are used to optimize the robots work, evaluate the quality of their work, and solve various tasks in control systems with robots. Calculating the full spectrum of LE is a complex problem. It relates to the identifiability recoverability and detectability of LE. The identifiability recoverability and detectability issues of Lyapunov exponents were not considered. This problem is relevant. We propose an approach to verify these characteristics for the linear dynamical system. It bases on the analysis of geometric frameworks (GF) that depends on the structural properties coefficient (SPC). The SPC reflects the change in Lyapunov exponents, and GF guarantees decision-making on the LE type. Chapter 9 obtains (1) conditions of fully detectable LE (these conditions correspond to the determination of an indicators complete set in the robot control system); (2) s-detectability conditions with level v-no recoverability if the system contains no recoverable lineals.

In Chapter 10, the number of vehicles on the roads worldwide grows annually. With the increase in the population of on-road vehicles comes the increase in traffic-related problems such as emissions, traffic accidents, traffic jams, and increased fuel consumption to mention but a few. The field of robotics presents the autonomous robotic platooning concept which presents a very promising future for the transportation industry in general and intelligent transportation industry, in particular. With autonomous platooning efficient road usage, reduced fuel consumption and emission are possible. This promising field, however, faces issues and challenges barring the deployment of autonomous platoons. In this chapter, we introduce the user to the concept of autonomous platooning, present the current state-of-the-art of autonomous robotic platoons, and issues that they still face, and ultimately then discuss their progress and future trends.

Chapter 11 shows the implementation of parametric accelerated life testing (ALT) as established way. It involves: (1) a product BX lifetime that X% of a system population is unsuccessful with ALT plan, (2) fatigue design, (3) ALTs with modifications, and (4) discernment as to whether the design(s) fulfills the objective BX lifetime. As a case study, pneumatic cylinder was investigated. The pneumatic cylinder was unsuccessful in a manufacturing line. To reproduce it, parametric ALT was carried out. At the first ALT, the metal seal made of nickel-iron alloy partially cracked and chipped with metal sound. As action plans, material of the seal was altered from metal to silicone rubber. At the second ALT, due to seal hardening and wear, the piston seal leaked gas. The failure of the silicone seal in the laboratory tests were alike to those returned from the marketplace., the seal material was changed from silicone rubber to (thermoset) polyurethane. For the third ALT, there were no issues. Finally, the lifetime of the pneumatic cylinder was manifested to have a B1 life 10 years.

The effectiveness of the welding process depends on some attributes, such as costs, the production time, and the quality of welding joints. This matter is more vital when dealing with the thick plates of steel for special structures in industry. To provide the quality when joining thick plates, especially with narrow gap and small chamfer angle, is it necessary to design the progressive process focusing on the main parameters, including the arc power, the translational velocity, and the rotational velocity of the tip. To evaluate the stability and the reliability of this technique, it must also perform the microstructure analysis in different zones of welding joint. In Chapter 12, previous publications related to this topic were gathered and analyzed. Particularly, several findings in our recent studies for improving this progressive technology in the manufacturing area were discussed in detail.

Chapter 13 talks about how artificial intelligence in the healthcare field has gotten the attention of professionals and researchers because the automated analysis of medical images can enhance diagnosis and therapy considerably. Deep Learning (DL) puts forwards tools to deal with this kind of problem. Data science is presumable to address significant changes in the study of cardiovascular imaging allowing change appreciably in the next years, encouraged by this learning. For medical professionals, it is fundamental to keep track of the development to make sure that it can have a significant impact on clinical practice. DL in medical image analysis is an interdisciplinary area that requires the support of medical experience and computer techniques. Nevertheless, the optimal clinical application of DL needs tricky formulation of problems, selection of most proper algorithms and data and balanced interpretation of outcomes. A current challenge comes from echocardiography. The need for human interpretation has restricted echocardiography's capability for precision medicine.

In Chapter 14, robotic-assisted surgery is a branch of engineering that develops robotic machines that allow the operator to perform surgery by maneuvering, from a distance, a robot that is not completely autonomous but capable of performing controlled maneuvers. It is a technique that has recently come into use, even in selected centers, and represents a further step in the field of minimally invasive surgery (MIS). It has basically the same indications but, at the moment, it is reserved for selected patients. Compared to traditional video-assisted surgery, it has some important differences. The surgeon is physically distant from the operating field and sits at a console, equipped with a monitor, from which, through a complex system, he controls the movement of the robotic arms. To these are fixed the various surgical tools, forceps, scissors, dissectors, which a team present at the operating table introduces into the cavity where the surgery is performed.

The domain of Flexible Robotics is gaining momentum in the recent past due to its widespread advantages despite major technological barricades, such as in-situ vibration and joint instability. Flexible Robotic System (FRS) pertains to unique designs of serial-chain assistive robots that exhibit the said inherent vibration impromptu. This chapter focuses on the control system algorithm of a prototype serial-chain FRS (PARv1.0), backed up by the indigenous firmware of the system controller and run-time software. Chapter 15 also brings out the parametric relationship of the control system hardware over the electromechanical output of the FRS. The design function architecture of the prototype multi-degrees-of-freedom FRS, highlighted in this chapter, is a rare combination of three revolute & one prismatic (RRRP) joints intended for pick and place applications, having a 1500 mm radial span reach in the horizontal plane. A prototype miniature gripper actuated by a mini-servomotor is another novel development of this FRS.

Mohamed Arezk Mellal
M'Hamed Bougara University, Algeria

Chapter 1
Basic Concepts of Manipulator Robot Control

Sara Benameur
M'Hamed Bougara University, Algeria

Sara Tadrist
M'Hamed Bougara University, Algeria

Mohamed Arezki Mellal
https://orcid.org/0000-0003-0667-8851
M'Hamed Bougara University, Algeria

Edward J. Williams
University of Michigan, Dearborn, USA

ABSTRACT

Robots presently occupy a large place in diverse activities. The robots currently in service are manipulator-type robots. They are often used in modern manufacturing processes to increase the production volume as well as to improve the quality of the product. This type of robot consists of a base, a carrier composed of segments, mobile joints expressed in the degree of freedom (rotoid and prismatic), actuator, sensor, and terminal organ. It is known that these systems have defects, such as insufficient accuracy, very slow reaction time, and instability. Furthermore, to adjust them, control systems are used: classic, dynamic, and adaptive. These commands need geometric, kinematic, and dynamic modeling of the studied systems. Most often, non-linear systems are transformed into linear systems using linearization methods. This chapter presents a short overview of robot manipulator control.

INTRODUCTION

Robotics is one of the sciences of the automaton, which represent techniques that allow the design and construction of automatic machines and robots. The robot is an automated device capable of handling objects or performing operations painful, tedious, and/or exhausting to the human being according to a fixed, modifiable or adaptable program. For the design, simulation or control of robots, it is necessary to have a model of their mechanism.

DOI: 10.4018/978-1-6684-5381-0.ch001

Robots can be classified into several types; in this chapter, we will focus on a single type called robotic manipulator. Several sectors of the industry are based on the use of robot manipulation (industrial) as is the case in the automotive industry or in shipbuilding (Sayers, 1999). They play an important role in reducing production costs, improving accuracy, quality, productivity and efficiency. The control of a rigid robotic manipulator is confronted with major difficulties such as non-linear, coupled behavior and time varying. In addition, the dynamic model of the system remains uncertain, particularly with regard to external disturbances and the uncertainty parameter, so that linear control methods are ill-adapted to strongly coupled behaviors, nonlinear and time-variables; and improvement thereof is the purpose of this work. A review of the literature on robotics can be found in Refs (Billard & Kragic, 2019; Cui & Trinkle, 2021).

The remainder of the chapter is organized as follows: Section 2 defines the manipulator robot. Section 3 describes the components of the manipulator robot. Section 4 illustrates modeling of these robots. Sections 5 and 6 discuss the control and linearization in manipulator robots. Section 7 illustrates an example. Finally, the last section concludes the chapter.

DEFINITION OF MANIPULATOR ROBOTS

A manipulator robot is a mechatronic device (combining mechanical, electronic, automation, and computer science) which automatically performs various tasks. It is a system designed to autonomously perform repetitive tasks in a fixed and ordered environment in which movements recorded in a memory are repeated cyclically (Isermann, 2005). It is position controlled, reprogrammable, and versatile with several degrees of freedom, capable of handling materials, parts, tools and specialized devices during the variable and programmed motion for execution. It is generally composed of two distinct subsets: a terminal organ and an articulated mechanical structure consisting of a set of solids, generally, one after the other where each solid is mobile relative to the previous one (see Figure 1) (Qin, 2013). This mobility is expressed in degrees of freedom.

Figure 1. Manipulator robots.

COMPONENTS OF MANIPULATOR ROBOTS

The different parts of a manipulator robot are illustrated in Figure 2.

Figure 2. Parts of a manipulator robot.

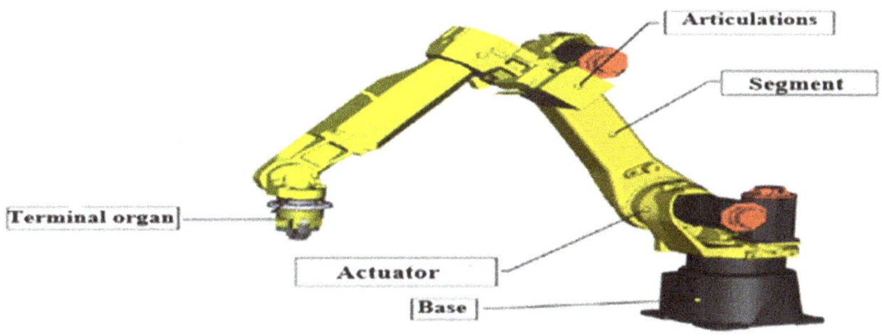

Base

The base of the manipulator is fixed at the workplace.

Support

This is the essential part of the articulated mechanical system (Bernardo, Sousa, & Gonçalves, 2022; Mello, Ribeiro, & Frizera-Neto, 2023). Its purpose is to put the terminal organ (end-effector) in a given location imposed by the task. It consists of:

Segments

The segments are rigid solid corps capable of movement relative to each other.

Articulations

The articulations are systems of bilateral connection that limit the relative possibilities of movements between two adjacent segments (regardless of the number of joint variables, the DOF should be less than or equal to six). Articulators are either of two types (see Figure 3):

- Rotoid articulator: it is of the pivot type, reducing the movement between two organs to a rotation around a common axis by the angle around it.
- Prismatic articulator: it is a slide type that reduces the movement between two organs to a translation along a common axis (the distance along this axis).

Figure 3. Rotoid and prismatic articulations.

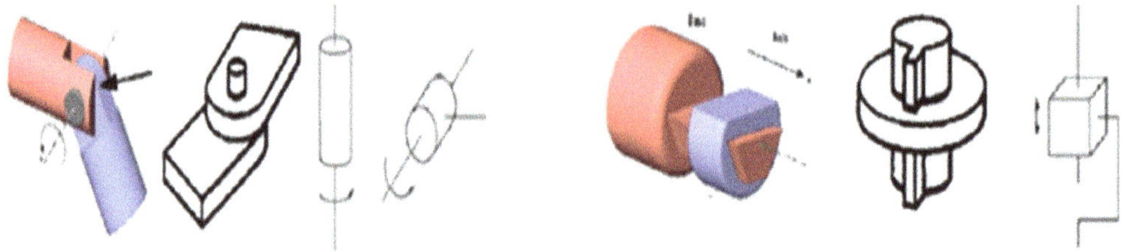

Actuator

The articulated mechanical structure comprises motors most often associated with transmissions; the set constitutes actuators. They frequently use motors, in which we find motors with electric switching, for the small robots of the stepper motors and for the robots manipulating very heavy loads the actuators are often hydraulic.

Terminal Organ (End-Effector)

It is the interface that allows the robot to interact with its environment. It can be multifunctional, equipped with several devices with different functions as it can also be nonfunctional but interchangeable. The number of terminal organs in a single robot depends on the number of arms (each arm has a deferential terminal organ of the other).

Sensors

Robotic sensors collect information about the situation of the robot such as the position and rapidity of joints or information about its environment through cameras and ultrasonic sensors.

MODELING OF MANIPULATOR ROBOTS

The control of robots or the simulation of their behavior requires the knowledge of geometric, dynamic and kinematic robot to be considered, each one relative to its own model. Obtaining these is not easy, the difficulty varies according to the complexity of the kinematics of the articulated chain, taking into account the number of degrees of freedom, the type of articulation and the fact that the chain can be opened simple, tree-shaped or closed. Details of modeling of robots can be found in the book (Siciliano & Khatib, 2008).

Direct and Inverse Geometric Models (DGM and IGM)

These models express the situation of the end effector as a function of the articular variables of the mechanism and vice versa (Boukens, 2017). The DGM is given as follows:

$$x_i = F(q_1, q_2, \ldots, q_n) \tag{1}$$

where q is the number of geometric variables and n is the number of degrees of freedom.

Then, the IGM allows us to calculate articular variables according to geometric variables. It is given by:

$$q = F^{-1}(x) \tag{2}$$

Modified Denavit-Hartenberg conversion: The homogeneous transformation is represented by the following matrix and the parameters are shown in Figure 4 (Guesbaya, 2012).

$$^{j-1}T_j = \begin{pmatrix} \cos(\theta_j) & -\sin(\theta_j) & 0 & \alpha_{j-1} \\ \sin(\theta_j)\cos(\alpha_{j-1}) & \cos(\theta_j)\cos(\alpha_{j-1}) & -\sin(\alpha_{j-1}) & -\alpha_j\sin(\alpha_{j-1}) \\ \sin(\theta_j)\sin(\alpha_{j-1}) & \cos(\theta_j) & \cos(\alpha_{j-1}) & \alpha_j\cos(\alpha_{j-1}) \\ 0 & 0 & 0 & 1 \end{pmatrix} \tag{3}$$

Figure 4. Parameters of the matrix (Guesbaya, 2012).

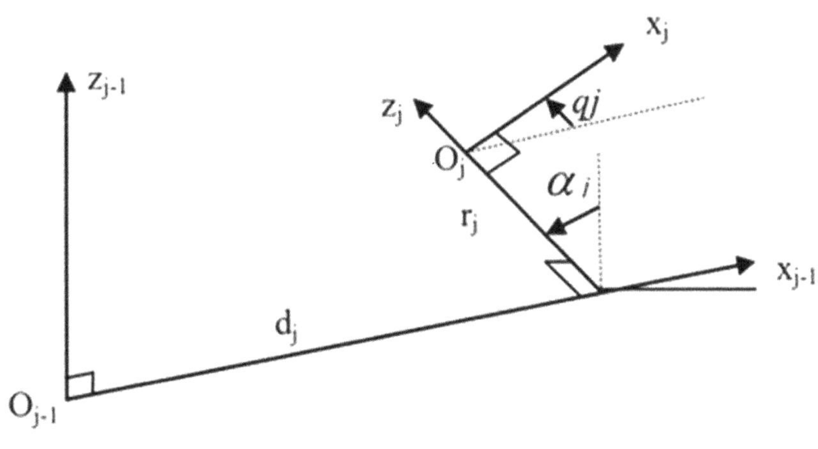

Direct and Inverse Kinematic Models (DKM and IKM)

These models express the speed of the terminal organ as a function of joint speeds and vice versa. The DKM is given as follows:

$$\frac{dx}{dt} = \frac{\partial F}{\partial q}\frac{dq}{dt} \tag{4}$$

where dx/dt is the position vector of the terminal member orientation, $\partial F/\partial q$=(J(q)) is the Jacobean matrix, and dq/dt is the articular velocity vector.

Therefore the MCI is given by:

$$\dot{q} = J^{-1}(q)\frac{dx}{dt} \tag{5}$$

Direct and Inverse Dynamic Models (DDM and IDM)

These models allow defining the robot's motion equations to establish the relationships between the torque or forces exerted by the actuators and the positions, speed and acceleration of the joints (Boukens, 2017). The DDM is written as follows:

$$\ddot{q} = f(q, \ddot{q}, \tau, f_e) \tag{6}$$

The IDM is written as follows:

$$\tau = f(q, \dot{q}, \ddot{q}, f_e) \tag{7}$$

where τ is the vector of torque/actuator forces, depending on whether the joint is rotoid or prismatic, q is the vector of the articular position, $\dot{q}$ is the vector of the articular velocity, $\ddot{q}$ is the vector of the articular acceleration, and f_e is the vector representing the external effort (force and moment) that the robot exerts on the environment.

The control of linear systems or systems approached by linear systems is a well-controlled domain that allows having controllers possessing properties of stability and optimality. However, the real world contradicts us by the fact that most systems are non-linear and dynamic with variable parameters.

CONTROLS OF ROBOTIC MANIPULATORS

Different techniques are used for the control of the manipulators; the mechanical design of these has an influence on the choice of control scheme. A robotic manipulator has a complex mechanical structure whose inertia in relation to the axes of the joints varies not only according to the charge but also according to the configuration of speeds and accelerations. Among the robots most used in industrial applications: conventional control, adaptive control and dynamic control.

Classical Command

A specification of a control loop imposes in a closed loop: stability, precision, speed, disturbances, etc. We know that the system may have defects in these parameters; it is often necessary to integrate in the system a corrector, so the objective is to improve one or more of these parameters without obviously doing it to the detriment of others. The classical control is the set of linear laws of type PID (proportional *P*, integral *I* and derivative *D*) with constant gain. To develop a PID control, it is necessary to consider each joint of the robot as an independent mechanism that can be linearized in an area of operation. The classic control is sufficient when the robot manipulator evolves without constraints of performance, rapidity and precision (see Figure 5). Some approaches are based on finding the values of PID using nature-inspired optimization algorithms (Mellal & Williams, 2017, 2018). The law corresponding to the PID command is (Siqueira, Terra, & Bergerman, 2011):

$$\tau(t)=k_p(q_d-q)+k_v(\dot{q}_d-\dot{q})+k_i\int_0^t(q_d-q)\,dt \tag{8}$$

where τ is the control torque applied to the joints, q_d is the desired angular position, q is the current angular position $((q_d-q)$ or q expresses the position error), q_d is the desired angular velocity, q is the current angular velocity $((q_d-q)$ or q expresses the velocity error vector. k_p, k_v and k_i are the the gains (k_{pj}, k_{vj} and k_{ij} of each joint).

Figure 5. Block diagram of the control of a robot with a PID controller.

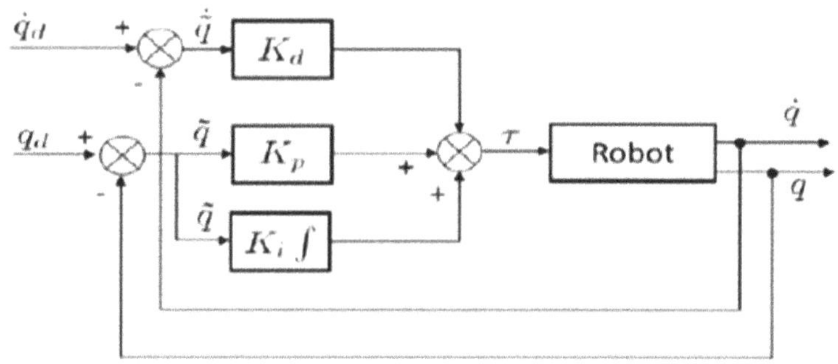

If the gravity forces are compensated mechanically or otherwise, the control law chosen is of type PD form:

$$\tau(t)=k_p(q_d-q)+k_v(\dot{q}_d-\dot{q}) \tag{9}$$

Dynamic Control (Control by Non-Linear Decoupling or Calculated Torque)

This type of technique considers all the joints and to decouple them we use the theory of non-linear decoupling; this theory utilizes the dynamic model of the robot for the calculation of control law, which leads to non-linear centralized control laws. Anticipation signals can be used to minimize non-linear effects. Dynamic control is not always the type of control required for good precision and stability. The implementation of this method requires the calculation of the dynamic model and knowledge of the numerical values of the inertial parameters (see Figure 6) (Gupta, 1997; Siqueira et al., 2011).

Equation model is given as follows:

$$\tau=D(q)\ddot{q}+H(q,\dot{q})+\tau f \tag{10}$$

Figure 6. Block diagram of the dynamic control.

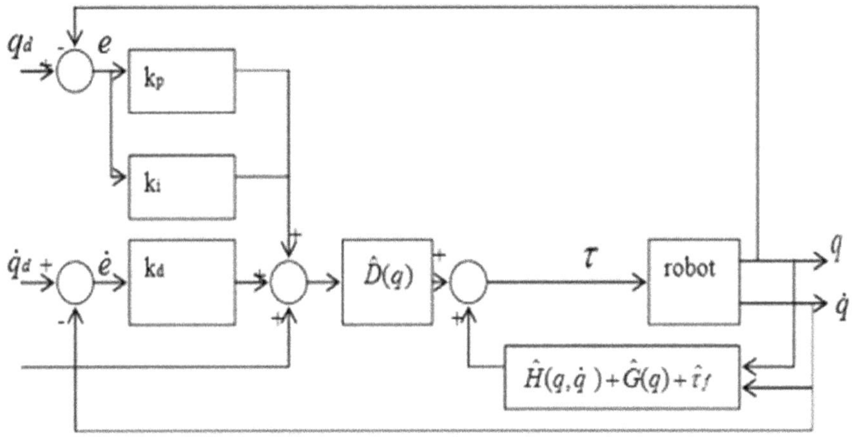

The control law will be given by:

$$\tau = \widehat{D}(q)u + \widehat{H}(q,\dot{q}) + \widehat{\tau_f} \tag{11}$$

with:

$$u = \ddot{q}_d + k_p(q_d - q) + k_d(\dot{q}_d - \dot{q}) + k_i\int_0^t (q_d - q)\,dt \tag{12}$$

and, $\widehat{D}, \widehat{H}, \widehat{\tau_f}$ esteemed from D, H, τ_f.

D: inertia matrix.
H: matrix of centrifugal and gravity terms.
τf: friction torque.

In case the parameters are correct, the equation becomes:

$$\ddot{e} + k_d\dot{e} + k_p e + k_i\int_0^t e(t)\,dt = 0 \tag{13}$$

With the equation of error $e = (q_d - q)$ it is therefore a linear and similarly decoupled equation.

Adaptive Command

Adaptive control is a set of strategies used for real-time automatic adjustment of control loop regulators to obtain or maintain a certain level of performance when the process parameters to be controlled are either unknown or time-variant. This type of technique is intended to correct control deficiencies using non-linear decoupling.

This type of scheme seeks to estimate or adjust the values of the parameters used in the calculation of the control law. In principle, an adaptive control system measures a certain performance index of the system to be ordered from the difference between the desired execution index and the measured execution index.

LINEARIZATION OF NON-LINEAR SYSTEMS (BY OFFSETTING NON-LINEARITIES)

Robots' manipulators are non-linear systems, so we need to focus on the linearization of non-linear systems. During linearization, non-linear effects are considered disruptive and thus neglected. The dynamics due to these non-linear effects are richer than those of linear systems.

We will see how to perform the linearization of a system presented by its non-linear differential equations using non-linear offset linearization.

We take the following nonlinear model:

$$a\ddot{x} + b\dot{x} + cx = f \tag{14}$$

The linear control law can be written as:

$$f = \alpha f' + \beta \tag{15}$$

We put: $\alpha = a$ and $\beta = b\dot{x} + cx$

We arrive at: $f' = \ddot{x}$ and we select:

$$f' = -k_p x - k_v \dot{x} \tag{16}$$

By replacing, we get:

$$a\ddot{x} + b\dot{x} + cx = \alpha\left(-k_p x - k_v \dot{x}\right) + \beta$$

$$\Rightarrow \alpha\ddot{x} + b\dot{x} + cx = \alpha\left(-k_p x - k_v \dot{x}\right) + b\dot{x} + cx$$

$$\Rightarrow \ddot{x} + k_v \dot{x} + k_p x = 0 \tag{17}$$

We finally arrive at the dynamic equation of a linear system. Just choose the k_p and k_v parameters to impose the desired poles and dynamic behavior.

EXAMPLE OF A ROBOTIC MANIPULATOR MODELING AND CONTROL APPLICATION

The robot studied in this example is a manipulator of 2 DOF (see Figure 7).

Figure 7. Robot's structure.

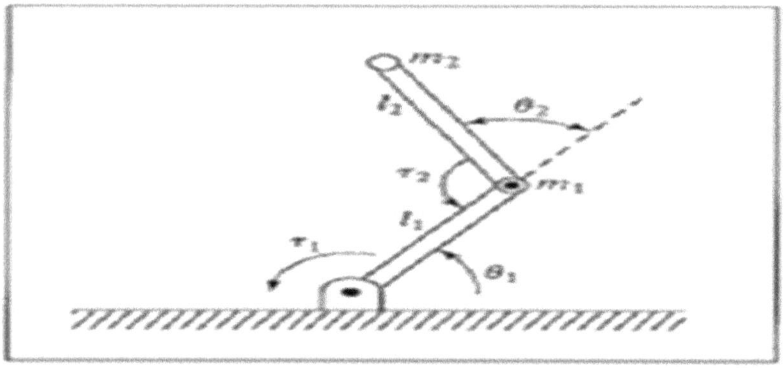

The parameters of this robot are as follows:

$m_1=m_2=m=1kg;\ l_1=l_2=l=1m;\ g=9.8m/s^2$

The dynamic arm model is given by:

$$\begin{bmatrix}\tau_1\\ \tau_2\end{bmatrix}=\begin{bmatrix} m_2l_2^{\,2}+2m_2l_1l_2C^2+(m_1+m_2)l_1^{\,2} & m_2l_2^{\,2}+m_2l_1l_2C_2\\ m_2l_1l_2C_2+m_2l_2^{\,2} & m_2l_2^{\,2}\end{bmatrix}\begin{bmatrix}\ddot{\theta}_1\\ \ddot{\theta}_2\end{bmatrix} +\begin{bmatrix} m_2l_1l_2S_2\dot{\theta}_2^{\,2}-2m_2l_1l_2S_2\dot{\theta}_2\dot{\theta}_1\\ 2m_2l_1l_2S_2\dot{\theta}_1\dot{\theta}_2\end{bmatrix}+\begin{bmatrix} m_2l_2C_{12}g+(m_1+m_2)l_1C_1g\\ m_2l_2C_{12}g\end{bmatrix} \tag{18}$$

With: $S_1=\cos\theta_1$; $S_1=\sin\theta_1$; $C_2=\cos\theta_2$; $S_2=sin\theta_2$; $C_{12}=\cos(\theta_1+\theta_2)$

We replace m and l with their value, we get:

$$\begin{bmatrix}\tau_1\\ \tau_2\end{bmatrix}=\begin{bmatrix}3+2C_2 & 1+C_2\\ C_2+1 & 1\end{bmatrix}\begin{bmatrix}\ddot{\theta}_1\\ \ddot{\theta}_2\end{bmatrix}+\begin{bmatrix}S_2\dot{\theta}_2^{\,2}-2S_2\dot{\theta}_2\dot{\theta}_1\\ 2S_2\dot{\theta}_1\dot{\theta}_2\end{bmatrix}+\begin{bmatrix}9.8C_{12}+19.6C_1\\ 9.8C_{12}\end{bmatrix}$$

This is the dynamic inverse arm model, since it gives the command vector as according to the acceleration, the direct model is given as follows:

$$\begin{bmatrix} \ddot{\theta}_1 \\ \ddot{\theta}_2 \end{bmatrix} = \begin{bmatrix} 3+2C_2 & 1+C_2 \\ C_2+1 & 1 \end{bmatrix}^{-1} * \left(\begin{bmatrix} \tau_1 \\ \tau_2 \end{bmatrix} - \begin{bmatrix} S^{2\dot{\theta}22} - 2S^{2\dot{\theta}2\dot{\theta}1} \\ 2S^{2\dot{\theta}^1\dot{\theta}^2} \end{bmatrix} - \begin{bmatrix} 9.8C_{12} + 19.6C_1 \\ 9.8C_{12} \end{bmatrix} \right) \tag{19}$$

CONCLUSION

Robotics can be defined as strategies and studies aimed at designing mechatronic systems (mechanical, computer and automatic), capable of replacing man in his motor, sensory and intellectual functions. The rapid development of the industry invoked the improvement of robotic manipulators. Then the robot must manipulate with great rapidity and increasing precision. This requires more suitable mechanical structures but also new and better control techniques. The structure and control of a robotic manipulator are dependent on the associated task. A slow task can be controlled by a conventional controller but for fast tasks a command based on the dynamic model becomes necessary. Sometimes when the parameters of the dynamic model are variants or male identifies one must resort to adaptive command or other commands.

REFERENCES

Bernardo, R., Sousa, J. M. C., & Gonçalves, P. J. S. (2022). Survey on robotic systems for internal logistics. *Journal of Manufacturing Systems*, *65*, 339–350. doi:10.1016/j.jmsy.2022.09.014

Billard, A., & Kragic, D. (2019). Trends and challenges in robot manipulation. *Science*, *364*(6446), eaat8414. doi:10.1126cience.aat8414 PMID:31221831

Boukens, M. (2017). *Contribution à la commande des robots manipulateurs mobiles* [*Contribution to the control of mobile manipulator robots.*]. University of Jijel.

Cui, J., & Trinkle, J. (2021). Toward next-generation learned robot manipulation. *Science Robotics*, *6*(54), 9461. doi:10.1126cirobotics.abd9461 PMID:34043539

Guesbaya, T. (n.d.). *Identification et commande de robot manipulateur rigide et flexible en utilisant les réseaux de neurones et la logique floue* [*Identification and control of rigid and flexible manipulator robots using neural networks and fuzzy logic.*]. University of Biskra.

Gupta, K. C. (1997). Solution manual for mechanics and control of robots. In Mechanical Engineering Series. doi:10.1007/978-1-4612-1840-1

Isermann, R. (2005). Mechatronic systems: Fundamentals. In *Mechatronic Systems*. Fundamentals. doi:10.1007/1-84628-259-4/COVER

Mellal, M. A., & Williams, E. J. (2017). The cuckoo optimization algorithm and its applications. In *Handbook of Neural Computation,* (pp. 269–277). Elsevier. doi:10.1016/B978-0-12-811318-9.00014-4

Mellal, M. A., & Williams, E. J. (2018). A survey on ant colony optimization, particle swarm optimization, and cuckoo algorithms. In *Handbook of Research on Emergent Applications of Optimization Algorithms*. IGI Global. doi:10.4018/978-1-5225-2990-3.ch002

Mello, R. C., Ribeiro, M. R. N., & Frizera-Neto, A. (2023). Implementing cloud robotics for practical applications. In Springer Tracts in Advanced Robotics. doi:10.1007/978-3-031-16908-3

Qin, J. (2013). *Commande hybride position/force robuste d'un robot manipulateur utilisé en usinageet/ ou en soudage*. [*Robust position/force hybrid control of a manipulator robot used in machining and/or welding*.] Ecole nationale supérieure d'arts et métiers - ENSAM.

Sayers, C. (1999). *Remote Control Robotics*. Remote Control Robotics. doi:10.1007/978-1-4612-1430-4

Siciliano, B., & Khatib, O. (2008). Springer handbook of robotics. In Springer Handbook of Robotics. doi:10.1007/978-3-540-30301-5

Siqueira, A. A. G., Terra, M. H., & Bergerman, M. (2011). Robust control of robots. In Robust Control of Robots. doi:10.1007/978-0-85729-898-0

Chapter 2
Machine Learning and Optimization Applications for Soft Robotics

Mehmet Mert İlman
https://orcid.org/0000-0001-7664-5217
Manisa Celal Bayar University, Manisa, Turkey

Pelin Yildirim Taser
https://orcid.org/0000-0002-5767-2700
Izmir Bakircay University, İzmir, Turkey

ABSTRACT

Due to their adaptability, flexibility, and deformability, soft robots have been widely studied in many areas. On the other hand, soft robots have some challenges in modeling, design, and control when compared to rigid robots, since the inherent features of soft materials may create complicated behaviors owing to non-linearity and hysteresis. To address these constraints, recent research has utilized different machine learning algorithms and meta-heuristic optimization techniques. First and foremost, the study looked at current breakthroughs and applications in the field of soft robots. Studies in the field are grouped under main headings such as modelling, design, and control. Fundamental issues and developed solutions were analyzed in this manner. Machine learning and meta-heuristic optimization-oriented methods created for various applications are highlighted in particular. At the same time, it is emphasized how the problems in each of the modeling, design, and control areas impact each other.

INTRODUCTION

As the use of robots grows, it may be necessary to redefine flexibility once more. While speed was one of the most essential factors for efficiency expected from robots confined to factories in the early stages of industrialization, flexibility could be considered as a margin of error. This approach may be appropriate for indoor applications. However, in outdoor applications, the opposite of this approach is mostly valid.

DOI: 10.4018/978-1-6684-5381-0.ch002

In fact, since the ideal conditions in the external environment disappear, the adaptation of traditional robots becomes very difficult with their cumbersome structure. When we look at the living nature, the reason for this is better understood. Because one of the features that ensures the existence of life even under the most extreme conditions is flexibility. Gerringer et al. (2017), for example, found that a fish species (hadal belt snail fish) can survive in the Mariana Trench at a depth of about 8 km and a pressure of around 800 atm. The adaptability brought about by flexibility enables the living thing to accept even the most difficult conditions (very high pressure, etc.) as new conditions, not as a disturbance effect, and to continue its standard behaviors such as feeding and reproduction. The development of the idea of benefiting from flexibility and deformation, inspired by living things with such examples, has led to the emergence of soft robots.

Figure 1. Classification of robots by materials and degrees of freedom (Trivedi, 2008)

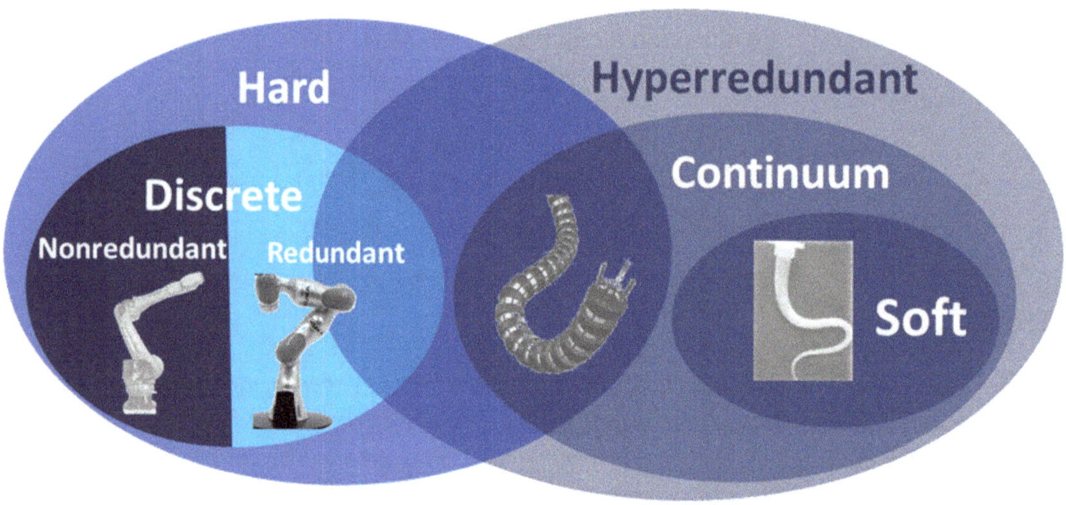

Soft robotic technologies have opened the door to a new era in robot design and development. However, these technological developments face new challenges and limitations due to the hyper-redundancy (Figure 1) and high degree of freedom caused by the flexible nature of the materials used. It also causes complex and unexpected behavior as a result of the nonlinear relationship between the input and output of the system.

The increasing complexity of soft robotic systems, which may be due to shape, material, actuation, and their complicated connection, renders conventional robot design methods inapplicable (Chen & Wang, 2020). The simulation and analytical tools required to effectively predict the complex mechanical behavior of soft robots are insufficient. In addition, there was a lack of effective optimization algorithms to automate the design process in the literature. To bridge this gap, optimization techniques have started to be implemented in the design of the soft robotics field.

On the other hand, the control challenges can be considered the final challenge to overcome before a soft robot can be utilized in the physical world. Although designs that use on-off logic, such as the soft robotic gripper, do not require control, this cannot be said for the other soft robot kinds. High-fidelity soft systems models have been effectively used to control soft systems, but this needs precise system identification and costly run-time calculation, hence restricting the applicability (Chin, Hellebrekers & Majidi, 2020). This high dimensionality is also a basic difficulty for soft sensing since many physical states might yield the same sensor readout once the higher dimensional data is converted. When traditional mathematical and statistical analytical models are deemed insufficient, data-driven machine learning (ML) control approaches have generated promising outcomes.

This chapter presents a comprehensive overview of ML-based studies and optimization methods in soft robotics, examines the current trends, and addresses the current limits of these algorithms. It also explains the application of ML algorithms and optimization techniques in soft robotics problems where the conventional methods continue to be ineffective. Furthermore, it analyses current ML-based studies in detail, covering learning types, tasks, and algorithms. The chapter also aims to provide an insight into researchers for their potential applications in this area. Finally, the advantages and some drawbacks of using ML techniques in soft robotics and future directions in this topic are provided in this chapter.

This chapter has three main sections in general. Soft robot applications and modeling challenges are discussed in the first section. The second section discusses design optimization strategies and challenges. Machine learning and control approaches are emphasized in the third section. Finally, some concluding remarks and future studies are presented in the Conclusion section.

SOFT ROBOT MODELING CHALLENGES

The application fields are expanding as the benefits of soft robots emerge. Modeling plays a fundamental role in further increasing this expansion. Soft robots must be precisely modelled in order to be used in both the design and control phases. Soft materials that give their name to soft robots are selected from materials such as new technology elastomeric materials, shape memory alloys/polymers, electroactive polymers, and hydrogel (Elango & Faudzi, 2015). This situation not only highlights the features of soft robots, but also complicates the modeling phase.

It is aimed to solve the modeling problem with discrete approaches such as finite-element-model (FEM) and lumped-parameter-model (LPM) as well as continuous techniques such as Cosserat rod theories. Gu et al. (2021) calculated bending and twisting deformations in three dimensions with the analytical model they discussed in their study. In the theoretical model created, minimum potential energy and continuum rod theories (CRT) were used. It is possible to compute the variable stiffness of pneumatically moving actuators in three dimensions thanks to the modeling used. Saunders et al. (2011) were able to calculate the displacements of the caterpillar bio-inspired crawling soft robot with acceptable accuracy, thanks to the discrete lumped-parameter model they used.

Analytical and numerical soft robot models can be compared in terms of practicality (applicability), speed (computation time), and accuracy. While selecting the modeling methods, the model that will provide at least two of these three features is selected in a way that will be most suitable according to the application. While selecting the modeling methods, the model that will provide at least two of these three features is chosen in a way that will be most suitable according to the application. Analytical models, for example, may be selected for speed and accuracy, whereas FEM from numerical approaches may be preferred for accuracy and practicality, or LPM may be preferred for practicality and speed. Reduced-order models (ROM) are strong candidates for providing all three of these features (Figure 2). Goury & Duriez (2018) demonstrated that the snapshot-proper orthogonal decomposition-based ROM technique they tested on detailed FEMs of diverse soft robot systems could boost the model's speed by more than

100 times without sacrificing accuracy. Figure 3 depicts the mesh structures used in the modeling of the octopus bio-inspired soft robot. When these models are compared in the study, it has been shown that ROM has close accuracy with the fine meshed full-order-model (FOM) and can perform faster and more accurate displacement calculations than coarse meshed FOM. Thanks to its advantages, ROM allows many control algorithms such as feedback linearization control and inverse Jacobian control to be used in real time (Sadati et al., 2020).

Figure 2. Main features of different modeling techniques

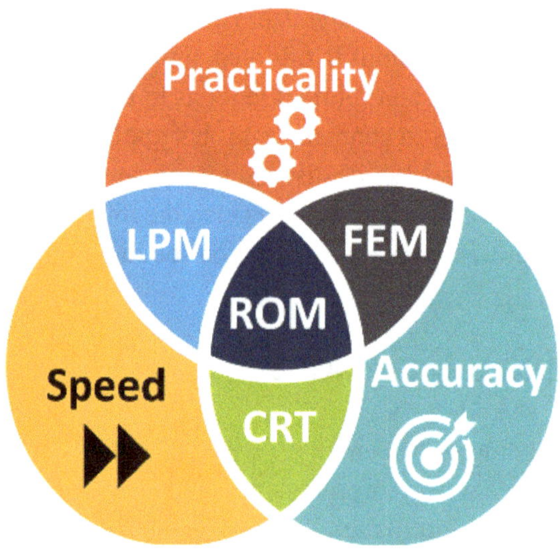

Figure 3. Reduced model (ROM) and full-order model (FOM) comparison of octopus bio-inspired soft robot manipulator a) coarse meshed FOM, b) fine meshed FOM, c) ROM derived from fine meshed FOM, d) comparison of strains during manipulation (Goury & Duriez, 2018)

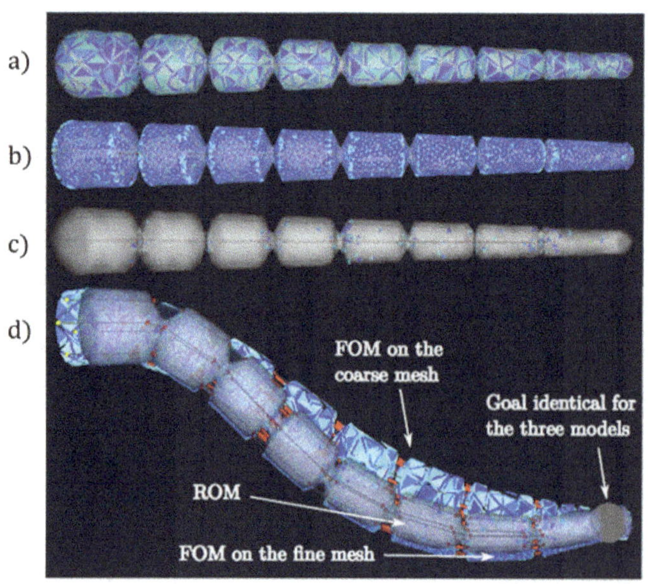

Furthermore, in recent years, the number of open-access simulation environments for soft robots produced employing the aforementioned modeling methodologies has increased. As examples of new open-access platforms designed for soft robots, programs utilizing volumetric FEM (voxel) technique (SOFA, ChainQueen) and programs using discrete rod model (Elastica, TMTDyn, SimSOFT, and SoRoSim) can be presented (Hiller & Lipson, 2013; Hu et al., 2019).

DESIGN OPTIMIZATION STRATEGIES AND CHALLENGES

It is feasible to generate substantially more complicated geometries (Hiller & Lipson, 2012; Joachimczak et al., 2012) and materials with varying mechanical characteristics using the additive manufacturing technology established for silicones (Wallin et al., 2020). With the elimination of traditional design constraints, soft robots can be designed to adapt to a wide variety of applications and environments. It is possible to diversify the applications such as swimming (Veenstra et al., 2018), crawling, manipulation, rehabilitation (Runge et al., 2017), exoskeleton, gripper. In addition, if necessary, it is possible to apply in field conditions, underwater (Veenstra et al., 2018) and extraplanetary environments (Naclerio et al., 2021).

However, because complex geometries include so many design parameters, achieving the optimal result by trial and error becomes increasingly challenging. Because as the design parameters increase, it becomes more difficult to find the optimum values of these parameters. In solving this problem, besides methods such as traditional topology optimization (Chen et al., 2018; Liu et al., 2018; Zhang et al., 2018; Zhang et al., 2019), currently genetic algorithm (GA, NSGA – Pagoli et al., 2020), Bayesian optimization (BO – Bhatia et al., 2022) and meta-heuristic algorithms such as compositional model generating network (CPPN-Neat – Methenitis et al., 2015) seem to be able to produce solutions. If the desired outcome can be attained through design optimization, the design phase can be shortened by standardizing it. This will also make it possible to produce more specialized robots as a result of increased bio-inspiration.

The modeling techniques and simulation platforms mentioned in the previous section also enable the iterative and rapid design required for optimization procedures. Two critical issues in design optimization are the design parameters, the objective function, and the chosen meta-heuristic optimization method. To give an example of the choices made in soft robot design optimizations in different applications in the literature (Figure 4): Bodily et al. (2017) performed the design optimization of the inflatable robot arm with the help of multi-objective objective genetic algorithm. For the design parameter, the lengths of the links, which are the sub-units of the arm, the total length and the mounting dimensions were evaluated together. Fitzgerald et al. (2021) performed the shape optimization of the granules of a soft robot jamming gripper using the NSGA-III (Non-dominated Sorting Genetic Algorithm - 3) algorithm. On the other hand, Pagoli et al. (2020) optimized the finger design on the finite element model for a soft robot gripper using the NSGA-II algorithm.

In the design optimization of soft robots, parameters such as the force applied by the end point, stiffness or strength can be used while determining the design criteria (objective function). In addition, the success of the soft robot in movements such as swimming, crawling and walking can be taken into account when determining the design criteria (Cheney et al., 2014; Methenitis et al., 2015; Rieffelet al., 2014; Veenstra et al., 2018).

Figure 4. a) A pneumatic soft gripper is composed of an inner chamber made of rubber and an outer layer made of Tango (Zhang et al., 2018), b) Module with Dragonskin 0030 accordion structure: module design, FEA analysis, and experimental study of onechamber actuation under 0.8 bar (Elsayed et al., 2014), c) A cabledriven soft finger is subject to topology optimization (Liu et al., 2018), d) an itrative study between experimental and numerical data. Referans girdi altında acquire edilen deneysel motion data (left) öncelikle system identification metodlarıyla modellenir (bottom right) ardından da trajectory optimizasyonu yapılarak iterative süreç tamamlanmış olur (Du et al., 2021), e) a soft robot manipulator with closed loop motion control (top), end-point position is measured via electromagnetic tracking probe and force torque sensor over the desired trajectory (bottom) (Lee et al., 2017)

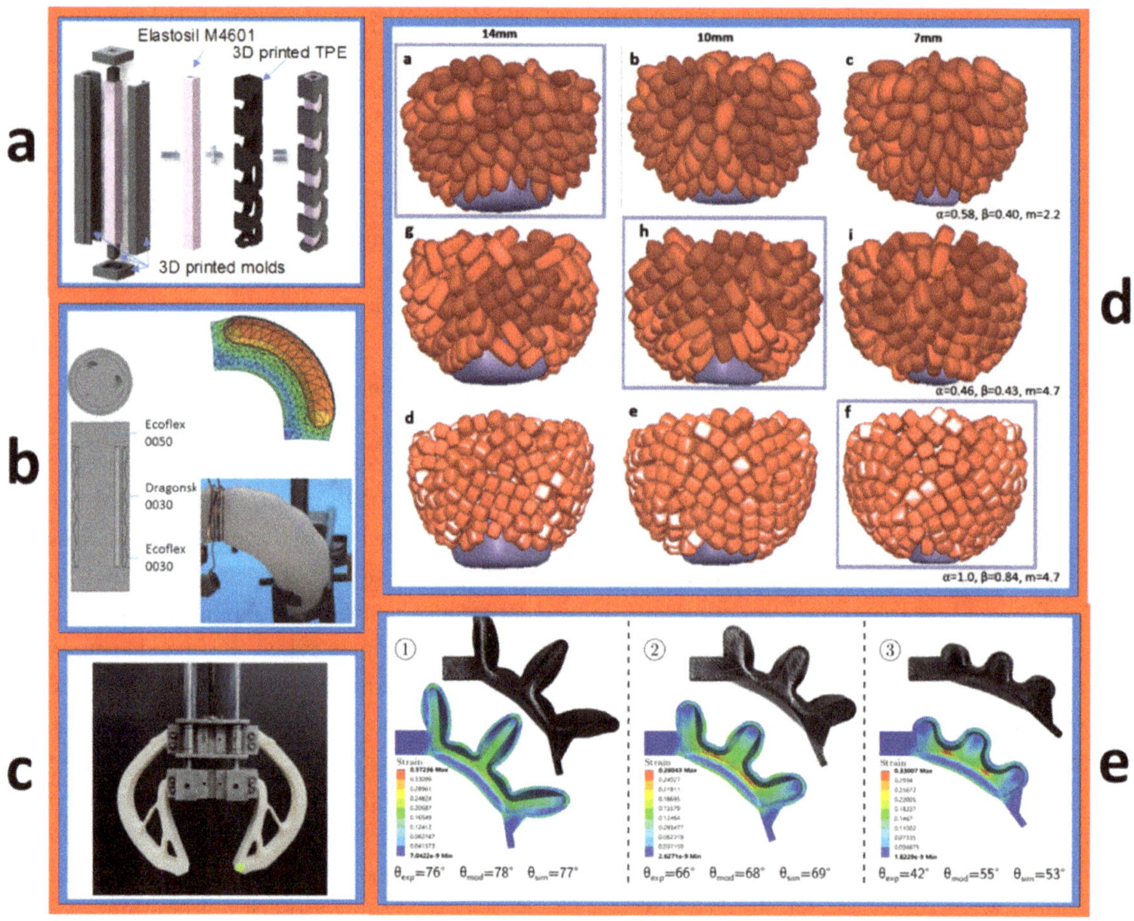

After examining at the studies, it is clear that design optimization procedures are used in conjunction with simulation tools. Therefore, it turns out that the simulation program is a prerequisite in meta-heuristic design optimization. As a result, incorporating meta-heuristic design optimization modules into soft robot simulation tools is crucial for speeding up research in the field. On this subject, Bhatia et al. (2021) developed a voxel-based benchmark that allows the use of meta-heuristic design optimization methods such as Bayesian optimization (BO), NSGA, and CPPN-Neat (Compositional Pattern Producing Network).

CONTROL CHALLENGES OF SOFT ROBOTICS

The difficulties in the control process can be considered as the last challenge to be overcome before using a soft robot in the physical world. While designs that use on-off logic, such as the soft robotic gripper, do not require control, this cannot be said for other types of soft robots. As a result, feedback or feedforward control applications are currently being investigated in both manipulation and gait applications to solve difficulties such as motion control/trajectory tracking (Thuruthel et al., 2016). In current reports, soft robot control applications appear to fall into two categories, namely: (i) model-based (white box) control (Della Santina et al., 2021; Manzano et al., 2021) and (ii) model-independent (black box) control. box) is the control (Rolf et al., 2015; Thuruthel et al., 2018).

In model-independent methods, direct learning/control can be achieved with machine learning-based approaches. When making these two classifications, it is assumed that the studies in which modeling techniques are used and the studies in which machine learning techniques are used are completely separate from each other. However, there are many studies in which this assumption is not valid. For example, in cases where traditional mathematical and statistical analytical models are deemed insufficient, machine learning-based models can be chosen instead of white box models (Jiang et al., 2017). Or, it is seen that machine learning can make control adaptive by integrating it into model-based control architectures (Lee et al., 2017). The goal may be to iteratively improve the model-based feedforward controller or to make the model-based feedback controller resistant to ambiguity. Likewise, on the contrary, model information can be used in model-free control applications (Huang et al., 2020). The goal here is to use the model to generate synthetic data for machine learning, or use it for pre-training and then fine-tune it with the test data. As a result, as hybrid research with machine learning applications and other methodologies becomes more common, it becomes more difficult to categorize machine learning methods separately.

Considering the control challenges, in the last decade, ML techniques have been preferred in the soft robotics field to understand relations of highly nonlinear, complex, and high-dimensional system parameters and to make predictions about the unknown feature values (Figure 5). Also, ML methods help soft robots adapt to unfamiliar environmental situations.

ML-based soft robotic control is an interdisciplinary area that can be considered as a synthesis of three distinct disciplines: computer science, statistics and math, and soft robotic control. As seen in Figure 6, the confluence of these three fields creates other subareas strongly connected to ML-based soft robotic control, including conventional software, traditional methods, and ML.

Machine Learning

ML is an area of artificial intelligence in which computers can self-learn based on past experiences (İlman et al., 2022). The main goal of this paradigm is to enable computers to make decisions independently, without any human intervention or support. The ML algorithm constructs a predictive model using some previous data in order to uncover hidden patterns in it and make predictions about future samples. There are three types of ML in the literature: supervised learning, unsupervised learning, and reinforcement learning, as shown in Figure 7.

Figure 5. a) Experimental setup for light lace sensors and tendon actuated soft mesh mounted in a solid structure (Manzano et al., 2021), b) Control system ftamework of honeycomb pneumatic network (HPN) (Jiang et al., 2017), c) Gestures for gesture recognition of soft robotic hands (Huang et al., 2020), d) An overview of the iterative, real-to-sim pipeline. Real-world motion data is captured from the robot for a given control input (left). This data is transferred to the differentiable simulator where system identication narrows the gap between reality (four green spheres bottom right) and simulation (four blue spheres bottom right) after which trajectory optimization is used to generate a new control sequence. (Du et al., 2021), e) Registration process of the predefined trajectory using an electromagnetic (EM) position tracking system. Blue line on the transparent sphere illustrates the tracking trajectory on the task space (top); Soft manipulator is commanded to follow the desired trajectory automatically. Its end-effector position is also measured by the tracking system to close the feedback loop under online learning control policy. Plastic rod actuated by a stepping motor pushes against the soft robot, generating external constrains. The contact force is monitored by a force/torque sensor (bottom) (Lee et al., 2017)

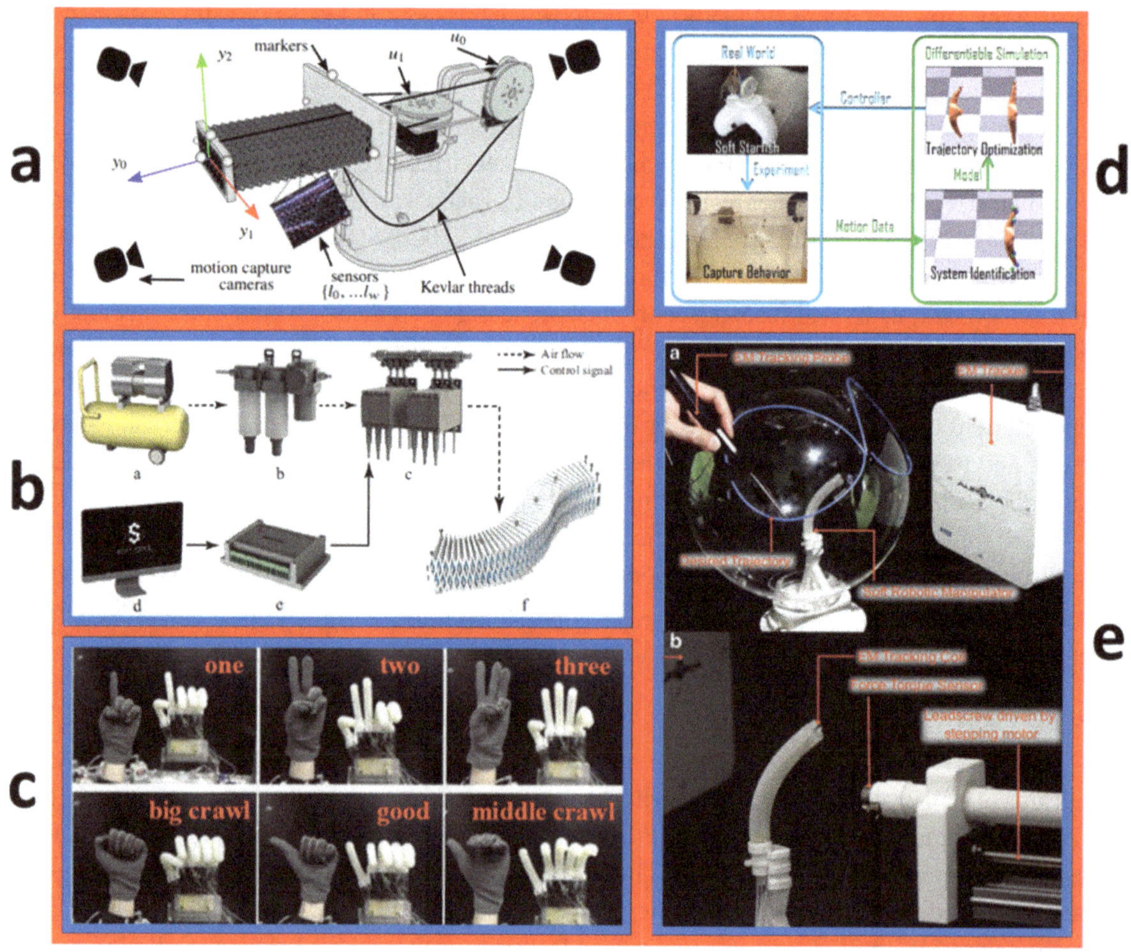

Figure 6. The main disciplines related to ML-based soft robotic control

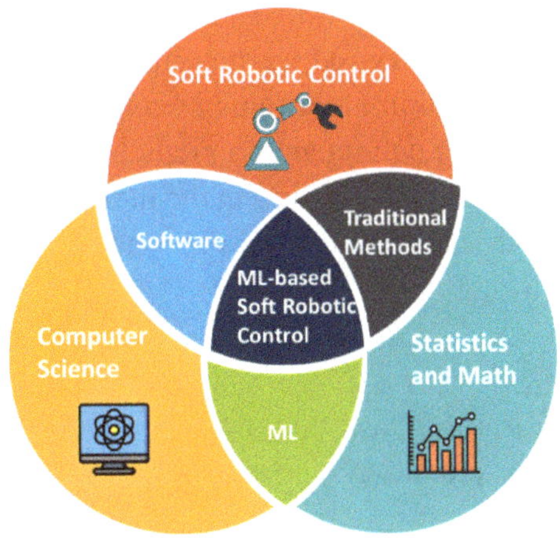

Figure 7. Types of machine learning

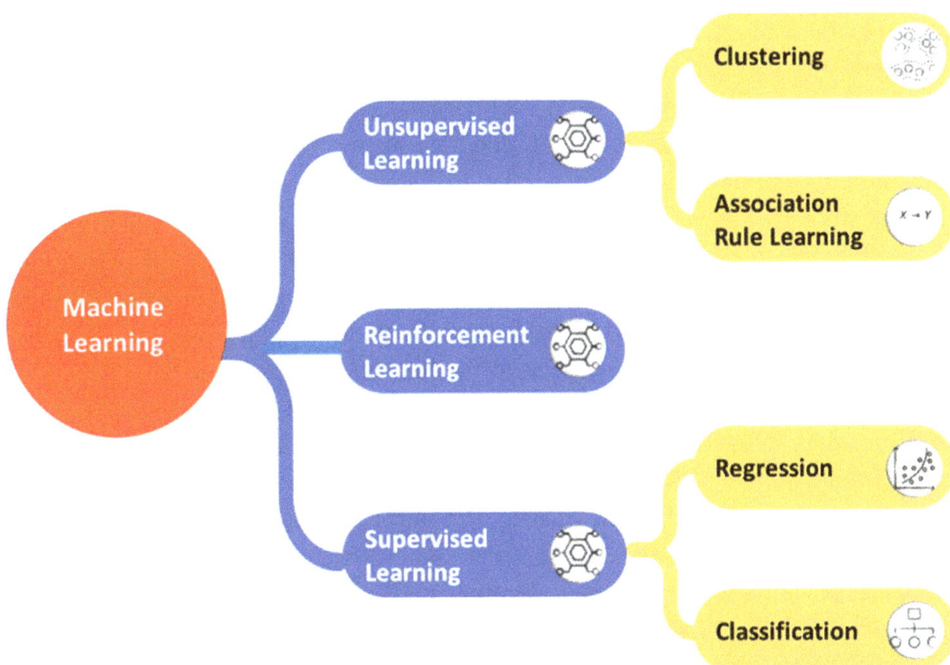

Supervised Learning

The supervised learning technique aims to train an ML model using pre-labeled data (Thuruthel et al., 2018). This constructed ML model tries to predict the unknown class label for a new instance using these already tagged data. The supervised learning technique differs in two tasks: classification and regression.

Classification

Classification is a commonly applied supervised learning method that assigns one of the predefined classes to new instances. The classification algorithms attempt to uncover correlations between the characteristics of samples in the training dataset and to predict an unknown target output given an input. The frequently preferred classification algorithms in the literature are Naive Bayes, Support Vector Machines, K-Nearest Neighbor, Artificial Neural Networks, and Decision Tree.

Regression

Regression is another supervised learning task that discovers correlations between data features and predicts continuous-valued target output based on the historical data. It is mostly used for prediction, forecasting, and time series modeling.

Supervised Learning Algorithms

- **Artificial Neural Network:** Artificial Neural Networks (ANNs) are supervised learners inspired by the biological nervous system of the human brain and try to predict the class label by processing the input values of the samples. In the ANNs, neuron, dendrite, nucleus, axon, and synapse structures in the biological nervous system are modeled as input, summation function, activation function, output, and weight, respectively. It is clearly seen from the literature that the most commonly applied ML algorithms in soft robotics studies are ANNs. In their studies, (Jiang et al., 2017) utilized neural networks in control algorithm for a multi-segment extensible HPN (honeycomb pneumatic network) arm. Machine learning is used to figure out the pressures from each segment's pose (from configuration space to actuation space) of an extensible soft arm in order to solve its inverse kinematics.
- **Support Vector Machines:** Support Vector Machines (SVM) is a binary supervised learning algorithm based on the concepts of statistical learning theory, decision planes, and risk minimization. In the SVM algorithm, each record in the training data is treated as a point in an n (feature number) dimensional space, and a hyperplane is created to separate the points on this space into two different classes. The purpose of the algorithm is to ensure that the generated hyperplane has the largest margin. Thus, excessive memorization of the system or poor classification performance can be prevented.
- **K-Nearest Neighbor:** As a non-parametric lazy learning classifier, k-Nearest Neighbor predicts the sample's class label based on a majority vote of its k neighbors who have particular class labels. The technique employs certain distance metrics such as Euclidean, Manhattan, Minkowski, and others to determine the k closest neighbors of a sample data point.
- **Decision Tree:** One of the most generally used ML approaches, a decision tree, is a basic combination of rules that uses a depth-first strategy to classify a new sample into one of the pre-defined classes. The decision tree that is formed has branches, internal nodes, and leaf nodes. They reflect the experimental dataset's attribute values, attributes, and classes, respectively. The output of any internal node of the tree is transferred as an input to another node in this method. The tree is traversed from the root node to one of the leaf nodes in order to categorize a new sample. In our past study (İlman et al., 2022), the machine learning paradigm was used to generalize an open-loop vibration control through parameter tuning

approach for multi-link flexible manipulators. In this study, the C4.5 (decision tree) algorithm was applied to a large experimental dataset, and its performance was compared to that of Naive Bayes (NB), Multilayer Perceptron (MLP), Support Vector Machine (SVM), and k-Nearest Neighbor (kNN), four other well-known traditional machine learning algorithms. The experimental results indicated that the C4.5 algorithm achieved the best classification performance with 92.25% and 83.75% accuracy rates.

- **Naive Bayes:** Naive Bayes is a statistical classification algorithm based on Bayes' theorem, calculating the probabilities of the values in each attribute of the sample to be classified in the training data and choosing the class label as the highest probability value. The biggest advantage of the algorithm is that it exhibits high accuracy classification performance on unbalanced and small numbers of data sets. Lepora et al. (2010) utilized Naive Bayes for texture classification by a tactile sensing of a moving robot.

Unsupervised Learning

Unsupervised learning technique learns patterns from unlabeled observation data with no previous knowledge of the output value (Kim et al., 2020). This technique employs two primary tasks: clustering and association rule learning.

Clustering

Clustering is a technique for grouping a collection of items into clusters based on their similarity. The similarity between items is determined using measures such as Manhattan, Euclidean, and Minkowski distances for numerical data and Jaccard's distance for categorical data. In the soft robotics literature, the studies were generally performed by combining classification and clustering algorithms to improve the performance of the constructed ML models. For example, Huang et al. (2020) utilized unsupervised learning algorithm of K-Means Clustering (KMC) in soft robotic hand application. They applied gesture recognition with parameters such as object shape, size, and weight are implemented weight (Figure 5-c).

Association Rule Learning

Association rule learning, which is also a form of unsupervised learning based on rules, is a well-researched technique for uncovering interesting correlations and common patterns among items in the datasets. Two primary user-defined criteria, such as support and confidence, are utilized to uncover hidden patterns in raw data. No association rule mining research has yet been conducted in this field to the best of our knowledge. Hidden, interesting, and useful rules and valuable relations between input variables of the soft robotic data can be discovered using association rule learning algorithms.

Reinforcement Learning

Reinforcement learning aims to construct actions or strategies for learning anticipated behaviors by developing reward functions (Schlagenhauf et al., 2018; Thuruthel et al., 2017; Zhang et al., 2017; Zhang et al., 2017). There are two types of reinforcement learning in the literature: positive and negative. Positive reinforcement learning augments existing ML models with positive behavior, increasing their likelihood of reproducing the outcomes they create. On the other hand, negative reinforcement learning introduces negative behavior in the form of punishment to encourage machine learning models to perform better. The main difference between reinforcement learning and supervised learning is that the learners receive only partial feedback on their predictions. Bu alanda yapılan çalışmalardan: Bhatia et al. (2021) controlled the different soft robots with different motions (walking, carrying, climbing, upstepping, throwing, catching) by a state-of-the-art RL algorithm named Proximal Policy Optimization (PPO). Rost & Schadle (2013) used Q-Learning and SARSA (State–action–reward–state–action) to optimize the gripping motion of a soft robot consisting of four fingers and 12 degrees of freedom.

CONCLUSION

Soft robotic technologies have ushered in a new era of robot design and development. These technological innovations and developments create new difficulties and limitations because of the material's structural compliance and viscoelasticity result in complicated and unexpected behaviors as a consequence of the nonlinear relation between the system's input and output. In addition, creep, drift, and high degrees-of-freedom (DOF) are factors that increase robot behaviors' complexity. In recent years, ML and optimization-based studies have been carried out in the field of soft robots as potential solutions to the drawbacks mentioned above. This chapter provides an overview of ML and optimization approaches specifically designed for soft robotics applications and a description of some experimental studies. It explains how ML and optimization methods can be used to solve an issue in the soft robotics field. In this chapter, it is shown that classification-based studies are more prevalent in soft robotics field than clustering-based researches. To our knowledge, no association rule learning studies have yet been conducted in this field.

In future work, association rule learning can be used for discovering interesting and valuable relations between input variables of the soft robotic data. Setting design goals, evaluating performance criteria, and producing and testing optimal designs are still open to innovation. New optimization methods (Grey Wolf Optimizer – Mirjalili et al., 2014, Corona Virus Optimization (CVO) – Salehan & Deldari, 2022 etc.) can be used to optimize design criteria during the design phase. In addition, these algorithms can be used to optimize the controller parameters in the control phase of the soft robot or hyperparameters in machine learning methods.

It would not be wrong to say that machine learning will become more common in soft robot control. It can be expected that especially hybrid control algorithms (deep reinforcement learning, neuro-fuzzy logic, etc.) will also give successful results in the control of soft robots. It is also possible to apply machine learning outside of soft control. For example, reduced-order-models of robots can also be made with Machine learning.

REFERENCES

Bhatia, J., Jackson, H., Tian, Y., Xu, J., & Matusik, W. (2021). Evolution Gym: A Large-Scale Benchmark for Evolving Soft Robots. *Advances in Neural Information Processing Systems, 34*, 2201–2214. https://proceedings.neurips.cc/paper/2021/hash/118921efba23fc329e6560b27861f0c2-Abstract.html

Bodily, D. M., Allen, T. F., & Killpack, M. D. (2017). Multi-objective design optimization of a soft, pneumatic robot. *IEEE International Conference on Robotics and Automation (ICRA)*, 1864–1871. 10.1109/ICRA.2017.7989218

Chen, F., & Wang, M. Y. (2020). Design Optimization of Soft Robots: A Review of the State of the Art. *IEEE Robotics & Automation Magazine*, *27*(4), 27–43. doi:10.1109/MRA.2020.3024280

Chen, F., Xu, W., Zhang, H., Wang, Y., Cao, J., Wang, M. Y., Ren, H., Zhu, J., & Zhang, Y. F. (2018). Topology Optimized Design, Fabrication, and Characterization of a Soft Cable-Driven Gripper. *IEEE Robotics and Automation Letters*, *3*(3), 2463–2470. doi:10.1109/LRA.2018.2800115

Cheney, N., MacCurdy, R., Clune, J., & Lipson, H. (2014). Unshackling evolution: Evolving soft robots with multiple materials and a powerful generative encoding. *ACM SIGEVOlution*, *7*(1), 11–23. doi:10.1145/2661735.2661737

Chin, K., Hellebrekers, T., & Majidi, C. (2020). Machine Learning for Soft Robotic Sensing and Control. *Advanced Intelligent Systems*, *2*(6), 1900171. doi:10.1002/aisy.201900171

Della Santina, C., Duriez, C., & Rus, D. (2021). *Model Based Control of Soft Robots: A Survey of the State of the Art and Open Challenges*

Demir, K. G., Zhang, Z., Yang, J., & Gu, G. X. (2020). Computational and Experimental Design Exploration of 3D-Printed Soft Pneumatic Actuators. *Advanced Intelligent Systems*, *2*(7), 2000013. doi:10.1002/aisy.202000013

Du, T., Hughes, J., Wah, S., Matusik, W., & Rus, D. (2021). Underwater Soft Robot Modeling and Control With Differentiable Simulation. *IEEE Robotics and Automation Letters*, *6*(3), 4994–5001. doi:10.1109/LRA.2021.3070305

Elango, N., & Faudzi, A. A. M. (2015). A review article: Investigations on soft materials for soft robot manipulations. *International Journal of Advanced Manufacturing Technology*, *80*(5), 1027–1037. doi:10.100700170-015-7085-3

Elsayed, Y., Vincensi, A., Lekakou, C., Geng, T., Saaj, C. M., Ranzani, T., Cianchetti, M., & Menciassi, A. (2014). Finite Element Analysis and Design Optimization of a Pneumatically Actuating Silicone Module for Robotic Surgery Applications. *Soft Robotics*, *1*(4), 255–262. doi:10.1089oro.2014.0016

Fitzgerald, S. G., Delaney, G. W., Howard, D., & Maire, F. (2021). Evolving soft robotic jamming grippers. *Proceedings of the Genetic and Evolutionary Computation Conference*, 102–110. 10.1145/3449639.3459331

George Thuruthel, T., Ansari, Y., Falotico, E., & Laschi, C. (2018). Control Strategies for Soft Robotic Manipulators: A Survey. *Soft Robotics*, *5*(2), 149–163. doi:10.1089oro.2017.0007 PMID:29297756

Goury, O., & Duriez, C. (2018). Fast, Generic, and Reliable Control and Simulation of Soft Robots Using Model Order Reduction. *IEEE Transactions on Robotics*, *34*(6), 1565–1576. doi:10.1109/TRO.2018.2861900

Gu, G., Wang, D., Ge, L., & Zhu, X. (2021). Analytical Modeling and Design of Generalized Pneu-Net Soft Actuators with Three-Dimensional Deformations. *Soft Robotics*, *8*(4), 462–477. doi:10.1089oro.2020.0039 PMID:32822253

Hiller, J., & Lipson, H. (2012). Automatic Design and Manufacture of Soft Robots. *IEEE Transactions on Robotics*, *28*(2), 457–466. doi:10.1109/TRO.2011.2172702

Hiller, J., & Lipson, H.HillerJonathan. (2014). Dynamic Simulation of Soft Multimaterial 3D-Printed Objects. *Soft Robotics*, *1*(1), 88–101. doi:10.1089oro.2013.0010

Hu, Y., Liu, J., Spielberg, A., Tenenbaum, J. B., Freeman, W. T., Wu, J., Rus, D., & Matusik, W. (2019). ChainQueen: A Real-Time Differentiable Physical Simulator for Soft Robotics. *2019 International Conference on Robotics and Automation (ICRA)*, 6265–6271. 10.1109/ICRA.2019.8794333

Huang, H., Lin, J., Wu, L., Fang, B., Wen, Z., & Sun, F. (2020). Machine learning-based multi-modal information perception for soft robotic hands. *Tsinghua Science and Technology*, *25*(2), 255–269. doi:10.26599/TST.2019.9010009

İlman, M. M., Yavuz, Ş., & Taser, P. Y. (2022). Generalized Input Preshaping Vibration Control Approach for Multi-Link Flexible Manipulators using Machine Intelligence. *Mechatronics*, *82*, 102735. doi:10.1016/j.mechatronics.2021.102735

Jiang, H., Wang, Z., Liu, X., Chen, X., Jin, Y., You, X., & Chen, X. (2017). A two-level approach for solving the inverse kinematics of an extensible soft arm considering viscoelastic behavior. *2017 IEEE International Conference on Robotics and Automation (ICRA)*, 6127–6133. 10.1109/ICRA.2017.7989727

Joachimczak, M., & Wróbel, B. (2012). Co-evolution of morphology and control of soft-bodied multicellular animats. *Proceedings of the 14th Annual Conference on Genetic and Evolutionary Computation*, 561–568. 10.1145/2330163.2330243

Lee, K.-H., Fu, D. K. C., Leong, M. C. W., Chow, M., Fu, H.-C., Althoefer, K., Sze, K. Y., Yeung, C.-K., & Kwok, K.-W. (2017). Nonparametric Online Learning Control for Soft Continuum Robot: An Enabling Technique for Effective Endoscopic Navigation. *Soft Robotics*, *4*(4), 324–337. doi:10.1089oro.2016.0065 PMID:29251567

Lepora, N. F., Evans, M., Fox, C. W., Diamond, M. E., Gurney, K., & Prescott, T. J. (2010). Naive Bayes texture classification applied to whisker data from a moving robot. *The 2010 International Joint Conference on Neural Networks (IJCNN)*, 1–8. 10.1109/IJCNN.2010.5596360

Liu, C.-H., Chen, T.-L., Chiu, C.-H., Hsu, M.-C., Chen, Y., Pai, T.-Y., Peng, W.-G., & Chiang, Y.-P. (2018). Optimal Design of a Soft Robotic Gripper for Grasping Unknown Objects. *Soft Robotics*, *5*(4), 452–465. doi:10.1089oro.2017.0121 PMID:29741987

Manzano, S. A., Xu, P., Ly, K., Shepherd, R., & Correll, N. (2021). High-Bandwidth Nonlinear Control for Soft Actuators with Recursive Network Models. In B. Siciliano, C. Laschi, & O. Khatib (Eds.), *Experimental Robotics,* (pp. 589–599). Springer International Publishing., doi:10.1007/978-3-030-71151-1_52

Methenitis, G., Hennes, D., Izzo, D., & Visser, A. (2015). Novelty Search for Soft Robotic Space Exploration. *Proceedings of the 2015 Annual Conference on Genetic and Evolutionary Computation*, 193–200. 10.1145/2739480.2754731

Mirjalili, S., Mirjalili, S. M., & Lewis, A. (2014). Grey Wolf Optimizer. *Advances in Engineering Software*, *69*, 46–61. doi:10.1016/j.advengsoft.2013.12.007

Naclerio, N. D., Karsai, A., Murray-Cooper, M., Ozkan-Aydin, Y., Aydin, E., Goldman, D. I., & Hawkes, E. W. (2021). Controlling subterranean forces enables a fast, steerable, burrowing soft robot. *Science Robotics*, *6*(55), eabe2922. doi:10.1126cirobotics.abe2922 PMID:34135117

Pagoli, A., Chapelle, F., Corrales-Ramon, J.-A., Mezouar, Y., & Lapusta, Y. (2020). Design and Optimization of a Dextrous Robotic Finger: Incorporating a Sliding, Rotating, and Soft-Bending Mechanism While Maximizing Dexterity and Minimizing Dimensions. *IEEE Robotics & Automation Magazine*, *27*(4), 56–64. doi:10.1109/MRA.2020.3024283

Rieffel, J., Knox, D., Smith, S., & Trimmer, B. (2014). Growing and Evolving Soft Robots. *Artificial Life*, *20*(1), 143–162. doi:10.1162/ARTL_a_00101 PMID:23373976

Rolf, M., Neumann, K., Queißer, J. F., Reinhart, R. F., Nordmann, A., & Steil, J. J. (2015). A multi-level control architecture for the bionic handling assistant. *Advanced Robotics*, *29*(13), 847–859. doi:10.1080/01691864.2015.1037793

Rost, A., & Schädle, S. (2013). The SLS-Generated Soft Robotic Hand—An Integrated Approach Using Additive Manufacturing and Reinforcement Learning. *12th International Conference on Machine Learning and Applications, 1*, 215–220. 10.1109/ICMLA.2013.44

Runge, G., Peters, J., & Raatz, A. (2017). Design optimization of soft pneumatic actuators using genetic algorithms. *IEEE International Conference on Robotics and Biomimetics (ROBIO)*, 393–400. 10.1109/ROBIO.2017.8324449

Sadati, S. M. H., Naghibi, S., Cruz, L., & Bergeles, C. (2020). *Reduced-Order Modeling and Model Order Reduction for Soft Robots*. doi:10.13140/RG.2.2.24115.86568

Salehan, A., & Deldari, A. (2022). Corona virus optimization (CVO): A novel optimization algorithm inspired from the Corona virus pandemic. *The Journal of Supercomputing*, *78*(4), 5712–5743. doi:10.100711227-021-04100-z PMID:34629744

Saunders, F., Golden, E., White, R. D., & Rife, J. (2011). Experimental verification of soft-robot gaits evolved using a lumped dynamic model. *Robotica*, *29*(6), 823–830. doi:10.1017/S0263574711000014

Schlagenhauf, C., Bauer, D., Chang, K.-H., King, J. P., Moro, D., Coros, S., & Pollard, N. (2018). Control of Tendon-Driven Soft Foam Robot Hands. *IEEE-RAS 18th International Conference on Humanoid Robots (Humanoids)*, 1–7. 10.1109/HUMANOIDS.2018.8624937

Thuruthel, T. G., Falotico, E., Cianchetti, M., Renda, F., & Laschi, C. (2016). Learning Global Inverse Statics Solution for a Redundant Soft Robot. *Proceedings of the 13th International Conference on Informatics in Control, Automation and Robotics*, 2, 303–310. 10.5220/0005979403030310

Thuruthel, T. G., Falotico, E., Renda, F., & Laschi, C. (2017). Learning dynamic models for open loop predictive control of soft robotic manipulators. *Bioinspiration &amp$\mathsemicolon$ Biomimetics, 12*(6), 066003. doi:10.1088/1748-3190/aa839f

Trivedi, D., Rahn, C. D., Kier, W. M., & Walker, I. D. (2008). Soft robotics: Biological inspiration, state of the art, and future research. *Applied Bionics and Biomechanics*, *5*(3), 99–117. doi:10.1155/2008/520417

Veenstra, F., Jørgensen, J., & Risi, S. (2018). Evolution of fin undulation on a physical knifefish-inspired soft robot. *Proceedings of the Genetic and Evolutionary Computation Conference*, 157–164. 10.1145/3205455.3205583

Zhang, H., Cao, R., Zilberstein, S., Wu, F., & Chen, X. (2017). Toward Effective Soft Robot Control via Reinforcement Learning. In Y. Huang, H. Wu, H. Liu, & Z. Yin (Eds.), *Intelligent Robotics and Applications,* (pp. 173–184). Springer International Publishing. doi:10.1007/978-3-319-65289-4_17

Zhang, H., Kumar, A. S., Chen, F., Fuh, J. Y. H., & Wang, M. Y. (2019). Topology Optimized Multi-material Soft Fingers for Applications on Grippers, Rehabilitation, and Artificial Hands. *IEEE/ASME Transactions on Mechatronics*, *24*(1), 120–131. doi:10.1109/TMECH.2018.2874067

Zhang, H., Kumar, A. S., Fuh, J. Y. H., & Wang, M. Y. (2018). Design and Development of a Topology-Optimized Three-Dimensional Printed Soft Gripper. *Soft Robotics*, *5*(5), 650–661. doi:10.1089oro.2017.0058 PMID:29985781

Zhang, M., Geng, X., Bruce, J., Caluwaerts, K., & Vespignani, M., SunSpiral, V., Abbeel, P., & Levine, S. (2017). Deep reinforcement learning for tensegrity robot locomotion. *2017 IEEE International Conference on Robotics and Automation (ICRA)*, 634–641. 10.1109/ICRA.2017.7989079

KEY TERMS AND DEFINITIONS

Association Rule Learning: Association rule learning, which is also a form of unsupervised learning based on rules, is a well-researched technique for uncovering interesting correlations and common patterns among items in the datasets.

Classification: Classification is a commonly applied supervised learning method that assigns one of the predefined classes to new instances.

Clustering: Clustering is a technique for grouping a collection of items into clusters based on their similarity.

Machine Learning: Machine Learning is an area of artificial intelligence in which computers can self-learn based on past experiences.

Meta-Heuristic Optimization: Meta-heuristic optimization is the collection of operations and models that employ randomness to optimize the candidates and discover the optimum solution.

Regression: Regression is another supervised learning task that discovers correlations between data features and predicts continuous-valued target output based on the historical data.

Reinforcement Learning: Reinforcement learning aims to construct actions or strategies for learning anticipated behaviors by developing reward functions.

Soft Robotics: Soft robotics is a branch of robotics that focuses on technologies that mimic the physical features of live beings.

Supervised Learning: The supervised learning technique aims to train an ML model using pre-labeled data.

Unsupervised Learning: Unsupervised learning technique learns patterns from unlabeled observation data with no previous knowledge of the output value.

Chapter 3
Perceiving the World With Sound:
An Overview to Robot Audition

Usama Saqib
https://orcid.org/0000-0001-9783-1887
GN Audio, Denmark

Robin Kerstens
https://orcid.org/0000-0002-5127-4947
University of Antwerp, Belgium

ABSTRACT

Robot perception is the ability of a robotic platform to perceive its environment by the means of sensor inputs, e.g., laser, IMU, motor encoders, and so on. Much like humans, robots are not limited to perceiving their environment through vision-based sensors, e.g., cameras. Robot perception, through the scope of this chapter, encompasses acoustic signal processing techniques to locate the presence of a sound source, e.g., human speaker, within an environment for human-robot interaction (HRI), that has gained great interest within scientific community. This chapter will serve as an introduction to acoustic signal processing within robotics, starting with passive acoustic localization and building up to contemporary active sensing methods, such as the usage of neural networks and spatial map generation. The origins of active acoustic localization, which finds its roots in biomimetics, are also discussed.

INTRODUCTION

The detection and localization of acoustic reflectors such as walls, objects or people within an environment is a popular topic within the area of robotics. Traditionally, camera and laser-based technologies are used to detect the presence of these landmarks to generate a spatial map of a 3D space and aid these robotic platforms in navigating within their environment. However, these light-based sensing modalities often face challenges such as a lack of light, overexposure (glare), the inability of detecting transparent

DOI: 10.4018/978-1-6684-5381-0.ch003

surfaces such as windows, false reflections, or their sensitivity to occlusion. These issues can be addressed when incorporating sound-based sensing modalities. Research in animal auditory system has inspired researchers to develop technologies to locate the presence of sound sources within an environment.

This chapter will serve as an introduction to robot audition, starting with the subdomains of robot audition and building up to more contemporary active sensing methods, such as the usage of Artificial Intelligence (AI) on recorded data to detect and track acoustic sources and for spatial map generation. The origins of active acoustic sensing, which finds its roots in biomimetics, are also discussed. This chapter is written as a reference for people working on robot perception using sound and wants to contribute to future works by bringing new challenges to the field of robot perception. The chapter will begin with an introduction to biomimicry in robotics, which aims to mimic an animal's auditory system to localize the position of sound sources in the nearby environment. Biomimicry facilitates intelligent designs in robots to achieve high performance and robustness when navigating between and localizing acoustic sources in a dynamic environment. Designers of such robots make use of new materials, sensors and actuators to provide high capabilities that allow robots to mimic biological processes such as hearing.

Furthermore, this chapter will review techniques in scientific literature associated with passive acoustic localization and active acoustic localization, which are the two important sub-domains of robot audition for SSL. Passive acoustic localization involves detecting sound generated by objects present in an environment while active acoustic localization techniques probe an environment with a known sound to detect the position of objects within an environment. Both sub-domains have their fair share of advantages and disadvantages. For example, active acoustic localization is useful in a quiet environment. This is normally the case when a robot explores an underground environment, such as caves, tunnels, and sewers. Bats, rats and even some aquatic mammals are known to use these techniques to navigate and hunt in complete darkness. These animals probe the environment with a unique sound, or call, and use acoustic echoes to distinguish flora and fauna, different types of animals/prey, and everything needed for their survival. Therefore, a discussion on the different types of probe signals that can be used in robotics to acquire spatial information from the environment is also an important highlight of this chapter. More specifically, analysis of additive white Gaussian noise (AWGN), coded emissions, and chirp signals will be discussed in detail. The application of spatial mapping using echolocation is also an important highlight of this chapter, which incorporate spatial filtering techniques, such as, beamforming techniques.

Finally, the chapter will review data-driven approaches to using contemporary methods such as neural networks for perceiving the environment in an artificially intelligent way. This is relatively a newer approach that combines physics-based model of sound with machine learning to teach robotic platforms to learn to classify and predict their surroundings.

BACKGROUND

Biomimicry in Robotics

The goal of robotics is to design an autonomous machine that behaves, reacts, and interacts with its surroundings much like a human does. That is, a robot should mimic the behavior and, in some cases, the physical appearance of humans. To fulfil this dream, researchers started studying humans and animal senses in an attempt to replicate the sensory organs, so that they can be used by robots. The five senses (seeing, hearing, touching, tasting, and smelling) have been a popular area of research and a source of inspiration for decades, as evident from a very large repository of scientific literature. By understanding and investigating human/animal senses, researchers can propose new techniques that are directly applicable to robotics such as image processing, computer vision, and audible systems, e.g., SSL techniques, while the remaining senses (touching, tasting, and smelling) are still an active area of research and beyond the scope of this chapter. Interested readers are advised to read (Ciui et al., 2018; Jie & Gunes, 2020; Kowadlo & Russell, 2008; Loutfi & Coradeschi, 2006; Shimazu et al., 2007).

Research in animal auditory systems has inspired researchers to develop technologies that locate the presence of sound sources within an environment. Robot audition is an area of research within robotics that aims to employ signal processing techniques to enable robots to locate and identify the source of a sound so that robots could effectively communicate with humans. For instance, in (Ono et al., 2003), an SSL structure is proposed that mimics the auditory system of a parasitoid fly. From the evolutionary perspective, two ears are useful to reduce cognitive load within the brain. Inspired from the animal auditory system, the sub-domain of robot audition called Binaural robot audition was defined. The common point among this subdomain is the use of two acoustic transducers. This falls under the sub-domain of passive acoustic localization.

Other examples of biomimicry include active acoustic localization, which is inspired by animals that probe the environment to obtain spatial information from the surroundings, such as the distance of an acoustic reflector, e.g., wall. This process is known as echolocation (Jones, 2005; Simmons & Stein, 1980; Sohl-Dickstein et al., 2015) and it is predominantly used by animals such as bats, rats, and dolphins to detect obstacles to aid their navigation in complete darkness (Edwards et al., 2009). According to research, echolocation is not limited to animals, but humans could also train themselves to use echolocation to navigate and detect obstacles (Norman et al., 2021). Compared to vision-based systems, echolocation-based systems can detect transparent or glass-like surfaces and can be used to work in the absence of light. Echolocation has been studied by researchers for several years and has led to the development of Sound Navigation and Ranging (SONAR) (Bienvenu, 2002; Mansour et al., 1988). SONAR is predominantly used in naval vessels to detect the presence of hostile ships, but as the technology matured and became easily accessible to the public, it is being used by the fishing industry to locate the presence of a school of fishes and used to monitor the ocean floor, as well as locate sunken ships.

Based on the above discussion, the ability of a robotic system to perceive its environment using echolocation, i.e., to acquire spatial information from the environment, is an interesting and challenging area of research. In the following subsection, the individual sub-domains within robot audition are discussed as shown in Figure 1.

Subdomains Within Robot Audition

Robot audition (Nakadai et al., 2000; Okuno & Nakadai, 2015) is an area of research within robotics that aims to employ signal processing techniques to provide robots with the ability to locate and identify the source of a sound such that robot could effectively communicate with a human. Moreover, sound-based modalities could complement existing state-of-the-art technologies in a dark or dusty environment to navigate while also avoiding potential collisions. Throughout literature, robot audition is divided into four main sub-domains:

- Sound source separation (SSS)
- Speech recognition

Figure 1. Sub-domains of Robot Audition

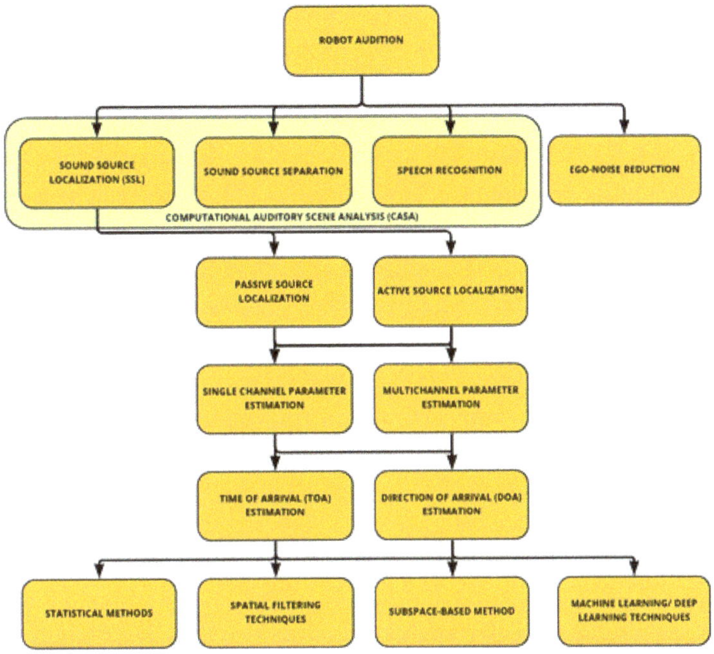

- Ego-noise reduction
- Sound source localization (SSL)

However, robot audition involves numerous challenges as the robots usually interact in dynamic, noisy, and reverberant environments, and they need to perform all computations in real-time on embedded hardware. SSL, sound source separation, and speech recognition and classification combine forms the basis of computational auditory scene analysis (CASA) as seen in Figure 1, which is another sub-domain of robot audition that has sparked interest from the scientific community.

Humans are capable of isolating audible sound sources from a mixture of sound sources (M. R. Pimpale et al., 2016). That is, a human can focus on a particular sound when multiple sound sources are simultaneously radiating within the environment. This is called the *cocktail party effect*. In robotics, SSS is a sub-domain of robot audition that aims to classify and distinguish audible signals to aid robotic systems intended for HRI to establish effective communication with human speakers (K.Duong et al., 2010; Mandel et al., 2009). Some approaches are based on a single-channel microphone that is formulated on the acoustic properties of sound, while the others require multiple microphones (Comon & Jutten, 2010).

Apart from SSS, a robotic system requires an understanding of a human speaker to enable effective HRI. Automatic speech recognition (ASR) is a research domain that makes use of algorithms that convert human speech into digital signals that are understood by robotic systems. Compared to voice recognition, a speech recognition system aims to translate human speech into text while voice recognition aims to recognize individual human voices. In a speech recognition system, audio signals are recorded by microphones which are first passed to a pre-processing stage that removes background noise from the signals. This is later passed to the feature extraction stage where the features such as statistical methods

(Shabani & Norouzi, 2016), spectral methods (Agrawal et al., 2010; Madan & Gupta, 2014), model-based methods (Lee et al., 1988; Russo et al., 2019), and pattern recognition methods (Chai et al., 2021; Jolad & Khanai, 2019; Seki et al., 2017). The features are then passed to the classification stage where algorithms are used to categorize speech (Barman & Deb, 2017; Spanias & Wu, 1991). Another important aspect of robot audition is reducing noise generated by the robotic system itself.

The term ego-noise is defined as the noise generated by the moving part of a robotic platform, e.g., the rotors of a drone. Ego-noise is a source of a problem in HRI systems as it corrupts the signals that prevent a robotic system from understanding human commands. The task of reducing ego-noise is further complicated because the ego-noise cannot be modeled as a static point source. Hence, traditional statistical methods cannot be applied directly to reduce ego-noise. However, there are attempts to simulate and model ego-noise of unmanned aerial vehicles (UAVs) (Insausti et al., 2020). Numerous techniques to reduce the ego-noise problem are suggested in the literature (Schmidt et al., 2020). One way to reduce the ego-noise of a robotic platform is to fuse the motor data information with a multi-channel dictionary algorithm (Schmidt et al., 2016). A multi-channel array ego-noise reduction algorithm such as beamforming is also proposed in drone audition as shown by the researchers in (Guzik et al., 2019; L. Wang & Cavallaro, 2017). Additionally, template matching (Ince et al., 2009) and data-driven methods for ego-noise reduction is also found in the literature (Ito et al., 2005; Schmidt et al., 2018). A relatively novel area of research is to utilize ego-noise constructively (Pico et al., 2016; U. Saqib & Jensen, 2021) to detect and estimate acoustic reflectors. The biggest sub-domain within robot audition is localizing sound sources present within an environment.

SSL techniques are further divided into two main categories: passive sound localization and active sound localization (Burgard et al., 1997; Schultheiss & Wagner, 1989). In passive sound localization, the sound source impinging the microphones from the environments are processed to estimate the distance and direction of the sound source, e.g., a robot detecting the presence of a human speaker. in contrast, active sound localization probes the environment with a known signal and the reflected sound signals are then recorded by the microphones for distance and direction estimation, e.g., echolocation employed by bats, active sonar in submarines, and radars (Heli & Abutalebi, 2011; Jeyasingh & Mohamed Ismail, 2018; Sasaki et al., 2005; T. Wang & Choy, 2015). There are two components to SSL that can be estimated to find the position of a sound source (Morikawa, 2014): the direction-of-arrival (DOA) information and the distance information which can be inferred from time-of-arrival (TOA) information of a sound source

SSL techniques are popularly used by the robotic community to enable robots to estimate the direction of human speakers. The popular techniques for direction estimation includes learning-based approaches, e.g., neural networks (Chazan et al., 2018; Deng et al., 2020; Nakadai et al., 2000; Pfeifenberger et al., 2017; Ramezanpour & Mosavi, 2020; Zhang, 2021), beamforming-based approaches, subspace-based approaches, tracking techniques such as Kalman filters and particle filters.

PASSIVE VS ACTIVE ACOUSTIC LOCALIZATION

Passive acoustic localization (PAL) is done when the sound sources are present within an environment and the job of the detecting system, e.g., robots, is to analyze the sound signals to determine the location of the object in question. Within robotics, PAL enables robots to detect the presence of human speakers and facilitate human-robot interaction (HRI). In scientific literature, the direction estimation includes learning-based approaches, e.g., neural networks (Chazan et al., 2018; Deng et al., 2020; Nakadai et al., 2000; Pfeifenberger et al., 2017; Ramezanpour & Mosavi, 2020), beamforming-based approaches, subspace-based approaches, tracking techniques such as Kalman filters and particle filters.

One major limitation of PAL is that it requires the presence of a sound source within an environment. In the event when there are no acoustic sources available then passive sound localization techniques do not work. In comparison, active sound localization probes the environment to acquire acoustic echoes for localization. Active acoustic localization (AAL) was inspired by animals, e.g., bats. Echolocation is a mechanism used by animals including humans to probe the environment with a known sound and receive a delayed version of the probe sound to infer the distance of acoustic reflectors, e.g., walls and obstacles. Researchers have studied and developed systems that utilize echolocation to locate objects (Quazi, 1981; Schultheiss & Wagner, 1989). This has also led to the development of sonar-based technologies that are used, notably, by naval vessels, especially submarines, to detect the presence of hostile ships.

Therefore, in this chapter, we define AAL as an approach that can enable a robot to probe an environment with a known sound, such that features about the environment can be extracted from acoustic echoes as shown in Figure 2. These features could be the distance to the acoustic reflectors, e.g., walls, objects, obstacles, etc. as the robot moves within an environment (Gutmann et al., n.d.). Compared to PAL, this approach does not require the presence of sound sources within an environment to enable room geometry estimation. Moreover, AAL can facilitate accurate localization of acoustic reflectors and the task of room geometry estimation becomes less complex (Saqib et al., 2020).

Figure 2. Active acoustic localization

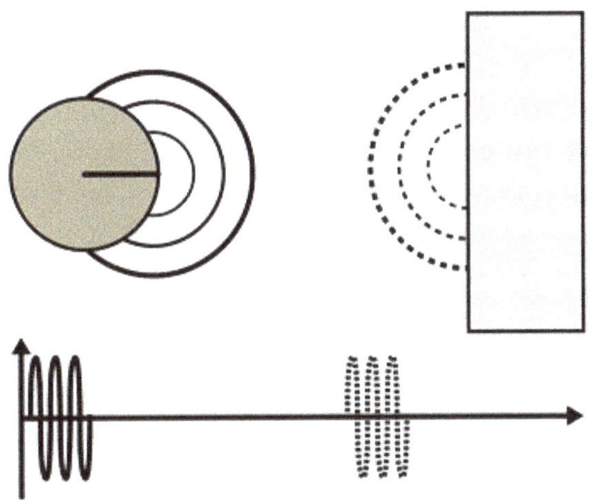

Within the context of robotics, exteroceptive sensors such as ultrasonic sensors are used to detect the presence of obstacles which helps robotic platforms to avoid collision and to estimate robot pose and navigation (Lim & Leonard, 2000; Moreira et al., 2019; M. Wang et al., 2005; Wu et al., 2016, 2015). AAL approach are found in scientific literature, e.g., the work performed by (Eliakim et al., 2018), a robotic platform was designed that probes the outdoor environment to navigate and classify floras. The robot is equipped with an ultrasonic emitter and receivers. The transducer emits a chirp sound to the environment while the receiver captures these reflected sounds for processing. The reflected sound is then processed to determine the TOA information of the acoustic echoes. The robot can classify the object around itself using an Artificial Neural Network (ANN) which makes the spatial map of the environment very rich. In another study, AAL was also used to construct a bat-inspired echolocation system that utilizes ultrasonic transducers to classify outdoor terrains. The authors in (Riopelle et al., 2018) trained an advance signal processing and machine learning technique to classify outdoor terrain with 97% accuracy. The echo data collected from grass, concrete and sand was then used to train a Support Vector Machine (SVM) classifier.

Table 1. Comparison between AAL and PAL techniques

	Passive acoustic localization	Active acoustic localization
Works in the absence of environmental sound?	No	Yes
Can be used to detect obstacles?	No	Yes
Can be used to map unknown environment?	No	Yes
Influenced by environmental noise?	Yes	No

Moreover, AAL were used in (Jan Steckel & Peremans, 2013). The authors take inspiration from RatSLAM (Milford et al., 2004), which is a navigation system based on a computational model of a rodent. With this work, the researchers successfully show that an autonomous system can navigate an environment purely using sonar sensors. This work is then extended by (Chen & Hu, 2019) to be applied on UAVs. Furthermore, the AAL approach is also useful to estimate the geometry of a room as shown in (Kreković et al., 2016). Here, the authors proposed two methods of detecting the shape of an environment using a robotic platform. The first method uses simple triangulation to estimate the virtual source which in turn is used to estimate the first-order early reflections while the second approach uses a Bayesian method to estimate the acoustic echoes. The process involves probing the environment and estimating the acoustic reflectors as the robot moves within an environment. AAL is also used in smartphone to estimate room geometry as shown in (Zhou et al., 2017). The authors use a standard smartphone to probe the environment using a chirp signal. In this work, the choice of frequency, pulse width and duration were empirically decided. The process involved in estimating the distance of an acoustic reflector using a standard cross-correlation technique. Additionally, the authors proposed a probabilistic mapping algorithm that enabled their setup to construct a spatial map of an environment. The comparison between AAL and PAL approach are summarized in Table 1.

Acoustic Echo Modeling

The sound propagation model depends on the position of the robotic platform as well as the properties of an environment. The two most used propagation models are 1) *near field* and 2) *far field*. In the near field, the observer is close to the radiating sound source and the sound originating from sources reaching the microphones are curved, assuming that there are no obstacles between the source and microphone. Moreover, it is also assumed that the reflections from the environment impinging the microphones do not exist. On the contrary, the far field assumption assumes that the observer is far away from the radiating sound source and the sound source that reaches each microphone within a microphone array are plane waves. If robot is moving within an indoor environment and probes the environment for acoustic reflector localization, then this is mathematically formulate as shown:

$$y(n) = h(n) * s(n) + v(n),$$

$$= x(n) + v(n) \tag{1}$$

where y(n) is the observed signal recorded by the microphone while h(n) and s(n) denote the acoustic impulse response (discussed next) and the probe signal, respectively. The background noise v(n) is usually represented as an additive white Gaussian noise (AWGN), which is useful to formulate a mathematically feasible model so that close form solutions to the estimators are found (Kay, 1993).

Figure 3. An example of acoustic impulse response (AIR)

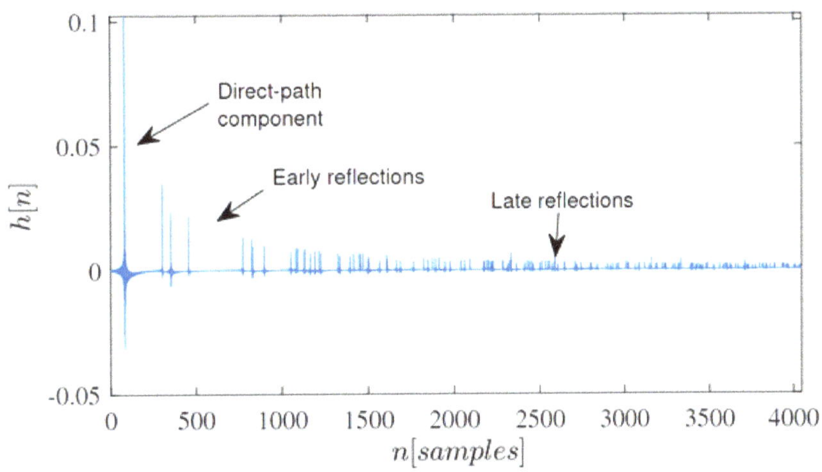

Acoustic Impulse Response (AIR)

The transfer function of an environment between the source and the microphone is known as acoustic *impulse response* (AIR) and is also known in the literature as *room impulse response* (RIR). The AIR is shown in Figure 3. AIR is divided into two distinct parts: the direct-path plus early reflections and a stochastic long tail that represents reverberation or late reflections (Moschioni, 2002). The direct-path component is the shortest distance that a sound takes while the early reflections represent the sound bouncing off an acoustic reflector, e.g., wall. The late reflections are sounds that bounce off multiple acoustic reflectors before reaching a microphone. The AIR structure is useful to formulate a model of sound for different applications. To detect and estimate the distance of an acoustic source within the environment, the direct-path component estimation is important. Moreover, early reflections give information about the geometry of the environment (Antonacci et al., 2010; Filos et al., n.d.; Gunel, 2002). Based on these observations, Eq. 1 can be decomposed as a sum of its direct-path and its reflections and expressing the transfer function between the emitter and receiver in terms of its gains and delay, the signal model can be rewritten as shown:

$$y(n)=\sum_{q=1}^{\infty}g_q s(n-\tau_q)+v(n) \tag{2}$$

where g_q is the gain or attenuation of the q-th reflection from the source to microphone and q is the TOA of the reflected signal while * represent the convolution operator. In our definition, the direct-path component corresponds to q = 1. Keeping the structure of AIR in mind, Eq. 1 can be rewritten as the sum of the first **R** reflections as shown:

$$\begin{aligned} y(n) &= \sum_{q=1}^{R}g_q s(n-\tau_q)+d(n)+v(n), \\ &= x(n)+v'(n) \end{aligned} \tag{3}$$

where d(n) is the stochastic and dense tail of the late reflections shown in Fig. Hence, if **N** samples are taken then Eq. 3 can be represented as shown:

$$\begin{aligned} \mathbf{y}(n) &= \sum_{q=1}^{R}g_q \mathbf{s}(n-\tau_q)+\mathbf{d}(n)+\mathbf{v}(n), \\ &= \mathbf{x}(n)+\mathbf{v}'(n) \end{aligned} \tag{4}$$

The task at hand is to estimate the TOA and DOA of the acoustic reflector, In the following sections, different techniques found in the literature for TOA estimations is discussed.

DISTANCE ESTIMATION OF AN ACOUSTIC REFLECTOR

Distance is usually inferred from the TOA information of a sound signal (Baba et al., 2017; Moore et al., 2013; Tervo & Tossavainen, 2012). As discussed earlier, the transfer function of an environment in Figure 3 is useful for estimating the TOA of a sound signal and location of acoustic reflectors. Traditionally, the extraction of TOA information of a sound source is done from the estimated AIR using a standard peak-picking approach. However, within the context of robot audition, extracting TOAs in this way is a time-consuming process because the AIR will have to be estimated every time a robot moves to a new location. Within robotics, ultrasonic sensors are used to estimate the distance between an acoustic reflector and robotic platform, but this method requires attaching robots with additional sensors which can increase the overall cost of robotic platforms.

Furthermore, the distance estimation could also be done by exploiting robot movement (Kreković et al., 2016; Nguyen et al., 2019; J Steckel & Peremans, 2012). In (Kreković et al., 2016), the robot probes the environment as it moves and by using simple trigonometry and later a Bayesian approach, the researchers were able to estimate the distance between the acoustic reflector and the robot. Similarly, in (Nguyen et al., 2019), the researchers proposed an algorithm that enables a robot equipped with a sound source and four microphones to follow a predefined path around an environment to estimate the geometry of an environment. In (Schillebeeckx et al., 2011; J Steckel & Peremans, 2012), the researchers propose a biomimetic sonar system based on bats' ability to smartly navigate an environment using sonar sensing alone. The authors based their analysis on the number of conditional entropy (Shannon, 2001) between the range data and the robot position.

PULSE-ECHO RANGING AND LOCALIZATION (SONAR)

While the previous methods explain how a passive system would interact with the environment, acoustics in robots is also found in an active sense where the system emits its own sound that will then be used to scan for reflections. An example of active sensing using sound is sonar. SOund NAvigation and Ranging has long been a popular method for systems to gain information about the nearby environment. There are multiple ways to implement a sonar system. One where the system has a continuous transmission, as is discussed in (Siciliano & Khatib, 2008), and one where short burst are being used. The latter method finds its origins in biology, as discussed earlier in the bio-mimicry section, systems will make use of sound signals to probe the environment for any obstacles. Knowing the signal that has been sent out, and the time it takes for the signal to return the sound transceiver, we can deduce a distance. Similar to when using radar methods, we can extract the object distance (Richards et al., 2010):

$$R = \frac{c\Delta T}{2} \tag{5}$$

Where c is the speed of the wave, for sound in the average case this is 343 m/s (Pierce & Beyer, 2005), ΔT is the time difference between the transmission of the signal, and the reception of the reflection. The halving of the value happens because the signals traverse the same distance twice (towards the object and back to the receiver).

In research, sonar is slowly increasing in popularity again. Sonar was often used in the early days of robotics as it does not require expensive hardware or complex computing platforms to perform basic tasks such as distance estimation in a single dimension. Systems like this are still ubiquitous, e.g., parking sensors are made using simple 1D transceivers. But now, with the ever-increasing computer power and efficiency of electronics, more complex sensors can be made that are able to scan an environment in 3D and offer reliable performance in scenarios where other sensing modalities often struggle (Jansen et al., 2020; Kerstens et al., 2019b). This is due to the long wavelength of sound, which is less impacted by dust, rain and small airborne particles when compared to the short wavelengths of light-based modalities such as LiDAR (Kutila et al., 2018). Of course, there are also drawbacks to using sonar in precise applications. The poor attenuation of sound through air works as a limiting factor on range (Pierce & Beyer, 2005), and depending on the size of the sensor, the system's Point-Spread-Function (PSF) (i.e., how a single point-source is registered by the sensor (Kerstens et al., 2019a) becomes larger as the distance to an object increases. This means that a reflecting object will appear to be increasingly larger the further it is positioned from the sensor.

RESOLUTION AND ACCURACY

To scan the environment using an active sensor, there are two main approaches. When a sensor performs its measurement, it does not directly know from which direction the received signal originates. In order to obtain this information, the sensor needs to be aimed in a certain direction, either mechanically or electronically. Mechanical steering is the most intuitive and can, in general, achieve higher quality measurements, as the sensor can be designed in such a way that the beam pattern is as narrow as possible in the frontal direction, cancelling out other directions. However, this comes with the inherent downside that mechanical movement is involved, which requires additional maintenance and accurate motors (Siciliano & Khatib, 2008). With the increasing amount of computing power, modern systems often rely more on electrical steering or scanning using phased arrays. Phased arrays rely on the presence of multiple receiving elements with a known distance between them. This distance can range from several millimeters to many meters. Popular examples of large active measuring systems are the over-the-horizon (early warning) RADAR systems, built by the Soviet Union in the 1980's, like the DUGA-1 RADAR where transmitter and receiver are separated by about 60 km (Litvinov, 2020). Or a more contemporary system that is in ongoing development is the Jindalee Operational Radar Network (JORN) in Australia, with stations separated by several hundreds of kilometers (Litvinov, 2020).[1]

When using sound, these extreme distances would be nonsensical as the wavelengths used in in-air robotics rarely exceed 0.02 meters (17 kHz). This is because designers often rely on frequencies outside the range of human hearing. Underwater sonar takes another approach and relies on infrasound, which is below human hearing having wavelengths approaching 34 meters (10 Hz) and longer.

To detect these pulses, several techniques can be used as well. In the early days a simple single frequency pulse would be transmitted, and, at the receiver side, the reflection would be band-pass filtered and fitted with a specific amplitude threshold. If the threshold was met, the system would see this as a legitimate reflection (Ureña et al., 2018). Very simple to implement, but prone to error, this method is no longer used in real applications. In modern applications it is common to send out a coded emission (García et al., 2015; Ureña et al., 2018) and create a filter that looks specifically for that coded signal. This approach is called matched filtering (Knapp & Carter, 1976) and proves to be very effective as a longer measuring signal can be compressed into a single peak that clearly indicates the TOA (Verellen et al., 2020). Also known as Pulse Compression, this process is now a popular way of working in many fields. Having these clear peaks in our recorded signal, also allows us to identify and distinguish signals that would otherwise overlap, and can also drastically improve the signal to noise ratio of our measurement (Richards et al., 2010) which in turn improves the likelihood of detection for smaller objects that would otherwise fail to reflect sufficient energy to meet the threshold. Once detected, the area where acoustic sensing shines is range resolution and accuracy. Having a longer wavelength compared to other RF- or light-based modalities helps the matched filter in achieving accurate results for TOA estimation.

The range resolution or a system is the smallest difference in range that can be measured by the system and is an important factor for autonomous platforms should they want to perform tasks such as autonomous navigation. The smallest difference in range can be determined using the following formula:

$$\Delta R = \frac{c\tau}{2} \tag{6}$$

Where ΔR is the range resolution of the system, two targets lying closer together than this distance will be seen as a single object. *c* being the speed of the travelling wave (343 m/s). τ is the pulse width of the unmodulated pulse used during the measurement (Richards et al., 2010).

When talking about resolution and accuracy, not only range is of importance, but one should also consider these in terms of direction or bearing. As previously mentioned, one of the downsides of sonar is that the size of the PSF increases with distance. If then at a certain distance two smaller objects are located close to each other, the system will fail to identify them as separate objects and will show a single reflection coming from the average direction of the two objects. When this happens, we can state that the two objects, and our sensor do not meet the Rayleigh criterion (Richards et al., 2010). These states:

$$\theta = 1.22\frac{\lambda}{D} \tag{7}$$

Where θ is the minimum resolvable angle expressed in radians, λ is the wavelength of our signal used during the measurement and D *is* the diameter of the sensor array. The Rayleigh criterion thus states that a smaller wavelength is preferable, along a large diameter array. This also explains the large distances between the over-the-horizon radar systems mentioned earlier. It is therefore also crucial to have a well-designed sensor, should the application require sufficient accuracy in both range and direction.

DESIGN ASPECTS

Array Geometry

As a short recapitulation: The array of an acoustic sensor is the portion of the sensor that features the (multiple) transmitting or receiving element(s). There are several ways in which the elements can be placed, all with their own benefits and disadvantages, but the concept behind every geometry is the same: Capturing multiple versions of the same signal, with predefined time (phase) differences between them. The ideal array would have an infinite number of elements and have a diameter capable of fitting everything, but of course designers need to take in account both restrictions in size and computing power. This in turn encourages the use of more complex designs and geometries to get the best performance possible for specific application.

Figure 4. Examples of different array topologies in which the elements are placed in a specific order. From left to right: Uniform Linear Array (ULA), Uniform Rectangular Array (URA), Uniform Circular Array (UCA), Random Array.

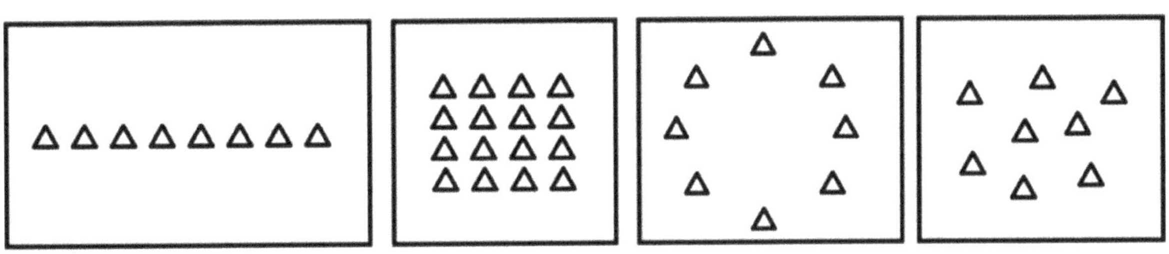

The figure above illustrates some commonly used array geometries, or topologies. From left to right, there is the Uniform Linear Array (ULA), Uniform Rectangular Array (URA), the Uniform Circular Array (UCA), and the random array. The ULA is a popular option because of its 1D characteristic, meaning it can easily be integrated into a system. It also has a broad aperture which, looking back at the discussion on the Rayleigh criterion, means it delivers an accurate response. An inherent downside of each 1D array topology is that the TDOA can only be measured in two dimensions: range and angle. For most applications this can be sufficient (e.g. autonomous navigation through a corridor) but more complex environments can benefit from having an additional dimension for its detections. When using 2D array topologies, such as the URA, UCA, or random array, we obtain TDOA information for both horizontal and vertical dimensions. This means we can position the obtained reflections with a certain range (average time delay), azimuth angle (horizontal plane TDOA), and elevation angle (vertical plane TDOA).

This is why a linear or circular microphone arrays are so often used to localize sound (Krim & Viberg, 1996). Recently, more sophisticated designs featuring pseudo random element placement (Kerstens et al., 2019b) or spherical microphone array designs are used as they offer SSL with an additional dimension that can offer better sound field capture (Meyer & Elko, 2002; 2004). Microphone array designs are particularly important if the aim is SSL. However, for application that require speech separation and noise reduction, other techniques can be used as well. More sophisticated methods for localization can be utilized such as subspace-based methods, which are a type of spatial filtering method (Kerstens et al., 2020; Moulines et al., 1995). Many of the problems in the field of microphone array signal processing seems to be similar to those encountered in the field of antenna array signal processing. However, when dealing with microphones and soundwaves, we have to take notice of some additional factors (Herbordt & Kellermann, 2003; Van Compernolle, 1990). These are mainly because audio/speech signals are broadband in nature, the reverberation of the environment is high, the number of microphones is often limited, and the acoustic signals are often non-stationary.

MULTI-PULSE SONAR

When using this kind of phased arrays, it is hard to skip the nomenclature of SIMO, MISO, MIMO. While being more common in radar literature (Richards et al., 2010), it can also be applied to sonar systems as there are many similarities between the two in terms of array design.

SISO, meaning Single Input Single Output is the most basic kind of system, and the term is therefore also not commonly used. SISO systems would be a basic transceiver, both sending and receiving a signal using the same element. A common example is the ubiquitous parking sensor.

The systems discussed in the previous section can be labeled as MISO - meaning Multiple Input, Single Output - having a single emitter and multiple receiving elements to increase accuracy. These systems are also very popular due to their simplicity, but if the designer want to increase the quality of measurements even further, it might be worthwhile to also add elements to the emitting part of the sensor, which brings us to MIMO systems, Multiple Input Multiple Output. Having multiple outputs allows the designer to add amplitude and complexity to the signals that are being emitted. The most common form of multi-pulse sonar are so-called sonar rings. These rely on several single transmitters that each look into a different direction and provide a single range value for the direction in which they are looking. Very effective and cheap to implement, this approach is very popular in many applications ranging from hobbyists at home to modern motorized vehicles to assist in terms of slow navigation tasks such as parking. When these emitters are all placed on the same surface, facing the same direction, the amount of energy being pushed into the environment drastically increases which results in a greater range and a higher sensitivity for nearby obstacles. In this case, we make a distinction between systems in which all elements transmit the same signal, or each signal sends out a unique, known, signal.

When using a set of known signals to "probe" the environment we can exploit this feature by treating each signal as a single measurement originating from a different, known, location. This technique is known as array virtualization, as we are now synthesizing a mathematically "larger" array than the one that is physically present and, as mentioned in the previous sections, a larger number of elements, spread out over a larger surface, results in measurements of a better quality. (Kerstens et al., 2022; Melvin & Scheer, 2013)

Probing Signals

The signal that is being used to scan the environment, i.e., being emitted by the sensor that will eventually record the reflections, is a factor that plays a crucial role in the performance of the sensor. Many books have been written on this topic (He et al., 2012; Levanon & Mozeson, 2004; Richards et al., 2010; Vaseghi, 2008, …) but the concepts are the following:

Figure 5. Different type of probe signals

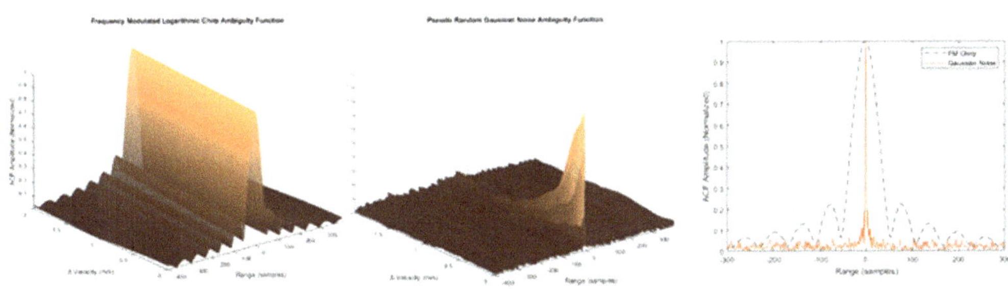

In essence, the more energy that is being emitted into the environment, the more energy the sensor will receive back. This total amount of energy is impacted by factors such as the duration of the pulse, the frequency content of the pulse, and of course the volume at which it is created. A silent 1 ms single frequency 40 kHz pulse will return less energy (e.g. illustrated in Figure 4 a), and will thus obtain less information about the environment, when compared to a loud 10 ms broadband chirp with a frequency content between 20 to 200 kHz (e.g. illustrated in Figure 4 b).

The frequency band that is used also plays an important role. Lower frequencies tend to be better at penetration and less troubled by attenuation than high frequency signals, while the Rayleigh criterion showed us that the higher frequency signals are a better fit for applications that require high resolutions. It is up to the designer of a system to then decide what is best for the application in mind or, which can also be done, fail to decide and use a very broadband signal that spans a very large range of frequencies. However, this would require a powerful sensor with a lot of computing power to process all the obtained information. For this debate, it might also be useful to look at nature, where bats have evolved to use a frequency range between 20 kHz and 80 kHz and are capable of high-speed navigation.

Coding of the signals can also be used to further enhance the TDOA estimation of the signal (e.g. illustrated in figure x c). Classical encoding schemes such as Kasami or Gold coding can help a system to greatly improve its localization capabilities (García et al., 2015). Aside from coding the signal, some researchers also employ "pseudo-random" Gaussian noise signals. The term "pseudo" is used, because the noise pulse is generated on beforehand, and reused for every measurement. This allows us to search for the unique, broadband, noise signal in the recordings, again resulting in a higher chance of detecting weaker reflections. Although, these advanced techniques are primarily used to really get peak performance and might not be necessary for basic navigation purposes.

Figure 6a. Examples of signals that can be used during an active measurement: A single frequency pulse.

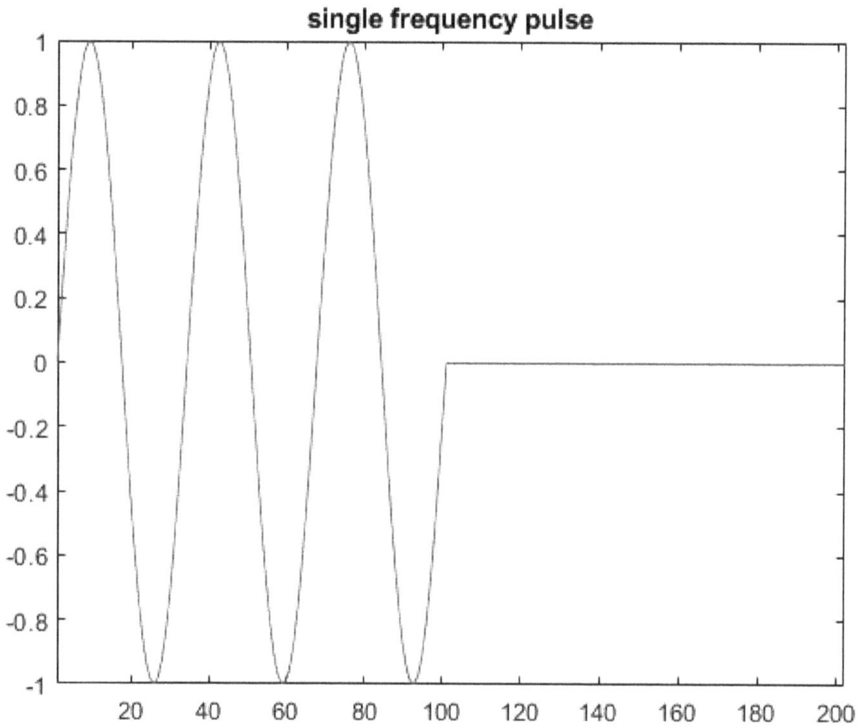

AMBIGUITY FUNCTIONS

It is not uncommon for active sensing systems to be used in a situation where the environment involves moving features, whether it is a robot driving around in a scene, or objects passing by a static setup. In both of these situations there will be a radial velocity component between the sensor and these objects. This difference in radial velocity will result in a Doppler shirt, a phenomenon in which a frequency scaling occurs depending on the radial velocity difference of the reflecting object (Richards et al., 2010). The resulting frequency shift can be represented by

$$f_d = \frac{2v_r}{\lambda} \tag{8}$$

Where v_r is the radial velocity component and λ is the wavelength of the signal used. The ambiguity function of a signal will show us how much the frequency shift of a signal impacts the matched filter performance of our system, or, in broader terms, how pulse detection performance is affected by the changing frequency (García et al., 2015). An ambiguity function can thus be seen as a collection of matched filter responses for a predetermined range of Doppler shifted versions of our probing signal.

Figure 6b. Examples of signals that can be used during an active measurement: A broadband pulse.

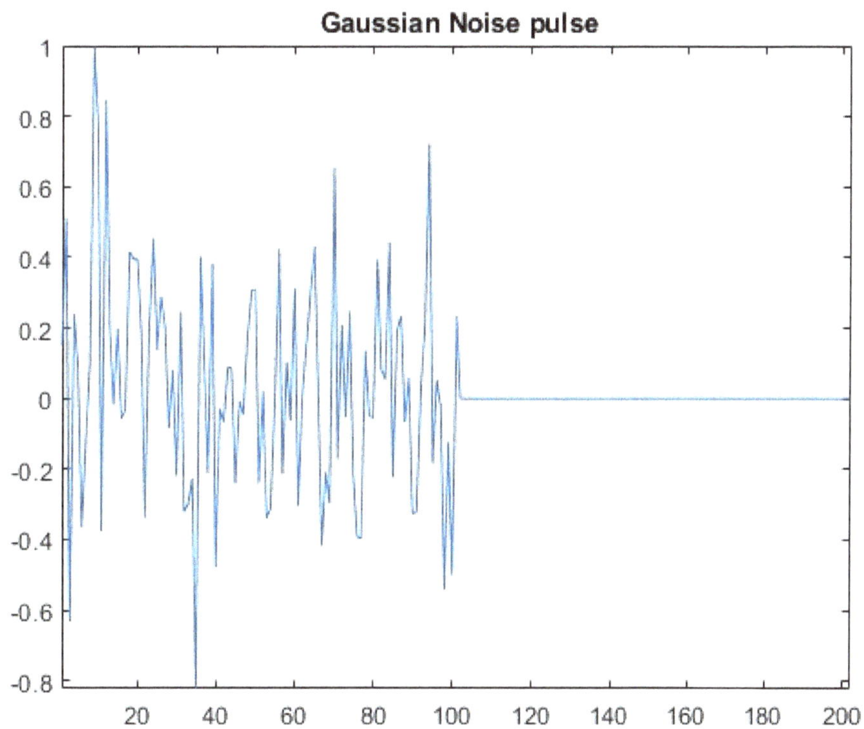

While the large wavelengths of sound will result in relatively small frequency shifts, it can be an important factor to keep in mind when designing signals. As seen in Figure 6 two examples: a frequency modulated logarithmic chirp signal, and a normal white Gaussian noise signal. The Gaussian noise signal is commonly used in active sensing, as it can contain a lot of energy in a shorter burst thanks to its large bandwidth. It also serves its purpose really well when it comes to accurately estimating distances of objects, as can be seen on the right of the Figure 4. It has a very sharp peak around the range where the object is located. However, if we take a look at the ambiguity function of the same signal, we can see that the matched filter, tuned to recognize this noise, is very sensitive to any frequency shift and once there is a radial velocity component present, measurement quality declines rapidly. When comparing to another type of broadband signal, the frequency modulated logarithmic chirp, the plot on the right shows

Figure 6c. Examples of signals that can be used during an active measurement: A single frequency pulse with phase coding.

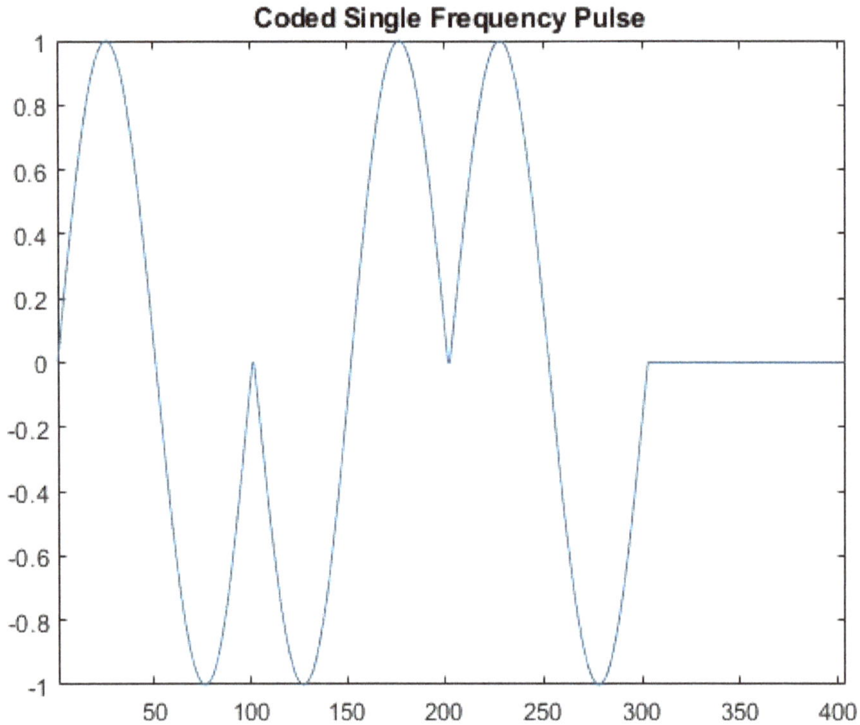

us that in terms of range estimating performance, the logarithmic chirp has less accurate results than the Gaussian noise. The situation changes when looking at the ambiguity function, where the logarithmic chirp hardly seems to suffer in terms of detection when movement is introduced. It should come as no surprise that in nature, bats use logarithmic chirps like this one to navigate in their environment. They travel at high speeds through dense environments and cannot afford to have their speed impact their quality of sensing.

NOTES ON NOISE

When perceiving the world using sound, noise will always be present and can pose as a real nuisance for the designer. Knowing where to look for different types of noise and how to handle it is therefore a valuable skill to have when working with these types of systems.

- **System noise:** Noise introduced by the system itself, can only be addressed by a better design, e.g. switching from analog to digital microphones (Kerstens et al., 2017) which will bypass any noise introduced by amplification. Using methods such as compressed sampling can make up for any samples that might have disappeared in the process and increase the quality of the recorded signals.

- **Atmospheric noise:** There might be a lot of ultrasonic noise present in the environment (e.g., pressurized air leaks, nearby construction works, …) and these might be hard to detect and inconsistent. Because of the higher wavelength, and the short distance high frequency acoustic waves can travel, this noise might be located in a limited region of the environment. To limit the impact these noise sources can have on our system it is useful to check out different ways to overcome noise.
- **Ghost/false reflections:** Technically this cannot be considered as "noise" but are rather artifacts of sensor design or signal processing. A matched filter might produce one large peak at the location of the object, and several periodically spaced smaller peaks. These can also be visible on the screen and be interpreted.
- **Multipath:** Also, these cannot really be seen as noise, but to the untrained eye these randomly occurring points at specific distances can be seen as false reflections. Multipath reflections in acoustic sensing are not as present as they are in RF applications. The large attenuation factor of air results in multipath reflections only posing issues in very close-range situations.

Identifying noise will help in the next step, which should be mitigating its effects on the system. Below is a short list of widespread coping methods that can be used with acoustic sensors:

- **Classic thresholding:** When using a small band signal, it might be useful to simply check how loud our reflection is compared to the other signals coming from the environment. This is by far the simplest method to implement but is prone to false triggering as many other sound sources might temporarily give a signal louder than our preset threshold.
- **CFAR:** Constant False Alarm Rate thresholding is a method that has been popular in RADAR for a while. Using CFAR, we apply an adaptive threshold level based on the value of neighboring cells, or samples. Short bursts of interference can be mitigated this way, and 1-D CFAR can be implemented with a relatively low computational need. More advanced systems, requiring 2-D CFAR operations are limited by the fact that every cell in the acoustic image needs to be analyzed independently, and because this is a serial process, framerates might be impacted.
- **Matched Filtering:** If a known measuring sequence is being used, it might be useful to design a filter that removes all information from a signal that does not "match" to know signal. This process is known as matched filtering. A matched filter keeps all portions of the sequence that "match" with the given sequence. A popular method is to use the GCC-PHAT (Generalized Cross Correlation – Phase Angle Transform) method (Knapp & Carter, 1976), which first transfers the signal into the frequency domain, multiplies it, and then transforms it back into the time domain. Mentioned briefly before, a matched filter is capable of detecting sounds in recordings where the SNR is far below 1, meaning it can not only increase your range resolution, but also support the operation of your system in high-noise environments. These reasons make it a very popular technique to be used.
- **Iteration:** If the possibility arises to perform multiple measurements of exactly the same scenario, the random noise will not be identical across the different frames. Performing multiple (near-) static measurements can then be used to average out the presence of noise in the environment. Techniques such as CLEAN-SC take this approach further and achieve SNR levels that are hard to match with any of the other techniques previously mentioned (Luesutthiviboon et al., 2019).

CONTEMPORARY METHODS

In recent development, data driven approaches to TOA and DOA estimation has attracted interest from the scientific community. Given enough data, machine learning (ML) can discover relationship between the features and the labels or could discover hidden relationship within a dataset (Bishop & Nasrabadi, 2006). Moreover, ML can discover complex models describing human speech and other acoustic phenomena. Applying ML especially deep learning (DL) within robotics is an active area of research (Pierson & Gashler, 2017). Data-driven approach is also used to estimate distance to acoustic reflectors. For example, features such as TOA are extracted from the observed signals at different distances, which are used to train an ANN for distance estimation as seen in (Kumarakulasingam & Agah, 2008). The input to the neural network is fed the coordinates of the microphone and features of the sound source are recorded at a different position within an environment. For example, in (Yiwere & Rhee, 2020), the researchers employ a convolutional recurrent neural network (CRNN) (Tang et al., 2015; Zuo et al., 2015) to learn features related to distance. This is done by converting the recorded audio signals into time-frequency representation i.e., the audio signals are transformed into mel-spectrogram (Stevens et al., 1937).

Within acoustics, ML and DL are used for SSL techniques, speech enhancement, classifying everyday sound from the environment. ML is classified into two main categories: supervised learning and unsupervised learning. In supervised learning, the aim is to learn a relationship between the input data and the labels associated with the data. In unsupervised learning, the labels are not available and the goal is to find hidden structure or relationship within the datasets (K. P. Murphy, 2012). For example, in the work of (Murray et al., 2009, 2006), the authors use a recurrent neural network (RNN) to estimate the next position of an acoustic source by first estimating DOA and feeding it to an RNN. Although ML especially DL offers a very powerful tool to SSL problem and to recognize speech for effective HRI but training a neural network for a specific task is a very computationally expensive process. This is why, most of the training is often done offline while the ML/DL train model is often used by the robotic platform to carry out its task.

APPLICATION OF SOUND AND MICROPHONE ARRAYS IN INDUSTRY AND FUTURE TRENDS

AAL has found its usefulness in robotics when detecting acoustic reflectors, e.g., walls, obstacles and people. By probing the environment with a known sound, robotic platform can use the reflected signals to from the environments to estimate the location of the acoustic reflectors. The use of acoustic sensors has certain advantages compared to traditional sensors technologies used in industries. For instance, camera-based technologies used in autonomous vehicles and drones can provide accurate information about the environment, but these technologies are susceptible to changing light condition and does not work in a completely dark environment as well as in in environments with limited visibility, e.g., fog. Moreover, camera-based technologies have limited field-of-view, limiting its use to detect objects around the corner. To increase the field-of-view, additional cameras are required. Camera and laser-based technologies are expensive compared to a good quality acoustic sensor, e.g., microphone. These limitations are addressed by incorporating acoustics sensors, e.g., microphone arrays as seen in Figure 5.

Figure 7. A robotic platform with AAL technique to detect objects in smoky environment

Additionally, acoustic sensors are also used to inspect water pipes and sewers. The work presented by (Worley et al., 2020) showcase a robotic platform to inspect water pipes for faults and leakage, which is useful for asset monitoring and repairing tasks for governmental agencies. Similarly, PAL techniques are also very popular within aerospace and automotive industries to study aerodynamics and to detect faults. For instance, acoustic cameras are very popular among such industries to detect the source of the noise originating from the moving part of a body of an aircraft or car (Erić, 2011). The basic principle behind acoustic camera is estimating the accurate location of a sound source from simultaneous recording from multiple microphones. As the sound travel in a fixed medium, e.g., air, at a fixed speed, it is recorded by each microphone at different time instances and at different intensity. One popular method for obtaining an acoustic image is by using beamforming techniques. DSB is a popular technique used to obtain an acoustic image by aligning the delayed time instances from multiple microphones. Furthermore, acoustic cameras are also used within robotics for search and rescue missions. For instance, in 2011 Tohoku Earthquake and Tsunami (Japan), Center for Robot Assisted Search and Rescue (CRASAR) deployed four ROVs with enhanced camera and acoustic cameras to inspect critical infrastructure and recover bodies from the affected areas (R. R. Murphy, 2012).

Traditionally, acoustic-based object detection and classification was only seen in navel devices to detect fisheries and hostile ships, but the vast adoption of acoustic localization techniques are also gaining popularities among autonomous vehicles. Some researchers such as (Deleforge, 2020) suggest using drones equipped with microphones array for search and rescue task. As discussed earlier, acoustic-based sensing is very useful in inspecting area that do not have access to light such as caves or sewers. Moreover, acoustic-based sensing is also useful to detect objects around the corner of the robotic platform because compared to camera-based technologies, acoustic sensors does not have limited field-of-view. However, acoustic sensing techniques are still an active area of research and more work on this domain is needed to produce product that could be used in industry.

CONCLUSION

This chapter is serves as an easy to grasp introduction to the concept of sound in robotics. Starting at the beginning, where research has drawn inspiration out of nature to use sound as a sensing method, through various topics such as passive and active sensing, Room Impulse Reponses, reflections in signals, noise, mapping sensitivity to object movement by means of an ambiguity function. The reader can use this as a guide to the many topics the field has to offer and use the referenced literature as a source for more information about the topics. Additional reading is also provided and features some of the more important works in the field, along with a few interesting reads for those interested in acoustic sensor design.

This research received no specific grant from any funding agency in the public, commercial, or not-for-profit sectors.

REFERENCES

Agrawal, S. S., Prakash, N., & Jain, A. (2010). Transformation of emotion based on acoustic features of intonation patterns for Hindi speech. *African Journal of Mathematics and Computer Science Research*, *3*(10), 255–266.

Antonacci, F., Sarti, A., & Tubaro, S. (2010). Geometric reconstruction of the environment from its response to multiple acoustic emissions. *IEEE International Conference on Acoustics, Speech and Signal Processing*, 2822–2825. 10.1109/ICASSP.2010.5496186

Baba, Y. E., Walther, A., & Habets, E. A. P. (2017). 3D room geometry inference based on room impulse response stacks. *IEEE/ACM Transactions on Audio, Speech, and Language Processing*, *26*(5), 857–872. doi:10.1109/TASLP.2017.2784298

Barman, T., & Deb, N. (2017). State of the art review of speech recognition using genetic algorithm. *IEEE International Conference on Power, Control, Signals and Instrumentation Engineering (ICPCSI)*, 2944–2946. 10.1109/ICPCSI.2017.8392264

Bienvenu, G. (2002). Signal sonar processing. *2002 14th International Conference on Digital Signal Processing Proceedings. DSP 2002 (Cat. No.02TH8628)*, *1*, 441–446. 10.1109/ICDSP.2002.1027914

Bishop, C. M., & Nasrabadi, N. M. (2006). Pattern recognition and machine learning, 4(4). Springer.

Burgard, W., Fox, D., & Thrun, S. (1997). Active mobile robot localization. *IJCAI (United States)*, 1346–1352.

Chai, L., Du, J., Liu, D. Y., Tu, Y. H., & Lee, C. H. (2021). Acoustic modeling for multi-array conversational speech recognition in the chime-6 challenge. *IEEE Spoken Language Technology Workshop (SLT)*, 912–918. 10.1109/SLT48900.2021.9383628

Chazan, S. E., Goldberger, J., & Gannot, S. (2018). DNN-based concurrent speakers detector and its application to speaker extraction with LCMV beamforming. *IEEE International Conference on Acoustics, Speech and Signal Processing (ICASSP)*, 6712–6716. 10.1109/ICASSP.2018.8462407

Ciui, B., Martin, A., Mishra, R. K., Nakagawa, T., Dawkins, T. J., Lyu, M., Cristea, C., & Wang, R. S. J. (2018). Chemical sensing at the robot fingertips: Toward automated taste discrimination in food samples. *ACS Sensors*, *3*(11), 2375–2384. doi:10.1021/acssensors.8b00778 PMID:30226368

Comon, P., & Jutten, C. (2010). *Handbook of Blind Source Separation: Independent component analysis and applications*. Academic press.

Deleforge, A. (2020). Drone audition for search and rescue: Datasets and challenges. *QUIET DRONES: International Symposium on UAV/UAS Noise*. https://hal.archives-ouvertes.fr/hal-03430293/

Deng, C., Song, H., Zhang, Y., Sha, Y., & Li, X. (2020). *DNN-based Mask Estimation Integrating Spectral and Spatial Features for Robust Beamforming*, 4647–4651.

Duong, K., Vincent, E., & Gribonval, R. (2010). Under-determined reverberant audio source separation using a full-rank spatial covariance model. *IEEE Transactions on Audio, Speech, and Language Processing*, *18*(7), 1830–1840. doi:10.1109/TASL.2010.2050716

Edwards, D. S., Allen, R., Papadopoulos, T., Rowan, D., Kim, S. Y., & Wilmot-Brown, L. (2009). Investigations of mammalian echolocation. *Annual International Conference of the IEEE Engineering in Medicine and Biology Society*, 7184–7187. 10.1109/IEMBS.2009.5335313

Eliakim, I., Cohen, Z., Kosa, G., & Yovel, Y. (2018). A fully autonomous terrestrial bat-like acoustic robot. *PLoS Computational Biology*, *14*(9), e1006406. doi:10.1371/journal.pcbi.1006406 PMID:30188901

Erić, M. M. (2011). Some research challenges of acoustic camera. *19th Telecommunications Forum, TELFOR Proceedings of Papers*, 1036–1039. 10.1109/TELFOR.2011.6143725

Filos, J., Habets, E. A. P., & Naylor, P. A. (n.d.). *A two-step approach to blindly infer room geometries*.

García, E., Paredes, J. A., Álvarez, F. J., Pérez, M. C., & García, J. J. (2015). Spreading sequences in active sensing: A review. *Signal Processing*, *106*, 88–105. doi:10.1016/j.sigpro.2014.07.002

Gunel, B. (2002). Room shape and size estimation using directional impulse response measurements. *Proc. Forum Acusticum Sevilla*.

Gutmann, J.-S., Burgard, W., Fox, D., & Konolige, K. (n.d.). An experimental comparison of localization methods. *Proceedings IEEE/RSJ International Conference on Intelligent Robots and Systems. Innovations in Theory, Practice and Applications (Cat. No.98CH36190)*, *2*(1), 736–743. 10.1109/IROS.1998.727280

Guzik, M., Kowalczyk, K., Woźniak, S., Fraś, M., Juros, K., Kaczor, D., & Walas, P. (2019). Acoustic source localization using drone-embedded microphone array. *The Journal of the Acoustical Society of America*, *146*(4), 3058–3059. doi:10.1121/1.5137614

He, H., Li, J., & Stoica, P. (2012). *Waveform design for active sensing systems: a computational approach*. Cambridge University Press. doi:10.1017/CBO9781139095174

Heli, H., & Abutalebi, H. R. (2011). Localization of multiple simultaneous sound sources in reverberant conditions using blind source separation methods. *International Symposium on Artificial Intelligence and Signal Processing (AISP)*, 1–5. 10.1109/AISP.2011.5960978

Herbordt, W., & Kellermann, W. (2003). Adaptive beamforming for audio signal acquisition. In *Adaptive Signal Processing,* (pp. 155–194). Springer. doi:10.1007/978-3-662-11028-7_6

Ince, G., Nakadai, K., Rodemann, T., Hasegawa, Y., Tsujino, H., & Imura, J. I. (2009). *Ego noise suppression of a robot using template subtraction*. 199–204.

Insausti, X., Hogstad, B. O., & Pätzold, M. (2020). Modelling and Simulation of Ego-Noise of Unmanned Aerial Vehicles. *IEEE 91st Vehicular Technology Conference (VTC2020-Spring)*, 1–5.

Ito, A., Kanayama, T., Suzuki, M., & Makino, S. (2005). Internal noise suppression for speech recognition by small robots. *Ninth European Conference on Speech Communication and Technology*. 10.21437/Interspeech.2005-260

Jansen, W., Laurijssen, D., Kerstens, R., Daems, W., & Steckel, J. (2020). *In-Air Imaging Sonar Sensor Network with Real-Time Processing Using GPUs*. Springer. doi:10.1007/978-3-030-33509-0_67

Jeyasingh, P., & Mohamed Ismail, M. (2018). Real-time Multi Source Speech Enhancement based on Sound Source Separation using Microphone Array. *Conference on Emerging Devices and Smart Systems (ICEDSS)*, 183–187. 10.1109/ICEDSS.2018.8544373

Jie, Z., & Gunes, H. (2020). Investigating Taste-liking with a Humanoid Robot Facilitator. *29th IEEE International Conference on Robot and Human Interactive Communication (RO-MAN)*, 131–136.

Jolad, B., & Khanai, R. (2019). An Art of Speech Recognition: A Review. *2nd International Conference on Signal Processing and Communication (ICSPC)*, 31–35.

Jones, G. (2005). Echolocation. *Current Biology*, *15*(13), R484–R488. doi:10.1016/j.cub.2005.06.051 PMID:16005275

Kay, S. M. (1993). *Fundamentals of statistical signal processing*. Prentice Hall PTR.

Kerstens, R., Jansen, W., & Steckel, J. (2022). *Radial Velocity Estimation for Multiple Objects using In-Air Sonar with MIMO Virtual Arrays*. IEEE. doi:10.1109/ACCESS.2022.3154395

Kerstens, R., Laurijssen, D., & Steckel, J. (2017). Low-cost One-bit MEMS Microphone Arrays for In-air Acoustic Imaging Using FPGA ' s. *IEEE SENSORS, 2017-. Decem, 1–3*, 1–3. doi:10.1109/ICSENS.2017.8234087

Kerstens, R., Laurijssen, D., & Steckel, J. (2019a). An Optimized Planar MIMO Array Approach to In-Air Synthetic Aperture Sonar. *IEEE Sensors Letters*, *3*(11), 1–4. doi:10.1109/LSENS.2019.2951061

Kerstens, R., Laurijssen, D., & Steckel, J. (2019b). eRTIS: A Fully Embedded Real Time 3D Imaging Sonar Sensor for Robotic Applications. *International Conference on Robotics and Automation (ICRA)*, 1438–1443. 10.1109/ICRA.2019.8794419

Kerstens, R., Verellen, T., & Steckel, J. (2020). Comparing Adaptations of MUSIC beamforming for 3D In-Air Sonar Applications with Coherent Source-spaces. *Berlin Beamforming Conference (BeBeC)*, 1–12.

Knapp, C., & Carter, G. (1976). The generalized correlation method for estimation of time delay. *IEEE Transactions on Acoustics, Speech, and Signal Processing*, *24*(4), 320–327. doi:10.1109/TASSP.1976.1162830

Kowadlo, G., & Russell, R. A. (2008). Robot odor localization: A taxonomy and survey. *The International Journal of Robotics Research, 27*(8), 869–894. doi:10.1177/0278364908095118

Kreković, M., Dokmanić, I., & Vetterli, M. (2016). *EchoSLAM: Simultaneous localization and mapping with acoustic echoes*, 11–15.

Krim, H., & Viberg, M. (1996). Two decades of array signal processing research: The parametric approach. *IEEE Signal Processing Magazine, 13*(4), 67–94. doi:10.1109/79.526899

Kumarakulasingam, P., & Agah, A. (2008). Neural network-based single sensor sound localization using a mobile robot. *Intelligent Automation \&. Soft Computing, 14*(1), 89–103.

Kutila, M., Pyykonen, P., Holzhuter, H., Colomb, M., & Duthon, P. (2018). Automotive LiDAR performance verification in fog and rain. *IEEE Conference on Intelligent Transportation Systems, Proceedings, ITSC,* 1695–1701. 10.1109/ITSC.2018.8569624

Lee, C. H., Soong, F. K., & Juang, B. H. (1988). *A segment model based approach to speech recognition*, 501–502.

Levanon, N., & Mozeson, E. (2004). *Radar Signals*. John Wiley & Sons. doi:10.1002/0471663085

Lim, J. H., & Leonard, J. J. (2000). Mobile robot relocation from echolocation constraints. *IEEE Transactions on Pattern Analysis and Machine Intelligence, 22*(9), 1035–1041. doi:10.1109/34.877524

Litvinov, S. (2020). Over-the-horizon detection radar ionosphere support system. *Journal of Physics: Conference Series, 1632*(1), 012021. doi:10.1088/1742-6596/1632/1/012021

Loutfi, A., & Coradeschi, S. (2006). Smell, think and act: A cognitive robot discriminating odours. *Autonomous Robots, 20*(3), 239–249. doi:10.100710514-006-7098-8

Luesutthiviboon, S., Malgoezar, A. M. N., Merino-Martinez, R., Snellen, M., Sijtsma, P., & Simons, D. G. (2019). Enhanced HR-CLEAN-SC for resolving multiple closely spaced sound sources. *International Journal of Aeroacoustics, 18*(4–5), 392–413. doi:10.1177/1475472X19852938

Madan, A., & Gupta, D. (2014). Speech feature extraction and classification: A comparative review. *International Journal of Computers and Applications, 90*(9).

Mandel, M. I., Weiss, R. J., & Ellis, D. P. W. (2009). Model-based expectation-maximization source separation and localization. *IEEE Transactions on Audio, Speech, and Language Processing, 18*(2), 382–394. doi:10.1109/TASL.2009.2029711

Mansour, M., Smith, B. V., & Noaks, D. R. (1988). Active sonar signal simulation. *IEE Colloquium on Simulation Techniques Applied to Sonar*, 2/1-2/4.

Melvin, W. L., & Scheer, J. A. (2013). Principles of modern radar: Advanced techniques. In Principles of Modern Radar: Advanced Techniques. Institution of Engineering and Technology. doi:10.1049/SBRA020E

Meyer, J., & Elko, G. (2002). A highly scalable spherical microphone array based on an orthonormal decomposition of the soundfield. *IEEE International Conference on Acoustics, Speech, and Signal Processing, 2*, II--1781. 10.1109/ICASSP.2002.1006109

Meyer, J., & Elko, G. W. (2004). Spherical microphone arrays for 3D sound recording. In *Audio signal processing for next-generation multimedia communication systems,* (pp. 67–89). Springer. doi:10.1007/1-4020-7769-6_3

Moore, A. H., Brookes, M., & Naylor, P. A. (2013). *Room geometry estimation from a single channel acoustic impulse response.*

Moreira, T., Lima, J., Costa, P., & Cunha, M. (2019). Low-cost Sonar based on the Echolocation. *ICINCO*, (1), 818–825.

Morikawa, D. (2014). Effect of Interaural Difference for Localization of Spatially Segregated Sound. *Tenth International Conference on Intelligent Information Hiding and Multimedia Signal Processing*, 602–605. 10.1109/IIH-MSP.2014.156

Moschioni, G. (2002). A new method for measurement of early sound reflections in theaters and halls. *IMTC/2002. Proceedings of the 19th IEEE Instrumentation and Measurement Technology Conference*, *1*, 425–430 vol.1.

Moulines, E., Duhamel, P., Cardoso, J.-F., & Mayrargue, S. (1995). Subspace methods for the blind identification of multichannel FIR filters. *IEEE Transactions on Signal Processing*, *43*(2), 516–525. doi:10.1109/78.348133

Murphy, K. P. (2012). *Machine learning: a probabilistic perspective.* MIT press.

Murphy, R. R. (2012). A decade of rescue robots. *IEEE International Conference on Intelligent Robots and Systems*, 5448–5449. 10.1109/IROS.2012.6386301

Murray, J. C., Erwin, H. R., & Wermter, S. (2009). Robotic sound-source localisation architecture using cross-correlation and recurrent neural networks. *Neural Networks*, *22*(2), 173–189. doi:10.1016/j.neunet.2009.01.013 PMID:19233613

Murray, J. C., Wermter, S., & Erwin, H. R. (2006). Bioinspired auditory sound localisation for improving the signal to noise ratio of socially interactive robots. *IEEE/RSJ International Conference on Intelligent Robots and Systems*, 1206–1211. 10.1109/IROS.2006.281855

Nakadai, K., Lourens, T., Okuno, H. G., & Kitano, H. (2000). Active audition for humanoid. *AAAI/IAAI*, 832–839.

Nguyen, L., Miro, J. V, & Qiu, X. (2019). *Can a Robot Hear the Shape and Dimensions of a Room?* 5346–5351.

Norman, L. J., Dodsworth, C. D., Foresteire, D., & Thaler, L. (2021). Human click-based echolocation: Effects of blindness and age, and real-life implications in a 10-week training program. *PLoS One*, *16*(6), e0252330. doi:10.1371/journal.pone.0252330 PMID:34077457

Okuno, H., & Nakadai, K. (2015). *Robot audition: Its rise and perspectives,* 5610–5614. doi:10.1109/ICASSP.2015.7179045

Ono, N., Saito, A., & Ando, S. (2003). Design and experiments of bio-mimicry sound source localization sensor with gimbal-supported circular diaphragm. *TRANSDUCERS '03. 12th International Conference on Solid-State Sensors, Actuators and Microsystems. Digest of Technical Papers, 1*, 935–938.

Pfeifenberger, L., Zöhrer, M., & Pernkopf, F. (2017). DNN-based speech mask estimation for eigenvector beamforming. *IEEE International Conference on Acoustics, Speech and Signal Processing (ICASSP)*, 66–70. 10.1109/ICASSP.2017.7952119

Pico, A., Schillaci, G., Hafner, V. V., & Lara, B. (2016). How do I sound like? forward models for robot ego-noise prediction. *Joint IEEE Int. Conf. on Development and Learning and Epigenetic Robotics*, 246–251. 10.1109/DEVLRN.2016.7846826

Pierce, A. D., & Beyer, R. T. (2005). Acoustics: An Introduction to Its Physical Principles and Applications. In *The Journal of the Acoustical Society of America*, *87*(4). Acoustical Society of America. https://asa.scitation.org/doi/10.1121/1.399390

Pierson, H. A., & Gashler, M. S. (2017). Deep learning in robotics: A review of recent research. *Advanced Robotics*, *31*(16), 821–835. doi:10.1080/01691864.2017.1365009

Pimpale, M. R., Therese, S., & Shinde, V. (2016). A survey on: Sound source separation methods. *International Journal (Toronto, Ont.)*, *3*(11), 580–584.

Quazi, A. (1981). An overview on the time delay estimate in active and passive systems for target localization. *IEEE Transactions on Acoustics, Speech, and Signal Processing*, *29*(3), 527–533. doi:10.1109/TASSP.1981.1163618

Ramezanpour, P., & Mosavi, M. R. (2020). DNN-based interference mitigation beamformer. *IET Radar, Sonar \& Navigation, 14*(11), 1788–1794.

Richards, M. A., Scheer, J. A., & Holm, W. A. (2010). Principles of modern radar: Basic principles. In Principles of Modern Radar: Basic Principles. Institution of Engineering and Technology. doi:10.1049/SBRA021E

Riopelle, N., Caspers, P., & Sofge, D. (2018). Terrain classification for autonomous vehicles using bat-inspired echolocation. *International Joint Conference on Neural Networks (IJCNN)*, 1–6. 10.1109/IJCNN.2018.8489046

Russo, M., Stella, M., Sikora, M., & Pekić, V. (2019). Robust cochlear-model-based speech recognition. *Computers*, *8*(1), 5. doi:10.3390/computers8010005

U. Saqib, A. D., & Jensen, J. R. (2021). *Detecting Acoustic Reflectors using a Robot's Ego-Noise*, 1–5.

Saqib, U., Gannot, S., & Jensen, J. R. (2020). Estimation of acoustic echoes using expectation-maximization methods. *EURASIP Journal on Audio, Speech, and Music Processing*, *2020*(1), 1–15. doi:10.118613636-020-00179-z

Sasaki, Y., Tamai, Y., Kagami, S., & Mizoguchi, H. (2005). 2D sound source localization on a mobile robot with a concentric microphone array. *IEEE International Conference on Systems, Man and Cybernetics*, *4*, 3528-3533. 10.1109/ICSMC.2005.1571694

Schillebeeckx, F., De Mey, F., Vanderelst, D., & Peremans, H. (2011). Biomimetic Sonar: Binaural 3D Localization using Artificial Bat Pinnae. *The International Journal of Robotics Research, 30*(8), 975–987. doi:10.1177/0278364910380474

Schmidt, A., W., HEinrich, W. L., & Kellermann, W. (2018). *A Novel Ego-Noise Suppression Algorithm for Acoustic Signal Enhancement in Autonomous Systems*. 6583–6587.

Schmidt, A., Deleforge, A., & Kellermann, W. (2016). *Ego-noise reduction using a motor data-guided multichannel dictionary*. 1281–1286.

Schmidt, A., Löllmann, H. W., & Kellermann, W. (2020). Acoustic self-awareness of autonomous systems in a world of sounds. *Proceedings of the IEEE, 108*(7), 1127–1149. doi:10.1109/JPROC.2020.2977372

Schultheiss, P. M., & Wagner, K. (1989). Active and passive localization: Similarities and differences. In *Underwater Acoustic Data Processing,* (pp. 215–232). Springer. doi:10.1007/978-94-009-2289-1_23

Seki, H., Yamamoto, K., & Nakagawa, S. (2017). *A deep neural network integrated with filterbank learning for speech recognition*. 5480–5484.

Shabani, S., & Norouzi, Y. (2016). Speech recognition using Principal Components Analysis and Neural Networks. *IEEE 8th International Conference on Intelligent Systems (IS)*, 90–95.

Shannon, C. E. (2001). A mathematical theory of communication. *Mobile Computing and Communications Review*, *5*(1), 3–55. doi:10.1145/584091.584093

Shimazu, H., Kobayashi, K., Hashimoto, A., & Kameoka, T. (2007). Tasting robot with an optical tongue: Real time examining and advice giving on food and drink. *Symposium on Human Interface and the Management of Information*, 950–957. 10.1007/978-3-540-73345-4_107

Siciliano, B., & Khatib, O. (2008). Springer Handbook of Robotics. In B. Siciliano & O. Khatib (Eds.), *Choice Reviews Online*. Springer Berlin Heidelberg. doi:10.1007/978-3-540-30301-5

Simmons, J. A., & Stein, R. A. (1980). Acoustic imaging in bat sonar: Echolocation signals and the evolution of echolocation. *Journal of Comparative Physiology*, *135*(1), 61–84. doi:10.1007/BF00660182

Sohl-Dickstein, J., Teng, S., Gaub, B. M., Rodgers, C. C., Li, C., DeWeese, M. R., & Harper, N. S. (2015). A Device for Human Ultrasonic Echolocation. *IEEE Transactions on Biomedical Engineering*, *62*(6), 1526–1534. doi:10.1109/TBME.2015.2393371 PMID:25608301

Spanias, A. S., & Wu, F. H. (1991). Speech coding and speech recognition technologies: a review. IEEE International Sympoisum on Circuits and Systems, 572–577.

Steckel, J., & Peremans, H. (2012). Biomimetic sonar for biomimetic SLAM. *Sensors (Basel)*, 1–4.

Steckel, J., & Peremans, H. (2013). BatSLAM: Simultaneous Localization and Mapping Using Biomimetic Sonar. *PLoS One*, *8*(1), e54076. doi:10.1371/journal.pone.0054076 PMID:23365647

Stevens, S. S., Volkmann, J., & Newman, E. B. (1937). A scale for the measurement of the psychological magnitude pitch. *The Journal of the Acoustical Society of America*, *8*(3), 185–190. doi:10.1121/1.1915893

Tang, D., Qin, B., & Liu, T. (2015). Document modeling with gated recurrent neural network for sentiment classification. *Proceedings of the Conference on Empirical Methods in Natural Language Processing*, 1422–1432. 10.18653/v1/D15-1167

Tervo, S., & Tossavainen, T. (2012). *3D room geometry estimation from measured impulse responses*, 513–516.

Ureña, J., Hernández, Á., García, J. J., Villadangos, J. M., Pérez, M. C., Gualda, D., Álvarez, F. J., & Aguilera, T. (2018). Acoustic Local Positioning with Encoded Emission Beacons. In *Proceedings of the IEEE,* 106(6), pp. 1042–1062. Institute of Electrical and Electronics Engineers Inc. 10.1109/JPROC.2018.2819938

Van Compernolle, D. (1990). Switching adaptive filters for enhancing noisy and reverberant speech from microphone array recordings. *International Conference on Acoustics, Speech, and Signal Processing*, 833–836. 10.1109/ICASSP.1990.115966

Vaseghi, S. V. (2008). Advanced Digital Signal Processing and Noise Reduction. In Advanced Digital Signal Processing and Noise Reduction: Fourth Edition. Wiley. doi:10.1002/9780470740156

Verellen, T., Kerstens, R., & Steckel, J. (2020). High-Resolution Ultrasound Sensing for Robotics Using Dense Microphone Arrays. *IEEE Access: Practical Innovations, Open Solutions*, *8*, 190083–190093. doi:10.1109/ACCESS.2020.3032177

Wang, L., & Cavallaro, A. (2017). Microphone-array ego-noise reduction algorithms for auditory micro aerial vehicles. *IEEE Sensors Journal*, *17*(8), 2447–2455. doi:10.1109/JSEN.2017.2669262

Wang, M., Tamimi, H., & Zell, A. (2005). Robot navigation using biosonar for natural landmark tracking. *International Symposium on Computational Intelligence in Robotics and Automation*, 3–7. 10.1109/CIRA.2005.1554246

Wang, T., & Choy, Y. (2015). An approach for sound sources localization and characterization using array of microphones. *International Conference on Noise and Fluctuations (ICNF)*, 1–4. 10.1109/ICNF.2015.7288571

Worley, R., Yu, Y., & Anderson, S. (2020). Acoustic Echo-Localization for Pipe Inspection Robots. *IEEE International Conference on Multisensor Fusion and Integration for Intelligent Systems (MFI)*, 160–165. 10.1109/MFI49285.2020.9235225

Wu, X., Abrahantes, M., & Edgington, M. (2016). MUSSE: A designed multi-ultrasonic-sensor system for echolocation on multiple robots. *Asia-Pacific Conference on Intelligent Robot Systems (ACIRS)*, 79–83. 10.1109/ACIRS.2016.7556192

Wu, X., D'Orazio, P., Edgington, M., & Abrahantes, M. (2015). Robotics echolocation test platform. *IEEE International Conference on Electro/Information Technology (EIT)*, 558–562. 10.1109/EIT.2015.7293397

Yiwere, M., & Rhee, E. J. (2020). Sound Source Distance Estimation Using Deep Learning: An Image Classification Approach. *Sensors (Basel)*, *20*(1), 172. doi:10.339020010172 PMID:31892213

Zhang, X. L. (2021). Deep ad-hoc beamforming. *Computer Speech \& Language, 68*, 101201.

Zhou, B., Elbadry, M., Gao, R., & Ye, F. (2017). BatMapper: Acoustic sensing based indoor floor plan construction using smartphones. *Proceedings of the 15th Annual International Conference on Mobile Systems, Applications, and Services*, 42–55. 10.1145/3081333.3081363

Zuo, Z., Shuai, B., Wang, G., Liu, X., Wang, X., Wang, B., & Chen, Y. (2015). Convolutional recurrent neural networks: Learning spatial dependencies for image representation. *Proceedings of the IEEE Conference on Computer Vision and Pattern Recognition Workshops*, 18–26. 10.1109/CVPRW.2015.7301268

ADDITIONAL READING

Garcia, E., Paredes, J. A., Álvarez, F. J., Pérez, M. C., & García, J. J. (2015). Spreading sequences in active sensing: A review. *Signal Processing*, *106*, 88–105. doi:10.1016/j.sigpro.2014.07.002

Kerstens, R. (2020). Advanced Array-based Sensors for Industrial In-Air Sonar Applications [Doctoral dissertation]. University of Antwerp.

Krim, H., & Viberg, M. (1996). Two decades of array signal processing research: The parametric approach. *IEEE Signal Processing Magazine*, *13*(4), 67–94. doi:10.1109/79.526899

Pierce, A. D. (2019). *Acoustics: an introduction to its physical principles and applications*. Springer. doi:10.1007/978-3-030-11214-1

Pierson, H. A., & Gashler, M. S. (2017). Deep learning in robotics: A review of recent research. *Advanced Robotics*, *31*(16), 821–835. doi:10.1080/01691864.2017.1365009

Rascon, C., & Meza, I. (2017). Localization of sound sources in robotics: A review. *Robotics and Autonomous Systems*, *96*, 184–210. doi:10.1016/j.robot.2017.07.011

Scheer, J., & Holm, W. A. (2010). Principles of modern radar.

Van Trees, H. L. (2004). Optimum array processing: Part IV of detection, estimation, and modulation theory. John Wiley & Sons.

KEY TERMS AND DEFINITIONS

AIR: – Acoustic impulse response
CASA: – Computational auditory scene analysis
CFAR: – Constant False Alarm Rate
DL: – Deep learning
DOA: – Direction of arrival
DSB: – Delay and sum beamformer
IPD: – Interaural phase difference
ITD: – interaural time difference
MIMO: – Multiple input and multiple output
MISO: – Multiple input and single output

ML: – Machine learning
MVDR: – Minimum variance distortionless response
PSF: – Point Spread Function
RIR: – Room impulse response
RNN: – Recurrent neural network
ROV: – Remotely operated vehicle
SIMO: – Single input and multiple output
SONAR: – Sound navigation and ranging
TDOA: – Time difference of arrival
TOA: – Time of arrival

ENDNOTE

1 *SONAR and RADAR systems share many design aspects, which is one of the reasons there can be a lot of cross-referencing between the two fields.*

Chapter 4

A Soft Robotic Gripper Material Study:

Effects of CNT Mixing Methodologies

Mehmet Mert İlman
https://orcid.org/0000-0001-7664-5217
Manisa Celal Bayar University, Manisa, Turkey

Hamza Taş
Manisa Celal Bayar University, Manisa, Turkey

ABSTRACT

Soft robots have gained superiority in outdoor applications compared to traditional robots today. This advantage is clearly due to bio-inspiration and evolving material technology. The objective of this research is to use nanotechnology to improve material qualities. For this, silicone named DragonSkin 20 (DS20), which can be employed in soft robot applications, was selected as the matrix material, while functionalized multi-walled-carbon-nanotube (MWCNT) was utilized as an additive. One of the parameters that determine the mechanical properties is the change of curing behavior. The choice of mixing technique, on the other hand, is very crucial since it affects the curing behavior. For this reason, the effects of not only the additive but also the various mixing techniques on the material behavior and curing time were reported as a result of the experiments. The results showed that the mixing methodologies plays an important role on the mechanical properties and curing time of neat and MWCNT reinforced silicone.

INTRODUCTION

Recently, robotic systems have gained importance and become widespread in parallel with technological evolution. They have made people's life easier in many ways, such as increasing safety, time and material savings, and productivity. Also, robots provide advantages in many applications like manufacturing, health care, agriculture, food preparation, military, and electronics (Marechal et al., 2021; Walker et al., 2020; El-Atab et al., 2020)

DOI: 10.4018/978-1-6684-5381-0.ch004

A robot is basically classified as hard (rigid) or soft based on its underlying materials (Alici, 2018). Soft robotics which is mostly inspired from biological systems and deals with the control, design, and production of robots consisting of stretchy and flexible materials as opposed to rigid links used in rigid-bodied robots is a subfield of robotics (Lee et al., 2017). While conventional rigid-bodied robots are actuated with rigid electrical motors, hydraulic pumps, or pneumatic compressors (Lee et al., 2017), the movement for soft robots is achieved by using variable-length tendons (Rus and Tolley, 2015) or by pneumatic, hydraulic or twist-and-coil actuators (Pawlowski et al., 2019). Soft robots have many advantages when compared to traditional rigid-bodied robots, such as adaptation to wearable devices, safe-human-machine interaction, and suitability for gripping systems. Besides, conventional rigid-bodied robots are made with aluminum, steel, stainless steel, titanium, etc. which are high stiffness materials. Accordingly, while conventional rigid-bodied robots have limited degrees of freedom, soft robots possess an infinite number of degrees of freedom. However, due to their flexible structure, soft robots are difficult to control because of their inherent deformations, such as twisting, bending, compressing, wrinkling, buckling, and so on (Walker et al., 2020).

Soft robots can be classified by their movements, such as crawling (Cao et al., 2017; Donatelli et al., 2017) and swimming (Marchese et al., 2014; Katzschmann et al., 2015; Ay et al., 2018), or their purpose of usages, such as rehabilitation (Chan et al., 2017; Polygerinos et al., 2015) and grippers (Glick et al. 2018; Zhu et al., 2017). Soft robots are inspired by various biological living creatures. For instance, soft robots that can move by crawling are inspired by animals such as worms, caterpillars, and snakes, whereas soft robots that can move by swimming are inspired by animals such as calamari, octopus, and ink fish. A soft robotic glove is an example of a soft robot used for rehabilitation purposes and used in people with movement disorders. The soft gripper is a new field of gripper technology that is used in pick and place applications by virtue of exploiting its compliance with object interaction. These advancements increase their importance day by day (Yetkin and Koca, 2021).

Recent advancements in rapid digital design and manufacturing tools have been enabled researchers to fabricate soft robotic systems by utilizing adaptable production techniques, including molding, additive manufacturing, thin-film manufacturing, shape deposition manufacturing, and bonding (Gul et al., 2018; Walker et al., 2020). The vast majority of soft robots are made of catalyzed polymers, such as silicone rubbers, which are prepared by mixing two components are generally cast into the mold for fabrication (Marechal et al., 2021).

Since soft robotic systems should perform flexible motion, silicone rubber is mostly used in soft robotics due to its soft and flexible structure. Dragon Skin™ series, Ecoflex™ series, Mold Star™ series, Rebound™ series, Body Double™ series, SORTA-Clear™ series and parts of Sylgard™ series are the most known silicone rubbers, and the characterization of these elastomers has been reported in the literature within the scope of soft robotics (Marechal et al., 2021; El Bana and Abbas, 2020; Presti et al., 2019; Zaltieri et al., 2020).Marechal et al. (2021) focused on the mechanical characterization of seventeen elastomers to aid the researcher to choose proper materials for their applications, namely SORTA-Clear™, Dragon Skin™ 10 MEDIUM, Dragon Skin™ 20, Dragon Skin™ 30, Dragon Skin™ FX-Pro, Dragon Skin™ FX-Pro+Slacker, Body Double™ SILK, Mold Star™ 16 FAST, Mold Star™ 20T, SOLOPLAST 150318, Psycho® Paint, PlatSil® Gel-10, RTV615, Rebound™ 25, Ecoflex™ 00–10, Ecoflex™ 00–30 and Ecoflex™ 00–50. El Bana and Abbas (2020) designed and fabricated a three-finger soft actuator gripper made of Dragon Skin™ 20, which can handle easily fragile and small objects. Presti et al., (2019) dealt with the manufacturing of a fiber Bragg gratings (FBG) based-flexible sensor encapsulated into Dragon Skin™ 20 silicone rubber. The sensor was fabricated with the intention of being utilized in the development of a wearable device for respiratory and cardiac monitoring. Zaltieri et al. (2020) presented a novel FBG based-wearable device for monitoring low back flexion/extension (F/E) movements. The flexible sensing element was fabricated by encapsulating the FBG into Dragon Skin™ 20 substrate.

One of the most recent techniques is the inclusion of fillers into the matrix material in order to improve the mechanical properties of the materials used in soft robotic applications. Nanoparticles, nanotubes, nanofibers, and nanowires are the forms of nanofillers (Natarajan et al., 2021). Carbon nanotubes in the form of single wall carbon nanotube (SWCNT), double wall carbon nanotube (DWCNT) and multiwall carbon nanotube (MWCNT) are one of the fillers gained recently an importance by the researchers because of their extremely high stiffness (Hanif et al., 2015). Golshahr et al. (2018) focused on the effects of MWCNT inclusion into the silicone on mechanical properties and reported that by adding MWCNT fillers into the silicone, the material properties were dramatically improved. Natarajan et al. (2021) presented the nonlinear material constants of MWCNT reinforced natural rubber (RMA 1) for soft robotic applications. To that end, the uniaxial tensile tests were carried out to RMA 1/MWCNT composites with the MWCNT loading of 5wt%, 10wt%, 20wt% and 30wt%. Hsiao et al. (2020) dealt with the carbon nanotube-integrated hydrogels. The findings of the tensile tests revealed that as the MWCNT loading increased, the Young modulus and ultimate tensile strength increased, but the ultimate strain before fracture decreased. On the contrary, Bokobza (2012) reported that Young's modulus and tensile strength of the MWCNT reinforced natural rubber were considerably lower than natural rubber. Besides, Montazeri et al. (2011) investigated the effects of sonication time and dispersing medium on the mechanical properties of MWCNT/epoxy composites. They reported that the increase in tensile strength with the increasing sonication time was followed by a decrease in tensile strength with longer sonication times. The common purpose of adding fillers into the polymers is to increase the mechanical properties. However, because of the agglomeration of fillers in the polymers, well-dispersing of fillers in the polymers is a challenge.

This study evaluates the effects of several literature-based mixing procedures on the mechanical properties and curing time of the end products. The compounds for the procedure are platinum-catalyzed liquid silicone DragonSkin20 (DS20) as a matrix material and COOH-functionalized carbon nanotubes (COOH-MWCNT) as an additive material. The reasons for the selection of these compounds are (i) DS20 product has the highest mechanical strength among Dragon Skin series, (ii) COOH-MWCNT can cause a higher increase in mechanical characteristics in mixtures with polymers compared to other functionalized MWCNTs (e.g. fluorinated-MWCNT). In order to determine the ideal additive ratio in the mixture in the literature, the processes are repeated using additive materials selected in different ratios. Based on the ideal values determined in these studies, a fixed additive material ratio (2%) is employed in this work. The followings are the mixing steps chosen in the study: (i) mechanical stirring, (ii) homogenization in ultrasonic bath, (iii) magnetic stirring. Different mixing techniques are determined with the combinations selected among these steps. The criteria chosen for the effect of different mixing techniques on the mixing result are (i) change in curing time, (ii) change in mechanical properties. Tensile tests are carried out with specimens created in accordance with ASTM D 412 standard for the relevant material in order to observe the change in mechanical properties. After the tensile tests, the modulus of elasticity, ultimate tensile strength, and maximum strain values are reported. The followings are some of the potential novelties that the study could bring to the literature:

1. mixing DS20 material, which is frequently used in soft robot production, with a nanoparticle additive material,

2. rather of selecting a single mixing approach, we present a comparison of various techniques,
3. for interpretation of data, both curing times and mechanical outcomes are taken into account,

This chapter has three main sections in general. Soft robot applications and manufacturing challenges are discussed in the first section. The fabrication and testing procedures of a soft robot are described in the second section. The third section discusses the effects of mixing methodologies on curing time and mechanical properties. Finally, some concluding remarks and future studies are presented in the Conclusion section.

MATERIALS AND METHODS

Materials

The basic compound used in many applications of soft robotic grippers in the industry is platinum cure liquid silicones. Therefore, in this study, DragonSkin20 (DS20, from Smooth-On) which is a trademark of a platinum cure liquid silicone was utilized as a matrix material. DS20 silicone is achieved by mixing Part A and Part B (1A:1B by volume or weight). While Part A is a silicone elastomer, Part B is a silicone elastomer crosslinker. The manufacturer's specifications of DS20 are given in Table 1.

Table 1. Some mechanical and physical properties of DS20

Mixed viscosity (ASTM D-2393) (Pa.s)	20
Pot life (ASTM D-2471) (min)	25
Cure time (hours)	4
Shore A hardness (ASTM D-2240)	20
Tensile strength (ASTM D-412) (MPa)	≈3.8
100% modulus (ASTM D-412) (MPa)	≈0.34
Elongation at break (ASTM D-412) (%)	620

(-COOH) Functionalized multi-walled carbon nanotubes (MWCNT) of which inside and outside diameter are 5–10 nm and 10–30 nm, respectively, were supplied by Nanografi Nano Technology, Germany. The content of (-COOH) is 1.5 wt% and the length of the MWCNT is 10–35 μm.

Acetone used to dilute Part A or Part B in this work has a molecular weight of 58.08 g/mol and a density of 0.791 g/cm^3 (179124, from Sigma Aldrich).

Fabrication Procedures

In this study, nine different fabrication procedures were employed to produce neat and MWCNT reinforced silicone rubber. In the first method (F_1), the required amount of Part A (40 g) and Part B (40 g) (1A:1B by volume or weight) is poured into the beaker and then mixed thoroughly by hand for three minutes using a glass rod. In the second and third procedures (F_2 and F_3), it was aimed to investigate whether pre-mechanical mixing of neat Part A (or neat Part B) effects the curing time and mechanical properties of neat DS20. Therefore, in the F_2, 40 g Part A was poured into the beaker and blended with a mechanical mixer (IKA-RW-20-digital) at 1500 rpm for 1 h. Next, 40 g of Part B was added into the beaker filled with pre-mechanically mixed Part A and mixed by hand for three minutes. In the F_3, unlike the F_2, pre-mechanical mixing was applied to Part B. In the fourth technique (F_4), with the purpose of determining the effects of pre-mixing of neat Part A with acetone on the curing time and mechanical properties of neat DS20, firstly 40 gr neat Part A and 50 ml acetone were stirred by using a magnetic stirrer (IKA-RH-basic-2) for 1 hour at 20°C. Afterwards, this mixture was kept in an ultrasonic bath (Mercury-Ultrasonic-Cleaner-12Lt) for 1 hour at 75°C for homogenization and to remove all acetone from Part A. During the homogenization, the weight of the mixture was measured. After being sure all of the acetone was removed from Part A, Part B was added and mixed by hand for three minutes. In the F_5, unlike the F_4, Part B was pre-mixed with the acetone. In the sixth and seventh procedures (F_6 and F_7), the effects of MWCNT added into the silicone rubber using mechanical mixing method on the mechanical properties and curing time of the silicone rubber were investigated. In the F_6, a 0.2wt% loading of MWCNT was added gradually into the Part A (Part B in the F_7) and blended with a mechanical mixer at 1500 rpm for 1 hour. Following that, Part B (Part A in the F_7) was added into Part A-MWCNT mixture (Part B-MWCNT mixture in the F_7). In the eighth method (F_8), first, 0.2wt% MWCNT was placed in 50 ml acetone and mixed by hand for ten minutes using a glass rod. Following this step, the acetone-MWCNT mixture was homogenized for 1 hour at 20°C in an ultrasonic bath. Next, Part A (Part B in the ninth procedure (F_9)) was included into the acetone-MWCNT mixture and stirred by using a magnetic stirrer for 1 hour at 20°C. Then, acetone-MWCNT-Part A mixture (acetone-MWCNT-Part B mixture in the F_9) was homogenized for 1 hour at 75°C in an ultrasonic bath in order to achieve a better homogenization and remove all acetone from the mixture. Part B (Part A in the F_9) was added into the MWCNT-Part A mixture (MWCNT-Part B) mixture and stirred by hand for three minutes after being sure all of the acetone had been removed from acetone-MWCNT-Part A mixture (acetone-MWCNT-Part B mixture in the F_9). Mixing procedures under the guidance of Part-A were given in Figure 1.

After each of mixing procedure, vacuum degassing was carried out to remove the entrapped gas from neat and MWCNT reinforced DS20 using a desiccator and a vacuum pump with a vacuum pressure of 700 mmHg for 5 minutes. All of the equipments used in this study were given in Figure 2.

Following the vacuum degassing process, mixtures were poured into the tensile test specimen mold made of polyamide 6 (PA6) (Figure 3), and vacuum degassing process was performed once again for 2 minutes. Lastly, degassed tensile test specimens were cured in an oven at 50°C for four hours. Dimensions of dumbbell-shaped test specimens with 2 mm thickness (Figure 4) were prepared according to method A of the ASTM D412 (1981) standard.

Tensile Test

A universal tensile testing machine (SHIMADZU AG-IS, Shimadzu Corp., Japan) fitted with a 5 kN load cell was used to conduct the tensile tests. Firstly, a dumbbell-shaped test specimen was attached to the upper and lower grippers originally developed for the tests. Since the tensile test specimens slipped or teared in the course of the tests when traditional grippers generally recommended for metallic materials were used, a specially developed gripper with a wavy-pattern was used (Figure 5). The specially developed gripper proved to have the best grip without any slippage or tearing. Then, the tensile load was gradually increased at a strain rate of 500 mm/min until the test sample breaks. Tests were conducted with five repetitions to ensure the reliability of the experiments. The crosshead displacement readings were used to calculate the global strain values. A camera setup or an extensometer to measure the local strain of the material can be more suitable, however, this feature is currently unavailable in the current setup. The recorded force-displacement data was used to calculate the material properties such as tensile strength, elongation at break and tensile modulus. The tensile test setup can be seen in Figure 5.

Figure 1. Mixing procedures under the guidance of Part A (Vacuum degassing and curing treatments are not included) a) 1st mixing procedure (F_1), b) 2nd mixing procedure (F_2), c) 4th mixing procedure (F_4), d) 6th mixing procedure (F_6), e) 8th mixing procedure (F_8)

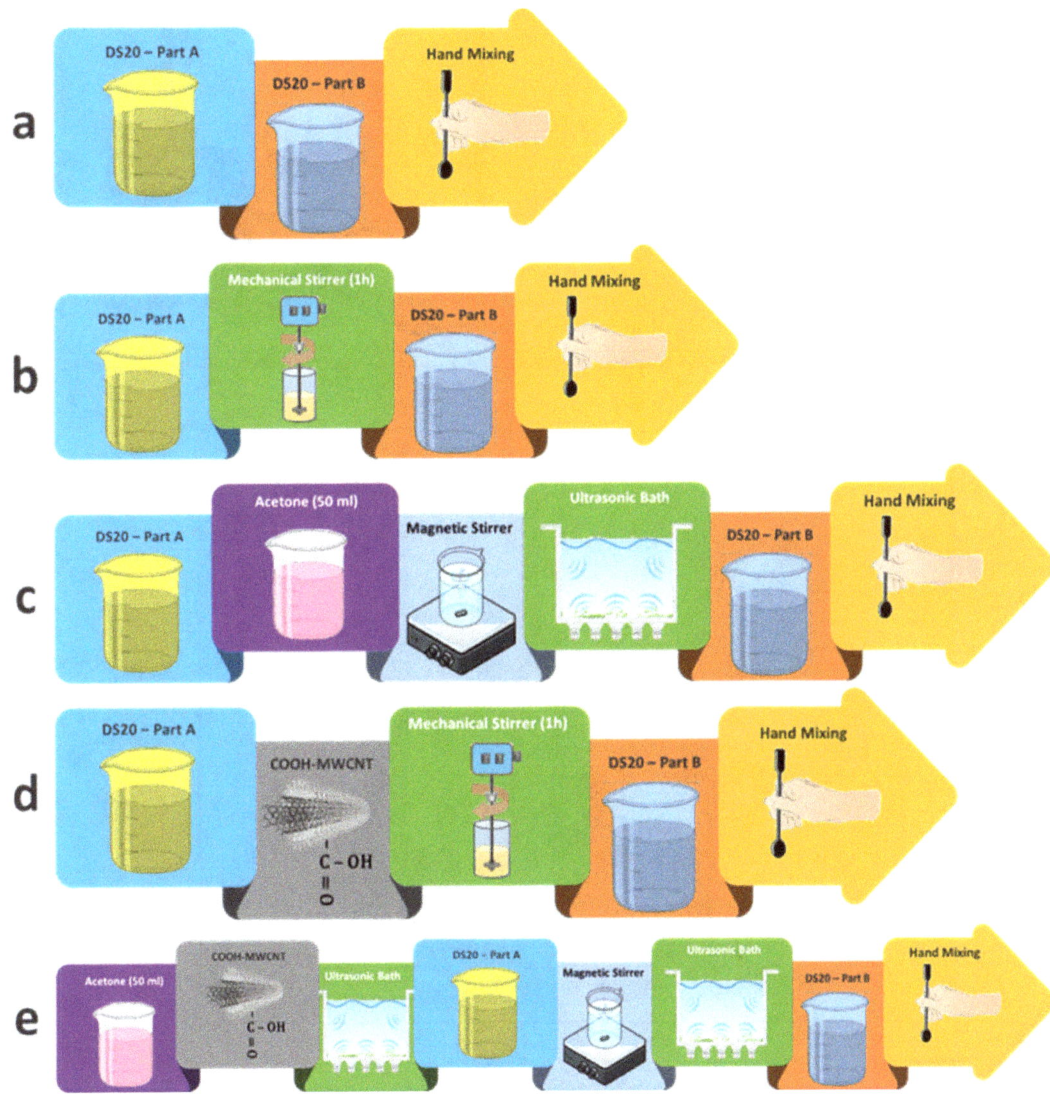

Figure 2. Pictures of experimental equipments a) Ultrasonic bath (Mercury-Ultrasonic-Cleaner-12Lt), b) Magnetic stirrer (IKA-RH-basic-2), c) Mechanical mixer (IKA-RW-20-digital), d) Desiccator and vacuum pump (Rocker-600) system

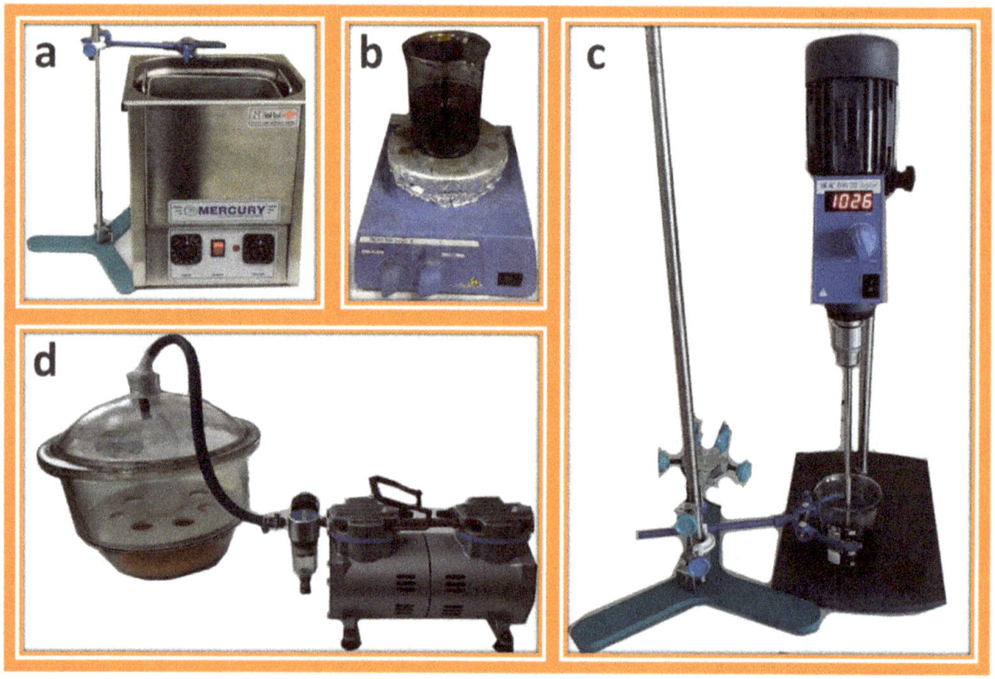

Figure 3. Pouring mixtures into the tensile test sample mold

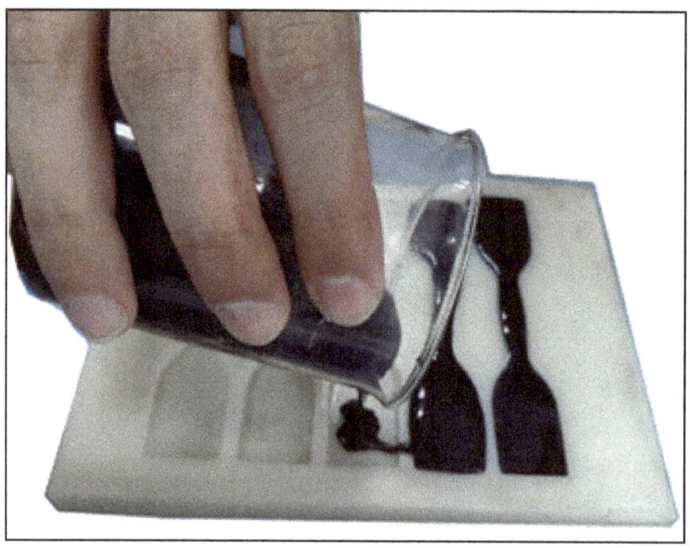

Figure 4. Dumbbell-shaped test specimen (Dimensions are in mm)

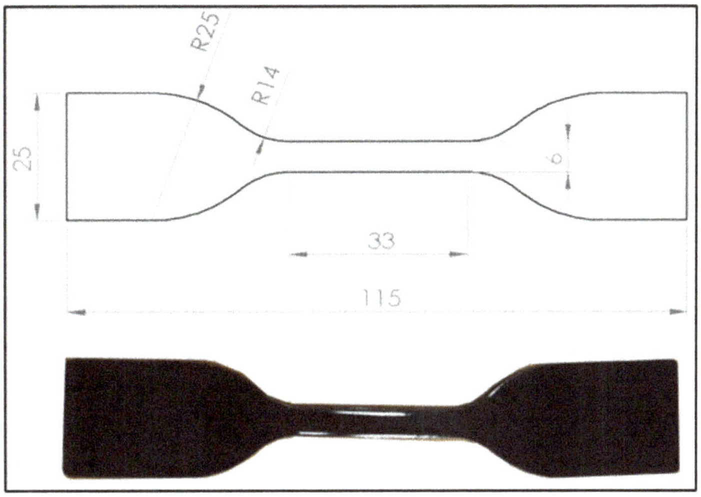

Figure 5. Dumbbell-shaped test specimen (Dimensions are in mm)

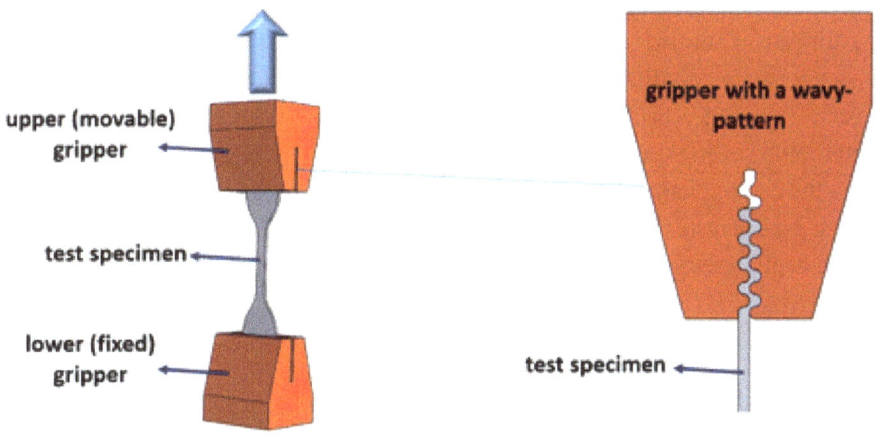

RESULTS AND DISCUSSIONS

The results of the different mixing and production methods shown in Figure 1 are discussed in this section. First of all, the effects of the mixtures on the curing times are investigated. Afterwards, the tensile test results of the produced specimens are reported.

Effects of Mixing Methodologies on Curing Time

It is observed that the mixing of the selected silicone material (DragonSkin 20) with functionalized-MWCNT significantly reduced curing time. When the catalog values of the matrix material (Dragon Skin, 2022) are examined, it is seen that the curing time is determined as four hours and the pot life as 25 minutes. Under normal conditions, it is very important that this period is not shortened, especially since finishing processes are applied throughout the pot life. However, all of the mixing procedures used showed a decrease in these durations. According to which compound is used first (Part-A or Part-B) in Table 2, it is reported if the curing time increases (↑), remains similar (≈), decreases (↓), or if a sudden curing occurs (↓↓↓↓) at the end of the related mixture options. When the Table 2 is examined, it is seen that the curing time decreases for each mixture option except F_1, F_2 and F_6, and instant curing occurs for some mixtures (F_3, F_5 and F_7). Observations obtained and their possible causes are listed with the following remarks:

Remark 1: *Instant Curing are Observed During Vacuuming Process*

The vacuuming process is very important in terms of increasing the quality of the material by providing homogenization. Two possible reasons for instant curings (F_3, F_5 and F_7) to occur at this stage: i) the curing time is generally considerably reduced during the processes and since the vacuuming process immediately after mixing (Figure 1), curing occurs during vacuuming, ii) the specimen tends to cure, but at the same time, sudden pressure changes (between 90 and 0 kPa) applied in the vacuum process further increase the curing of the system.

Remark 2. *Instant Curing Occurs when Part-B is the Leading Compound*

When the results are examined, it is observed that the material, which normally has a curing time of 4 hours, cured in approximately the same shortened time (10 minutes) as a result of the mixtures in F_3 and F_5. It is observed that the specimen produced in the F_5 process gelled under vacuum and is not in the fluidity to be molded even though it is not fully cured. For this reason, a tensile specimens could not be created. When the catalog values of the matrix material (Dragon Skin, 2022) are examined, it is seen that Part-B is a curing-agent (crosslinker). Considering this feature, it can be said that the curing time decreases with the heating due to increased kinetic energy in long-term mixtures.

Remark 3. *While Part-B is the Leading Compound, Instant Curing Occurs Except When CNT is Added (F_9)*

Looking at the table, it is seen that when acetone, magnetic stirrer and ultrasonic bath processes are applied together, it causes instant curing. However, the absence of immediate curing when CNT is included in the process can be interpreted as the effect of CNT increasing the curing time in the system. It is thought that the reason for the instant curing is the heat treatment applied during the evaporation of the acetone added from the mixture to ensure easy mixing. Therefore, it can be said that the added CNT acts to reduce the effect during this process.

Table 2. Effect of mixing couples on curing times

Leading Compound of Silicone	Mechanical Stirrer	Acetone + Magnetic Stirrer + Ultrasonic Bath	COOH-MWCNT + Mechanical Stirrer	COOH-MWCNT + Acetone + Magnetic Stirrer + Ultrasonic Bath
Part-A	≈	↓	≈	↓
Part-B	↓↓↓↓	↓↓↓↓	↓↓↓↓	↓

Effects of Mixing Methodologies on Mechanical Properties

Figure 6 is to compare the mechanical properties of the specimens produced as a result of the mixtures made. When the figure is examined, it is seen that some operations (F_3, F_5 and F_7) have no results. This is because, as explained in the previous section, the casting process could not be carried out because instant curing occurs as a result of the mixtures involved. The tensile tests described in the previous sections are applied for the specimens that can be produced. As test data, elasticity module, ultimate tensile strength and elongation percentage at break parameters are calculated. In order for the data to be statistically significant, five samples are prepared for each mixing method (there are 9 mixing methods in total). Observations obtained and their possible causes are listed with the following remarks:

Remark 4. *The F_1 Mix and Catalog Values are Coherent.*

The F_1 coded mixture is prepared by the method specified in the catalogue. Therefore, the close results can be considered as a kind of confirmation for the matrix material used.

Remark 5. *Although the Results do not change much in the F_2 mixture, the mechanical properties decreased when CNT was included (F_6).*

In F_2 coded mixture, unlike mixture-F_1, it is the mixing of the matrix material by a mechanical mixer (1500 rpm) for one hour. Since Part-A compound is not a curing agent, it is observed that the mechanical properties do not change much after this mixture. On top of that, when this mixing method is used to mix CNT and silicone, it is observed that the mechanical strength decreases. The reason for this is that mechanical mixing cannot ensure a homogeneous mixing of the CNT, regardless of the curing time. In Figure 7-a, it is seen that nano-materials, which are reinforcing elements, form agglomerates and exhibit an uneven distribution in the silicon matrix. Therefore, agglomerated CNTs caused a decrease in mechanical strength.

Remark 6. *The Mechanical Properties Increased when the CNT is Included (F_8), Although the Results Decreased Slightly in the F_4 Mixture*

The effect of the addition of acetone, which reduces the density and facilitates mixing in the F_4 coded mixture, and then evaporation with the help of magnetic stirrer and ultrasonic bath, on the silicon material is investigated. As a result, as it is said in the previous section, the curing time of the matrix material exposed to heat treatment is shortened, and therefore, mechanical properties decreased due to air bubbles. However, when the same mixture is used to mix silicon and CNT, it is observed that the mechanical properties increased compared to the previous situation (F_4). The reason is, although this mixing method reduces the mechanical properties of silicon, the mechanical properties increase again, as it allows the nano reinforcement material to be mixed homogeneously.

Remark 7. *The Mechanical Strength of the Mixture F_9 (The Only Produceable Sample When Part-B is the Leading Compound) Decreased.*

Figure 6. Results of each fabricated specimens with different mixtures, a) modulus of elasticity, b) ultimate tensile strength, c) elongation percentage at break

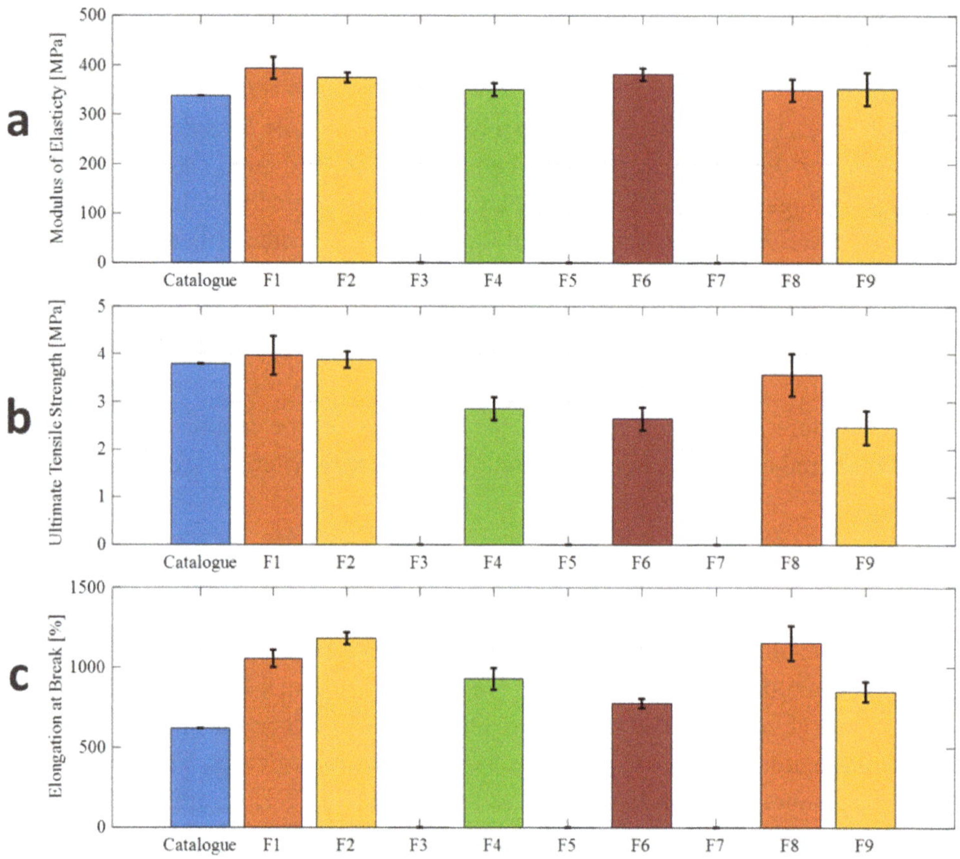

Figure 7. Pictures from fabricated specimen, a) specimen from mixture-F_8, b) specimen from mixture-F_9, c) specimen in tensile test

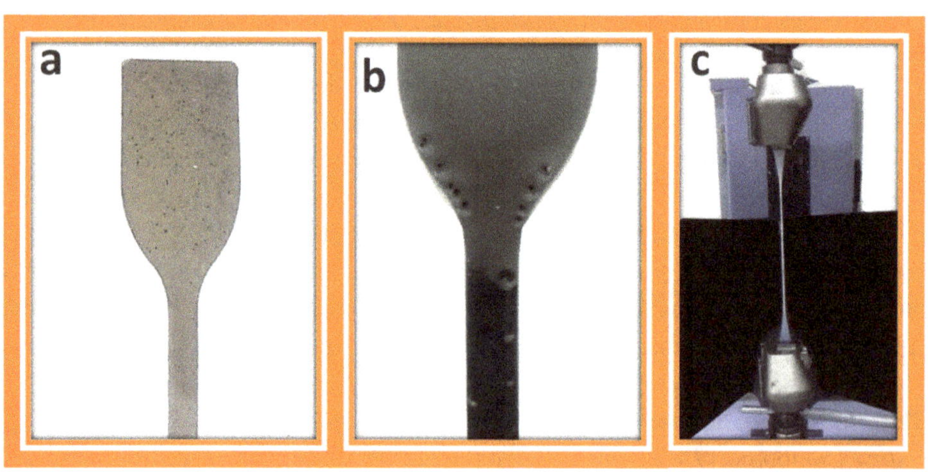

As said in the previous section, the addition of CNT prevented immediate curing, unlike the mixture in F_7. When the results obtained from the sample produced in this way are examined, it is seen that the mechanical properties are decreased compared to the catalog data and F_1. The reason for this is that sufficient vacuuming process could not be performed due to the large reduction in curing time currently available. Therefore, the air bubbles cannot leave the specimen as the material is curing while leaving. The structure in question is depicted in Figure 7-b. As a result, it can be said that the mechanical strength of the mixture decreases.

CONCLUSION

It is aimed to increase the mechanical strength of the material without changing the elasticity of the material much by adding a functionalized-MWCNT to the DS20 silicone material, which was chosen as the matrix material because it has the highest ultimate tensile strength in the product group (DragonSkin). Numerically, the expectations before the study are that the ultimate tensile strength value increases, the modulus of elasticity value remains the same, and the elongation at break value increases. However, different results than the targeted results were achieved. It would not be wrong to say it is because the matrix material is more sensitive to mixing processes than it is because of the additive used. In order to achieve the targeted improvement in future studies, the sensitivity of the selected matrix materials to mixing processes should also be considered. In addition, different processes can be applied to eliminate this sensitivity. Case studies may be in the form of reducing exposure to heat treatment or applying techniques such as homogenizers that will shorten the mixing time.

REFERENCES

Alici, G. (2018). Softer is Harder: What Differentiates Soft Robotics from Hard Robotics? *MRS Advances*, *3*(28), 1557–1568. doi:10.1557/adv.2018.159

ASTM D412. (1981). Standard test methods for vulcanized rubber and thermoplastic elastomers–tension. *Annual Book of ASTM Standards*, Philadelphia.

Ay, M., Korkmaz, D., Ozmen Koca, G., Bal, C., Akpolat, Z. H., & Bingol, M. C. (2018). Mechatronic Design and Manufacturing of the Intelligent Robotic Fish for Bio-Inspired Swimming Modes. *Electronics (Basel)*, *7*(7), 118. doi:10.3390/electronics7070118

Bokobza, L. (2012). Multiwall carbon nanotube-filled natural rubber: Electrical and mechanical properties. *Express Polymer Letters*, *6*(3), 213–223. doi:10.3144/expresspolymlett.2012.24

Cao, Y., Liu, Y., Chen, Y., Zhu, L., Yan, Y., & Chen, X. (2017). A novel slithering locomotion mechanism for a snake-like soft robot. *Journal of the Mechanics and Physics of Solids*, *99*, 304–320. doi:10.1016/j.jmps.2016.11.019

Chan, Y. H., Tse, Z., & Ren, H. (2017). Design evolution and pilot study for a kirigami-inspired flexible and soft anthropomorphic robotic hand. *18th International Conference on Advanced Robotics (ICAR)*, 432–437. 10.1109/ICAR.2017.8023645

Donatelli, C. M., Serlin, Z. T., Echols-Jones, P., Scibelli, A. E., Cohen, A., Musca, J.-M., Rozen-Levy, S., Buckingham, D., White, R., & Trimmer, B. A. (2017). Soft foam robot with caterpillar-inspired gait regimes for terrestrial locomotion. *2017 IEEE/RSJ International Conference on Intelligent Robots and Systems (IROS)*, 476–481. 10.1109/IROS.2017.8202196

Dragon Skin. (2022). *Dragon Skin™ 20 Product Information*. Smooth-On, Inc. https://www.smooth-on.com/products/dragon-skin-20/

El-Atab, N., Mishra, R. B., Al-Modaf, F., Joharji, L., Alsharif, A. A., Alamoudi, H., Diaz, M., Qaiser, N., & Hussain, M. M. (2020). Soft Actuators for Soft Robotic Applications: A Review. *Advanced Intelligent Systems*, *2*(10), 2000128. doi:10.1002/aisy.202000128

El Bana, H. G., & Abbas, A. (2020, February). A novel design of the utilisation of soft grippers in loading and unloading applications. *International Conference on Innovative Trends in Communication and Computer Engineering (ITCE)*, 158-163. 10.1109/ITCE48509.2020.9047779

Glick, P., Suresh, S. A., Ruffatto, D., Cutkosky, M., Tolley, M. T., & Parness, A. (2018). A Soft Robotic Gripper With Gecko-Inspired Adhesive. *IEEE Robotics and Automation Letters*, *3*(2), 903–910. doi:10.1109/LRA.2018.2792688

Golshahr, A., Natarajan, E., Santhosh, M. S., Sasikumar, R., Ramesh, S., & Durairaj, R. (2018). Multi wall carbon nanotube reinforced silicone for aerospace applications. *International Journal of Mechanical and Production Engineering Research Development*, *8*(4), 743–752. doi:10.24247/ijmperdaug201880

Gul, J. Z., Sajid, M., Rehman, M. M., Siddiqui, G. U., Shah, I., Kim, K.-H., Lee, J.-W., & Choi, K. H. (2018). 3D printing for soft robotics – a review. *Science and Technology of Advanced Materials*, *19*(1), 243–262. doi:10.1080/14686996.2018.1431862 PMID:29707065

Hanif, W. W., Risby, M. S., & Noor, M. M. (2015). Influence of carbon nanotube inclusion on the fracture toughness and ballistic resistance of twaron/epoxy composite panels. *Procedia Engineering*, *114*, 118–123. doi:10.1016/j.proeng.2015.08.049

Hsiao, L. Y., Jing, L., Li, K., Yang, H., Li, Y., & Chen, P. Y. (2020). Carbon nanotube-integrated conductive hydrogels as multifunctional robotic skin. *Carbon*, *161*, 784–793. doi:10.1016/j.carbon.2020.01.109

Katzschmann, R. K., Marchese, A. D., & Rus, D. (2016). Hydraulic Autonomous Soft Robotic Fish for 3D Swimming. In M. A. Hsieh, O. Khatib, & V. Kumar (Eds.), *Experimental Robotics: The 14th International Symposium on Experimental Robotics,* (pp. 405–420). Springer International Publishing. 10.1007/978-3-319-23778-7_27

Lee, C., Kim, M., Kim, Y. J., Hong, N., Ryu, S., Kim, H. J., & Kim, S. (2017). Soft robot review. *International Journal of Control, Automation, and Systems*, *15*(1), 3–15. doi:10.100712555-016-0462-3

Marchese, A. D., Onal, C. D., & Rus, D. (2014). Autonomous Soft Robotic Fish Capable of Escape Maneuvers Using Fluidic Elastomer Actuators. *Soft Robotics*, *1*(1), 75–87. doi:10.1089oro.2013.0009 PMID:27625912

Marechal, L., Balland, P., Lindenroth, L., Petrou, F., Kontovounisios, C., & Bello, F. (2021). Toward a Common Framework and Database of Materials for Soft Robotics. *Soft Robotics*, *8*(3), 284–297. doi:10.1089oro.2019.0115 PMID:32589507

Montazeri, A., Montazeri, N., Pourshamsian, K., & Tcharkhtchi, A. (2011). The effect of sonication time and dispersing medium on the mechanical properties of multiwalled carbon nanotube (MWCNT)/ epoxy composite. *International Journal of Polymer Analysis and Characterization*, *16*(7), 465–476. do i:10.1080/1023666X.2011.600517

Natarajan, E., Hassan, C. S., Kit, A. C., Santhosh, M. S., Ramesh, S., & Sasikumar, R. (2021). Modeling of multiwall carbon nanotubes reinforced natural rubber for soft robotic applications–A comprehensive presentation. *Materials Today: Proceedings*, *46*, 3251–3258. doi:10.1016/j.matpr.2020.11.293

Pawlowski, B., Sun, J., Xu, J., Liu, Y., & Zhao, J. (2019). Modeling of Soft Robots Actuated by Twisted-and-Coiled Actuators. *IEEE/ASME Transactions on Mechatronics*, *24*(1), 5–15. doi:10.1109/ TMECH.2018.2873014

Polygerinos, P., Galloway, K. C., Sanan, S., Herman, M., & Walsh, C. J. (2015). EMG controlled soft robotic glove for assistance during activities of daily living. *IEEE International Conference on Rehabilitation Robotics (ICORR)*, 55–60. 10.1109/ICORR.2015.7281175

Presti, D. L., Massaroni, C., D'Abbraccio, J., Massari, L., Caponero, M., Longo, U. G., Formica, D., Oddo, C., & Schena, E. (2019). Wearable system based on flexible FBG for respiratory and cardiac monitoring. *Sensors Journal*, *19*(17), 7391–7398. doi:10.1109/JSEN.2019.2916320

Rus, D., & Tolley, M. T. (2015). Design, fabrication and control of soft robots. *Nature*, *521*(7553), 467–475. doi:10.1038/nature14543 PMID:26017446

Walker, J., Zidek, T., Harbel, C., Yoon, S., Strickland, F. S., Kumar, S., & Shin, M. (2020). Soft Robotics: A Review of Recent Developments of Pneumatic Soft Actuators. *Actuators*, *9*(1), 3. doi:10.3390/ act9010003

Yetki̇n, S., & Koca, G. O. (2021). Esnek Robotların Tasarım, Kontrol ve İmalat Çalışmaları. *International Journal of Engineering Research and Development*, *13*(1), 74–86. doi:10.29137/umagd.706041

Zaltieri, M., Massaroni, C., Presti, D. L., Bravi, M., Sabbadini, R., Miccinilli, S., Sterzi, S., Formica, D., & Schena, E. (2020). A wearable device based on a fiber bragg grating sensor for low back movements monitoring. *Sensors (Basel)*, *20*(14), 3825. doi:10.339020143825 PMID:32659958

Zhu, M., Wang, Z., Hirai, S., & Kawamura, S. (2017). Design and fabrication of a soft-bodied gripper with integrated curvature sensors. *24th International Conference on Mechatronics and Machine Vision in Practice (M2VIP)*, 1–6. 10.1109/M2VIP.2017.8211521

Chapter 5
Active Vibration Control of a Flexible Manipulator With Velocity Feedback

Şahin Yavuz
https://orcid.org/0000-0001-9007-772X
Dokuz Eylül University, Turkey

Doğukan Akgöl
Dokuz Eylül University, Turkey

Gökçe Naz Biricik
Dokuz Eylül University, Turkey

ABSTRACT

In this study, active vibration control of a single-link flexible manipulator has been performed. Residual vibrations affect the end-point positioning accuracy of the manipulator. To control residual vibration amplitudes, closed loop control scheme has been effectuated. Velocity feedback of the desired endpoint has been carried out and traditional proportional-integral (PI) Control has been applied. The mathematical model of the flexible manipulator has been established by the finite element method (FEM). The FEM results in modal analysis have been verified by the experimental results and then, the transient analysis has been realized by the Newmark method. Different velocity profiles have been applied to the manipulator to demonstrate the effect of PI controller on vibration amplitudes. The root mean square values and steady-state time have been calculated presented for different cases. It is observed from the results that the residual vibration amplitudes of the manipulators are successfully suppressed by the proportional-integral control gains.

DOI: 10.4018/978-1-6684-5381-0.ch005

INTRODUCTION

Robot manipulators with rigid links have been widely used on industrial applications. These links are generally heavy to achieve high accuracy and repeatability during high-speed operations. Developments in the field of materials science and manufacturing technologies, flexible manipulator concepts have started to drawn attention and have attracted the interest of many researchers.

The most common usage example of flexible manipulators is robot manipulators. Flexible and light-weighted manipulators show many advantages over rigid and bulky ones. Flexible manipulators can be driven by smaller driving actuators and accordingly have less energy consumption levels. However, these and similar advantages reveal disadvantages in their advantages due to lightness. Residual vibrations occur due to high flexibility. Residual vibrations due to flexibility affects the settling time and end point accuracy of the robot manipulators while operating with high speeds. Robot manipulator action time increases accordingly residual vibrations as the flexible manipulator need to be allowed to settle at the end of the application.

Developing dynamic models of flexible manipulators can be constructed by the finite element method or analytical methods (Benosman & LeVey, 2004). The governing differential equations of a dynamic system can be solved by different numerical methods (Fung, 1997; Haido et al., 2010; Owren & Simonsen, 1995; Zhang et al., 1999). However, developing a proper dynamic model plays an important role in the controlling or suppression of the residual vibrations of flexible manipulators. Conspicuously, since the model-based control system performance depends on the dynamic model accuracy highly.

Elimination of residual vibrations at the end of the manipulation action by using different control strategies is a challenging problem for many researchers (Dwivedy & Eberhard, 2006; Oguamanam et al., 2006; Vakil et al., 2011). It is well known that flexible manipulators can be deformed easily due to their high flexibility. Thus, suppression of the residual vibrations come into prominence, which have destructive consequences on effectiveness and accuracy.

Residual vibration amplitudes can be eliminated by using two main strategies, passive and active control schemes. One of them is the passive control strategy stated in other words feed-forward control, which deals with the open loop system can be achieved by using techniques like motion input command shaping methods or trajectory planning methods (Conker et al., 2016). In addition, several studies can be found about different vibration control approaches with counter springs to eliminate residual vibrations of the flexible manipulators (Tang et al., 2021). This method aims to eliminate residual vibrations under favour of additional spring elements which have been attached to a flexible manipulator with opposite angles to each other to suppress or reduce the vibration amplitudes in both directions.

In trajectory planning methods, specifying the trajectory of the joint angle is the main target by using a combination of polynomial and cycloidal functions. The coefficients of the trajectory function can be adjusted by using different algorithms such as genetic algorithm (Kojima & Hiruma, 2006).

However, among these feed-forward control (passive control) strategies, motion command shaping or in other words input shaping methods take an important place in the field of vibration control of flexible manipulators and have drawn attention of many researchers (Blackburn et al., 2010; Chan et al., 2003; Dharne & Jayasuriya, 2007; Gürleyük & Cinal, 2007; Kim & Singhose, 2010; Singhose & Pao, 1997; Singhose et al., 2000).

Active control strategy (feed-back control) is the other method which is used to eliminate vibrations. Active control requires sensors to be equipped with thus, can be expensive or difficult to implement. Besides that, to achieve active control properly it can require sufficient computing power to evaluate the feedback signal that comes from sensing system and decide suitable control input to eliminate vibration. For this purpose, different control strategies and tuning algorithms have been used like Artificial Neural Networks (ANN) or Fuzzy Neural Networks (FNN) (Qiu et al., 2019). Active control methods require the closed loop model and an external actuator. To generate feed-back signal for controller, piezoelectric elements are widely used for flexible manipulators to eliminate the residual vibrations (Gaudenzi et al., 2000; Hassan et al., 2007; Xɨanmin et al., 2002). Ma et al. (2022) proposed motion control method on a flexible manipulator to reduce residual vibrations on the deceleration stage of the movement. After verifying the numerical model created by running a number of experiments, they analyzed the vibration damping effects. This study showed that when the acceleration/deceleration time of the felxible manipulator is a positive full multiple of its natural period, the residual vibrations can be eliminated effectively. Shang et al. (2022) obtained mathematical model of a dual-flexible manipulator by considering bearing friction. After that the controller parameters are obtained by using the adaptive law of the PI controller. This study showed that adaptive PI controller control method is more appropriate for conditions with longer flexiable loads

A 4-bar linkage system with piezoelectric actuators and sensors was numerically simulated for active control of the mechanism the reduced model controller, the classical and the robust H¥ controller (Xianmin et al., 2002). Piezoceramic actuators were used in simulation and experiments to reduce vibration of a flexible one-link manipulator with a model-based predictive controller (Xianmin et al., 2002). Piezo patches were used in a study by Gaudenzi et al. (2000) to reduce vibration in an active cantilever beam. For the control method, researchers employed a single-input single-output feedback closed loop control system.

In this study, numerical analysis is used to examine the active vibration management of a single-link flexible steel manipulator with a payload. The finite element method is used to build the mathematical model of a manipulator (FEM). The experimental findings support the FEM modal analysis conclusions. The Newmark approach is then used to implement the transient analysis. The manipulator is subjected to two different velocity profiles, such as trapezoidal and triangular. The control procedure is integrated into the Newmark system for active vibration control. It uses the PI (proportional and integral) control gain. After motion is complete, active control is initiated. For each scenario, the reduction ratios and root mean square (RMS) values derived from the residual vibration signals are provided. The results show that the PI control gain effectively suppresses the flexible manipulator's residual vibration amplitudes.

MODELING OF SINGLE LINK MANIPULATOR BASED ON FINITE ELEMENT METHOD

Finite element method (FEM) is used to construct the mathematical model of the flexible manipulator [23]. Modeling and dynamic analysis are performed in MATLAB. The line model and FE model of flexible manipulator are shown in Figure 1. There is a revolute joint at the point O between OB beam (Member-2) and the frame (Member-1). The manipulator is rotated by the servo motor located at O. The mass of the servo motor is on the frame at O. The payload mass located at the point B has a translational inertia (m_L) and rotational inertia (I_L).

Member-2 rotates with the instantaneous angular displacement θ_2, which changes with respect to time. The manipulator length of Member-2 indicates as L_2=OB.

Figure 1. (a) Model, (b) FE model, (c) starting and stopping locations of the manipulator.

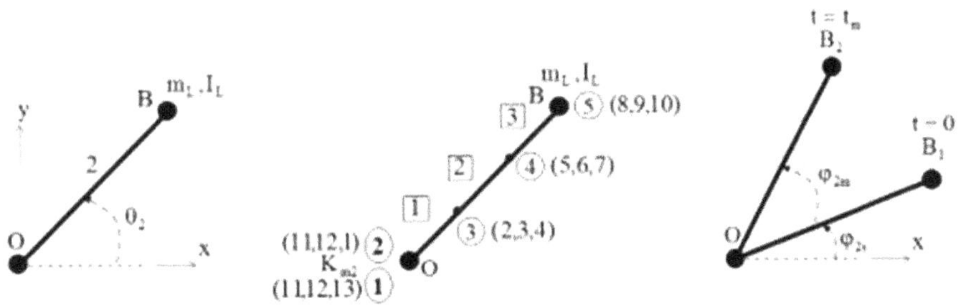

The finite element number for the manipulator has been chosen as 3 for explanation. In Figure 1 (b), the numbers in circles indicate the node numbers. The numbers in squares indicate each divided element (FE-n). The numbers in parentheses indicate degrees of freedom (DOF) of each node. Since 2D analysis is considered, each node has 2 translational and 1 rotational DOF. For example, Node-3 and Node-4 are at both end of FE-2. Each FE-n has local coordinates x_n, y_n and z_n. Here, x_n is from Node-n to Node-(n+1). z_n is always parallel to global z axis. The instantaneous angle of orientation for z_2 is γ_2, $\gamma_2 = \theta_2$. Beam FE's and their parameters are shown in Table 1.

Table 1. Beam FE's on the system

FE-	On Member-	FE Nodes	Length	γ_n	Identification numbers for displacements at nodes
1	2	2,3	h_2	θ_2	11,12,1,2,3,4
2	2	3,4	h_2	θ_2	2,3,4,5,6,7
3	2	4,5	h_2	θ_2	5,6,7,8,9,10

The displacement ($\mathbf{d}_{eln}$), force ($\mathbf{f}_{eln}$), stiffness ($\mathbf{k}_{eln}$), and mass ($\mathbf{m}_{eln}$) matrices in local coordinates of a finite element (FE-n) are given below (Bathe, 2014). The node numbers j and k are at the local origin at the far end of FE-n, respectively.

$$k_{eln} = \begin{bmatrix} \frac{A_nE_n}{h_n} & 0 & 0 & -\frac{A_nE_n}{h_n} & 0 & 0 \\ 0 & \frac{12E_nI_n}{h_n^3} & \frac{6E_nI_n}{h_n^2} & 0 & \frac{-12E_nI_n}{h_n^3} & \frac{6E_nI_n}{h_n^2} \\ 0 & \frac{6E_nI_n}{h_n^2} & \frac{4E_nI_n}{h_n} & 0 & \frac{-6E_nI_n}{h_n^2} & \frac{2E_nI_n}{h_n} \\ -\frac{A_nE_n}{h_n} & 0 & 0 & \frac{A_nE_n}{h_n} & 0 & 0 \\ 0 & \frac{-12E_nI_n}{h_n^3} & \frac{-6E_nI_n}{h_n^2} & 0 & \frac{12E_nI_n}{h_n^3} & \frac{-6E_nI_n}{h_n^2} \\ 0 & \frac{6E_nI_n}{h_n^2} & \frac{2E_nI_n}{h_n} & 0 & \frac{-6E_nI_n}{h_n^2} & \frac{4E_nI_n}{h_n} \end{bmatrix} \quad d_{eln} = \begin{Bmatrix} u_{jn} \\ v_{jn} \\ h_n r_{jn} \\ u_{kn} \\ v_{kn} \\ h_n r_{kn} \end{Bmatrix} \tag{1}$$

$$f_{eln} = \begin{Bmatrix} F_{jnx'} + q_{nx'}\frac{h_n}{2} \\ F_{jny'} + q_{ny'}\frac{h_n}{2} \\ T_{jn} + q_{ny'}\frac{h_n^2}{12} \\ F_{knx'} + q_{nx'}\frac{h_n}{2} \\ F_{kny'} + q_{ny'}\frac{h_n}{2} \\ T_{kn} + q_{ny'}\frac{h_n^2}{12} \end{Bmatrix} \quad m_{eln} = \frac{ÁA_nh_n}{420}\begin{bmatrix} 140 & 0 & 0 & 70 & 0 & 0 \\ 0 & 156 & 22h_n & 0 & 54 & -13h_n \\ 0 & 22h_n & 4h_n^2 & 0 & 13h_n & -3h_n^2 \\ 70 & 0 & 0 & 140 & 0 & 0 \\ 0 & 54 & 13h_n & 0 & 156 & -22h_n \\ 0 & -13h_n & -3h_n^2 & 0 & -22h_n & 4h_n^2 \end{bmatrix} \tag{2}$$

Here, h_n is the length of FE-n. A_n is the cross-sectional area. Where m=j or k, the nodal displacements at Node-m are u_{mn} and v_{mn} in the x_n and y_n directions respectively. The cross section's flexural rotation at Node-m is r_{mn}. $F_{mnx'}$ and $F_{mny'}$, respectively, are the external load forces at Node-m in the x_n and y_n directions. T_{mn} represents the external bending moment at Node-m. In the x_n and y_n directions, the distributed external loads on the FE-n are, respectively, $q_{nx'}$ and $q_{ny'}$. En is the elastic modulus, I_n is the cross-sectional bending moment of inertia, and n is the density.

The displacement ($\mathbf{d}_{egn}$), force ($\mathbf{f}_{egn}$), stiffness ($\mathbf{k}_{egn}$), and mass ($\mathbf{m}_{egn}$) matrices in global coordinates of FE-n are given below (Bathe, 2014).

$$d_{egn}=T_n d_{eln} \quad k_{egn} = T_n^T k_{eln} T_n \quad f_{egn}=T_n f_{eln} \quad m_{egn} = T_n^T m_{eln} T_n \tag{3}$$

where, $\mathbf{T}_n$ is the transformation matrix and $\mathbf{T}_n^T$ is the transpose of $\mathbf{T}_n$. The transformation matrix is given as

$s_n = \sin \gamma_n$

$$T_n = \begin{bmatrix} c_n & s_n & 0 & 0 & 0 & 0 \\ -s_n & c_n & 0 & 0 & 0 & 0 \\ 0 & 0 & 1 & 0 & 0 & 0 \\ 0 & 0 & 0 & c_n & s_n & 0 \\ 0 & 0 & 0 & -s_n & c_n & 0 \\ 0 & 0 & 0 & 0 & 0 & 1 \end{bmatrix} \tag{4}$$

$c_n = \cos \gamma_n$

Node-1 and Node-2 are coincident in Figure 1 (b), but their flexural rotations are different due to the revolute joint at O. The value of rotational spring between Node-1 and Node-2 (K_{m2}) can be found in chosen servo motor datasheet. There is a mass representing the payload at Node-5. The Node-1 is fixed. So, $d_{s11}=0$, $d_{s12}=0$, and $d_{s13}=0$.

The mathematical model of the system is obtained as

$$m_s\ddot{d}_s + c_s\dot{d}_s + k_s d_s = f_s \tag{5}$$

Here, the system's matrices for mass, damping, stiffness, displacement, and force are denoted as $\mathbf{m}_s$, $\mathbf{c}_s$, $\mathbf{k}_s$, $\mathbf{d}_s$ and $\mathbf{f}_s$ respectively. For the setup shown in Figure 1(b), the sizes of the ds and fs are 10x1, while the sizes of the ms, cs, and ks are 10x10 (b). For instance, the displacement of Node-4 in the y direction is represented as $\mathbf{d}_s(6,1)= d_{s6}$. The external force acting in the y direction at Node-4 is represented as $\mathbf{f}_s(6,1)= f_{s6}$.

Global FE matrices are assembled with a size of 6x6 to obtain the mass ($\mathbf{m}_s$) and the system stiffness ($\mathbf{k}_s$) matrices. For example,

$$\mathbf{k}_s(6,5) = \mathbf{k}_{eg2}(5,4) + \mathbf{k}_{eg3}(2,1) \text{ and } \mathbf{m}_s(6,5) = \mathbf{m}_{eg2}(5,4) + \mathbf{m}_{eg3}(2,1) \tag{6}$$

The index (6,5) exists in FE-2 and FE-3 as given in Table 1. The index (6,5) is the sum of (5,4) for the FE-2 matrix, and (2,1) for FE-3 matrix.

Considering the kinetic energy, m_L and I_L are added to the system mass matrix as the following

$$\mathbf{m}_s(8,8)=\mathbf{m}_{eg3}(4,4)+m_L,\ \mathbf{m}_s(9,9)=\mathbf{m}_{eg3}(5,5)+m_L,\ \mathbf{m}_s(10,10)=\mathbf{m}_{eg3}(6,6)+I_L \tag{7}$$

Considering the potential energy, K_{m2} is added to the system stiffness matrix as the following

$$\mathbf{k}_s(1,1)=\mathbf{k}_{eg1}(3,3)+K_{m2} \tag{8}$$

- **Damping:** The damping matrix is calculated as the Rayleigh damping as given below.

$$c_s = \eta m_s + \beta k_s \tag{9}$$

where, η and β are damping coefficients (Rao, 2011).

- **Motion:** As seen in Figure 1(c), the manipulator's trapezoidal motion profile is taken into consideration. Manipulator moves from a starting position (OB1) at time t=0 to an end position (OB2) at time $t=t_m$. The motion time is denoted by tm. The initial angular position is given as $\theta_2=\varphi_{2s}$. The angular location is represented as $\theta_2= \varphi_{2s}+ \varphi_{2m}$ at $t=t_m$, at stopping time.

The motor maintains a velocity profile which is trapezoidal. The motor rotation, φ_{2m}, is determined from the area under the velocity curve.

Figure 2. The angular trapezoidal velocity profile of the motor.

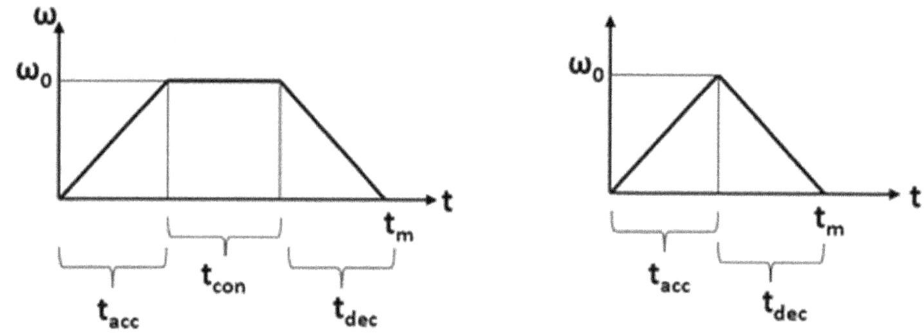

To consider the distributed inertia forces, the mid-point accelerations of the FEs are calculated. For the calculation, the rigid body kinematics is considered. Let L_{nm} represent the distance between the point O and the midpoint of the FE-n on Member-2. The midpoint's position vector is therefore expressed as $R_{nm} = L_{nm}e^{i\theta_2}$, in complex form, where $i=\sqrt{-1}$. The x component of $\mathbf{R}_{nm}$ is real, whereas the y component is imaginary. The acceleration is given by the second derivative of $\mathbf{R}_{nm}$ as $a_{nm} = \left(i\ddot{\theta}_2 - \dot{\theta}_2^2\right)L_{nm}e^{i\theta_2}$. The imaginary component of the $\mathbf{a}_{nm}$ is a_{nmy}, and its real component is a_{nmx}.

Considering the velocity profile of the motor, the following equations can be obtained.

$$\omega_{02} = \frac{\phi_{2m}}{\left(0.5t_{acc} + t_{con} + 0.5t_{dec}\right)} \tag{10}$$

$$\dot{\theta}_2 = \frac{\omega_{02}t}{t_m} \quad \ddot{\theta}_2 = \frac{\omega_{02}}{t_{acc}} \quad \text{for } 0 \le t < t_{acc} \tag{11}$$

$$\dot{\theta}_2 = \omega_{02} \quad \ddot{\theta}_2 = 0 \quad \text{for } t_{acc} \le t < t_m\text{-}t_{dec} \tag{12}$$

$$\dot{\theta}_2 = \frac{\omega_{02}\left(t_m - t\right)}{\left(t_m - t_{\text{dec}}\right)} \quad \ddot{\theta}_2 = -\frac{\omega_{02}}{t_{dec}} \quad \text{for } t_m\text{-}t_{dec} \le t < t_m \tag{13}$$

$$\dot{\theta}_2 = \ddot{\theta}_2 = 0 \quad \text{for } t \ge t_m \tag{14}$$

- **Forces:** There are gravity forces and distributed inertia forces. Let the distributed forces for FE-n be q_{nx} and q_{ny}. The following can be written:

$$q_{nx} = -(\rho_n A_n h_n)\text{ anmx and } q_{ny} = -(\rho_n A_n h_n)(a_{nmy} + g) \text{ where } g=9.81 \text{ m/s}^2. \tag{15}$$

For instance, the distributed forces for FE-2 and FE-3 result in the nodal forces for Node-6. Therefore, the components of the nodal force vector, fs, corresponding to Node-6, can be expressed as follows.

$$\mathbf{f}_s(5,1) = q_{2x}h_2/2 + q_{3x}h_3/2,\ \mathbf{f}_s(6,1) = q_{2y}h_2/2 + q_{2y}h_2/2,\ \mathbf{f}_s(7,1) = q_{2y}h_2^2/12 + q_{3y}h_3^2/12. \tag{16}$$

The inertia forces due to the payload mass are added for $\mathbf{f}_s(8,1)$, $\mathbf{f}_s(9,1)$ and $\mathbf{f}_s(10,1)$.

- **Modal Analysis:** Modal analyses are performed by MATLAB to identify the flexible manipulators' undamped natural frequencies. For the single link manipulator, MATLAB solves the following eigenvalues equation that is derived from the mathematical model based on the FEM.

$$|-\omega^2 m_2 + k_s| = 0 \tag{17}$$

MATLAB determines the eigenvalues ω_i in rad/s. Let f_1 be the manipulators' first natural frequency. Following that, $T_1 = 1/f_1$ is used to get the period T_1 that corresponds to the first natural frequency. The motion parameters t_{acc}, t_{con} and t_{dec} are defined using the half of the period, $T_{1h} = 0.5T_1$.

NUMERICAL VALUES

At point O, a revolute joint is defined for the manipulator. The global origin is at O. The manipulators are actuated by a servo motor, which rotates around z axis is used in the experiments, which is the global coordinate of the flexible link. The rotational spring constant K_m is defined at the point O for the revolute joint since the servo-motor has a joint flexibility.

As shown in Table 2, the value of K_m is provided. A pay-load at B, where vibration responses are collected, is thought of as the wireless accelerometer sensor that will be added to the experimental system.

The Finite Element model of the planar manipulator has 100 elements and 301 nodes. The geometrical and material properties of the manipulators are given in Table 2.

Table 2. Properties of the manipulator

Description	Value	Description	Value
Modulus of elasticity	E = 200 GPa	Density	ρ =7800 g/m^3
Length	L = 500 mm	Width	b = 20 mm
Thickness	h=3.2 mm	Cross section area	A = 64 mm^2
Distance of the payload from the end point	$d_{payload}$ =15 mm	Inertia of cross section	I = 54.61 mm^4
Distance of the receiving point from the center of the payload	d_{sensor} =55 mm	Newmark amplitude decaying factor	γ =0.005
Number of finite elements	n_e=100	Rayleigh damping coefficient	β=3.8 x10^{-4}
Motor rotational spring constants	K_{m2} = 16000 Nm/rad	Time step	Δt =0.01 s
Weight of the sensor	m_s = 0.054 kg	Weight of the payload	m_L=0.62 kg

EXPERIMENTAL SYSTEM

The experimental setup used in this study is shown in Figure 3.

Figure 3. Experimental system, (a) planar manipulator, (b) pc-based motion and measurement system.

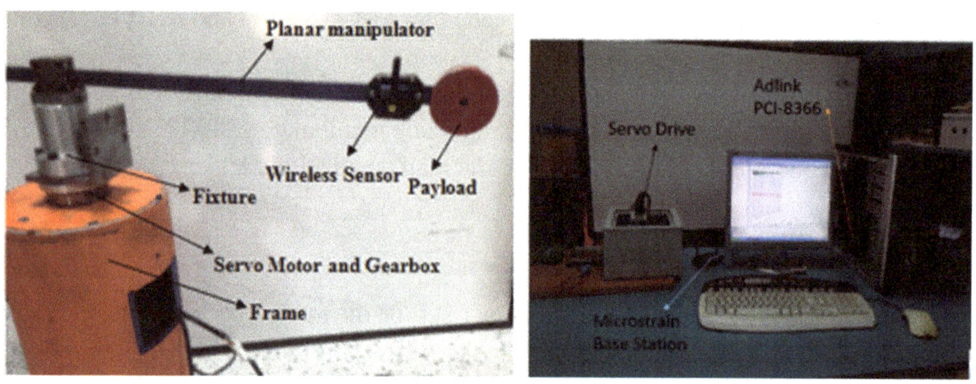

To conduct the experiments, the manipulator is created. The experimental set-up includes a wireless accelerometer sensor (WAS), a wireless data acquisition system (WDA), a PC, a servo motor and servodriver, a motion control card, and a single-link flexible manipulator. In the experiments, the WAS is located at the end point of the manipulators. Since it affects the dynamic response of the manipulators, the WAS is also treated as additional payload in the simulations.

It uses a 200 W Mitsubishi Electric servo motor and driver (Model HC-KFS23B/MR-J2S-20A). The motor is driven by a harmonic drive gearbox, model HFUC-32-100/100, with a gear ratio of 100. Adlink PCI-8366, a motion control card for computers, is employed. The SSCNET network connects the motion control card and driver in serial fashion. The Adlink-ActiveX component is used to program the driver using Visual Basic commands. Readily accessible in the market are the motion control units that generate pulses in accordance with the profile shown in Figure 2.

At the end point of the manipulators, acceleration signals are measured using MicroStrain WDA. The three main elements of the WDA system are the WAS, a USB base station to receive and transmit data to a host, and software to run the system and record data. The WAS combines triaxial accelerometers and uses embedded accelerometers to sense vibrations in three directions. 617 Hz has been chosen as the sample rate. The planar manipulator's low-pass filter is set at and 5 Hz.

SIMULATION AND EXPERIMENTAL RESULTS

The single-link flexible manipulator is considered in the numerical and experimental analyses. The natural frequencies of the manipulators corresponding to the first three modes are listed in Table 3.

Table 3. Natural frequencies of the manipulator

Order of Frequency	Flexible manipulator	
	FEM (Hz)	Experiment (Hz)
1.	3.166	3.163
2.	-	17.850
3.	48.250	43.080

The mode in out of plane corresponds to the second natural frequency of the planar manipulator as determined by modeling and experiment. The value of its second mode is not shown in the table because the FEM solution yields results in plane.

The trapezoidal and triangular velocity motion profiles are used to study various stopping positions of the manipulators, which are specified by the vector $\mathbf{q}_p = [\theta_s, \theta_m]$. Since s's value represents starting position and is unimportant in a resting state, it is always assumed to be zero. The angular position's incremental value is shown by the symbol m. The vector $\mathbf{q}_m = [t_{acc}, t_{con}, t_{dec}, t_m]$ specifies the time parameters of the trapezoidal velocity profile. Three time parameters are selected, and the equation below is used to compute the third time parameter.

$$t_m = t_{acc} + t_{con} + t_{dec} \tag{18}$$

In $\mathbf{q}_m$, the calculated parameter is denoted by a "*". Unless otherwise specified, the units for angles and time parameters are degrees and seconds, respectively. The vector $\mathbf{q}_m = [t_{acc}, 0, t_{dec}, t_m]$ defines the triangular velocity profile, which can be thought of as a specific case of a trapezoidal motion profile. The triangular velocity profile's constant time parameter is $t_{con}=0$.

From a resting position (OB1) at time t=0 to a halting position (OB2) at time $t=t_m$, the manipulator moves.

ACTIVE VIBRATION CONTROL OF THE MANIPULATOR

It is known that, flexibility causes residual vibrations after motion is done. The aim of this study is to reduce the end point vibrations of the manipulator with active control action. MATLAB is used to construct the finite element models of the system and to integrate the control strategy into Newmark method. The closed loop block diagram is shown in Figure 4.

Figure 4. Closed loop block diagram

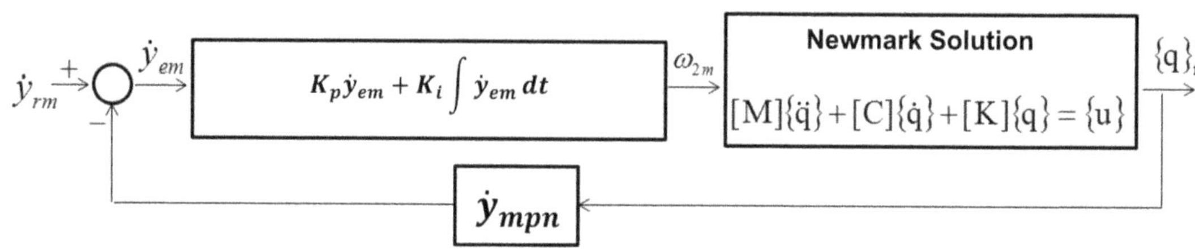

Since the flexible beam is driven by a servo motor in velocity mode, average velocity feedback is used in the closed loop block diagram. Here, $\dot{y}_{rm}$ is the reference velocity signal at the time step m. It is desired to reduce the residual vibrations of the flexible beam, therefore reference velocity signal is taken as zero at each step. The feedback velocity data is defined as $\dot{y}_{mpn}$ at the y-direction of the measurement point at the time step n. $\dot{y}_{erm}$ is the error velocity data at the time step m and m=n+1. K_p and K_i are proportional and integral control action constants, respectively. Trapezoidal and triangular velocity profiles are applied to the flexible beam. Closed loop control is activated after motion is done and the instantaneous velocity at the measurement point is subtracted from the reference velocity input to find the error signal value at a time step in the Newmark solution. The proportional control action is applied to find the actuator signal value in the time step. This input value is used to find the acceleration of the end point for the subsequent time step. The process is continued with the time interval Δt until the steady-state value is approximately reached. A time step, Δt, is chosen for the solution as $\Delta t<1/(20f_1)$, where f_1 is the first natural frequency considered. $\{q\}_m$ is the value of the translations at x and y directions and the rotation about z axis for each nodes at the time step m. $\omega 2_m$ is the control input at the time step m.

Arbitrary trapezoidal and triangular velocity profiles with the stopping position $\theta_m= 90^o$ are applied to the flexible manipulator and the effect of control gains is investigated in active vibration control. Examples signals of vibration responses obtained by Newmark are shown in Figure 5 and Figure 6.

The window of the vibration signals for $t_m<t<t_m+t_{res}$ is considered to analyze residual vibrations, where t_{res} is the chosen time for residual vibrations. The time value is taken as $t_{res}= 4s$ for the window. The root mean square (RMS) value of the windowed signal and steady-state time (Tss) are computed for the cases. The computed RMS and Tss values for the cases are listed in Table 4.

Figure 5. Vibration responses for $\theta_m = 90^o$ and $\mathbf{q}_m$ = [,0, 0.1, 1]*

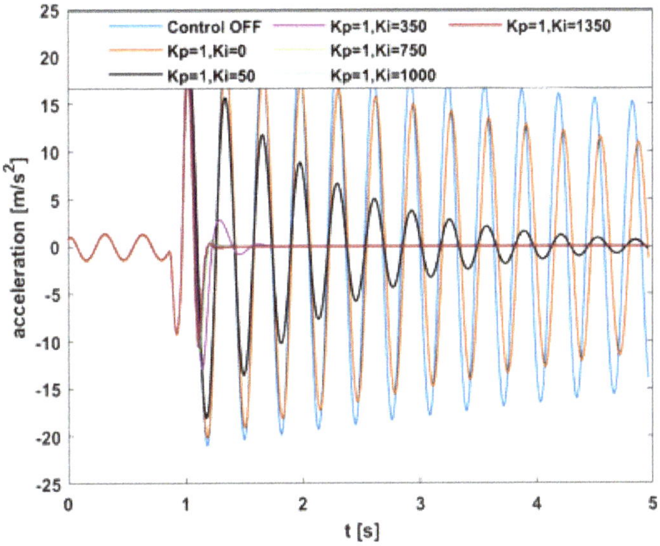

Figure 6. Vibration responses for $\theta_m = 90^o$ and $\mathbf{q}_m$ = [,0.8, 0.1, 1]*

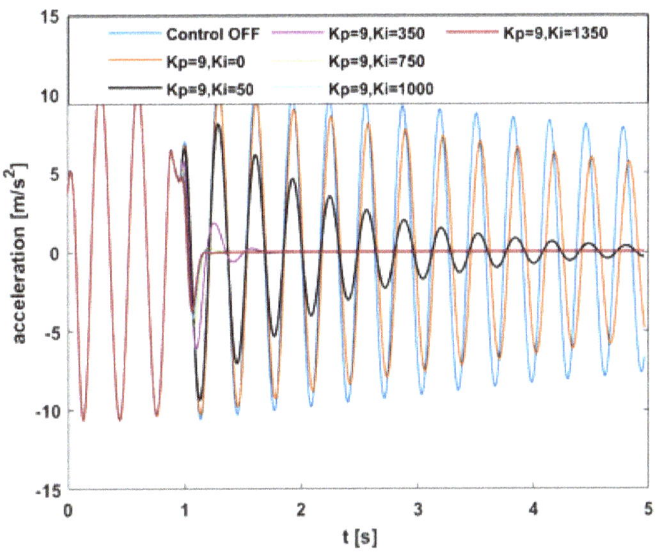

In case of the triangular velocity profile with $\mathbf{q}_m$=[*, 0, 0.1, 1]; the RMS value for Control-OFF result is 12.79 and the reductions are 13.92%, 55.04%, 77.72%, 80.53%, 81.39%, 81.94% for K_p= 1 and Ki=0, 50, 350, 750, 1000, 1350, respectively. For Kp=1, the steady state time decreases from 67.49 s to 39.60, 1.81, 1.85, 1.4, 1.35 and 1.25 s, respectively. The RMS value for Control-OFF result is 12.79 and the reductions are 42.38%,62.86%, 79.28%, 80.77%, for K_p= 5 and Ki=0, 50, 350, 750, respectively. For K_p=5, the steady state time decreases from 67.49 s to 15.25, 6.7, 1.92, 1.42, respectively. For given proportional control constant, it is observed that integral control gains have no effect on the desired control such as K_i=1000 and 1350.

Table 4. Case results for the manipulator

Motion parameters	K_P	Ki	RMS (m/s²) Control-OFF	Tss Control-OFF	RMS (m/s²) Control-ON	Tss Control-ON	Reduction %
*, 0, 0.1, 1]	1	0	12.79	67.49	11.01	39.60	13.92
		50			5.75	1.81	55.04
		350			2.85	1.85	77.72
		750			2.49	1.4	80.53
		1000			2.38	1.35	81.39
		1350			2.31	1.25	81.94
	5	0			7.37	15.25	42.38
		50			4.75	6.7	62.86
		350			2.65	1.92	79.28
		750			2.46	1.42	80.77
		1000			-	-	-
		1350			-	-	-
[*, 0, 0.05, 1]	1	0	14.81	49.89	12.70	28.43	14.25
		50			6.70	7.81	54.76
		350			3.42	1.89	76.91
		750			3.02	1.43	79.61
		1000			2.90	1.39	80.42
		1350			2.78	1.34	81.23
	5	0			8.34	12.97	43.69
		50			5.42	6.28	63.4
		350			3.17	1.99	78.6
		750			3.25	1.47	78.06
		1000			-	-	-
		1350			-	-	-
[*, 0.8, 0.1, 1]	1	0	6.49	75.66	5.55	39.05	14.48
		50			2.77	8.17	57.32
		350			0.95	1.85	85.36
		750			0.63	1.43	90.29
		1000			0.54	1.26	91.68
		1350			0.47	1.28	92.76
	5	0			3.69	17.18	43.14
		50			2.24	7.20	65.49
		350			0.84	1.98	87.06
		750			0.58	1.38	91.06
		1000			0.51	1.23	92.14
		1350			0.45	1.40	93.07
	9	0			2.88	12.57	55.62
		50			1.92	6.84	70.42
		350			0.76	1.24	88.29
		750			0.54	1.13	91.68
		1000			0.48	1.28	92.6
		1350			0.43	1.17	93.37

continues on following page

Table 4. Continued

Motion parameters	K_p	Ki	RMS (m/s^2) Control-OFF	Tss Control-OFF	RMS (m/s^2) Control-ON	Tss Control-ON	Reduction %
[*, 0.9, 0.05, 1]	1	0	2.25	66.20	1.94	39.51	13.78
		50			1.00	8.21	55.56
		350			0.42	2.31	81.33
		750			0.28	1.35	87.56
		1000			0.24	1.23	89.33
		1350			0.19	1.21	91.56
	5	0			1.28	15.89	43.11
		50			0.82	10.10	63.56
		350			0.38	2.38	83.11
		750			0.26	1.41	88.44
		1000			0.22	1.27	90.22
		1350			0.18	1.21	92.0
	9	0			0.99	19.43	56.0
		50			0.70	8.4	68.89
		350			0.35	2.39	84.44
		750			0.25	1.48	88.89
		1000			0.22	1.27	90.22
		1350			0.18	1.23	92.0

In case of the triangular velocity profile with $\mathbf{q}_m$=[*, 0, 0.05, 1]; the RMS value for Control-OFF result is 14.81 and the reductions are 14.25%, 54.76% and 76.91%, 79.61%, 80.42%, 81.23%, for K_p= 1 and Ki=0, 50, 350, 750, 1000, 1350 respectively. For Kp=1, the steady state time decreases from 49.89 s to 28.43, 7.81, 1.89, 1.43, 1.39 and 1.34 and respectively.

The RMS value for Control-OFF result is 14.81 and the reductions are 43.69%, 63.4%, 78.6%, 78.06%, for K_p= 5 and Ki=0, 50, 350, 750, respectively. For K_p=5, the steady state time decreases from 49.89 s to 12.97, 6.28, 1.99, 1.47, respectively. For given proportional control constant, it is observed that integral control gains have no effect on the desired control such as K_i=1000 and 1350.

In case of the trapezoidal velocity profile with $\mathbf{q}_m$=[*, 0.8, 0.1, 1]; the RMS value for Control-OFF result is 6.49 and the reductions are 14.48%, 57.32%, 85.36%, 90.29%, 91.68%, 92.76% for K_p= 1 and Ki=0,50,350,750,1000,1350 respectively. For K_p=1, the steady state time decreases from 75.66 s to 39.05, 8.17, 1.85, 1.43, 1.26, 1.28 respectively. The RMS value for Control-OFF result is 6.49 and the reductions are 43.14%, 65.49%, 87.06%, 91.06%, 92.14%, 93.07% for K_p= 5 and Ki=0,50,350,750,1000,1350 respectively. For K_p=5, the steady state time decreases from 75.66 s to 17.18, 7.20, 1.98, 1.38, 1.23,1.40 respectively. the RMS value for Control-OFF result is 6.49 and the reductions are 55.62%, 70.42%, 88.29%, 91.68%, 92.6%, 93.37% for K_p= 9 and Ki=0,50,350,750,1000,1350 respectively. For K_p=9, the steady state time decreases from 75.66 s to 12.57, 6.84, 1.24, 1.13, 1.28, 1.17 respectively.

In case of the trapezoidal velocity profile with $\mathbf{q}_m$=[*, 0.9, 0.05, 1]; the RMS value for Control-OFF result is 2.25 and the reductions are 13.78%, 55.56%, 81.33%, 87.56%, 89.33%, 91.56% for K_p= 1 and Ki=0,50,350,750,1000,1350 respectively. For K_p=1, the steady state time decreases from 66.20 s to 39.51, 8.21, 2.31, 1.35, 1.23, 1.21 respectively.

the RMS value for Control-OFF result is 2.25 and the reductions are 43.11%, 63.56%, 83.11%, 88.44%, 90.22% and 92.0% for K_p = 5 and Ki=0,50,350,750,1000,1350 respectively. For K_p=5, the steady state time decreases from 66.20 s to 15.89, 10.10, 2.38, 1.41, 1.27, 1.21 respectively the RMS value for Control-OFF result is 2.25 and the reductions are 56.0%, 68.89%, 84.44%, 88.89%, 90.22%, 92.0% for K_p= 9 and Ki=0,50,350,750,1000,1350 respectively. For K_p=9, the steady state time decreases from 66.20 s to 19.43, 8.4, 2.39, 1.48, 1.27, 1.23 respectively.

It is observed from Table 4 that; residual vibration amplitudes can be successfully suppressed and steady-state time can dramatically decrease by applying PI control for all velocity profile inputs. The reduction ratio of residual vibration amplitudes is up to 92% and steady-state time decreases from 75.66 s to 1.17s.

CONCLUSION

Robot manipulators have a wide area of usage on industry. However, these manipulators are bulky and consumes more energy than flexible robot manipulators which have smaller actuators and consumes less energy. Besides these advantages of flexible manipulators, residual vibrations occur due to their high flexibility especially on high-speed motions. Active vibration control is important to control these vibrations due to the flexibility.

Robot arms and flexible mechanical systems in general can be modeled with finite element method and Newmark method can be used to solve mathematical model of such systems.

The finite element model of a rotating flexible manipulator can also be created in MATLAB and transient results can be observed by the Newmark method. In this study, triangular and trapezoidal velocity profiles have been applied to the manipulator. To control residual vibration amplitudes, closed loop control scheme has been effectuated. Velocity feedback of the desired endpoint has been carried out and traditional Proportional-Integral (PI) Control has been applied. The root mean square values and steady-state time have been calculated presented for different cases. It is observed from the results that the residual vibration amplitudes of the manipulators are successfully suppressed, and steady-state time decreased by the proportional-integral control gain.

REFERENCES

Bathe, K. J. (2014). *Finite Element Procedures,* (2nd ed.). Prentice-Hall.

Benosman, M., & Le Vey, G. (2004). Control of flexible manipulators: A survey. *Robotica, 22*(5), 533–545. doi:10.1017/S0263574703005642

Blackburn, D., Singhose, W., Kitchen, J., Patrangenaru, V., Lawrence, J., Kamoi, T., & Taura, A. (2010). Command shaping for nonlinear crane dynamics. *Journal of Vibration and Control, 16*(4), 477–501. doi:10.1177/1077546309106142

Chan, T., Godbole, K., & Hou, E. (2003). Optimal input shaper design for high-speed robotic workcells. *Journal of Vibration and Control, 9*(12), 1359–1376. doi:10.1177/1077546304031165

Conker, C., Yavuz, H., & Bilgic, H. H. (2016). A review of command shaping techniques for elimination of residual vibrations in flexible-joint manipulators. *Journal of Vibroengineering*, *18*(5), 2947–2958. doi:10.21595/jve.2016.16725

Dharne, A. G., & Jayasuriya, S. (2007). Robust adaptive control of residual vibration in point-to-point motion of flexible bodies. *Journal of Vibration and Control*, *13*(7), 951–968. doi:10.1177/1077546307078830

Dwivedy, S. K., & Eberhard, P. (2006). Dynamic analysis of flexible manipulators, a literature review. *Mechanism and Machine Theory*, *41*(7), 749–777. doi:10.1016/j.mechmachtheory.2006.01.014

Fung, T. C. (1997). Unconditionally stable higher-order Newmark methods by sub-stepping procedure. *Computer Methods in Applied Mechanics and Engineering*, *147*(1-2), 61–84. doi:10.1016/S0045-7825(96)01243-1

Gaudenzi, P., Carbonaro, R., & Benzi, E. (2000). Control of beam vibrations by means of piezoelectric devices: Theory and experiments. *Composite Structures*, *50*(4), 373–379. doi:10.1016/S0263-8223(00)00114-8

Gürleyük, S. S., & Cinal, Ş. (2007). Robust three-impulse sequence input shaper design. *Journal of Vibration and Control*, *13*(12), 1807–1818. doi:10.1177/1077546307080012

Haido, J. H., Bakar, B. A., Abdul-Razzak, A. A., & Jayaprakash, J. (2010). Dynamic response simulation for reinforced concrete slabs. *Simulation Modelling Practice and Theory*, *18*(6), 696–711. doi:10.1016/j.simpat.2010.01.011

Hassan, M., Dubay, R., Li, C., & Wang, R. (2007). Active vibration control of a flexible one-link manipulator using a multivariable predictive controller. *Mechatronics*, *17*(6), 311–323. doi:10.1016/j.mechatronics.2007.02.004

Kim, D., & Singhose, W. (2010). Performance studies of human operators driving double-pendulum bridge cranes. *Control Engineering Practice*, *18*(6), 567–576. doi:10.1016/j.conengprac.2010.01.011

Kojima, H., & Hiruma, T. (2006). Evolutionary Learning Acquisition of Optimal Joint Angle Trajectories of Flexible Robot Arm. J. *Robotics Mechatronics*, *18*(1), 103–110. doi:10.20965/jrm.2006.p0103

Ma, P., Xia, R., Wang, X., Zhang, X., Królczyk, G., Gardoni, P., & Li, Z. (2022). An active control method for vibration reduction of a single-link flexible manipulator. *Journal of Low Frequency Noise, Vibration and Active Control*, *41*(4), 14613484221094982. doi:10.1177/14613484221094982

Oguamanam, D., Bosnjak, S., & Zrnic, N. (2006). On the dynamic modelling of flexible manipulators. *FME Trans.*, *34*, 231–237.

Owren, B., & Simonsen, H. H. (1995). Alternative integration methods for problems in structural dynamics. *Computer Methods in Applied Mechanics and Engineering*, *122*(1-2), 1–10. doi:10.1016/0045-7825(94)00717-2

Qiu, Z. C., Li, C., & Zhang, X. M. (2019). Experimental study on active vibration control for a kind of two-link flexible manipulator. *Mechanical Systems and Signal Processing*, *118*, 623–644. doi:10.1016/j.ymssp.2018.09.001

Rao, S. (2011). *Mechanical Vibrations,* (5th ed.). Prantice-Hall.

Shang, D., Li, X., Yin, M., & Li, F. (2022). Vibration Suppression Method for Dual-flexible Manipulator Considering Bearing Friction Based on Adaptive PI Control Strategy with Nonlinear Disturbance Observer.

Singhose, W., & Pao, L. (1997). A comparison of input shaping and time-optimal flexible-body control. *Control Engineering Practice*, *5*(4), 459–467. doi:10.1016/S0967-0661(97)00025-7

Singhose, W., Porter, L., Kenison, M., & Kriikku, E. (2000). Effects of hoisting on the input shaping control of gantry cranes. *Control Engineering Practice*, *8*(10), 1159–1165. doi:10.1016/S0967-0661(00)00054-X

Tang, L., Gouttefarde, M., Sun, H., Yin, L., & Zhou, C. (2021). Dynamic modelling and vibration suppression of a single-link flexible manipulator with two cables. *Mechanism and Machine Theory*, *162*, 104347. doi:10.1016/j.mechmachtheory.2021.104347

Vakil, M., Fotouhi, R., Nikiforuk, P. N., & Heidari, F. (2011). A study of the free vibration of flexible-link flexible-joint manipulators. *Proceedings of the Institution of Mechanical Engineers. Part C, Journal of Mechanical Engineering Science*, *225*(6), 1361–1371. doi:10.1177/0954406211399517

Xianmin, Z., Changjian, S., & Erdman, A. G. (2002). Active vibration controller design and comparison study of flexible linkage mechanism systems. *Mechanism and Machine Theory*, *37*(9), 985–997. doi:10.1016/S0094-114X(02)00025-3

Zhang, L., Zu, J. W., & Zheng, Z. (1999). The stochastic Newmark algorithm for random analysis of multi-degree-of-freedom nonlinear systems. *Computers & Structures*, *70*(5), 557–568. doi:10.1016/S0045-7949(98)00203-X

Chapter 6
Advances of the Robotics Technology in Modern Minimally Invasive Surgery

Ranjit Barua
https://orcid.org/0000-0003-2236-3876
Omdayal Group of Institutions, India

Sumit Bhowmik
Omdayal Group of Institutions, India

Arghya Dey
Omdayal Group of Institutions, India

Jaydeep Mondal
College of Medicine and Sagore Dutta Hospital, India

ABSTRACT

Currently, numerous robotically controlled steerable needle-based methods have been offered, which can precisely maintain needle tip deflection to monitor premeditated curved paths inside soft tissue. Percutaneous interventions via MIS methods can deliver patients with improved consequences and earlier recovery times than traditional or open surgeries. Precise needle insertions are important for effective procedures, and dynamically piloted needles can intensify system exactness. Current robotic technology can overcome individuals' limitations and upturn the attainment of minimally invasive percutaneous treatment. Needle insertion can also be applied as a "satellite access" area, and parallel calculus can be pushed or broken into the renal pelvis. The main access is applied for the removal of stones, in that way avoiding the formation of various channels and efficiently caring for kidney function. Here, the authors will present the present state-of-the-art insertion needles for MIS, concentrating on the limitations and challenges still open for their application in the medical area.

DOI: 10.4018/978-1-6684-5381-0.ch006

INTRODUCTION

The advances in medicinal and clinical performance and modern surgical machinery have continuously developed at the same time. Surgical robotics application in the last few years has succeeded in rapid progress and has been extensively functional in the modern surgical area. The varieties of surgical robotics include orthopedics, neurosurgery, vascular interventional, laparoscopic, prosthetics, exoskeleton robotics, capsule robotics, and rehabilitation robotics. Main technical investigation in advanced robotics in medical science is now very important. The rise of advanced machinery made additional developments in minimal access or invasive surgery (MIS) conceivable. Robotic surgery and telerobotics surgery successfully addressed the restrictions of thoracoscopic, and laparoscopic processes, therefore modernizing the surgery through minimal access. Robotic surgery is estimated to endure to involve a rising portion of advanced surgery. They are computer-controlled self-powered devices that can be planned to help in the arranging and guidance of surgical tools, allowing the surgeon to perform more intricate tasks. The moderately latest signs of progress in endoscopic machinery have permitted the recognition of the MIS or nominally form of surgery (Moit et al., 2019). The signs of progress in robotics assist accurate surgical procedures that are habitually combined with medical image control competence. This opportunity has determined the additional progress of equipment to pay for the exclusive difficulties created by this fresh arrangement and to progress the presentation and increase the possibility of the measures which can be achieved. Surgical robotics technology has been an essential factor of this progress because of the greatly appropriate appearances that a machinelike (programmed)arrangement can significance, with extremely advanced automated compliance and the aptitude to suit operative purposes in surgical robots for doctors to achieve harmless and truthful operations (Zhang et al., 2020). Furthermore, linking the robot-controlled interferences with recognizing and medicinal imaging expertise can importantly progress the existing evidence and thus advantage to confirm that MIS remains to improve accepted and stay at the emphasis of modern medical technology progress (Lee et al., 2011) (Moit et al., 2019) (Barua et al., 2022). Founded on the nature of invasiveness of surgical or operating techniques, there are normally three key techniques: (a) traditional surgery or operation, (b) minimally invasive processes, and (c) noninvasive processes. MIS is planned to deliver excessive assistance to the patient on traditional surgery (open) by reducing avoidable trauma-affected in the procedure of accomplishing a medical process. Also, a smaller amount of disturbance, blood loss, pain, and trauma, these modifications make possible a speedy recovery time and decreased the possibility of difficulties (Wagner et al., 2002). Though, it is also well recognized that this method brings several consistent problems to the medical team performing it. These problems are because of the very restricted workspace, dedicated tools needing additional staff teaching and adjustment for usage, and importantly reduced visual and touch info. Even though the above-mentioned disadvantages, MIS has sustained improvement acceptance and is to be extensively used (Mack et al., 2001) (Barua et al., 2022). A major issue in this sustained acceptance is the conforming growth of surgical tools envisioned for practice in MIS (Barua et al., 2022). Surgical robotic platforms are most appropriate because of their promising features. Such features contain high correctness, repeatability, and the opportunity of planning specific apparatuses that can be useful for precise events and vital organs. Also, modern surgical robots can include sensors to reappearance trace and force recognition and can also be joined with advanced imaging knowledge to permit sovereign, semiautonomous, or teleoperated mechanism, which can recover surgical presentation and the possibility of minimally invasive surgery. Furthermore, by joining evolving imaging methods and non-invasive modalities, cost-effective, more exact, and convenient treatment apparatuses can be made potential. Medical robotics has been in usage for almost 30 years; the first generation of medical robots was applied as tool holders and positioners, formerly the improvement of dynamic medical robots in the initial 1990s (Davies et al., 2000). Even with the fact that this knowledge has been nearby for three decades, medical robots have not been broadly accepted because of restrictions of control and the high risk of their uses. The associated procedures and standards, and in specific for dynamic systems, which are envisioned to be powered during a technique, are essentially challenging and as such, the standard growth time is very long in this area. There has also been the problem of recognition from the medical community and surgeons, who still opinion with disbelief any technology that would make a separation of the surgeon from the surgery (Schmidt et al., 2021). A modern trend for medical robotics and devices being advanced for use with minimally invasive surgery and (NIS) non-invasive surgery is to additional rise the capacity of uses with the design of greatly function-specific devices that can be mutual with medical imaging technology. A new tendency for this technique is the expansion of systems to return touch and force information to help in surgery or to deliver more info for diagnosis commitments (Mamone et al., 2019). This paper defines the improvement of modern surgical robotic methods for minimally invasive surgery MIS, the move to not as much autonomous but more form-precise ways, and the increase in investigation and progress of haptics knowledge in modern surgical robotics.

NEEDLE-STEERING METHODS

Needles normally used in therapy can be categorized as either asymmetric or symmetric; the former consist of beveled tipped needles, whereas the latter is prismatically or conically (Eslami et al., 2013). Clinicians who practiced needle insertion identify that beveled tip needles tend to bend as an effect of their tip asymmetry, in contrast with symmetrically tipped needles (Leibinger et al., 2016). As shown in Figure 1, the beveled shape creates forces at the needle–tissue boundary that imparts a bending needle force throughout insertion. The resulting bending may be reduced either by raising the needle stiffness or by rotating it, in a drill-like motion, during insertion (Casanova et al., 2014). With the indifference to conventional needle treatments that seek to boundary the bending motion, surgical robotics investigation tends to highlight and achieve this distortion for novel methods of treatment delivery, the method referred to as needle navigation. One of the core methods in needle navigation is created on the asymmetry of the needle tip, which contains either a beveled termination of Figure 1B or a combination of a beveled tip and pre-curved section to the needle as shown in Figure 1C.

Figure 1. a) Conical, b) Beveled shape needle insertion process, and c) Beveled needle deflection during insertion (Audette et al., 2020).

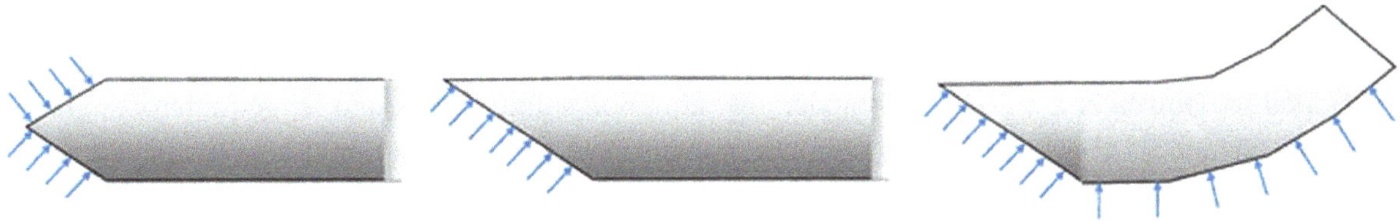

A B C

HUMAN-CONTROLLED ROBOT-ASSISTED SURGERY

The idea of tele-surgery was initiated differently from that of a controlled surgical robot. The first tele-robotic surgical technology was established in the early 1990s under the Defense Advanced Research Projects Agency (DARPA) which was an advanced combat casualty care program (Lucas et al., 2005). More of these advanced and sophisticated cases are the ZEUS of Computer Motion Inc. (acquired by Intuitive Surgical in 2003), the "da Vinci robot" of Intuitive Surgical Inc., and, Titan Medical Inc's Amadeus System. In the practice of traditional MIS (minimally invasive surgery) techniques in these ways, various robotic arms control the endoscopic tools, the arms are inflexible and cable-driven, and the additional arm controls a laparoscopic camera. The surgeon drives the robot from a console by two hand masters for the tools and pedal controls or voice control (ZEUS) for the laparoscope arm. With the "da Vinci" system, the surgeon can monitor the surgical position from side to side with a viewfinder, which creates pseudo-3D images. The "da Vinci" system shown in Figure 2 also includes "endo-wrists," a special feature that provides two extra degrees of freedom (DOF) at the point of interference and significantly ease the relief of use, mainly for cutting and suturing processes (McBeth et al., 2004). Other causes of MI (minimally invasive) processes executed by the "da Vinci" system contain urological surgery, cardiac surgery, and prostatectomy (Cleary et al., 2006). Endoscopic and laparoscopic procedures are at the moment controlled by multiport access approaches, which mimic the traditional hand-eye coordination with an instrument-stereovision arrangement. Those methods need multiple ports in the patient's body for the MIS (minimally invasive surgery) apparatus insertions. For lesser invasive methods, surgeons are increasingly opting for single-port access (SPA) minimally invasive surgery, with the help of robotic arms and computer assistive strategies in the operating room. Single-port access (SPA) can lessen problems related to multiport incisions (Bucher et al., 2008) (Borowski et al., 2011), although single-port access (SPA) poses important faces for the design of endoscopic instruments as a result of the maneuverability restraints over a single port (Park et al., 2011).

WIRELESS MIS (MINIMALLY INVASIVE SURGERY) ROBOTIC SYSTEMS

Wireless robotic arrangements that can influence far privileged the human body have to turn into a innovative division in minimally invasive systems. Several examples of these methods can be established in use for gastrointestinal (GI) events. The intestine offerings a long and complicated atmosphere in gastrointestinal (GI) endoscopic techniques. By performing the traditional gastrointestinal (GI) procedures, patients complain of discomfort and pain, and clinicians face technical problems complicated in steering the apparatuses. As such, robotic strategies for example the PillCam established by Given Imaging Inc. are small and thin, can be inserted into the gastrointestinal (GI) tract, and planned under their individual decision. The PillCam also contains a mini camera, LEDs, and radio frequencies transmitter (RF) but is transported by exhausting the body's digestive method. The "inchworm" design (Figure 3), developed by Quirini et al., 2007, is a moderately common solution although legged microbots have also been engaged. Other popular actuation mechanisms have also been explored for example the ones driven by outer magnetic fields, offered by Ciuti et al. 2010, Wang et al., 2011, and Lien et al., 2009, or hybrid-driven by external/ internal actuators (Simi et al., 2010).

Figure 2. Endo-wrists end effectors (McBeth et al., 2004).

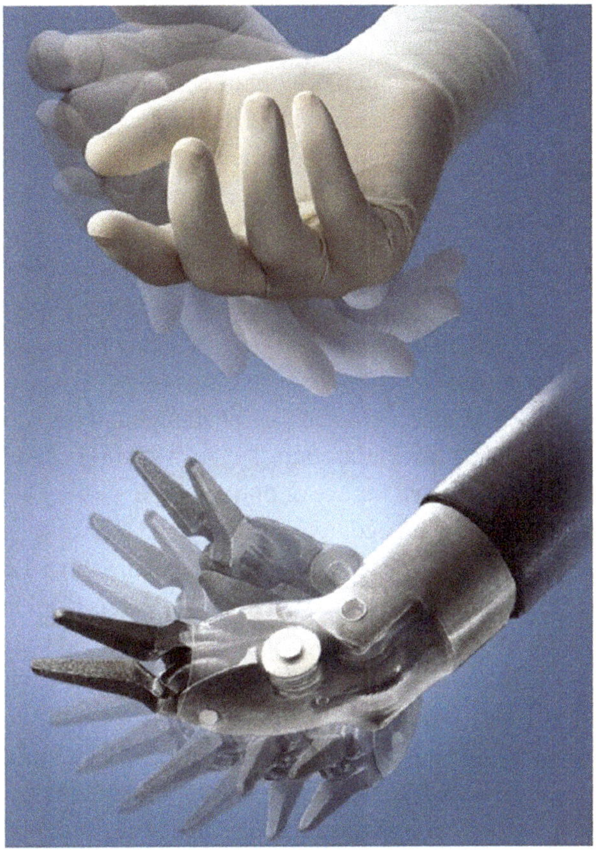

Figure 3. Inchworm capsule robot (Prototype of 12-legged endoscopic capsular robot) (Valdastri et al., 2009).

MICROSURGERY BY HUMAN-CONTROLLED ROBOTS

Dynamic control of the needle can intensify engagement accuracy and simultaneously dropping risk, e.g., by evading difficulties or critical areas in tissue that lie in between the entrance point and object. The inset of the needle is tough as the displacement and deformation of the vital organs are the essentials in surgical performance. Initiating concentrated tissue motion and deformation throughout needle insertion can intensify the correctness and henceforth the achievement of a medical process. If the tissue structure has changed by a smaller amount because of the needle existence introduced, surgeons can depend on advanced assurance on internal scans developed pre-operatively. To reduce tissue trauma, and permit further difficult path ensuing, a flexible surgical type of needle is emerging, approximately concentrating on steerable, extended, and vigorous devices. Tip navigation has the benefit of a follow-the-leader track, where the shaft monitors the tip route, and steerability is mainly without air by the inset complexity. Though the needle–tissue contact forces are challenging to perfection and forecasting, and non-holonomic restraints make difficulties in the control of such strategies. At Northwestern University, a microsurgical manipulator was developed for eye or optical surgery which is mostly used as a surgical robot (Jensen et al., 1997). This surgical procedure is programmed to execute a micropipette by a hypodermic surgical needle through the lumen. The technique is done through a Trackball master which can target retinal blood vessels to an inner diameter of 20 microns. The Johns Hopkins University steady-hand robot for eye surgery highlights motion scaling over tremor filtering and has a placing resolution of 5 to 10 microns (Figure 4) (Üneri et al., 2010). The University of Washington developed a 7-DOF cable-actuated manipulator RAVEN which is controlled by the PHANToM haptic control interface (Sensable Technologies Inc.), containing wrist joints similar to the "da Vinci" System, and has the possibility for haptic feedback (King et al., 2009). At ETH Zurich, a wireless system has been developed to deliver drugs into the retinal blood vessels (Dogangil et al., 2008). This is a "microbot" and involves only one cut made in the sclera wall. A telerobotic system famous as Heartlander has been made at Carnegie Mellon University, which uses suction pads to attach itself to the epicardium and propels itself by applying small onboard piezoelectric motors (Patronik et al., 2011).

SURGICAL ROBOTS WITH PARALLEL MECHANISMS

For the use in medical techniques, the advancement of various designs of parallel robots has been configured. Brandt et al. designed a compressed robot CRIGOS for surgery in image-guided orthopedics which included a parallel robot along with a principle of computation for arranging the surgical intrusions and parallel stage controlling (Brandt et al., 1999). Tanikawa et al., 1999 developed dexterous micro-manipulation, whose arrangement was based on parallel apparatus and used for execution microsurgery among various other works. Merlet established a MIPS i.e. a microrobot having a parallel manipulator. The 3-DOF scheme endorsed the adequate placing of a surgical instrument which was used as a dynamic wrist at the tip of the endoscope (Merlet et al., 2002). Shoham et al. 2003 produced a small robot named MARS, for surgical practice having a mechanism of the 6-DOF parallel cylinder and it has the ability for being applied in various surgical procedures having trauma and spine surgery, where precise placement and positioning of a portable surgical robot for necessary. Maurin et al., 2004 developed a robotic stage having parallel 5-DOF for CT-guided percutaneous methods having automatic patient-to-image needles registration and force feedback. Tsai et al., 2007 produced a parallel surgical robot for the exact boring of the skull for stereotactic neurosurgical operation. The "da Vinci" robotic system is shown in figure 5. As pointed out earlier that the main limitation of the parallel mechanisms is that the partial workstation is connected to the serial link robots. For the neurosurgical processes, the area of working lies above the skull exterior which is positioned as one robot side. This unequal workspace was examined by Hsu and Tsai and found the perfect qualified places of the parallel mechanism and the skull to express the exact workspace on the skull (Tsai et al., 2007).

Figure 4. Retinal microsurgery (Üneri et al., 2010).

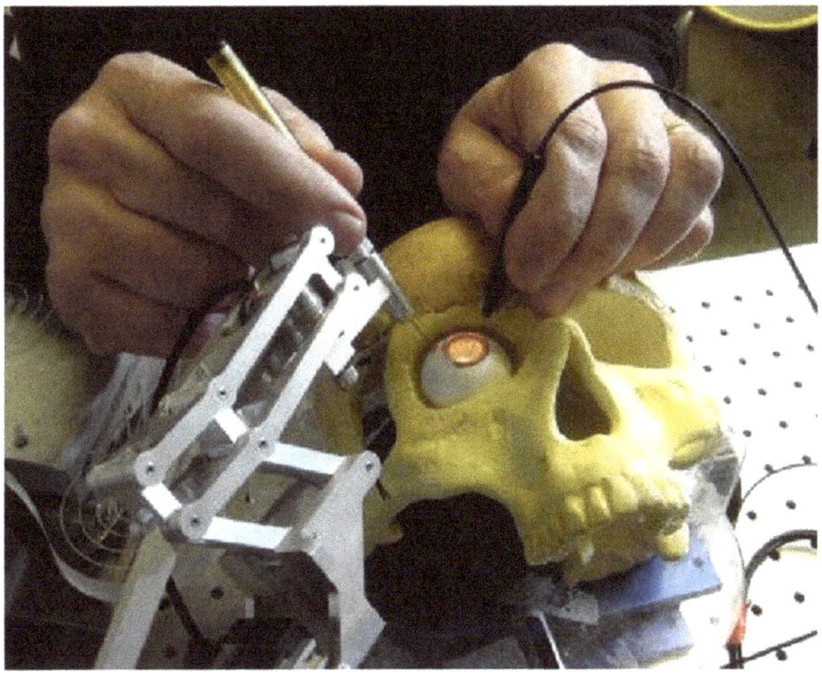

TELE-ROBOTIC SURGERY

With the increased use of teleoperated MIS, the haptic feedback response has become a significant area for examination. This kind of feedback in the clinic can rectify the surgeon's knowledge of telepresence, which completely leads to a better presentation. Validation forcefully proposes that the capability to debate haptic response to current robotic surgical setups for example "da Vinci" can donate expressively to innocent cardiac surgical activities by these multifaceted structures (Cavusoglu et al., 2001). The inadequacy of haptic response in the present robotic systems is an important handicap in the presentation of the officially additional complex and complex surgical tasks. These works are essential for specializations like heart surgery [Okamura et al., 2004] and the absence of a haptic response can create an indeterminate force level by the medical staff (Wagner et al., 2002). An amount of surgical haptics that gives a provisional between master-slave telesurgery and independent medical robotics is the practice of "active restraints" (Davies et al., 2006) or "virtual fixtures" (Payandeh et al., 2002). These machines do not energetically control the devices as given, somewhat the surgeon's determination authority is given, though the machine can give forces that are controlled immediately as the prearranged workspace boundary is reached. This method improves the robot-controlled surgery as well as the techniques of MIS, simultaneously allowing the total control as well as the decision to carry on with the surgeon (Davies et al., 2000). In the Imperial College, the "Acrobat" is a model of this robot. Another area of research that has grasped an important growth in attractiveness is the growth of haptic abilities for telerobotic approaches, for example, the "da Vinci".

Figure 5. "da Vinci" robotic system (Liu et al., 2017).

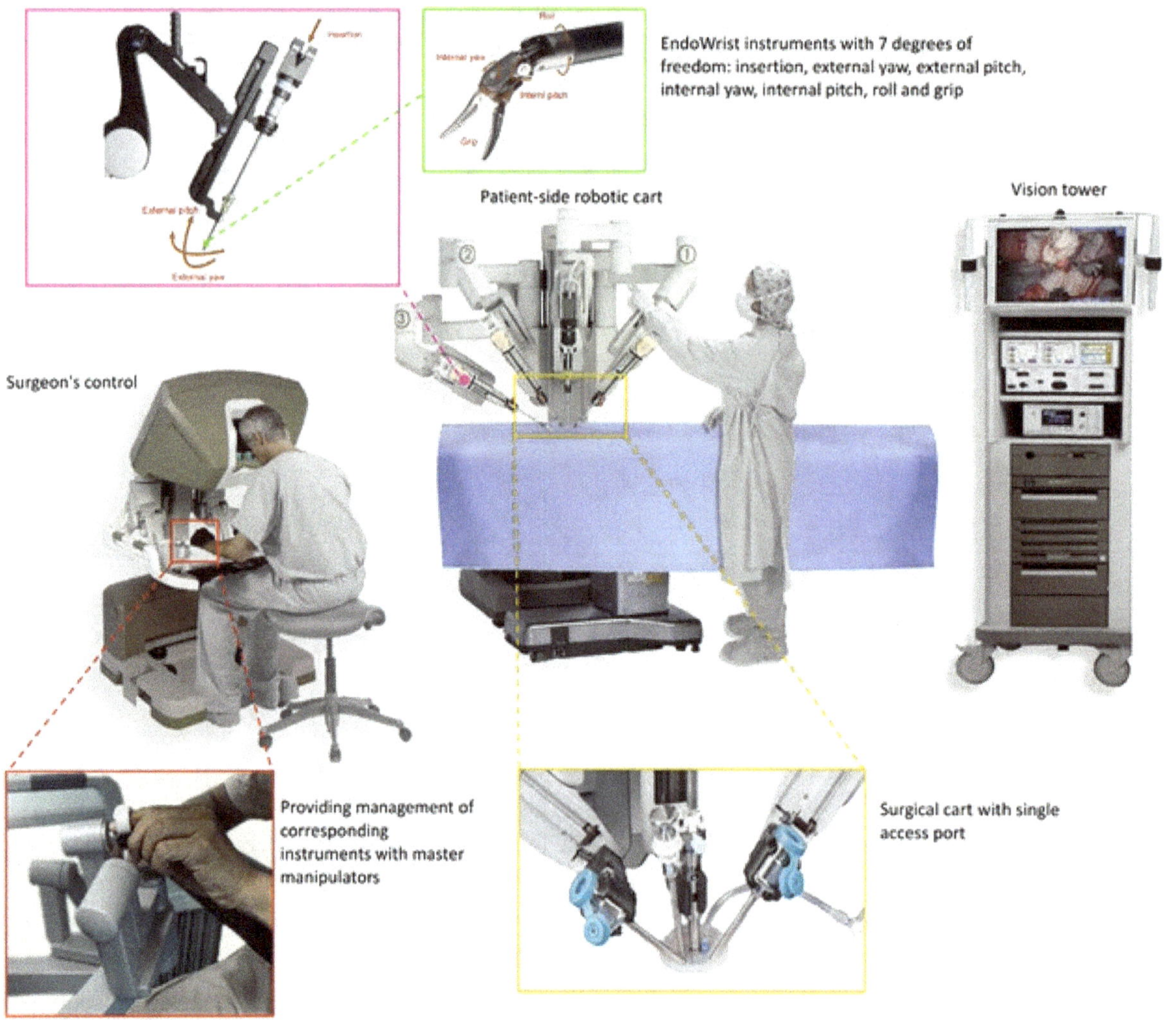

SURGICAL SIMULATION AND TRAINING

The haptic surgical simulators' progress is found out by the boundaries of the orthodox approaches to surgical preparation. Typically, the technique of training is in dual stages: in the first stage, the surgeon training structure from manuscripts and extra graphical supports; this is later "hands-on" workout in the operation (OT) chamber or by simulation manikins or corpse separation (Cooper et al., 2008). The trouble with the orthodox form of workout is that the availability of patients or corpses to struggle with is changeable and inadequate (Liu et al., 2003). Real human anatomical stuff and many morphologies can be created by simulations. For this purpose, since the 90s virtual reality (VR) simulations are being used (Burdea, 1996). Also, for assistance in cure planning and analysis, surgical simulators can be used (Deurloo et al., 2001) (Barua et al., 2022).

CONCLUSION

For complex surgeries and tough medical connections, modern robots are very much useful which enable these complicated tasks which cannot be done with bare hands. The development of surgical stages which are robotically controlled is hugely changed from the period of pre-electronics to create medical procedures inclusive, more consistent, safer, and faster. In operating rooms, the use of robotic manipulators is useful and suitable because of the current changes in computer programming, controls, and mechanical designs. Robotics has also operated to recuperate robots' skills over version, overriding sensory information to respond to switching conditions, along with autonomy. An additional benefit is that machines could be taken into account with immeasurable dissimilar morphologies to a procedure having a particular working topology; the physicians, equally, are alike to common-using machines suitable to numerous occasions but flawless to nothing. As well as the technological and historical developments in robotics, many surgical propensities have also affected the importance of research. The main issues comprise the rising significance of non-invasive and marginally invasive approaches and the accessibility of the 3D image information. Nevertheless, the mentioned productive character, still there are many duties opposite medical robotics. One of the main limitations for any robotic procedure is the inability to dynamically procedure different basics of workout meaningful decisions, complete qualitative intellectual and information. This is the trouble of artificial intelligence (AI) and is one of the important inspirations in the inconsistency of self-governing surgery currently.

ACKNOWLEDGMENT

The authors would like to thank Mechanical Engineering Dept. and Advanced Robotics lab of ODGI, College of Medicine & Sagore Dutta Hospital (CMSDH), Kamarhati Kolkata, and thanks to Mrs. Nibedita Bardhan for language proof reading.

REFERENCES

Audette, M. A., Bordas, S., & Blatt, J. E. (2020). Robotically Steered Needles: A Survey of Neurosurgical Applications and Technical Innovations. *Robotic Surgery (Auckland)*, *7*, 1–23. doi:10.2147/RSRR.S224446 PMID:32258180

Barua, R., & Das, S. (2022). Improvements of Virtual and Augmented Reality for Advanced Treatments in Urology. In *Emerging Advancements for Virtual and Augmented Reality in Healthcare,* (pp. 117–131). IGI Global. doi:10.4018/978-1-7998-8371-5.ch008

Barua, R., Das, S., Datta, P., & RoyChowdhury, A. (2022). Computational FEM Application on Percutaneous Nephrolithotomy (PCNL) Minimum Invasive Surgery Through Needle Insertion Process. In P. Pain, S. Banerjee, & G. Bose (Ed.), *Advances in Computational Approaches in Biomechanics,* (pp. 210-222). IGI Global. doi:10.4018/978-1-7998-9078-2.ch013

Barua, R., Das, S., Datta, S., Datta, P., & Chowdhury, A. R. (2022). Analysis of surgical needle insertion modeling and viscoelastic tissue material interaction for minimally invasive surgery (MIS). *Materials Today: Proceedings*, *57*, 259–264. doi:10.1016/j.matpr.2022.02.498

Barua, R., Das, S., Datta, S., Roy Chowdhury, A., & Datta, P. (2022). Experimental Study of the Robotically Controlled Surgical Needle Insertion for Analysis of the Minimum Invasive Process. In *Emergent Converging Technologies and Biomedical Systems,* (pp. 473–482). Springer. doi:10.1007/978-981-16-8774-7_38

Borowski, D. W., Kanakala, V., Agarwal, A. K., Tabaqchali, M. A., Garg, D. K., & Gill, T. S. (2011). Single-port access laparoscopic reversal of Hartmann operation. *Diseases of the Colon and Rectum*, *54*(8), 1053–1056. doi:10.1097/DCR.0b013e318221aa13 PMID:21730797

Brandt, G., Zimolong, A., Carrat, L., Merloz, P., Staudte, H. W., Lavallée, S., Radermacher, K., & Rau, G. (1999). CRIGOS: a compact robot for image-guided orthopedic surgery. *IEEE transactions on information technology in biomedicine: a publication of the IEEE Engineering in Medicine and Biology Society, 3*(4), 252–260.

Bucher, P., Pugin, F., & Morel, P. (2008). Single port access laparoscopic right hemicolectomy. *International Journal of Colorectal Disease*, *23*(10), 1013–1016. doi:10.100700384-008-0519-8 PMID:18607608

Burdea, G. C. (1996). *Force and Touch Feedback for Virtual Reality*. John Wiley & Sons.

Casanova, F., Carney, P. R., & Sarntinoranont, M. (2014). Effect of needle insertion speed on tissue injury, stress, and backflow distribution for convection-enhanced delivery in the rat brain. *PLoS One*, *9*(4), e94919. doi:10.1371/journal.pone.0094919 PMID:24776986

Cavusoglu, M. C., Villanueva, I., & Tendick, F. (2001, November). Workspace analysis of robotic manipulators for a teleoperated suturing task. In *Proceedings IEEE/RSJ International Conference on Intelligent Robots and Systems. Expanding the Societal Role of Robotics in the the Next Millennium (Cat. No. 01CH37180),* (Vol. 4, pp. 2234-2239). IEEE. 10.1109/IROS.2001.976402

Ciuti, G., Valdastri, P., Menciassi, A., & Dario, P. (2010). Robotic magnetic steering and locomotion of capsule endoscope for diagnostic and surgical endoluminal procedures. *Robotica*, *28*(2), 199–207. doi:10.1017/S0263574709990361

Cleary, K., Melzer, A., Watson, V., Kronreif, G., & Stoianovici, D. (2006). Interventional robotic systems: Applications and technology state-of-the-art. *Minimally Invasive Therapy & Allied Technologies*, *15*(2), 101–113. doi:10.1080/13645700600674179 PMID:16754193

Cooper, J. B., & Taqueti, V. (2008). A brief history of the development of mannequin simulators for clinical education and training. *Postgraduate Medical Journal*, *84*(997), 563–570. doi:10.1136/qshc.2004.009886 PMID:19103813

Davies, B. (2000). A review of robotics in surgery. *Proceedings of the Institution of Mechanical Engineers. Part H, Journal of Engineering in Medicine*, *214*(1), 129–140. doi:10.1243/0954411001535309 PMID:10718057

Davies, B. (2000). A review of robotics in surgery. *Proceedings of the Institution of Mechanical Engineers. Part H, Journal of Engineering in Medicine*, *214*(1), 129–140. doi:10.1243/0954411001535309 PMID:10718057

Davies, B., Jakopec, M., Harris, S. J., Baena, F. R. Y., Barrett, A., Evangelidis, A., Gomes, P., Henckel, J., & Cobb, J. (2006). Active-constraint robotics for surgery. *Proceedings of the IEEE*, *94*(9), 1696–1704. doi:10.1109/JPROC.2006.880680

Deurloo, E. E., Gilhuijs, K. G., Schultze Kool, L. J., & Muller, S. H. (2001). Displacement of breast tissue and needle deviations during stereotactic procedures. *Investigative Radiology*, *36*(6), 347–353. doi:10.1097/00004424-200106000-00008 PMID:11410756

Dogangil, G., Ergeneman, O., Abbott, J. J., Pané, S., Hall, H., Muntwyler, S., & Nelson, B. J. (2008, September). Toward targeted retinal drug delivery with wireless magnetic microrobots. In *IEEE/RSJ International Conference on Intelligent Robots and Systems,* (pp. 1921-1926). IEEE. 10.1109/IROS.2008.4650768

Eslami, S., Fischer, G. S., Song, S. E., Tokuda, J., Hata, N., Tempany, C. M., & Iordachita, I. (2013). Towards Clinically Optimized MRI-guided Surgical Manipulator for Minimally Invasive Prostate Percutaneous Interventions: Constructive Design. *International Conference on Robotics and Automation*, *20132*, 1228–1233. PMID:24683502

Jensen, P. S., Grace, K. W., Attariwala, R., Colgate, J. E., & Glucksberg, M. R. (1997). Toward robot-assisted vascular microsurgery in the retina. *Graefes Archive for Clinical and Experimental Ophthalmology*, *235*(11), 696–701. doi:10.1007/BF01880668 PMID:9407227

King, H. H., Tadano, K., Donlin, R., Friedman, D., Lum, M. J., Asch, V., Wang, C., Kawashima, K., & Hannaford, B. (2009, June). Preliminary protocol for interoperable telesurgery. In *International Conference on Advanced Robotics,* (pp. 1-6). IEEE.

Lee, J. Y., Mucksavage, P., Sundaram, C. P., & McDougall, E. M. (2011). Best practices for robotic surgery training and credentialing. *The Journal of Urology*, *185*(4), 1191–1197. doi:10.1016/j.juro.2010.11.067 PMID:21334030

Leibinger, A., Oldfield, M. J., Rodriguez, Y., & Baena, F. (2016). Minimally disruptive needle insertion: A biologically inspired solution. *Interface Focus*, *6*(3), 20150107. doi:10.1098/rsfs.2015.0107 PMID:27274797

Lien, G. S., Liu, C. W., Teng, M. T., & Huang, Y. M. (2009, May). Integration of two optical image modules and locomotion functions in capsule endoscope applications. In *IEEE 13th International Symposium on Consumer Electronics,* (pp. 828-829). IEEE.

Liu, A., Tendick, F., Cleary, K., & Kaufmann, C. (2003). A survey of surgical simulation: Applications, technology, and education. *Presence (Cambridge, Mass.)*, *12*(6), 599–614. doi:10.1162/105474603322955905

Liu, H. H., Li, L. J., Shi, B., Xu, C. W., & Luo, E. (2017). Robotic surgical systems in maxillofacial surgery: A review. *International Journal of Oral Science*, *9*(2), 63–73. doi:10.1038/ijos.2017.24 PMID:28660906

Lucas, M. A., Phys, F., & Krstic, A. R. (2005). Unmanned tele-operated robots as medical support on the battlefield. *ADF Health, 6*, 34-8.

Mack, M. J. (2001). Minimally invasive and robotic surgery. *Journal of the American Medical Association*, *285*(5), 568–572. doi:10.1001/jama.285.5.568 PMID:11176860

Mamone, V., Condino, S., Cutolo, F., Tamadon, I., Menciassi, A., Murzi, M., Ferrari, M., & Ferrari, V. (2019). Low-Computational Cost Stitching Method in a Three-Eyed Endoscope. *Journal of Healthcare Engineering*, *2019*, 5613931. doi:10.1155/2019/5613931 PMID:31316742

Maurin, B., Gangloff, J., Bayle, B., De Mathelin, M., Piccin, O., Zanne, P., Doignon, C., Soler, L., & Gangi, A. (2004, September). A parallel robotic system with force sensors for percutaneous procedures under CT-guidance. In *International Conference on Medical Image Computing and Computer-Assisted Intervention,* (pp. 176-183). Springer, Berlin, Heidelberg. 10.1007/978-3-540-30136-3_23

McBeth, P. B., Louw, D. F., Rizun, P. R., & Sutherland, G. R. (2004). Robotics in neurosurgery. *American Journal of Surgery*, *188*(4), 68–75. doi:10.1016/j.amjsurg.2004.08.004 PMID:15476655

Merlet, J. P. (2002, May). Optimal design for the micro parallel robot MIPS. In *Proceedings IEEE International Conference on Robotics and Automation (Cat. No. 02CH37292),* 2, pp. 1149-1154. IEEE. 10.1109/ROBOT.2002.1014698

Moit, H., Dwyer, A., De Sutter, M., Heinzel, S., & Crawford, D. (2019). A Standardized Robotic Training Curriculum in a General Surgery Program. *JSLS: Journal of the Society of Laparoendoscopic Surgeons, 23*(4), e2019.00045. doi:10.4293/JSLS.2019.00045

Okamura, A. M. (2004). Methods for haptic feedback in teleoperated robot-assisted surgery. *Industrial Robot. International Journal (Toronto, Ont.).*

Park, H. S., Kim, T. J., Song, T., Kim, M. K., Lee, Y. Y., Choi, C. H., Lee, J.-W., Kim, B.-G., & Bae, D. S. (2011). Single-port access (SPA) laparoscopic surgery in gynecology: A surgeon's experience with an initial 200 cases. *European Journal of Obstetrics, Gynecology, and Reproductive Biology*, *154*(1), 81–84. doi:10.1016/j.ejogrb.2010.09.004 PMID:21056527

Patronik, N. A., Zenati, M. A., & Riviere, C. N. (2011). A study ex vivo of the effect of epicardial fat on the heartlander robotic crawler. In *5th European Conference of the International Federation for Medical and Biological Engineering,* (pp. 227-230). Springer, Berlin, Heidelberg. 10.1007/978-3-642-23508-5_60

Payandeh, S., & Stanisic, Z. (2002). On application of virtual fixtures as an aid for telemanipulation and training. *Proceedings 10th Symposium on Haptic Interfaces for Virtual Environment and Teleoperator Systems. HAPTICS*, 18-23. 10.1109/HAPTIC.2002.998936

Quirini, M., Webster, R. J., Menciassi, A., & Dario, P. (2007, April). Design of a pill-sized 12-legged endoscopic capsule robot. In *Proceedings IEEE International Conference on Robotics and Automation,* (pp. 1856-1862). IEEE. 10.1109/ROBOT.2007.363592

Salicrú, S., Gil-Moreno, A., Montero, A., Roure, M., Pérez-Benavente, A., & Xercavins, J. (2011). Laparoscopic radical hysterectomy with pelvic lymphadenectomy in early invasive cervical cancer. *Journal of Minimally Invasive Gynecology, 18*(5), 555–568. doi:10.1016/j.jmig.2011.05.003 PMID:21783432

Schmidt, T., Babic, B., Bruns, C. J., & Fuchs, H. F. (2021). ChirurgischeTherapie des Ösophaguskarzinoms – neueTechnologien, moderneKonzepte [Surgical treatment of esophageal cancer-New technologies, modern concepts]. *Der Chirurg; Zeitschrift fur alleGebiete der operativenMedizen, 92*(12), 1100–1106.

Shoham, M., Burman, M., Zehavi, E., Joskowicz, L., Batkilin, E., & Kunicher, Y. (2003). Bone-mounted miniature robot for surgical procedures: Concept and clinical applications. *IEEE Transactions on Robotics and Automation, 19*(5), 893–901. doi:10.1109/TRA.2003.817075

Simi, M., Valdastri, P., Quaglia, C., Menciassi, A., & Dario, P. (2010). Design, fabrication, and testing of a capsule with hybrid locomotion for gastrointestinal tract exploration. *IEEE/ASME Transactions on Mechatronics, 15*(2), 170–180. doi:10.1109/TMECH.2010.2041244

Tanikawa, T., & Arai, T. (1999). Development of a micro-manipulation system having a two-fingered micro-hand. *IEEE Transactions on Robotics and Automation, 15*(1), 152–162. doi:10.1109/70.744610

Tsai, T. C., & Hsu, Y. L. (2007). Development of a parallel surgical robot with automatic bone drilling carriage for stereotactic neurosurgery. *Biomedical Engineering: Applications, Basis and Communications, 19*(04), 269–277.

Üneri, A., Balicki, M. A., Handa, J., Gehlbach, P., Taylor, R. H., & Iordachita, I. (2010, September). New steady-hand eye robot with micro-force sensing for vitreoretinal surgery. In *3rd IEEE RAS & EMBS International Conference on Biomedical Robotics and Biomechatronics,* (pp. 814-819). IEEE. 10.1109/BIOROB.2010.5625991

Valdastri, P., Webster, R. J., Quaglia, C., Quirini, M., Menciassi, A., & Dario, P. (2009). A new mechanism for mesoscale legged locomotion in compliant tubular environments. *IEEE Transactions on Robotics, 25*(5), 1047–1057. doi:10.1109/TRO.2009.2014127

Wagner, C. R., Lederman, S. J., & Howe, R. D. (2002). A tactile shape display using RC servomotors. In *Proceedings 10th Symposium on Haptic Interfaces for Virtual Environment and Teleoperator Systems. HAPTICS,* (pp. 354-355). IEEE. 10.1109/HAPTIC.2002.998981

Wagner, C. R., Stylopoulos, N., & Howe, R. D. (2002). The role of force feedback in surgery: analysis of blunt dissection. *Proceedings 10th Symposium on Haptic Interfaces for Virtual Environment and Teleoperator Systems. HAPTICS*, 68-74. 10.1109/HAPTIC.2002.998943

Wang, X., & Meng, M. Q. (2011). Perspective of active capsule endoscope: Actuation and localisation. *International Journal of Mechatronics and Automation, 1*(1), 38–45. doi:10.1504/IJMA.2011.039154

Zhang, D., Liu, J., Zhang, L., & Yang, G. Z. (2020). Hamlyn CRM: A compact master manipulator for surgical robot remote control. *International Journal of Computer Assisted Radiology and Surgery*, *15*(3), 503–514. doi:10.100711548-019-02112-y PMID:31956954

ADDITIONAL READING

At present healthcare sector has developed a lot as both the medical science and engineering field are working together for the betterment of this sector. Recently minimally invasive surgery (MIS) gained huge popularity as the trauma period of the patients, post-operation is less and fatality rate is also lesser in percentage. Robotics surgery is like a boon for the surgeon, it works like the helping hand of the surgeon. Bio-robotic engineers are focusing on developing more user-friendly surgical robots, so that modern invasive can execute smoothly.

KEY TERMS AND DEFINITIONS

Minimally Invasive Surgery (MIS): An invasive surgery, surgeons practice a variety of methods to operate with a minimal damage to patient's body than with traditional surgery.

Robot-Assisted Needle Insertion: A procedure of needle insertion with the help of surgical robot.

Soft Tissue: A biological components mainly in nonlinear characteristics (viscoelastic) in nature found in living organism.

Surgical Needle: A needle with different dia. and different geometrical tip (bevel, conical, and blunt) used in clinical or different MIS surgical purpose.

Tissue Disruption: A phenomenon occurs when needle tries to insert the tissue.

Chapter 7
An Automatic Approach to Control Wheelchair Movement for Rehabilitation Using Electroencephalogram

Darunjeet Bag
Maulana Abul Kalam Azad University of Technology, West Bengal, India

Ahona Ghosh
https://orcid.org/0000-0003-0498-285X
Maulana Abul Kalam Azad University of Technology, West Bengal, India

Sriparna Saha
Maulana Abul Kalam Azad University of Technology, West Bengal, India

ABSTRACT

Modern science and technology development enables humans to control devices using their brains. brain computer interface (BCI) system using EEG (Electroencephalogram) is a non-invasive method that uses brain signals to control robots. Brain control robots can assist people with disabilities to improve their livelihood. In our proposed work to control wheelchair movement for rehabilitation using EEG, electrodes are first placed on the subject's scalp to acquire brain signal activity. After that, the signals got filtered using fast fourier transformation (FFT) method, and then features got extracted using power spectral density (PSD). Random forest is used for the classification of wheelchair movement. For this purpose, a publicly available dataset from Kaggle is used, and an average accuracy of 96.79 is achieved. The proposed architecture has outperformed all the existing ones in its concerned domain; thus, it is suitable, cost-effective, and flexible for the users, which also helps maintain user privacy.

DOI: 10.4018/978-1-6684-5381-0.ch007

INTRODUCTION

A spinal cord injury can cause a person to lose control over his motor control function. In addition to lower limbs, this injury can also affect the upper limb. Brain-computer interfaces (BCI) and robotic machines can be very helpful for people with this type of injury. Robotics can also improve the life of a normal human being. Communication within our body is done using electrical signals. BCI system enables humans to communicate with associated devices. BCI can be very useful for patients with critical injury and rehabilitation. They can regain their ability using robotics.

Electroencephalography (EEG) measures electrical activity within our brain. The EEG is used to study electrical activity in our brain to diagnose neurological states and different disorders like head injury, epilepsy, sleep disorder, brain tumor (Li et al., 2016), etc. EEG is widely used in medical science. There are two methods to record EEG signals, invasive and non-invasive. An EEG-based BCI system uses a non-invasive method to record EEG signals for communication with the machine. This technology is vital for people with disabilities. Robotic movement is possible using EEG-based BCI, where disabled people do not have to rely on someone every time. Also, BCI technology has various applications in education, neuromarketing, advertising, and games. The mostly used EEG-based BCI technologies are steady-state visually evoked potentials, P300, event-related desynchronization, and event-related synchronization (Lazarou et al., 2018). Data acquired through EEG-based BCI can further be classified using different types of machine learning methods to classify different actions of the user.

Our work aims to develop a classification scheme for wheelchair movement. For our work, we have used EEG data for hand movement of persons belonging to different age groups from Kaggle. The EEG signal is passed through Fast Fourier Transformation (FFT) at preprocessing step to acquire the desired frequency range. Then, Power Spectral Density is used to get features from that signal. Then, a random Forest algorithm is used to classify this movement. Hyperparameter tuning has been performed to identify the best combination of parameter values of the classifier, and the outcome has also been compared with the other existing classifiers. The applicability of the proposed approach in controlling wheelchair movement for real-time rehabilitation purposes has been checked by evaluating its performance based on accuracy, precision, recall, and error rate metrics (M & M.N, 2015). It is also effective in terms of robustness, computational time, and reliability.

The chapter outline is structured like the following. First, the related works section has described the related works in this domain, and their benefits and shortcomings have also been analyzed. Second, the proposed methodology section has presented the working mechanism of different steps followed in our proposed framework. Third, the experimental result and performance evaluation have been discussed in the experimental outcome session. Finally, the concluding statements and possible future directions have been discussed in the conclusion section.

RELATED WORKS

Various researchers have researched our proposed framework (Ghosh & Saha, 2020; Saha & Ghosh, 2019). Here we have discussed some of the related works. Edla et al. (2018) use Neurosky Mindwave mobile as an EEG recording device. It is a non-invasive device that connects wirelessly with a computer or mobile using Bluetooth. This device consists of two electrodes placed on the forehead at the Fp1 location according to the 10-20 system and clip-on-ear as a reference electrode, respectively. They intend to classify two different states of the human mind: concentration and meditation. First, raw EEG signals of 8 different frequency bands are collected from the Fp1 channel: low gamma, high gamma, high alpha, low alpha, high beta, low beta, delta, and beta. After that, these raw EEG signals are preprocessed. For preprocessing, a single chip is presented in that Neurosky Mindwave device. Then useful features are extracted from that signal. First, the statistical tool is used for signal features, such as mean, standard deviation, and difference of the highest and the lowest value. After that, classification is done using a random forest machine-learning algorithm. It is an ensemble learning approach which applies a random amount of decision trees to develop the final prediction. They achieved an accuracy of 75% on predicting the class of the mental state.

The authors Cantillo-Negrete et al. (2018) applied a brain-computer interface using motor imagery (Herman et al., 2008). The imaginary movement of a limb creates it without actual muscle activation. However, people with disabilities can still imagine the movement of their paralyzed limbs. This creates electrical activity in their brain, especially in the frequency band of alpha and beta in the sensory-motor cortex of their brain. The author uses samples of eight healthy subjects and six patients having stroke disabilities for their study. For EEG signal recording, a g.USBamp bio-signal amplifier is used. It has a sampling rate of 256 Hz. It consists of 11 electrodes placed in C3, C4, Cz, T3, T4, F3, F4, Fz, P3, P4, and Pz locations using a 10-20 system. Signals were recorded using a visual action shown on a monitor to perform motor imagery. After recording, these EEG signals were divided using the temporal filtering method to obtain six frequency sub-bands. Each sub-band is 4 Hz broad between ranges 8-32 Hz. After that spatial filter is computed for each sub-band using a common spatial pattern (CSP) algorithm. Next, particle swarm optimization (PSO) selects useful features. After that linear discriminant analysis classifier is used to classify two motor imagery classes; for their work, a robotic hand orthosis is used, a 3D printed frame of polylactic acid. This orthosis is connected with the BCI using Bluetooth. It has four actuators; each can perform flexion and extension. Their offline performance has achieved an average of 74.1 ± 11% for stroke patients and 76.2 ± 7.6% for healthy people.

Imran et al. (2014) implement a method to classify EEG signals while performing motor imagery tasks. They classify two activities: right hand and right foot movement. For their work, they used dataset IVa of BCI competition III. It is a publicly available dataset. This dataset consists of an EEG recording of five healthy subjects. One hundred eighteen electrodes were placed using a 10-20 system to record EEG data. There are 140 trails of each subject doing motor imagery tasks of the right hand and right foot. First, the data is collected at a sampling rate of 1000 Hz; then, it is down-sampled at 100 Hz. Then this data was filtered using the bandpass filter at the range of .005-200 Hz. After that, discrete wavelet transform (DWT) is applied to extract features. DWT has the advantage of low cost and simpler implementation. The signal coefficients obtained from these filters are called detail and approximate coefficients. For their work, approximate coefficients are used. These coefficients are normalized between the range 0 and 1. After performing DWT at each signal, four statistical features are selected. Those features are the maximum value of wavelet coefficients, the variance of wavelet coefficients, wavelet domain energy, and wavelet domain entropy. The next principal component analysis was applied to decrease the dimension of these features. After that, these data are fed into the k-nearest neighbor (kNN) algorithm for classification. The author shows an average classification accuracy of 78.26% using this method.

Kant et al. (2019) projected an approach to classifying hand movement using EEG signals recorded during motor imagery tasks. Their study used the dataset of the BCI 2003 competition, also named Graz2003 data. This dataset consists of an EEG recording of a healthy female volunteer of age 25. Signals were recorded by showing left or right cues on the screen. The task was to imagine the movement of the hand in the left or right way, depending on the cue. The dataset contains 280 trails. One hundred forty trails are used for training the system, and 140 are used for testing. An electrode was placed at C3, Cz, and C4 using a 10-20 system, where the sampling rate was 128 Hz. The alpha frequency band is used as a feature. Discrete wavelet transformation is a feature extraction method on the EEG signal. First, DWT decomposes the signal into approximate and detailed coefficients. Then, the approximate coefficient is further decomposed to the 3rd level to extract alpha band signals. After that, some statistical features of that alpha band signal (Sebastián-Romagosa et al., 2020) are calculated: Mean, Variance, Energy, Shannon entropy, Log energy entropy, Kurtosis, and Skewness. These features are used to classify the signal. Support Vector Machine (SVM) and kNN are used for classification purposes. The author shows that the best classifying accuracy is achieved using the Shannon entropy feature at 86.4% using SVM as a classifier.

Table 1. Recent literature review

References	A device used for EEG	Filtering and feature extraction	Classification algorithm	Classifications	Accuracy
(Shedeed et al., 2013)	Emotive Epoc headset at 128 Hz	5th order Butterworth filter then WT, FFT and PCA	Multi-Layer Perceptron using sigmoid function at learning rate 0.03	Three different arm movements, namely, close arm, open arm, and close hand	91.1% average accuracy
(Gandhi et al., 2014)	g.USBamp at 256 Hz sampling rate	Recurrent Quantum Neural Network (RQNN)	Intelligent adaptive user interface (iAUI), a method proposed by the author	Two classes are left or right-hand movement	Accuracy is measured in time and number of commands to completion of the task
(Li et al., 2016)	Emotive EPOC neuroheadset with 128 Hz sampling rate	online K-means clustering algorithm	PCA is used for the classification	Four classes, forward, backward, left, and right	Accuracy 86%
(Ofner et al., 2019)	Uses MRCP and g.USBamps devise with sampling rate at 256 Hz	4th order Butterworth filter then PCA	Shrinkage linear discriminant analysis (LDA)	Five classes for hand movement	68.4% accuracy
(Shao et al., 2020)	Brain Product at 500 Hz sampling rate	Wavelength transformation (WT)	Canonical correlation analysis (CCA)	Four classes top, bottom, left, and right	The average accuracy of 89.92%
(Suresh et al., 2019)	Neurosky Mindwave EEG headset to capture brain signals at 512 Hz	Fast Fourier Transformation (FFT)	LDA, SVM, and KNN	Two class classification if a person is trying to solve quick math or doing nothing	100% using SVM
(Korovesis et al., 2019)	OpenBCI at 200 Hz sampling rate	Butterworth IIR band pass filter and FFT algorithm	Multilayer Perceptron (MLP) with ReLU function	Four classes forward, reverse, left, right	92.1%
(Song et al., 2020)	BrainCap MR at 500 Hz sampling rate	4th order Butterworth bandpass filter then xDAWN algorithm	SVM, LDA, MLP	Two class classification	SVM with 90.57%

Bousseta et al. (2018) proposed a method for controlling a robotic arm using EEG based brain-computer interface. They acquire EEG signals using an EMOTIV EPOC headset consisting of 14 electrodes. The signal was sampled at 128 Hz. Their experiment used four volunteers between 20 and 29 age without any prior neurological or psychological disorder. Among them, three were male and one female. There was four class of movements for the robotic arms. During the training phase, instructions were displayed to control that robotic arm. These instructions were arrows in four directions. First, the user has to imagine a right-hand movement for a right arrow display to move the arm right. Next, the left arrow was a movement of the left hand to move the arm to the left. Then, for the up arrow, the user has to imagine the movement of both arms to control the elbow of the robotic arm to point upward. Finally, for the down arrow, the user has to imagine the movement of feet to control the elbow of the robotic arm to point down. Data acquired from the electrodes AF3, AF4, F3, and F4 are used later. After that, a Butterworth filter of 50 Hz is used to remove noise. Then useful features are extracted using Fourier transform combined with Principal Component Analysis to select frequency between 8-22 Hz. Then these features were classified using SVM. The SVM classifier used in this method uses Radial Base Function as its kernel. Using this method, they have achieved an average accuracy above 85%. Still, very simple tasks were considered in the study. In reality, more complex tasks like successive action execution or action stopping based on the detection of a searched object are required for motor-impaired people. Table 1 presents a comparative study of the existing state of the art.

Figure 1. Flowchart of Robotic component movement using EEG signal

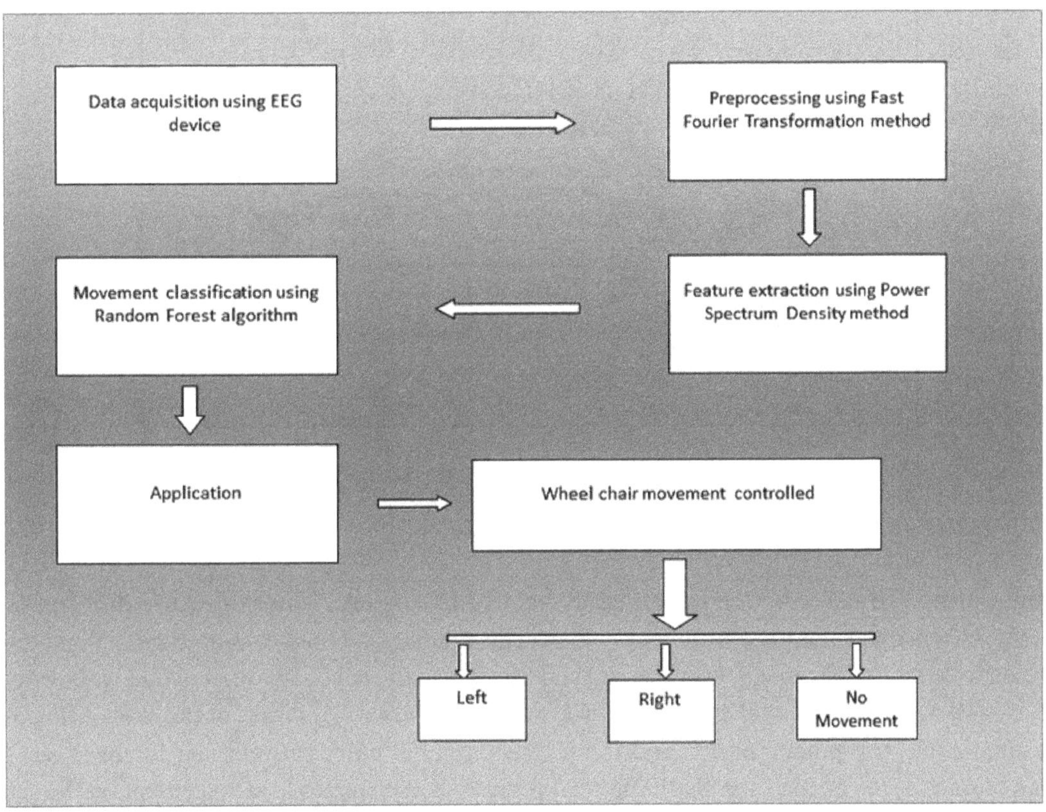

PROPOSED METHODOLOGY

Our intuitive approach to controlling wheelchair movement for rehabilitation has been divided into three parts. First, in the proposed framework shown in Figure 1, data has been collected from EEG; then the preprocessing has been done, feature extraction of power spectral density has been performed, and then using Random Forest classifier, the wheelchair movement has been classified.

Introduction to Electroencephalogram

The human brain is made of billions of neurons—these neuron cells of our nervous system exchange information as electrical or chemical signals along its axon. There are three types of neurons. First are the sensory neuron cells. They pick information from senses and send a signal to the neural system. Then motor neurons send a signal from the brain to the rest of the body using a spinal cord. The third type of neuron is the interneuron. They are located in the central nervous system. It connects the central nervous system with sensory and motor neurons. Figure 2 shows the structure of a single neuron. Brain signals are detected between 1-100 Hz. These signals are divided into five categories, shown in Figure 3: alpha, beta, delta, theta, and gamma, depending on their frequencies.

Figure 2. Structure of a single neuron

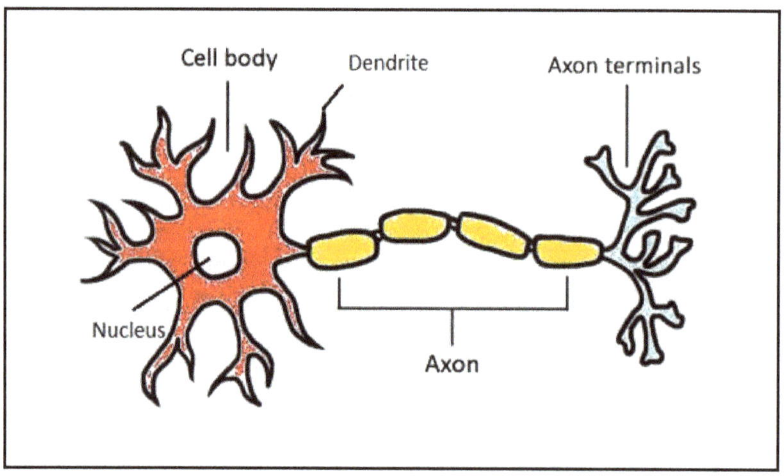

10-20 System of Standard EEG

Electrodes placed on the scalp are important to record EEG signals. This is because different human brain lobes are responsible for different activities. The standard method to record EEG signals is the 10-20 electrode system (Terms, 1961). The 10 and 20 signify the space between two neighboring electrodes. It is 10 or 20% of the total distance between the front-back or right-left of the skull. The position is measured using two points' nasion and Inion. Nasion is the point between the forehead and nose, whereas Inion is at the back of the skull. Figure 4 represents the electrode arrangement on the scalp of

21 electrodes. Each location uses a letter along with a number for identification. The letter corresponds to the lobe, and the number corresponds to the hemisphere of electrode location. The letters F, T, C, P, and O represent Frontal, Temporal, Central, Parietal and Occipital. Letter "z" refers to the midline of the head. Even numbered electrodes refer to electrodes placed on the right hemisphere, and the odd number represents the left hemisphere.

Figure 3. Different types of EEG signals

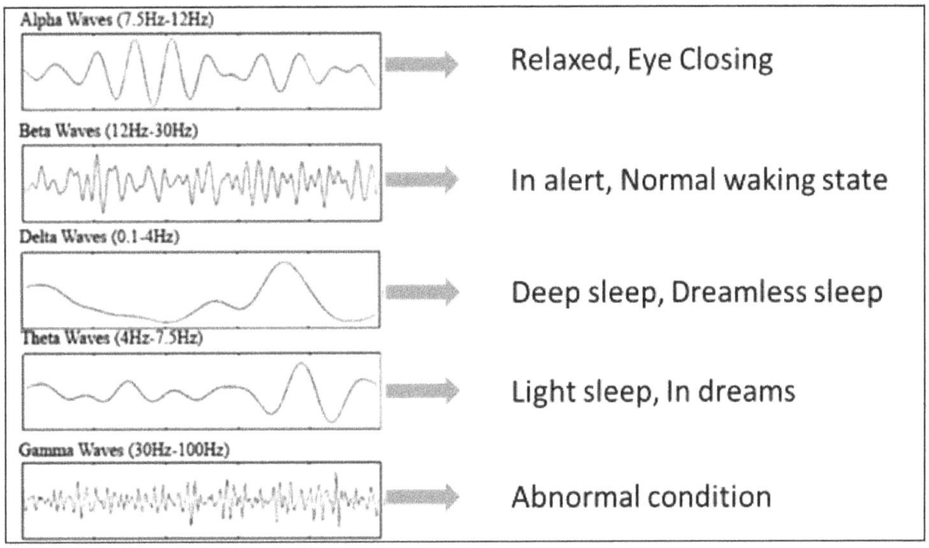

Figure 4. 10-20 Electrode System

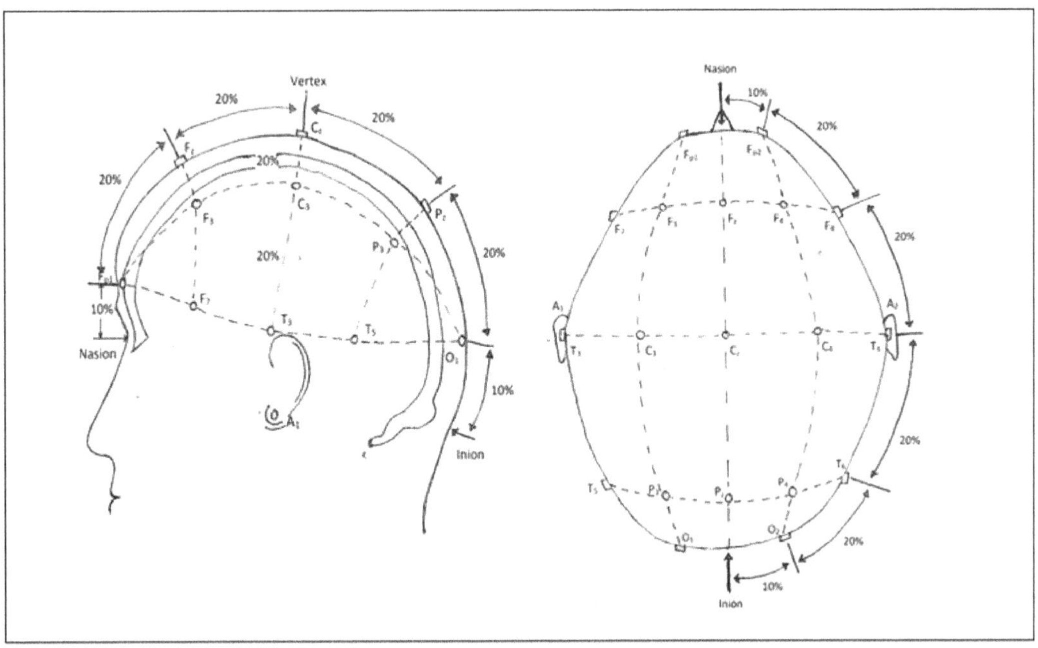

Electrodes

EEG signals are collected using some electrodes placed on the scalp. The chosen electrodes are low-cost and stable with low impedance. Commonly used electrodes are made using silver or silver chloride. To improve conductivity between electrode and scalp, the gel is used. However, applying gel to each electrode and placing them on the scalp is troublesome. Therefore, dry electrodes introduced by Emotive System Inc. or NeuroSky Inc (Jenita Amali Rani & Umamakeswari, 2015) can be used instead of wet electrodes. Nowadays, various data acquisition devices have different shapes and numbers of electrodes. The most commonly used devices are Emotive Epoc, g.USBamp and g.HIamp, which is made by g.tech medical engineering. Epoc headset uses 14 dry electrodes, whereas g.tech devices use 256 (Aljalal et al., 2020).

Lobes

The most complex organ in the human body is the human brain. It contains billions of nerves communicating with trillions of connections called synapses. In addition, there are specialized areas in the human brain for specific work. The human brain can further be divided into right and left hemispheres. Each hemisphere comprises four different lobes: frontal, parietal, occipital, and temporal. And every lobe is responsible for some purposes within the human body (Shipp, 2007). Figure 5 has outlined the positions of the lobes within the human brain.

Figure 5. Different lobes in the human brain

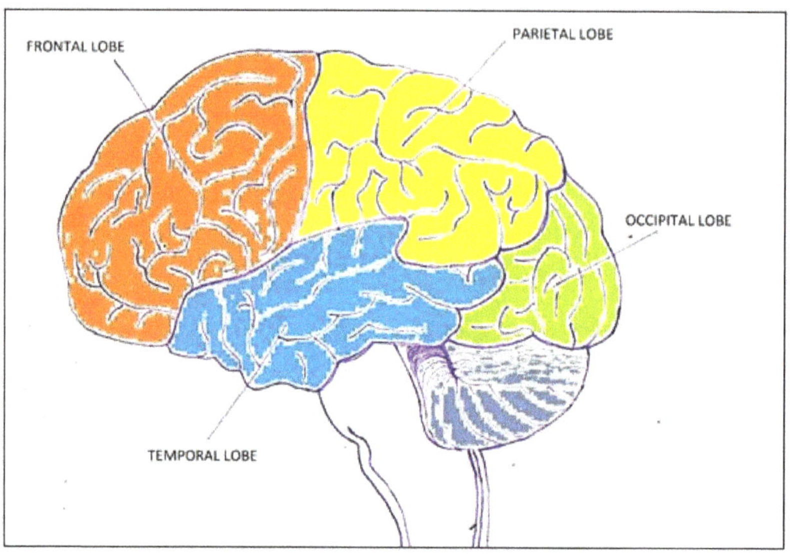

- **Frontal lobe:** The frontal lobe is the largest. It is responsible for speech, language, emotion, planning, movement, and personality (Chanda et al., 2022). In addition, this lobe has an area named Broca's area. This area is responsible for speech production. In addition, the frontal lobe has the motor cortex dealing with different movement coordination and planning.

- **Parietal lobe:** The parietal lobe is positioned behind the frontal lobe. It contains the sensory cortex. It deals with the sensory information throughout the human body—sensations like touch, pain (Ghosh et al., 2020), etc.
- **Occipital lobe:** The occipital lobe is the smallest. It resides at the back of the human brain. It processes the visual activity of the human brain using the visual cortex. This information helps us recognize we compare objects.
- **Temporal Lobe:** The temporal lobe is the second largest. It is at the side of the human head above the ears. The temporal lobe deals with speech, perception, and memory. With the help of the primary auditory cortex, it processes and translates all sounds. EEG signals generated from each lobe of two hemispheres are recorded for our purpose.

Preprocessing Using Fast Fourier Transform

Fast Fourier Transformation (FFT) is the optimized implementation of Discrete Fourier Transformation (DFT). FFT is faster than DFT because it takes fewer computations. FFT converts l length time-domain digital signal x_t into frequency domain x_f. It converts a signal into spectral components and provides frequency spectrum analysis. In other words, the original signal is divided into sub-spectral components (Ingle & Proakis, 2008) (Patel et al., 2021), which gives us information about the signal frequencies, and we can select the desired ones. The converted frequency domain signal can be represented by

$$x_f[j] = \sum_{i=0}^{l-1} (x_t[i] e^{-jki2\Pi/l}) \quad (1)$$

where j=0, 1, 2, …, (l-1).

If the Fourier Transform gets separated into even and odd indexed sub-sequences such that

i=2p if even

=2p+1 if odd (2)

where p = 1, 2, …, $\frac{l}{2}-1$.

Then the transformed signal of the frequency domain can be represented by

$$x_f[j] = \sum_{p=0}^{l/2-1} (x_t[2p] e^{-jk(2p)2\Pi/l}) + x_f(j)$$
$$= \sum_{p=0}^{l/2-1} (x_t[2p+1] e^{-jk(2p+1)2\Pi/l}) + x_f(j)$$

$$x_f[j] = \sum_{p=0}^{l/2-1} (x_t[2p]e^{-jk(2p)2\Pi/l}) + x_f(j)$$
$$= e^{-jk2\Pi/l} \sum_{p=0}^{l/2-1} (x_t[2p+1]e^{-jk(2p)2\Pi/l})$$

$$x_f[j] = \sum_{p=0}^{l/2-1} (x_t[2p]e^{-jk(2p)4\Pi/l}) + x_f(j)$$
$$= e^{-jk2\Pi/l} \sum_{p=0}^{l/2-1} (x_t[2p+1]e^{-jk(2p)4\Pi/l})$$

$$x_f(j) = x_{even}(j) + e^{-jk2\Pi/l} x_{odd(j)} \tag{3}$$

Feature Extraction Using Power Spectral Density

Our proposed model uses Power Spectral Density abbreviated as PSD in the feature extraction step. It is the most commonly used method in signal processing. PSD detects the presence of power in the signal. It shows the strength distribution of signal power with frequency commonly expressed in watts per hertz (W/Hz). Among different methods of calculating PSD, the welch method is used in our study. It is based on periodogram spectrum estimation, which occurs due to the conversion of the signal from the time domain to the frequency domain. Welch's method enhances Bartlett's method for reduced noise (Bin Ng et al., 2019). The Welch Power Spectrum, i.e., the periodogram mean for each interval, can be expressed as

$$P_{welch}(f) = \frac{1}{MU} \left| \sum_{n=0}^{M-1} x_i(n) w(n) e^{-j2\Pi f} \right|^2 \tag{4}$$

where $x_i(n)$ denotes the signal vector having M number of elements, which means the number of recordings of each channel in the ith trial, $w(n)$ is the windowing function, i.e., the hamming window, and U is the normalization constant.

Classification Using Random Forest Classifier

As the name suggests, Random Forest comprises a random number of decision trees. It is very flexible and possesses very high accuracy. It is an ensemble learning approach that uses several decision trees to generate the final output. The ensemble learning approach uses multiple learning algorithms to improve its performance. The Random Forest approach builds multiple decision trees on data samples and obtains the result. Then it chooses the majority of these results and selects them as the final class.

Each decision tree in the random forest is designed using some chosen attributes or features. There are many approaches the attribute selection. The classifier uses the Gini index measure to select attributes (Pal, 2005). It measures impurity among attributes concerning their classes. For example, let F is a training set; now, for a randomly selected case, its Gini index for a class *Cm* can be written as:

$$\sum\sum_{m \neq n}(f(Cm,F)/|F|)(f(Cn,F)/|F|) \quad (5)$$

Here f (*Cm*, F)/ |F| is the probability that the selected instance belongs to a class *Cm*. First, the user decides the number of trees in the classifier. Then, every node uses those selected features to grow a tree. Figure 6 shows the working mechanism of the classifier.

Figure 6. Random Forest Classification approach

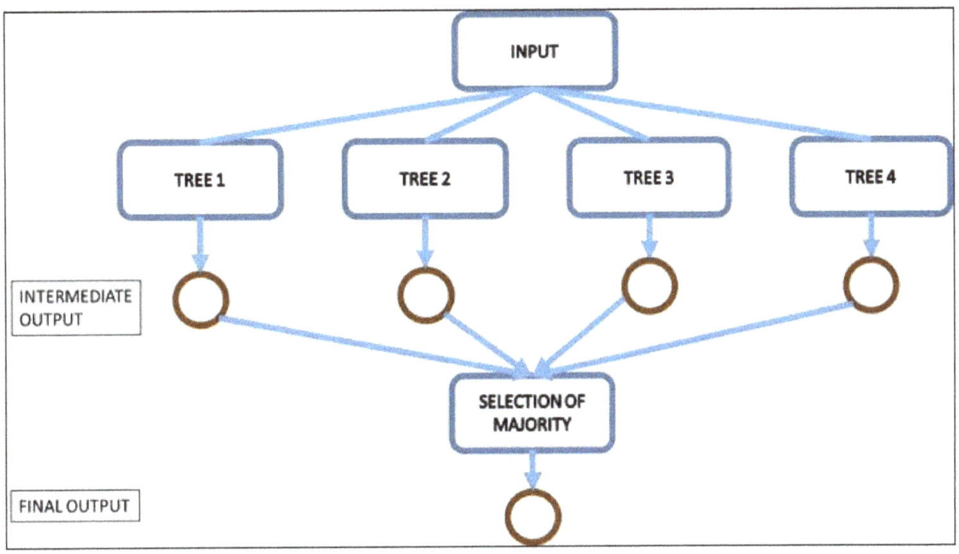

EXPERIMENTAL RESULT

This section has discussed the data collection procedure, the experimental outcome, and the applicability of our proposed approach. It has been tested by comparing it with some existing literature in terms of accuracy, precision, recall, average error rate, and F1 score.

Dataset Description

For our work, we have taken a dataset of EEG data from hand movement from Kaggle (EEG data, 2022). To capture EEG data, they used an EMOTIV EPOC +14 device. As the name suggests, this device has 14 electrode channels, namely, AF3, F7, F3, FC5, T7, P7, O1, O2, P8, T8, FC6, F4, F8, and AF4. A twenty-five-minute data collecting protocol was built to control the virtual object, with multiple cycles that handle different conditions when the user tries to control the object. Various scenarios have been presented as part of the protocol so that the machine learning component can generalize brainwave behavior concerning object control commands independently of the environment. During signal acquisition, participants were shown images for corresponding motor action. A right arrow signifies a movement in the right direction, a left arrow signifies a movement in the left direction, and a circle signifies no movement. The start of data acquisition is indicated using an audible beep until that participant was with his eyes closed. The dataset contains 0 for movement in the left hand, 1 for movement in the right hand, and 2 for no movement.

Preprocessing Result Using Fast Fourier Transform

Four user data were collected. The collected data is needed to be preprocessed for later use. The 14-channel EEG equipment collected data at 128-Hz and created a 16-bit 128╳14 matrix. This data is passed through the Fast Fourier Transform (FFT) algorithm between 0 to 30 Hz. This results in a 30╳14 matrix. After that, the weighted and arithmetic average of every waveform is measured. This creates a matrix with dimensions 14╳4╳2. Finally, the dataset consists of 14 device channels with four wave classifications.

Figure 7. Four waveforms of AF3 channel

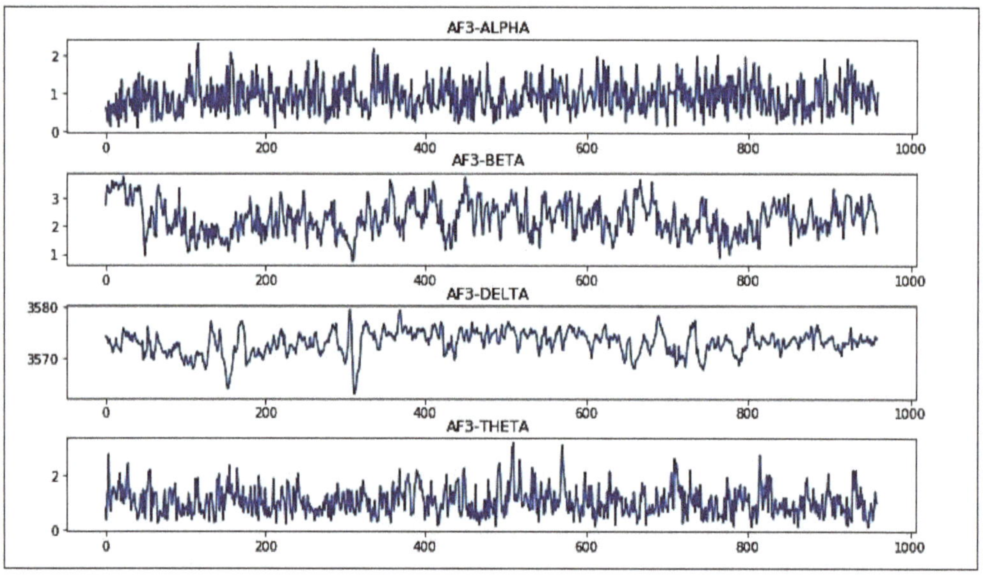

Each wave is included with a weighted and arithmetic mean. A sheet is not feasible to give waveform output of all the channels; only different waveforms from channel AF3 are shown in Figure 7.

Feature Extraction Outcome Using Power Spectral Density

After processing these data, useful features are extracted from the data. We have used this signal's power spectral density (PSD) for our work. We have to use the Welch method to extract PSD values. Figure 8 shows the PSD outcome for four different waveforms of the AF3 channel.

Figure 8. PSD of Four waveforms of AF3 channel

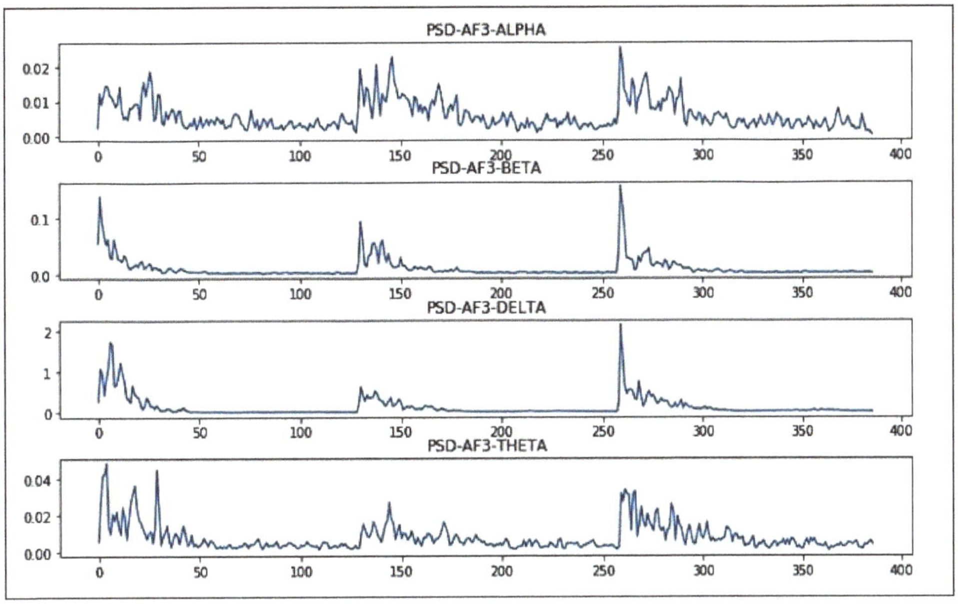

Random Forest Classification Outcome

For classification, the Random Forest algorithm is chosen. The training and testing ratio has been defined as 3:1. Hyperparameter tuning has been performed based on the five parameters, namely the number of trees in the forest, the maximum number of features needed to split a node, the maximum number of layers at each decision tree, the minimum amount of data points placed in a node before splitting the node, the minimum number of data points allowed in a leaf node, and data points sampling method (with/without replacement). For applying Randomized Search CV, we first created a parameter grid to sample from during fitting. The 'bootstrap' value was chosen among True and False, maximum depth was considered as [10, 20, 30, 40, 50, 60, 70, 80, 90, 100, None], for the maximum features parameter, two values were considered, i.e., ['auto', 'sqrt'], minimum samples leaf value was considered among [1, 2, 4], minimum samples split were selected among [2, 5, 10], the number of estimators parameter was varied among [200, 400, 600, 800, 1000, 1200, 1400, 1600, 1800, 2000]. The algorithm has been led to choose different parameters for each iteration. There are a total of 4320 settings: $2\times12\times2\times3\times3\times10$. The advantage of a random search is that we do not test each combination but rather sample a large range of values at random. The essential inputs in Randomized Search CV are the number of iterations, which specifies how many combinations to try, and the number of folds to utilize for cross-validation (100 and 3 have been used here, respectively) (Fatihah Rahmadayana & Yuliant Sibaroni, 2021). More iterations will cover a larger search space, and more cross-validation folds will lower the odds of overfitting, but each will take longer to run. The best parameters for fitting the random search were identified, and the base model was compared to the best random search model to check whether the random search could improvise the performance. We achieved an improvement in the accuracy of 0.8%. Our work has three classes: right movement, left movement, and no movement. There are four users' data available. As shown in Figure 9, the model's best accuracy for user A is 96.58%, user B is 99.14%, user C is 93.16%, and user D is 98.29%. Overall average accuracy is 96.79%.

Figure 9. Accuracy comparison of users

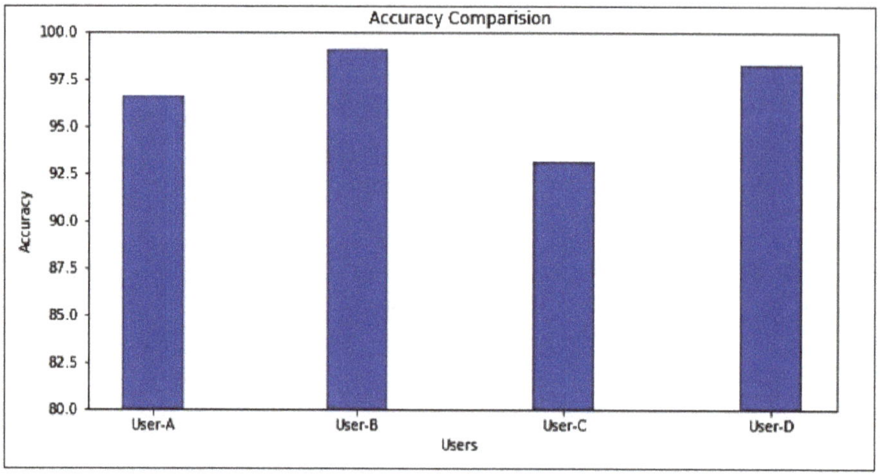

Performance Evaluation

A classification report is generated for testing our model. This report measures the quality of prediction of our model. A certain measure aspect is considered for performance, such as precision, recall, F1-score, and support. These measurements are true and false positive, true and false negative. True positive abbreviated as TP, is when the real instance is positive, and the classifier also predicts it as positive. True negative abbreviated TN, is when the real value is negative, and the classifier also identifies it as negative. False-positive (FP) is when the real value is negative, but the classifier identifies it as positive. False-negative (FN) is when the real instance is positive, but the classifier identifies negative. Accuracy is used for the evaluation of classification problems. It is measured as the number of true instances among the total instances.

$$\text{Accuracy} = (TP+TN) / (TP+TN+FP+FN) \quad (6)$$

Precision is used to show how many are correctly predicted among that class. Precision is the ratio between True positive and total positive values and can be measured using

$$\text{Precision} = TP / (TP+FP) \quad (7)$$

Recall measures the proportion of actual positives which are classified correctly. It is measured as

$$\text{Recall} = TP / (TP+FN) \quad (8)$$

The F1- score is a harmonic average of recall and precision. An F1 score is a number between the range of 0 and 1. The best one is 1.0, while the worst one is 0.0.

$$\text{F1-score} = 2 * (\text{Precision} * \text{Recall}) / (\text{Precision} + \text{Recall}) \quad (9)$$

Support value shows us several occurrences of that class in our dataset. Classification reports for four users are given in Figure 10. A confusion matrix is another way of representing the performance of any machine learning model. The confusion matrix shows the set of actual and predicted classes in tabular form. It gives an insight into the model, and errors are identified easily. For example, the value X_{mn} (where *m* and *n* represent the row and column number, respectively) in the confusion matrix shows that the case belongs to m, classified as *n*. Hence the diagonal values of the confusion matrix are the ones properly classified. At the same time, the other non-diagonal elements are misclassified. The confusion matrix for the four users is shown in Figure 11.

Figure 10. Classification report of four users

Accuracy on training set is : 1.0
Accuracy on test set is : 0.9658119658119658

	precision	recall	f1-score	support
0	0.97	1.00	0.99	38
1	1.00	0.93	0.97	46
2	0.91	0.97	0.94	33
accuracy			0.97	117
macro avg	0.96	0.97	0.96	117
weighted avg	0.97	0.97	0.97	117

User (A)

Accuracy on training set is : 1.0
Accuracy on test set is : 0.9829059829059829

	precision	recall	f1-score	support
0	0.97	0.97	0.97	36
1	0.97	0.97	0.97	39
2	1.00	1.00	1.00	42
accuracy			0.98	117
macro avg	0.98	0.98	0.98	117
weighted avg	0.98	0.98	0.98	117

User (B)

Accuracy on training set is : 1.0
Accuracy on test set is : 0.9914529914529915

	precision	recall	f1-score	support
0	1.00	0.98	0.99	45
1	0.97	1.00	0.99	36
2	1.00	1.00	1.00	36
accuracy			0.99	117
macro avg	0.99	0.99	0.99	117
weighted avg	0.99	0.99	0.99	117

User (C)

Accuracy on training set is : 1.0
Accuracy on test set is : 0.9316239316239316

	precision	recall	f1-score	support
0	0.97	0.92	0.94	37
1	0.90	1.00	0.95	44
2	0.94	0.86	0.90	36
accuracy			0.93	117
macro avg	0.94	0.93	0.93	117
weighted avg	0.93	0.93	0.93	117

User (D)

Comparison With Other Classifiers

This section compares the proposed Random Forest based approach toward wheelchair movement classification with the existing benchmarks, i.e., kNN, Feed Forward Neural Network (FFNN), and SVM. Four classification metrics have been considered for comparison to assess the effectiveness and applicability of the proposed one. The comparison graphs have been presented in Figure 12, where it is observed that the proposed work has benefitted from the hyperparameter tuning and performed better than all the others. However, GridSearchCV can be employed in the future to check for the possibility of more improvement (Shekar & Dagnew, 2019). The main difference is that in grid search, the combinations get defined by the user and trained in the model, whereas in RandomizedSearchCV, the classifier chooses the combinations randomly.

Figure 11. Confusion matrix of four users

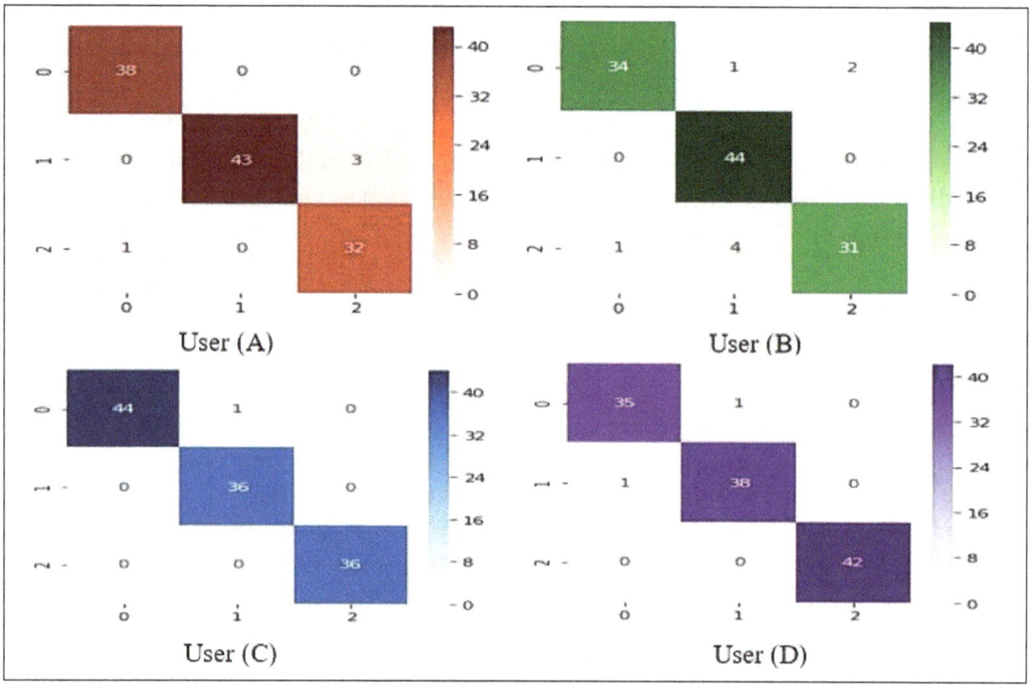

Figure 12. Comparison of proposed RF with kNN, FFNN, and SVM

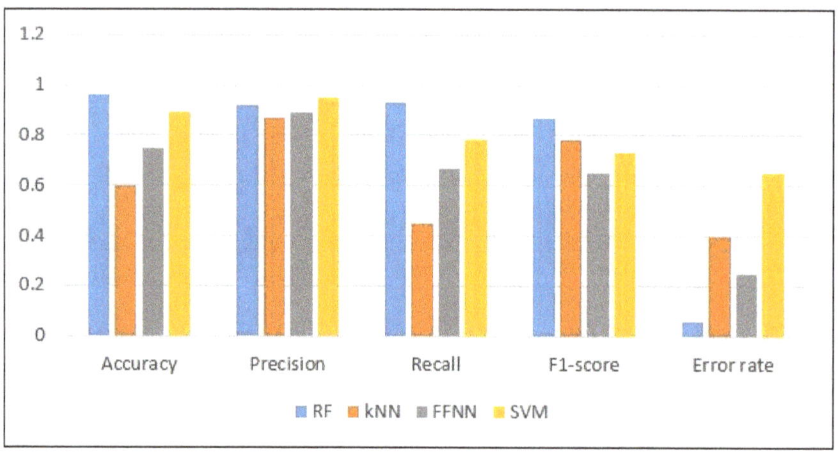

Statistical Analysis

Table 2 shows the mean and standard deviation mentioned within the bracket of the best values of the performance metrics for each algorithm's 25 different runs. In addition, the means obtained from each metric obtained by the best and second-best algorithms have also been presented using the paired t-test. Finally, the statistical significance level of the difference between the means of the best two algorithms has been presented in Table 2's last column, where the '+' denotes the t-value of 49 degrees of freedom is statistically significant at a 0.05 level of significance by two-tailed test, and the '–' denotes that the difference of means is not statistically significant. The sample size for all of the t-tests is set as 25. The best one has been marked in bold.

Table 2. Comparison of different EEG-based movement recognition algorithms for 25 Runs

Performance metric	RF	kNN	FFNN	SVM	Statistical significance
Accuracy	**0.96 (0.161)**	0.60 (0.531)	0.75 (0.657)	0.89 (0.645)	+
Precision	0.92 (0.134)	0.87 (0.634)	0.89 (0.453)	**0.95 (0.564)**	-
Recall	**0.93 (0.168)**	0.45 (0.324)	0.67 (0.498)	0.78 (0.654)	+
F1 score	**0.87 (0.019)**	0.78 (0.121)	0.65 (0.056)	0.73 (0.109)	+
Mean error rate	**0.06 (0.013)**	0.40 (0.232)	0.25 (0.143)	0.65 (0.435)	+

Performance Evaluation

The current approach toward wheelchair movement control for motor rehabilitation has been evaluated by comparing its performance with related works in the concerned domain. Figure 13 shows the comparison with (Hossain et al., 2015) (Baig et al., 2015) and (Zhang et al., 2020) in terms of overall accuracy, computational time, and the ratio between these two, where it has been observed that the achieved accuracy is the highest, computational time in seconds is the lowest. Furthermore, the ratio of these two is the highest for our proposed random forest classifier compared to the others. From the result, it is observed that the proposed method outperforms the existing EEG-based activity recognition frameworks. The proposed approach has eventually proved to be suitable in real-time.

Figure 13. Comparison with state-of-the-art frameworks

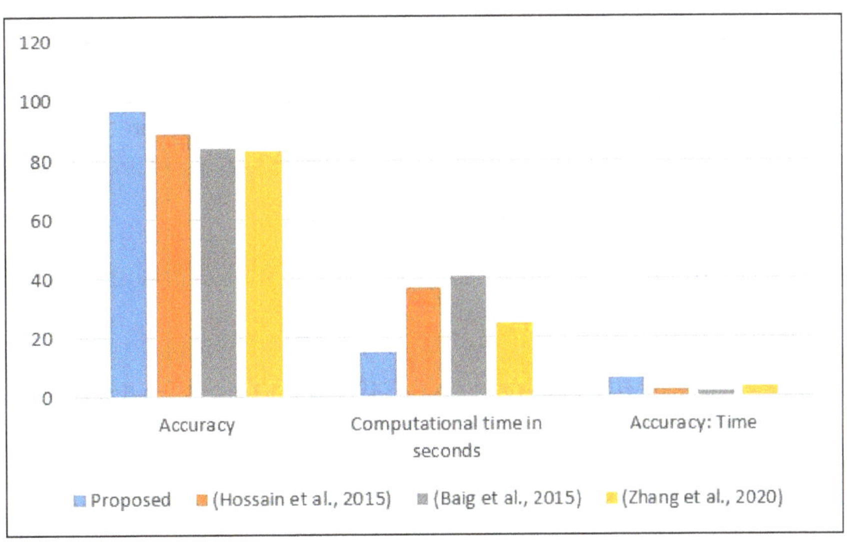

Movement of Wheelchair Prototype

Since the classification performance was satisfactory enough to implement it into the real-world scenario, a mobile robot TurtleBot3 Waffle Pi as a prototype of a wheelchair walked along the track to the destination controlled by the EEG signal. During the robot's travel, the software recorded each direction conversion information as an array, including the robot's walking time, steering direction, and steering angle elements. The running track and return track arrays, respectively, for left and right movement, are represented by

$$\begin{pmatrix} x_1 & y_1 & z_1 \\ x_2 & y_2 & z_2 \\ & \vdots & \\ x_{n-1} & y_{n-1} & z_{n-1} \\ x_n & y_n & z_n \end{pmatrix} \text{ and } \begin{pmatrix} x_1 & -y_1 & z_1 \\ x_1 & -y_1 & z_1 \\ & \vdots & \\ x_1 & -y_1 & z_1 \\ x_1 & -y_1 & z_1 \end{pmatrix}$$

where x, y, and z are the robot walking time, steering direction (1: turn to the left and -1: turn to the right), and the steering angle, respectively. The experimental paradigm has been presented in Figure 14.

CONCLUSION AND FUTURE SCOPE

Wheelchair control using EEG brain signals will help elderly or disabled people lead their lives with fewer difficulties for rehabilitation purposes. Here we studied some papers on EEG-based BCI systems and their use in robotics. We presented detailed comparisons between them. BCI system is at its beginning stage. It needs more improvement on noise reduction to make the wheelchair system more accurate. For our work, we have relied on an open data set that contains data with three classes corresponding to three motor actions. The advantages may be summed up as follows:

1. **Robustness analysis:** When a classifier fits the training data extremely closely, it performs well on the training data but not on the validation data, resulting in overfitting. Table 2 shows the average error rate numbers, which show that the suggested approach does not produce the same results.
2. **Benefits:** The proposed work's major purpose is to detect the wheelchair activity of people of various ages and provide a viable rehabilitative path for those affected. This goal appears to be achievable with the proposed effort.
3. **Effectiveness and efficiency:** The suggested model's effectiveness and efficiency have been compared to current standard alternatives and detailed in the article.
4. **Computational time:** On an Intel® Core TM i5-7200U @ 2.50GHz 2.70GHz CPU with 8GB RAM running Python 3.7.10 version, the needed time for successfully recognizing an unknown emotion is in the range of 15.02 seconds.
5. **Diversity:** The proposed approach has been tried on various people with different physical characteristics and genders.

6. **Reliability:** The suggested framework of wheelchair movement control for motor imagery-based rehabilitation is reliable in a real-time environment.

For future work, we can include more motor action in this system. Here EEG signals are preprocessed using Fast Fourier Transformation. After that, feature extraction is done using the PSD method. Then we classified the data using the Random Forest algorithm and achieved an average accuracy of 96.79. We can use the neural network in the future for classification to improve accuracy. For future work, we can use this method as an online classifier to classify wheelchair movements in real-time.

Figure 14. Pictorial illustration of the experimental paradigm

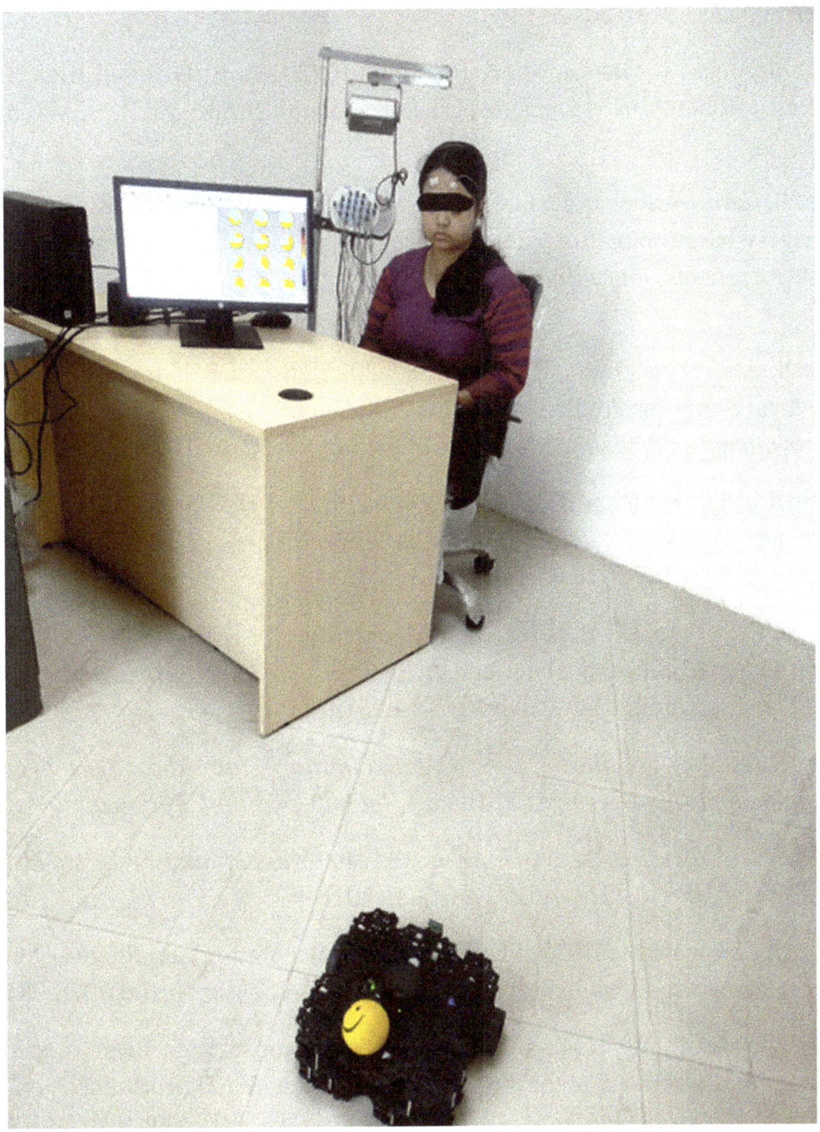

REFERENCES

Aljalal, M., Ibrahim, S., Djemal, R., & Ko, W. (2020). Comprehensive review on brain-controlled mobile robots and robotic arms based on electroencephalography signals. *Intelligent Service Robotics, 13*(4), 539–563. doi:10.100711370-020-00328-5

Baig, M. Z., Javed, E., Ayaz, Y., Afzal, W., Gillani, S. O., Naveed, M., & Jamil, M. (2015). Classification of left/right hand movement from EEG signal by intelligent algorithms. *ISCAIE 2014 - 2014 IEEE Symposium on Computer Applications and Industrial Electronics*, 163–168. 10.1109/ISCAIE.2014.7010230

Bin Ng, W., Saidatul, A., Chong, Y. F., & Ibrahim, Z. (2019). PSD-Based Features Extraction for EEG Signal during Typing Task. *IOP Conference Series. Materials Science and Engineering, 557*(1), 012032. doi:10.1088/1757-899X/557/1/012032

Bousseta, R., El Ouakouak, I., Gharbi, M., & Regragui, F. (2018). EEG Based Brain Computer Interface for Controlling a Robot Arm Movement Through Thought. *IRBM, 39*(2), 129–135. doi:10.1016/j.irbm.2018.02.001

Cantillo-Negrete, J., Carino-Escobar, R. I., Carrillo-Mora, P., Elias-Vinas, D., & Gutierrez-Martinez, J. (2018). Motor imagery-based brain-computer interface coupled to a robotic hand orthosis aimed for neurorehabilitation of stroke patients. *Journal of Healthcare Engineering, 2018*, 1–10. doi:10.1155/2018/1624637 PMID:29849992

Chanda, K., Ghosh, A., Dey, S., Bose, R., & Roy, S. (2022). Smart Self-Immolation Prediction Techniques: An Analytical Study for Predicting Suicidal Tendencies Using Machine Learning Algorithms. *EAI/Springer Innovations in Communication and Computing*, 69–91. doi:10.1007/978-3-030-71485-7_4

Edla, D. R., Mangalorekar, K., Dhavalikar, G., & Dodia, S. (2018). Classification of EEG data for human mental state analysis using Random Forest Classifier. *Procedia Computer Science, 132*(Iccids), 1523–1532. doi:10.1016/j.procs.2018.05.116

Gandhi, V., Prasad, G., Coyle, D., Behera, L., & McGinnity, T. M. (2014). EEG-Based mobile robot control through an adaptive brain-robot interface. *IEEE Transactions on Systems, Man, and Cybernetics. Systems, 44*(9), 1278–1285. doi:10.1109/TSMC.2014.2313317

Ghosh, A., & Dey, S. (2021). *"Sensing the Mind": An Exploratory Study About Sensors Used in E-Health and M-Health Applications for Diagnosis of Mental Health Condition*. Springer.

Ghosh, A., & Saha, S. (2020). *Interactive Game-Based Motor Rehabilitation Using Hybrid Sensor Architecture*, 312–337. doi:10.4018/978-1-5225-9643-1.ch015

Ghosh, A., Saha, S., & Konar, A. (2020). Fuzzy Posture Matching for Pain Recovery Using Yoga. *Advances in Intelligent Systems and Computing, 999*(January), 957–967. doi:10.1007/978-981-13-9042-5_82

Herman, P., Prasad, G., McGinnity, T. M., & Coyle, D. (2008). Comparative analysis of spectral approaches to feature extraction for EEG-based motor imagery classification. *IEEE Transactions on Neural Systems and Rehabilitation Engineering, 16*(4), 317–326. doi:10.1109/TNSRE.2008.926694 PMID:18701380

Hossain, A. B. M. A., Rahman, M. W., & Riheen, M. A. (2015). Left and right hand movements EEG signals classification using wavelet transform and probabilistic neural network. *Iranian Journal of Electrical and Computer Engineering*, *5*(1), 92–101. doi:10.11591/ijece.v5i1.pp92-101

Imran, S. M., Talukdar, M. T. F., Sakib, S. K., Pathan, N. S., & Fattah, S. A. (2014). Motor imagery EEG signal classification scheme based on wavelet domain statistical features. *1st International Conference on Electrical Engineering and Information and Communication Technology, ICEEICT*. 10.1109/ICEEICT.2014.6919172

Ingle, V. K., & Proakis, J. G. (2008). Digital Signal Processing Using MATLAB, (3rd Edition). In Electronics {&} Communications Engineering Journal, 5(1).

Jenita Amali Rani, B., & Umamakeswari, A. (2015). Electroencephalogram-based Brain Controlled Robotic Wheelchair. *Indian Journal of Science and Technology*, *8*(S9), 188. doi:10.17485/ijst/2015/v8iS9/65580

Kant, P., Hazarika, J., & Laskar, S. H. (2019). Wavelet transform based approach for EEG feature selection of motor imagery data for braincomputer interfaces. *Proceedings of the 3rd International Conference on Inventive Systems and Control, ICISC, Icisc*, 101–105. 10.1109/ICISC44355.2019.9036445

Korovesis, N., Kandris, D., Koulouras, G., & Alexandridis, A. (2019). Robot motion control via an eeg-based brain–computer interface by using neural networks and alpha brainwaves. *Electronics (Switzerland)*, *8*(12), 1–15. doi:10.3390/electronics8121387

Lazarou, I., Nikolopoulos, S., Petrantonakis, P. C., Kompatsiaris, I., & Tsolaki, M. (2018). EEG-based brain–computer interfaces for communication and rehabilitation of people with motor impairment: A novel approach of the 21st century. *Frontiers in Human Neuroscience*, *12*(January), 1–18. doi:10.3389/fnhum.2018.00014 PMID:29472849

Li, Y., Zhou, G., Graham, D., & Holtzhauer, A. (2016). Towards an EEG-based brain-computer interface for online robot control. *Multimedia Tools and Applications*, *75*(13), 7999–8017. doi:10.100711042-015-2717-z

M, H., & M.N, S. (2015). A Review on Evaluation Metrics for Data Classification Evaluations. *International Journal of Data Mining & Knowledge Management Process*, *5*(2), 01–11. doi:10.5121/ijdkp.2015.5201

Ofner, P., Schwarz, A., Pereira, J., Wyss, D., Wildburger, R., & Müller-Putz, G. R. (2019). Attempted Arm and Hand Movements can be Decoded from Low-Frequency EEG from Persons with Spinal Cord Injury. *Scientific Reports*, *9*(1), 1–15. doi:10.103841598-019-43594-9 PMID:31073142

Pal, M. (2005). Random forest classifier for remote sensing classification. *International Journal of Remote Sensing*, *26*(1), 217–222. doi:10.1080/01431160412331269698

Patel, J., Khaked, A. A., Patel, J., & Patel, J. (2021). Heart Disease Prediction Using Machine Learning. *Lecture Notes in Networks and Systems, 203 LNNS*(6), 653–665. doi:10.1007/978-981-16-0733-2_46

Rahmadayana, F., & Sibaroni, Y. (2021). Sentiment Analysis of Work from Home Activity using SVM with Randomized Search Optimization. [Rekayasa Sistem Dan Teknologi Informasi]. *Jurnal RESTI*, *5*(5), 936–942. doi:10.29207/resti.v5i5.3457

Saha, S., & Ghosh, A. (2019). Rehabilitation using neighbor-cluster based matching inducing artificial bee colony optimization. *IEEE 16th India Council International Conference, INDICON Symposium Proceedings*, 5–8. 10.1109/INDICON47234.2019.9028975

Sebastián-Romagosa, M., Udina, E., Ortner, R., Dinarès-Ferran, J., Cho, W., Murovec, N., Matencio-Peralba, C., Sieghartsleitner, S., Allison, B. Z., & Guger, C. (2020). EEG Biomarkers Related With the Functional State of Stroke Patients. *Frontiers in Neuroscience*, *14*(July), 1–16. doi:10.3389/fnins.2020.00582 PMID:32733182

Shao, L., Zhang, L., Belkacem, A. N., Zhang, Y., Chen, X., Li, J., Liu, H., & Minati, L. (2020). EEG-Controlled Wall-Crawling Cleaning Robot Using SSVEP-Based Brain-Computer Interface. *Journal of Healthcare Engineering*, *2020*, 1–11. doi:10.1155/2020/6968713 PMID:32399166

Shedeed, H. A., Issa, M. F., & El-Sayed, S. M. (2013). Brain EEG signal processing for controlling a robotic arm. *Proceedings, 8th International Conference on Computer Engineering and Systems, ICCES*, 152–157. 10.1109/ICCES.2013.6707191

Shekar, B. H., & Dagnew, G. (2019). Grid search-based hyperparameter tuning and classification of microarray cancer data. *2nd International Conference on Advanced Computational and Communication Paradigms, ICACCP*, 1–8. 10.1109/ICACCP.2019.8882943

Shipp, S. (2007). Structure and function of the cerebral cortex. *Current Biology*, *17*(12), R443–R449. doi:10.1016/j.cub.2007.03.044 PMID:17580069

Song, Y., Cai, S., Yang, L., Li, G., Wu, W., & Xie, L. (2020). A Practical EEG-Based Human-Machine Interface to Online Control an Upper-Limb Assist Robot. *Frontiers in Neurorobotics*, *14*(July), 1–13. doi:10.3389/fnbot.2020.00032 PMID:32754025

Suresh, A., Laha, D., Gaba, D., & Bhambri, S. (2019). The classification of EEG signal using different machine learning techniques for BCI application. In Robot Intelligence Technology and Applications, 5(January). Springer Singapore. doi:10.1007/978-981-13-7780-8

Terms, F. (1961). The Ten Twenty Electrode System: International Federation of Societies for Electroencephalography and Clinical Neurophysiology. *The American Journal of EEG Technology*, *1*(1), 13–19. doi:10.1080/00029238.1961.11080571

Zhang, G., Davoodnia, V., Sepas-Moghaddam, A., Zhang, Y., & Etemad, A. (2020). Classification of Hand Movements from EEG Using a Deep Attention-Based LSTM Network. *IEEE Sensors Journal*, *20*(6), 3113–3122. doi:10.1109/JSEN.2019.2956998

Chapter 8
Design and Implementation of the FOIMC Using the PSO Algorithm for the Coupled Tank System

Sateesh Kumar Vavilala
Vignan's Lara Institute of Technology and Science, India

Vinopraba Thirumavalavan
National Institute of Technology, Puducherry, India

Radhakrishnan T. K.
National Institute of Technology, Tiruchirapalli, India

Sivakumaran N.
National Institute of Technology, Tiruchrapalli, India

ABSTRACT

In this work, a fractional order internal model controller (FOIMC) is developed for the level control of a coupled tank system. The parameters of the FOIMC are optimized by the particle swarm optimization (PSO) algorithm. The controller parameters obtained by the PSO algorithm, are compared with the gray wolf optimizer (GWO) and the whale optimization algorithms (WOA), and the PSO algorithm gave better results. The performance of the proposed FOIMC is compared with the state of the art. The experimental results obtained from the laboratory setup of the cylindrical tank system using the proposed FOIMC are also presented. The step responses from the simulation and the experimental results, indicated that the proposed FOIMC has good time domain performance than the state of the art. The proposed controller showed better robustness to a change in the operating point. The novelty of the present work is, for the first time, a comparative study of latest swarm algorithms is done for the tuning of FOIMC parameters.

DOI: 10.4018/978-1-6684-5381-0.ch008

INTRODUCTION

The coupled tank system has two modes of operation: interacting and noninteracting. In the interacting mode, the tanks are connected in cascade physically, and in the noninteracting mode the tanks are not connected physically. In this work interacting coupled tank system is considered as the case study. The level control problem of the coupled tank system usually involves, changing the level of the tank from the one value to another value, or keeping the level constant at one value. Many works in the control literature discussed the level control problem. Patel and Shah (2018) discussed a fault tolerant controller for the level control of the coupled tank system using the artificial intelligence (AI) techniques. The controller parameters are tuned using the neural networks. Sathish et al. (2017) developed a linear quadratic regulator (LQR) for the multi-input multi-output (MIMO) model of the cylindrical three tank system level control (Kumar et. al, 2017). Singla et al. (2018) proposed a modified proportional integrator differentiator (PID) controller for the coupled tank system. Mahapatro et al. (2019) discussed the design and implementation of a decentralized proportional integrator (PI) controller for the level control of the coupled tank system. A reduced order model of the coupled tank system is obtained based on the frequency response fitting. Gouta et al. (2019) discussed the design of an adaptive high gain observer-based controller design, for the coupled tank system. A generalized predictive controller is used to minimize the multi stage objective function. Dumlu and Ayten (2018) proposed an adaptive sliding mode controller for the coupled tank system level control. The design combines the feedforward action with the adaptive feedback. Other popular controllers for the level control of the coupled tank system are conventional PID controller, internal model control (IMC), Model predictive controller (MPC), and adaptive controller etc.

Among the controllers available for the coupled tank system level control, the IMC has the advantages like stability, and minimum number of the parameters to be tuned. Some of the recent literature works on the IMC design are presented here. Li and Zhu (2019) developed an IMC- PID based controller for the speed control of the servo motor using an extended state observer. Pati and Negi (2019) designed a two degrees of freedom IMC for the magnetic levitation system. The controller parameters are tuned using the PSO algorithm. Vadamalu and Beidl (2019) designed an adaptive IMC for the internal combustion engine's torsional oscillations control. The controller reduces the disturbances at various operating speeds. Tasoujian et al. (2019) discussed the design of a robust IMC for the blood pressure regulation. An IMC-PID controller is developed including the uncertainty, with a time varying model. Li and Bing (2019) proposed a Smith predictor-based IMC for a second order time delay system.

The fractional order control became popular in the past three decades for the engineering, economic, biological problems. The fractional order proportional integrator differentiator (FOPID) is proposed by Podlubny (1994). Later many authors extended the fractional order concept to the popular controllers like IMC, MPC, and adaptive control etc. Vinopraba et al. (2012) discussed the design of a two degrees of freedom (2-DOF) FOIMC based on the closed loop system bandwidth. The FOIMC is the fractional order version of the IMC. Rodrigues et al. (2019) proposes an analytical design of the FOIMC for a FOPDT system. The controller parameters are obtained from optimizing an objective function. Erhan et al. (2019) proposed a FOPID controller for the higher order process. The controller parameters are tuned using the frequency domain specifications. Rayalla et al. (2019) discussed the design of the FOPID for a FOS using the higher order fractional filter based FOIMC. The controller parameters are tuned using the maximum sensitivity by minimizing the integral absolute Error (IAE). Wen et al. (2020) developed a FOIMC for the force/ position control of a parallel robot. Chekari et al. (2019), proposed a 2-DOF FOIMC satisfying the disturbance rejection and servo tracking. The peak overshoot of the servo response is high. Nath et al. (2019) further discussed IMC-PID based fractional controller. Model order reduction from the higher order system to a FOPDT model, is developed. Wu et al. (2019) discussed the Smith predictor based fractional order proportional integrator (FOPI) for the higher order system. A multi objective genetic algorithm (GA) is used to tune the parameters of the controller.

After careful selection of the controller configuration, the next step is to tune the controller parameters based on the given specifications. The specifications may be time domain or frequency domain. The time domain specifications include rise time, settling time, peak time, peak overshoot and steady state error. The frequency domain specifications include phase margin and gain margin. The parameters of the IMC/ FOIMC can be tuned using, heuristic algorithms, numerical methods, empirical formulae or analytical procedures and the soft computing techniques like the neural networks, fuzzy logic, GA.

The analytical and numerical tuning methods are time consuming and complex, as they depend on the operating personnel's experience and, at times use trial and error-based procedure. Lakshmanprabu and Sabura Banu (2019) proposed an IMC based PID controller for the two-tank system using the bat optimization algorithm (BOA). Muresan et al. (2021) discussed the design of the FOIMC for time delay system. A novel approximation method for the time delay is used to tune the controller parameters. Arya and Chakrabarty (2020) discussed the design of the fractional order controller for the fractional order time delay system. A novel analytical method is proposed to tune the controller parameters based on the frequency domain specifications. Jain et al. (2020) designed a FOIMC for the buck boost converter. The controller parameters are tuned using the frequency domain specifications. Bettayab and Rachid (2014) discussed the design of a FOIMC based on the fractional filter. An empirical formula based on the frequency domain specifications, are used to tune the controller parameters.

The swarm intelligence techniques like the particle swarm optimization (PSO) algorithm proposed by Kennedy and Eberhart (1995), are becoming popular for the optimization problems and the controller tuning. The PSO algorithm uses a social communication among the swarm to get the global minimum value of the objective function. Some of the recent swarm algorithms are the grey wolf optimizer (GWO) and the whale optimization algorithms (WOA). Mirjalili et al. (2016) developed the GWO algorithm. Mirjalili and Lewis (2016) developed the WOA. The WOA and GWO algorithms have an exploration and an exploitation phase incorporated in the algorithms, which increase the randomness of the optimization process. Abo Hammour et al. (2018) discussed the tuning of the integer and the fractional order PID controllers using the PSO algorithm. A heat solid process with a fractional order transfer function, a higher order fractional order system (FOS), and a ship's roll angle control are considered as the case studies. Tavakoli (2018) developed a PI controller for the system with long dead time. A constraint-based optimization is used to tune the controller parameters. The performance of the FOIMC controller proposed in this work, is compared with the performance of the state-of-the-art controllers (Abo-Hammour, 2018 and Tavakoli, 2018). The objective function that should be minimized in the algorithms, is constructed based on the goal of the controller, i. e. if it is required to improve the time domain specifications, then the objective function should consist of the time domain specifications, or else if it is required to improve the frequency domain specifications, then the objective function should consist of the frequency domain specifications.

Latest literature on the tuning methods using algorithms like antlion algorithm, Big-Bang, Big-Crunch algorithm, Lion swarm optimization, Hunger games search optimizer etc., (Fathy A. etal.,2022, Wang et al., 2022., Ghamari S.M. et al., 2022) are developed for the FOPID controller. There is no study of such algorithms for the FOIMC.

To simulate the fractional order controller in MATLAB software, toolboxes like fractional order transfer function (fotf), fractional order modelling and control (FOMCON), CRONE, and NINTEGER, etc. are available. Among these toolboxes, FOMCON is suitable for the present application. The FOMCON is a toolbox developed by the Teplijakov et al. (2011), and it is used for the identification and control of the fractional and integer order systems.

For the level control problem, none of the existing literature applied the PSO algorithm-based tuning of the FOIMC parameters. To the author's best knowledge, this is the only work proposing a PSO algorithm-based tuning for obtaining the controller parameters of the fractional filter based FOIMC. The authors Vavilala and Vinopraba (2018) presented the design and simulation of the FOIMC using the PSO algorithm, for a conical tank system case study. The current work developed the design and simulation of the FOIMC for the coupled tank system case study, and experimental validation. Simulations in this work, are implemented using the MATLAB/ SIMULINK software with the FOMCON toolbox.

The main contributions of this work are:

1. Comparison of the different heuristic algorithms for the tuning of the controller parameters, and experimental implementation of the proposed controller.
2. The stability and robustness of the proposed controller are analyzed.

The novelty of the present work is; for the first time a comparative study of the latest swarm algorithms is done for the tuning of the FOIMC parameters. Recent literature works on the swarm algorithms have focused on the FOPID controller only, since it is the extension of the industry popular PID controller. A few works like Saxena S. et al. (2022), have discussed the Big-Bang, Big-Crunch algorithm for FOIMC, but a comparative study of the different algorithms is not made.

The contents of this chapter are organized as follows. The second section presents the problem formulation. The third section discusses the design of the FOIMC using three swarm intelligence algorithms; PSO, GWO, WOA and the fourth section discusses the dynamics of the coupled tank nonlinear system as a case study and the presents the results obtained (both simulation and experimental), fifth section gives the future direction of the research work, and the sixth section concludes the results obtained.

Problem Formulation Using the FOIMC

Problem statement: For the FOPDT model of the coupled tank system, a fractional filter-based IMC should be designed, which minimizes the integral square error (ISE) of the closed loop system, using the swarm intelligence techniques like the PSO algorithm, GWO algorithm, and WOA.

The controller objectives are given below:

1. To achieve good transient and steady state step responses $M_p \leq 5\%$ and $t_s \leq 1500$ sec).
2. To achieve a satisfactory servo tracking and disturbance rejection properties simultaneously.
3. To obtain a stable and robust controller.

Figure 1 shows the tuning procedure employed in this work using the swarm intelligence techniques. Consider the FOPDT model of the plant given by Eq. (1).

$$G(s) = \frac{Ke^{-Ls}}{Ts+1} \tag{1}$$

The exponential term involving the delay time *L*, is expanded using the first order Taylor series, and Eq. (1) can be rewritten as Eq. (2).

$$G(s) = \frac{-KLs+K}{Ts+1} \tag{2}$$

Figure 1. Tuning procedure employed to obtain the parameters

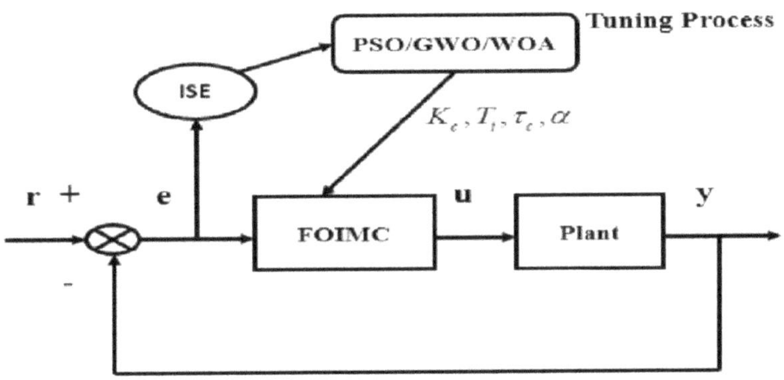

It is difficult to design the controller for systems with larger time constants and longer time delays. The time delay indicates the delay in the output after input is applied to the system. Time constant represents the rise time of the system's response.

The structure of the FOIMC implemented in this work is selected as Eq. (3), since it is suitable for a wide variety of systems like the FOPDT, the second order plus delay time (SOPDT), an integrator, and an integrator with delay etc. (Bettayab & Rachid, 2014).

$$C_{FOIMC}(s) = \left(\frac{1}{1+\tau_c s^{\alpha}}\right) K_c \left(1+\frac{1}{T_i s}\right) \tag{3}$$

The error *e(t)*, and integral square error (ISE) are given by Eq. (4) and (5).

$$e(t) = r(t) - y(t) \tag{4}$$

$$ISE = \int_0^t \left(r(\tau) - y(\tau)\right)^2 d\tau \tag{5}$$

The ISE is a time domain integral performance index used for analyzing the performance of the controller and the controller parameter optimization. The set point *r* in Figure 1, is a step input of magnitude '*A*', therefore the Laplace transform of *r* '*R(s)*' is *A/s*. From the feedback dynamics of the closed loop system, the error signal *E(s)* is obtained as Eq. (6). Using the Eq. (5), (6) the controller parameters are obtained using the evolutionary algorithms like PSO/ WOA/ GWO.

$$E(s)=\frac{R(s)}{1+G(s)C_{FOIMC}(s)}=\frac{a/s}{1+\left(\left(\frac{-KLs+K}{Ts+1}\right)\left(\frac{1}{1+\tau_c s^{\alpha}}\right)K_c\left(1+\frac{1}{T_i s}\right)\right)} \tag{6}$$

Since the presented case study is a commonly used FOPDT system, the controller design method and evolutionary algorithms are applicable for a generalized class of systems. The objective function considered for the optimization can include complex terms like, integral error indices and time domain indices, hence the evolutionary algorithms are useful in complex scenarios also.

DESIGN OF THE FOIMC

Basics of the IMC

The IMC is used when a good model of the plant is available. The IMC is an inherently stable controller. Figure 2 shows the IMC configuration, and Figure 3 shows the equivalent block diagram of the IMC in the unity feedback system configuration. The performance of the IMC depends on the accuracy of the model and proper choice of the time constant of the IMC. The only disadvantage is it is difficult to handle open loop unstable systems.

R(s) = Reference input
Q(s) = Internal model controller
Y(s) = Plant output
E(s) = *R(s)-Y(s)*
G(s) = Plant transfer function
$\tilde{G}(s)$ = Model of the plant
D(s) = Disturbance at the output side
U(s) = Controller output
$\tilde{Y}(s)$ = Model output

$$V(s)=Y(s)-\tilde{Y}(s)$$

C(s) = Controller transfer function

Figure 2. Block diagram of IMC

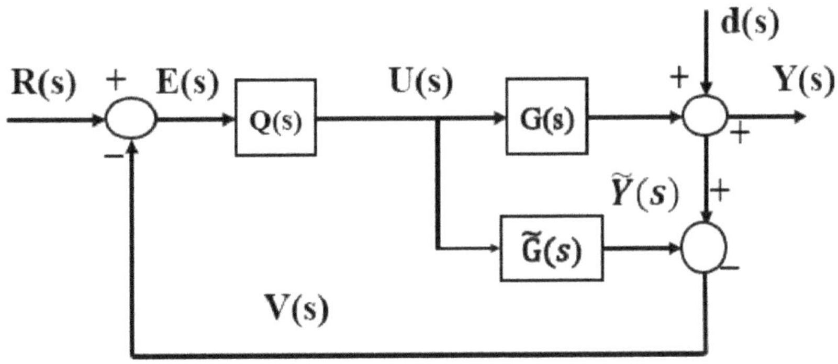

Figure 3. Block Diagram of control system

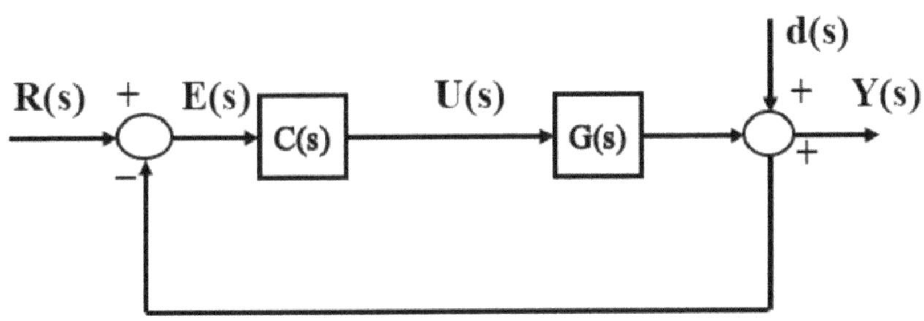

Assuming that the plant transfer function and model of the plant are same, the relation between *C(s)*, and *Q(s)* is given by Eq. (7).

$$C(s) = \frac{Q(s)}{1 - Q(s)\tilde{G}(s)} \tag{7}$$

Basics of the Fractional Calculus

The fractional calculus is 300 years old, but its applications became popular in the past three to four decades only. There are three definitions available for the fractional differentiation; Reimann-Liouville, Caputo, and Grunwald-Letnikov (G-L). Among these definitions, the Caputo definition is used in the control and engineering applications. The G-L definition is more amenable to the digital computer implementation. The Caputo definition of the fractional order derivative, used in this work is given by Eq. (8).

$$D_t^{\alpha} f(t) = \frac{1}{\Gamma(n-\alpha)} \int_0^t \frac{f^n(\tau)}{(t-\tau)^{\alpha-n+1}} d\tau \tag{8}$$

In this definition, the function $f(t)$ is first differentiated and then integrated. The fractional derivatives are nonlocal and are suitable to represent the systems involving the distributed parameters. The fractional powers in the FOPID controller give more flexibility to the integrator and the differentiator of the PID controller. A general form of the FOPID controller is represented as Eq. (9).

$$C(s) = K_p + \frac{K_i}{s^{\lambda}} + K_d s^{\mu} \tag{9}$$

where K_p, K_i, and K_d are proportional, integral and derivative gains respectively. The fractional order is extended to all the major controller configurations. One among them is the FOIMC based on a fractional filter. The configuration of the FOIMC used in this work is based on the ideal Bode loop transfer function given by Eq. (10).

$$L(s) = \frac{1}{\tau_c s^{\lambda}} \tag{10}$$

This configuration has the advantages of isodamping property, and wider stability margins (Bettayab & Rachid, 2014). The structure of the FOIMC is given by Eq. (3). The fractional filter in the Eq. (3) is, $\frac{1}{1+\tau_c s^{\alpha}}$, where α is the fractional order of the filter. In this work, the control parameters are tuned and optimized by the PSO algorithm. The only disadvantage with the FOC is it increases the complexity of the controller design.

PSO Based Tuning

As discussed in the Introduction, the heuristic algorithms are becoming popular nowadays for the tuning of the controller parameters. The PSO algorithm is popular for the optimization problem with or without constraints. The optimization process using the PSO algorithm is governed by Eq. (11) and (12).

$$V_i(k+1) = \omega V_i(k) + c_1(p_i(k) - x_i(k)) + c_2(g(k) - x_i(k)) \tag{11}$$

$$x_i(k+1) = x_i(k) + V_i(k+1) \tag{12}$$

where,

$V_i(k+1)$ = Velocity of the particle at time $(k+1)$
$V_i(k)$ = Velocity of the particle at time k
ω = Inertia weight
c_1 = Cognitive learning rate
c_2 = Social learning rate
$g(k)$ = Global best position
$p_i(k)$ = Particle best position

$x_i(k)$ = Position of the particle at time k
$x_i(k+1)$ = Position of the particle at time $(k+1)$

Initially, the particle's positions will be set to the lower bound, and as the iteration progresses, the particle's positions and velocity are updated as per Eq. (11) and (12). After completion of the maximum number of iterations, the best particle position and minimum value of the objective function are displayed. The details of the different parameters used in the PSO algorithm are given Table 1. Figure 4 shows the flowchart of the FOIMC controller parameter tuning process. Swarm algorithms solve the nonlinear optimization problems also with or without constraints. The disadvantage with swarm algorithms is, the design person need not have deep knowledge of the system, but still, he can be successful in the controller design.

Table 1. Details of the PSO algorithm

Parameter	No. of iterations	c_1	c_2	Population size	ω_{max}	ω_{min}
Value	10	2	2	5	0.9	0.4

In the optimization process, an objective function will be developed based on the desired specifications. This objective function will be minimized during the optimization. The objective function can have either the time domain specifications like steady state error, peak overshoot, ISE, IAE etc., or the frequency domain specifications like phase margin, gain margin etc. or an algebraic combination of both the time domain and frequency domain specifications.

In the present work, ISE is chosen as the objective function and is shown in Eq. (5). The convergence of the objective function for the PSO algorithm is shown in Figure 5. The upper and lower bounds of the FOIMC controller parameters given in PSO coding are shown in Eq. (13).

$$\begin{aligned} &5 \le K_c \le 50 \\ &500 \le T_i \le 1000 \\ &1 \le \tau c \le 50 \\ &0.01 \le \alpha \le 1.99 \end{aligned} \tag{13}$$

The optimization used in this work is offline. The controller parameters are obtained from running the MATLAB code of the evolutionary algorithm and then the controller parameters are plugged in the SIMULINK file implementing the controller. Hence the system stability will not affect the search process. This offline search method is followed because, it is difficult to get the level measurements using the sensors from the coupled tank system quickly to suit online search (due to its dead time and slow dynamics).

Before finalizing the algorithm to be used for optimizing the controller parameters, a comparison of the three popular and recent algorithms, PSO, WOA and GWO is made. The coding of the PSO, WOA, and GWO algorithms is done in the MATLAB. The controller parameters obtained using the different algorithms are listed Eq. (14)- (16).

Figure 4. Flowchart of the PSO algorithm

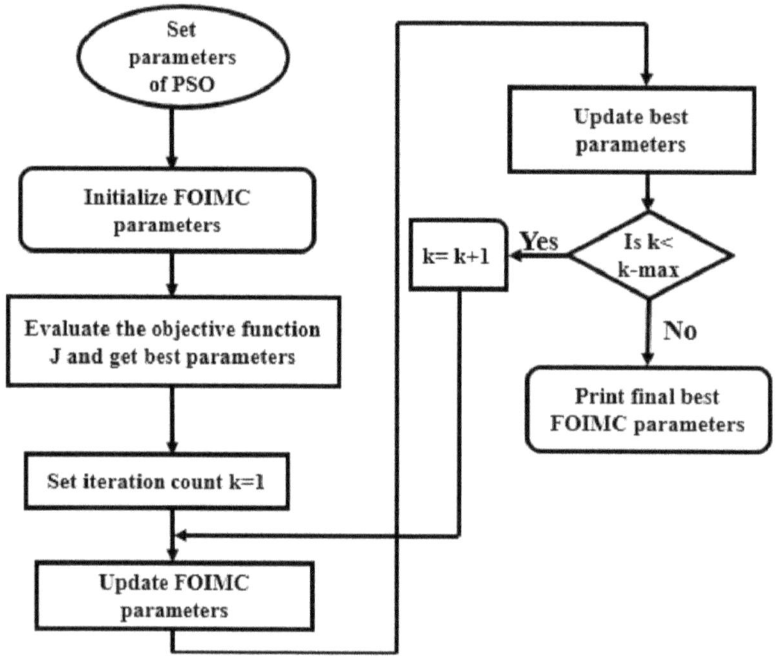

$$C_{FOIMC-PSO}(s)=\left(\frac{1}{1+s^{1.9}}\right)7.9\left(1+\frac{1}{790s}\right) \tag{14}$$

$$C_{FOIMC-WOA}(s)=\left(\frac{1}{1+0.29s^{1.99}}\right)8.06\left(1+\frac{1}{806s}\right) \tag{15}$$

$$C_{FOIMC-GWO}(s)=\left(\frac{1}{1+3.98s^{1.99}}\right)50\left(1+\frac{1}{347s}\right) \tag{16}$$

Figure 5. Convergence of the objective function using the PSO algorithm- Offline

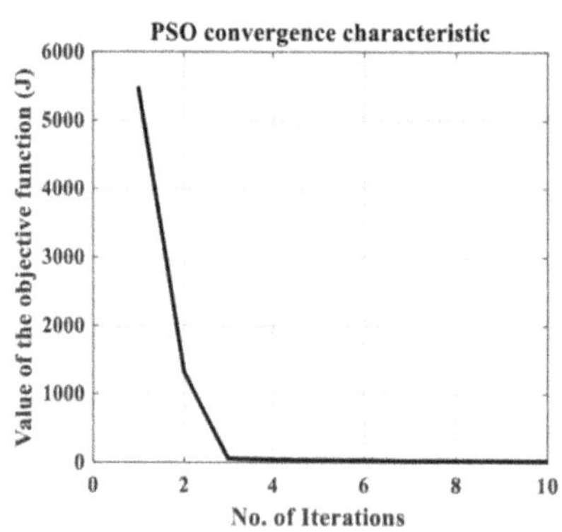

In Figure 6, the step responses of the closed loop system with the controllers tuned by the PSO, WOA and GWO algorithms (Eq. (14)- (16)) are compared.

From Figure 6 it is also observed that the FOIMC tuned using the GWO algorithm showed high frequency transients in the step response. Hence while implementing such control signal on the experimental setup the actuator gets damaged.

GWO Based Tuning

GWO is a swarm based meta heuristics algorithm Mirjalili et al. (2014). Four types of grey wolves alpha, beta, gamma, delta are available. This algorithm mimics the predator-prey configuration. The hunting by predator consists of two phases exploration and exploitation. The mathematical model of the GWO algorithm is represented by;

$$\vec{D} = \left|\vec{C}.\overrightarrow{X_p}(t) - \vec{X}(t)\right| \tag{17}$$

$$\vec{X}(t+1) = \overrightarrow{X_p}(t) - \vec{A}.\vec{D} \tag{18}$$

where,

T= Present iteration
$\overrightarrow{A},\vec{C}$ = Coefficient vectors
$\overrightarrow{X_p}$ = Position vector of prey
$\vec{X}$ = Position vector of grey wolf

The vectors $\vec{A}$ and $\vec{C}$ are updated as follows,

$$\vec{A} = 2\vec{a}.\overrightarrow{r_1} - \vec{a}$$

$$\vec{C} = 2.\overrightarrow{r_2}$$

Where r_1 and r_2 are random vectors in [0,1] and $\vec{a}$ decreases linearly from 2 to 0. WOA (Mirjalili & Lewis, 2016) is also a meta heuristics algorithm with mathematical model given by Eq. (17)- (18). It has also an exploitation and exploration phases.

In Table 2, the integral time domain performance indices like ISE, IAE, integral time absolute error (ITAE) and integral time square error (ITSE) and the time domain specifications like the rise time, the settling time, the steady state error and the peak overshoot obtained from Figure 6, are compared. From Table 2, it can be observed that the PSO algorithm based tuning gives better controller performance compared to the WOA and GWO algorithms. Hence for the simulation, and the experimental works, the PSO algorithm is finalized, and in the next section, results obtained from the PSO algorithm will only be discussed.

Figure 6. Comparison of the different algorithms- Simulation

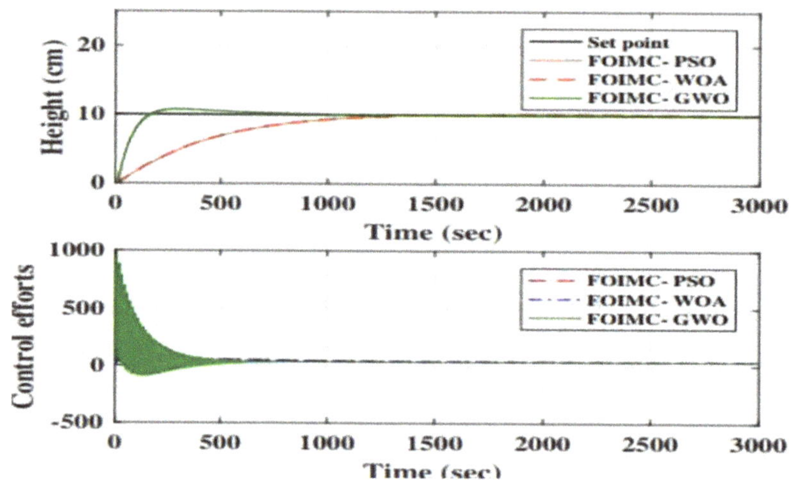

Table 2. Time domain comparison of the algorithms- Simulation

Parameter	PSO	WOA	GWO
ISE	2.45 10^4	2.7 10^4	3.06 10^4
IAE	4.32 10^3	4.76 10^3	7.89 10^3
ITAE	1.97 10^6	2.45 10^6	9.96 10^3
ITSE	1.22 10^8	1.35 10^8	1.53 10^8
Rise time (sec)	900	900	500
Settling time (sec)	1300	3600	5000
Peak overshoot (%)	2	5	30
Steady state error	0	0	0

SIMULATION RESULTS

Case Study: Coupled Tank System

In this work, a cylindrical coupled tank system in the interacting mode is taken as a case study. The interacting mode consists of two cylindrical shaped tanks connected physically, as shown in Figure 7. The control objective for the experimental setup is to regulate the height of the tank two, by changing the inlet flow rate of the tank one. The transfer function of this problem is represented as, $\frac{h_{2s}}{q_{ins}}$. The transfer function of the level control problem in the interacting mode of the coupled tank system is of second order. The experimental laboratory setup of the coupled tank system is shown in Figure 8. Figure 9 shows the plot of the heights of the tanks and the inlet flow rate. Figure 9 is also referred to as the open loop input and output data, which gives the dynamics of the system.

Figure 7. Block diagram of the coupled tank system

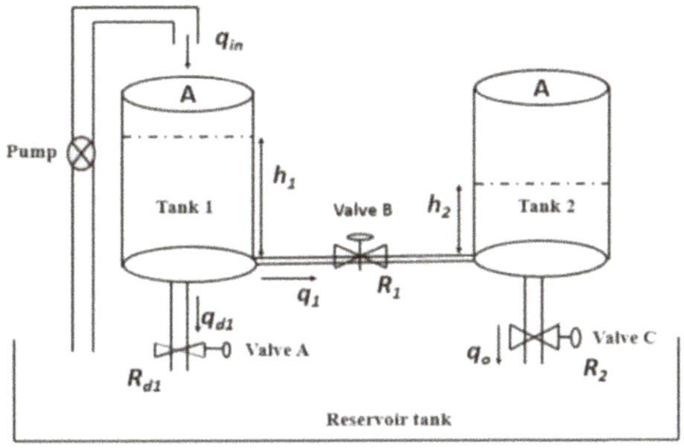

In Figure 7,

q_{in} = Inlet flow rate
q_1 = Flow rate from the tank one to tank two
q_o = Outlet flow rate from the tank two
h_1 = Height of the tank one
h_2 = Height of the tank two
R_1 = Resistance of valve *B*, between the two tanks
R_2 = Resistance of the outlet valve *C*, at the second tank
D = Diameter of the cross-sectional area of each tank
A = Cross sectional area of each tank
R_{d1} = Disturbance flow rate from tank one
R_{d2} = Resistance of the valve A, at tank one

Figure 8. Experimental setup of the coupled tank system

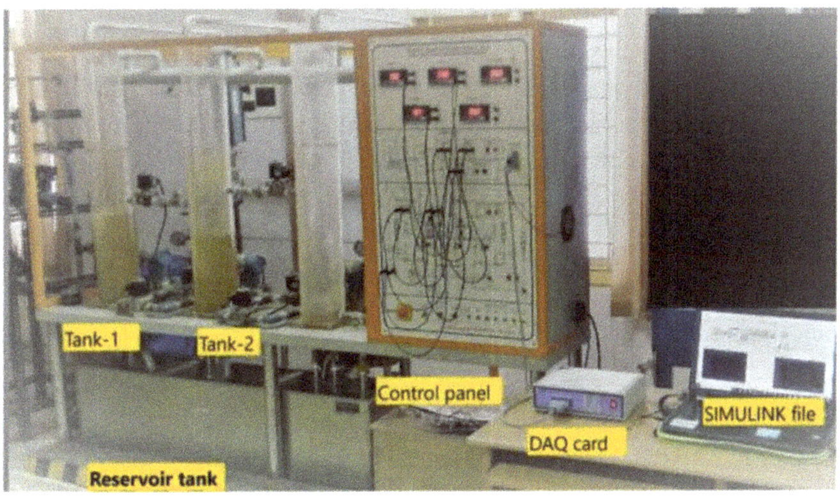

Figure 9. Open loop test results of the level control- Experimental

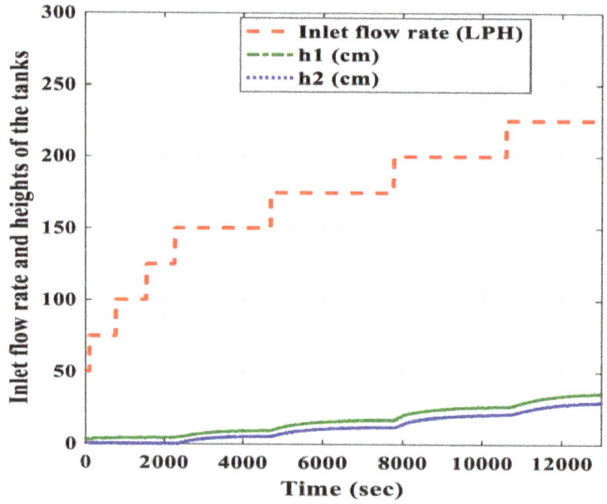

The experimental test data shows that a FOPDT transfer function is suitable for the coupled tank system. The open loop test results of Figure 9 are obtained by varying the inlet flow rate from 50 LPH to 275 LPH in steps of 25 LPH. The desired operating range of the tank two is (5-15) cm. From the plot, a FOPDT model of the level control system in the range (5-15) cm is obtained as Eq. (19).

$$\frac{h_2(s)}{q_{in}(s)} = \frac{0.268e^{-10s}}{975s+1} \tag{19}$$

The Eq. (7) is used in the design and tuning of the controller parameters. The nominal physical parameters of the cylindrical tank system are shown in Table 1, where H is the maximum height of the tank. The front control panel consists of the power supplies to different components, the interconnections between data acquisition (DAQ) and the system. The DAQ system consists of the analog and digital input and output channels and acts as an interface between the computer and the front control panel. The DAQ converts the voltage signals into (4-20) mA current signals. The control valve position ranges from (0-100) %, which corresponds to the current to pressure converter output of (3-15) psi. The diaphragm valve corresponds to a (0-100) % change, and finally, the inlet flow rate has a range of (44- 400) LPH. A 1/3 HP air compressor maintains air pressure of the total system between (15-25) psi.

To compare the performance of the proposed FOIMC controller, two state of the art controllers are considered. The first controller considered is an integer order proportional integrator (IOPI) Abo-Hammour (2018), denoted as Ref-1.

Table 3. Parameters of the cylindrical tank

Parameter	H (cm)	R_1 (ohm)	R_2 (ohm)	D (mm)	Q_{in} (LPH)
Nominal value	100	1.5	1.7	92	20

A phase margin of 60þ is selected, and the Ref-1 controller obtained is given by Eq. (20).

$$C_{IOPI-Ref-1}(s) = 2.095 + \frac{0.0122}{s} \quad (20)$$

$$C_{FOPID-Ref-2}(s) = 21.6 + \frac{3.86}{s^{0.47}} + 10s^{0.01} \quad (21)$$

The second controller considered is a FOPID controller and denoted as Ref-2 Tavakoli (2018). Using the PSO algorithm, the controller Ref-2 obtained is given in Eq. (21).

Any controller designed for the real time system should be stable. The stability investigation of the proposed FOIMC is shown in Figure 10. The stability test is conducted using the FOMCON toolbox. Figure 10 shows the poles of the primary Reimann sheet, and from the results of the FOMCON, the system is found to be stable with a fractional order of 0.1.

The desired height (set point) of the second tank is 10 cm (in the operating range of (5-15) cm). In the closed loop system (of Figure 3), $R(s)$ $(= 10/s)$ is a step input, and the controller transfer function $C(s)$ is given by Eq. (15), plant transfer function $G(s)$ is given by Eq. (7), and $Y(s)$ is the controlled height of the tank two. The MATLAB/ SIMULINK is used for the simulations. The FOMCON toolbox is used for implementing the fractional controllers and fractional transfer functions in the SIMULINK. The FOMCON toolbox uses Nelder-Mead optimization method. The step responses of the system with the proposed controller, Ref-1 and Ref- 2 are given in Figure 11. The time domain specifications and indices obtained from the step responses of the controllers are compared in Table 4.

From Figure 11 and Table 4, the proposed FOIMC has the 92% lesser peak overshoot than Ref-1 and 86.66% lesser peak overshoot than Ref- 2. The time domain performance indices like IAE and ITAE are lesser for the proposed controller compared Ref-1 and Ref- 2. Hence from the peak overshoot, and integral indices, the proposed FOIMC is efficient than the other controllers. Even though the Ref- 2 showed the least rise time and settling time compared to other controllers, it did not give zero steady state error. In the level control applications to avoid spillage of the liquid, there is a stringent condition on the peak overshoot.

The controller action has a saturation limit of (0-100) %. If the saturation limit of the controller is omitted, then the actuator gets damaged. Figure 11 also shows the control efforts, which indicate the energy consumed by the controllers in the process of the control action. From Figure 11, it can be seen that the proposed controller has its control effort within the limits and the control effort is the least among all the controllers. The control efforts of Ref-1 and Ref- 2 are hitting the saturation limits, which is not a desirable feature. Again, the efficiency of the proposed controller is reiterated with less control energy consumed by the proposed controller.

From the above discussion, it can be observed that the proposed controller showed better transient and steady state time domain performances. The better performance of the proposed controller is once again proved in the transient and steady state responses. For checking the disturbance rejection capability of the controller, the regulatory responses are studied. A step disturbance of magnitude 5 cm ($d(s)= 5/s$) is applied at time 8000 sec. Figure 12 shows the regulatory responses (disturbance rejection) of the system with different controllers. From Figure 12, it is seen that the proposed controller resumes back to the 10 cm level quickly when compared to other controllers. Figure 13 shows the servo or tracking responses of the controllers. The servo response indicates the ability of the controller to track changes in the level (both increment and decrement) satisfactorily. The set point is varied from 5 cm to 20 cm at 8000 sec and varied from 20 cm to 15 cm at 16000 sec. The servo responses in Figure 13 indicate the better performance of the proposed controller, both in terms of the level control and the control effort. Normally a controller giving good servo response may not give good regulatory response, but the proposed controller gave better servo and regulatory responses than the other controllers.

Figure 10. Primary Reimann sheet of the closed loop system

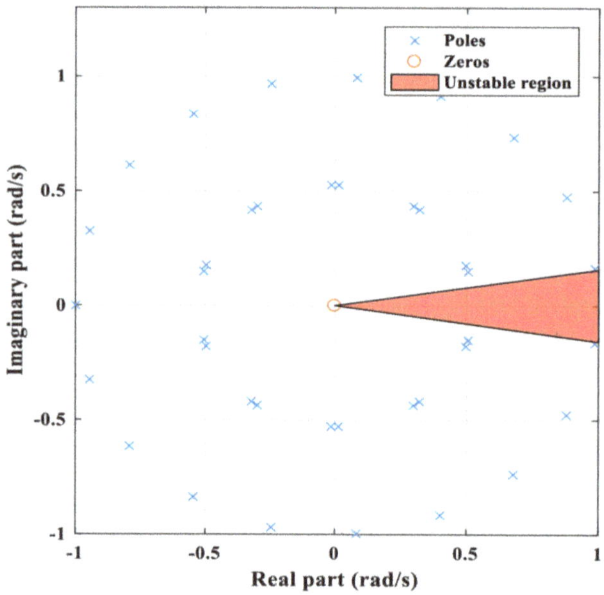

Figure 11. Step responses with controllers- Simulation

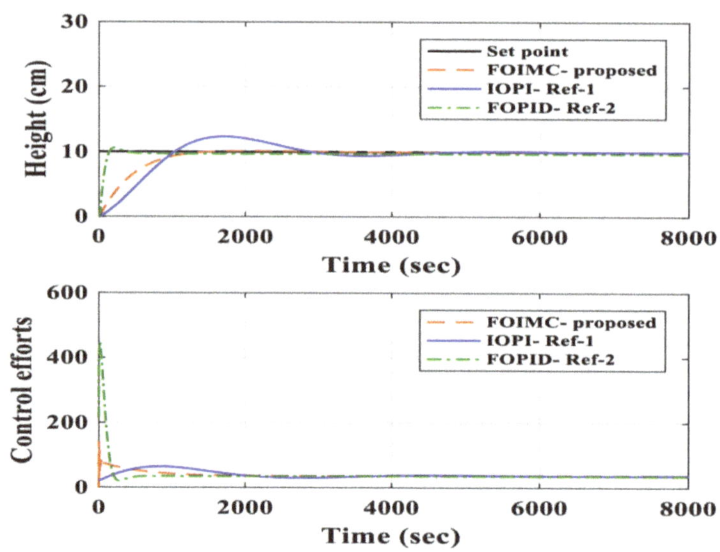

The controller has to accommodate the changes (be robust) in the plant parameters, changes in the set point, changes in the operating region. In this work, the controller's robustness is verified by operating it in an adjacent operating region of (16-30) cm. The transfer function of the new region is given by Eq. (22). A set point of 30 cm is chosen in the new adjacent operating region of (16-30) cm. The step responses of the closed loop system with the new transfer function given by Eq. (22) are presented in Figure 14.

$$\frac{h_2(s)}{q_{in}(s)} = \frac{0.8e^{-30s}}{1095s+1} \tag{22}$$

For Ref- 2, the steady state error remains the same as 0.25 and for the Ref- 1 the peak overshoot has increased from 25% to 40%. The proposed controller showed an increase in the overshoot from 2% to 16.66%. The rise time of the proposed controller is reduced from 900 sec to 500 sec and the settling time reduced is from 3600 sec to 2600 sec. From this discussion, the proposed controller showed better robustness.

From the controller equations (14), (20), (21), Ref-2 gives lesser peak overshoot than Ref-1, since Ref-2 has differentiator term. The rise time of Ref-2 is lesser than Ref-1, since Ref-2 has higher proportional gain than Ref-1. The steady state error is zero for both the Ref-1 and Ref-2 since an integrator is included in the controllers.

Comparing the proposed and Ref-1 controllers, it is observed that the proposed controller has the least peak overshoot, since the proposed controller has lesser integral gain. The proposed controller showed good robustness to change in the set point and change in the controller parameters due to its fractional nature. The proposed controller has good integral indices because of the least settling time and rise time, as the integral time domain indices depend on the time and error. The plant considered has different transfer functions in different zones of the operating height. A controller designed for one zone will not be effective in the other zone. The proposed controller gave satisfactory step responses in two different adjacent zones.

Figure 12. Regulatory responses with controllers- Simulation

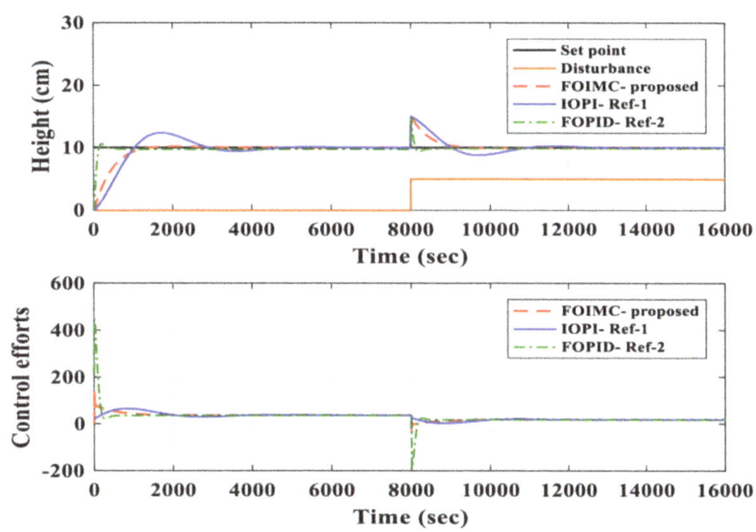

Figure 13. Servo responses of the controllers- Simulation

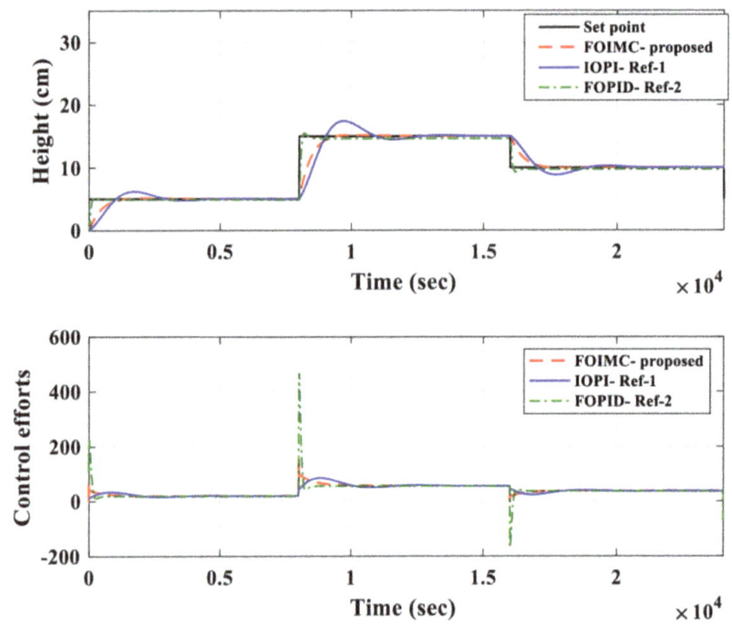

Figure 14. Robustness verification- Simulation

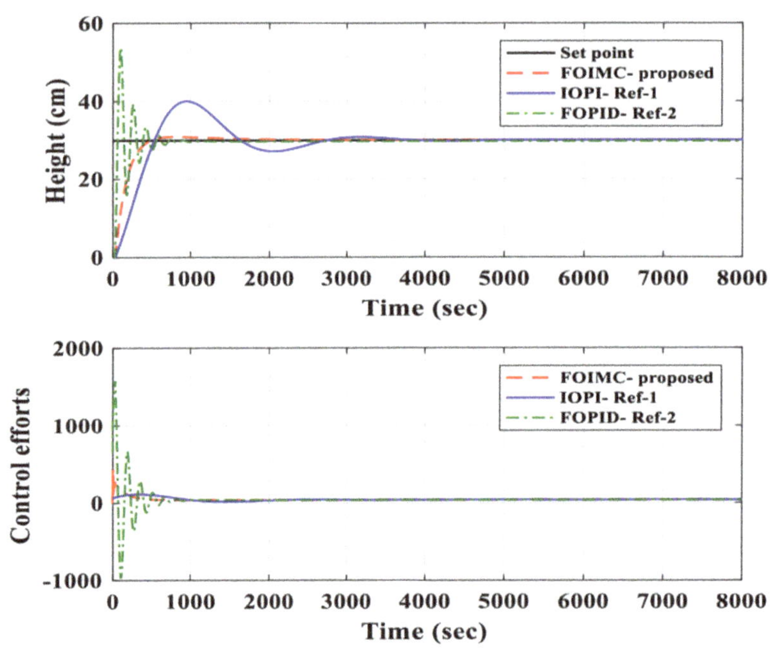

Table 4. Time domain specifications of the controllers- Simulation

Parameter	FOIMC (proposed)	Ref- 1	Ref- 2
ISE	$8.754\ 10^3$	$8.754\ 10^3$	$4.36\ 10^3$
IAE	$4.27\ 10^4$	$4.2\ 10^4$	$1.52\ 10^4$
ITAE	$1.002\ 10^6$	$1.98\ 10^6$	$8.907\ 10^6$
ITSE	$3.366\ 10^8$	$1.773\ 10^8$	$3.774\ 10^7$
Control energy	$1.338\ 10^7$	$1.38\ 10^7$	$2.383\ 10^7$
Rise time (sec)	900	970	120
Settling time (sec)	1300	2000	2000
Peak overshoot (%)	2	25	15
Steady state error	0	0	0.25

Experimental Results

The proposed controller and Ref-1 are implemented on a laboratory scale experimental setup of the coupled tank system. The details of the experimental setup are given in subsection 4.1. Figure 15 shows the step responses of the controllers and the control efforts implemented on the experimental setup.

The experimental step responses are tabulated in Table 5. The sampling time used for the simulation and experimental results is 0.01 sec (1 sec= 100 samples). The regulatory responses and the corresponding control efforts are presented in Figure 16. The disturbances of 2.5 cm and 7.5 cm are applied to the FOIMC at 3100 sec and Ref-1 at 2500 sec respectively. The regulatory responses of the controllers settled back to the set point and rejected the disturbance. The disturbance is applied by changing the position of the outlet valve C in the experimental setup. For verifying the robustness of the controllers, the set point is chosen as 30 cm, and automatically the real time system picks up the transfer function given by Eq. (20). The robustness verification responses and the corresponding controller efforts are shown in Figure 17. Figure 17 shows that the proposed controller has good robustness to a change in the setpoint in the experimental responses also.

From the Table 5, it can be seen that the experimental step response of the FOIMC has lesser peak overshoot, lesser steady state error and lesser settling time when compared to the Ref-1 response. The experimental responses also, confirm the superiority of the proposed FOIMC and they are close to the simulations results (evident from Table 4 and Table 5).

Overall observations from the present work are:

1. Increasing the fractional filter time constant τ_c, increases the rise time and the settling times. Increasing the fractional filter time constant and does not affect the peak overshoot.
2. A lower value of the fractional order α of the order 0.1 results in sluggish response with high rise time.
3. All the controller objectives set in the Section 2 are achieved by the proposed controller.

Table 5. Time domain specifications of the controllers from step responses- Experimental

Parameter	FOIMC (proposed)	Ref- 1
Rise time (sec)	2200	690
Settling time (sec)	3394	3500
Peak overshoot (%)	0	25
Steady state error	0	1.5

Figure 15. Step responses- Experimental

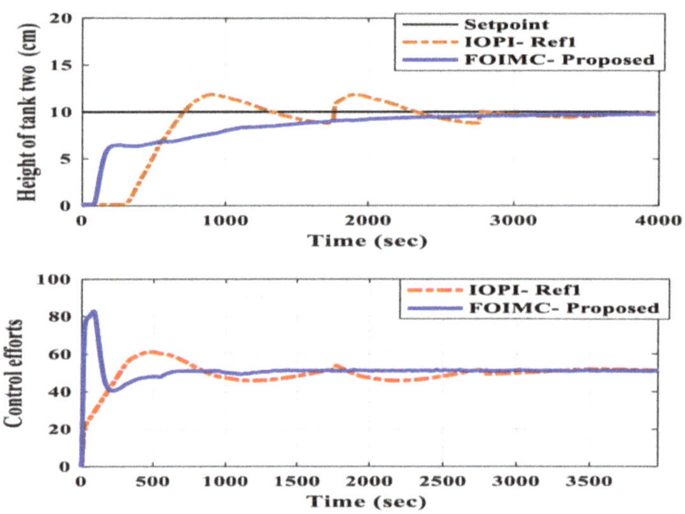

Figure 16. Regulatory responses- Experimental

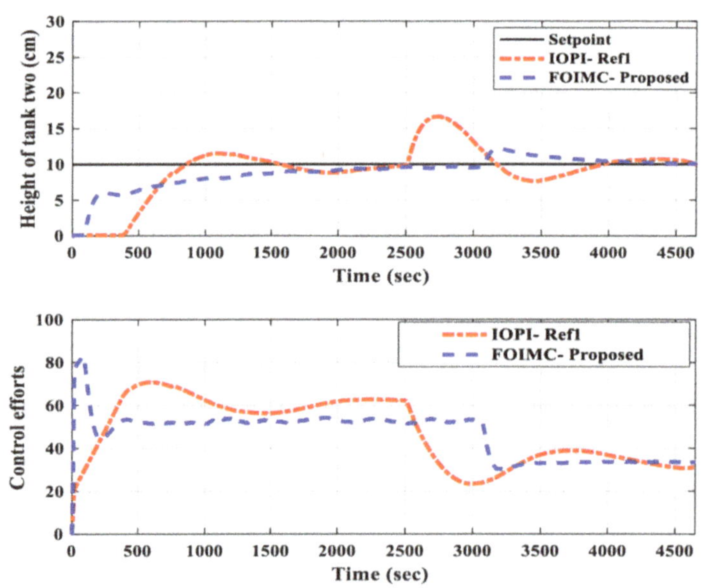

Figure 17. Robustness verification- Experimental

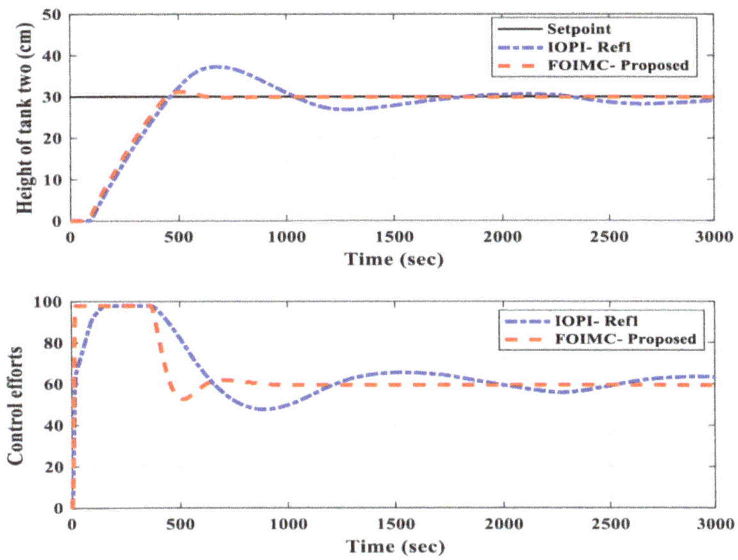

FUTURE RESEARCH DIRECTIONS

The latest trends in the tuning of the controller parameters are using AI techniques like, neural networks, fuzzy logic, adaptive neural fuzzy inference system (ANFIS), (or) new swarm intelligence-based algorithms like, marine predators' algorithm, ant colony optimization etc. In future a combination of these techniques might be used for the tuning of the controller parameters.

In line with the focus of the book's theme, "Design and Control Advances in Robotics", the present chapter considers the design and implementation of the advanced controller like FOIMC. The proposed controller tuning method is suitable for any piece-wise linear model of the nonlinear system's control. The PSO algorithm can be used for minimizing a variety of complex objective functions. The limitations of the present tuning method are it is easy to fall into local minima, low convergence rate in the iterative process. The implementation issues include the proper selection of sampling time period and applying identical disturbance for different controller implementation. Future research opportunities include the design and implementation on a variety of process control systems, robotics, and hybrid electric vehicle's power converters etc.

CONCLUSION

In this chapter, the fractional filter based FOIMC tuned by the PSO algorithm, is proposed for the level control of the coupled tank system. The performance of the PSO algorithm is compared with the GWO and WOA algorithms, and the PSO algorithm gave better time domain performance. The step responses, regulatory responses, and the servo responses of the proposed controller and the state of the art are compared. The time domain performance indices like the rise time, peak overshoot, and the IAE of the proposed controller are better than the state of the art. The control effort of the proposed controller is less than the other controllers. The control energy of the proposed controller is reduced by 3.14 when compared to Ref-1 and reduced by 78% when compared to Ref-2. The rise time of the proposed controller is reduced by 7.77% when compared to Ref-1. The settling time of the proposed controller is reduced by 53.86 when compared to Ref-1 and Ref-2. The peak overshoot is reduced by 11.5 time compared to Ref-1 and reduced by 6.5 times compared to Ref-2.

The stability of the closed loop system with the proposed controller is analyzed, and the system is found to be stable. The proposed controller showed better robustness to the change in the set point of the system. The proposed controller and the state of the art are implemented experimentally on the laboratory scale cylindrical coupled tank system. The experimental results are in tune with the simulation results and confirm the superiority of the proposed controller. In this chapter an attempt is made to compare the efficiency of the different swarm algorithms and the simulation results are verified on a hardware model. In the future, the fractional robust and the fractional optimal controllers can be developed for the level control of the coupled tank system along with an experimental implementation. The novelty of the present work is, for the first time a comparative study of latest swarm intelligence algorithms is done for the tuning of the controller parameters of the FOIMC.

REFERENCES

Abo-Hammour, Z. S., Alkayyali, Khasawneh, M. J., Hussam, I., Al Saadeh, Mohammed (2018). On the design of the integer and fractional PID controller using particle swarm optimization, In *Proceedings of International Conference on Fractional Differentiation and its Applications (ICFDA)*. 10.2139srn.3277839

Arya, P. P., & Chakrabarty, S. (2020). A robust internal model based fractional order controller for fractional order plus time delay processes. *IEEE Control Systems Letters*, *4*(4), 862–867. doi:10.1109/LCSYS.2020.2994606

Bettayab, M., & Rachid, M. (2014). IMC PID fractional order filter controllers design for integer order systems. *ISA Transactions*, *53*(5), 1620–1628. doi:10.1016/j.isatra.2014.05.007 PMID:24957276

Chekari, T., Rachid, M., & Bettayab, M. (2019). Improved internal model control proportional integral derivative fractional order multiloop controller design for non-integer multivariable systems. *Journal of Dynamic Systems, Measurement, and Control, ASME*, *141*(1), 011014. doi:10.1115/1.4041353

Dumlu, A., & Ayten, K. K. (2018). A combined control approach for industrial process systems using feed-forward and adaptive action based on second order sliding mode controller design, *Transactions of the Institute of Measurement and Control. Sage (Atlanta, Ga.)*, 1–12.

Fathy, A., Yousri, D., Rezk, H., Thanikanti, S., & Hasaniern, H. M. (2022). A robust fractional order PID controller based load frequency control using modified hunger games optimizer. *Energies*, *15*(361), 361. doi:10.3390/en15010361

Ghamari, S. M., Narm, H. G., & Mollaee, H. (2022). Fractional order fuzzy PID controller design on buck converter with antlion optimization algorithm. *IET Control Theory & Applications*, *16*(3), 340–352. doi:10.1049/cth2.12230

Gouta, H., Said, S. H., Turki, A., & Sahli, F. (2019). Experimental sensor less control for a coupled tank system using high gain adaptive observer and nonlinear generalized predictive strategy. *ISA Transactions*, *87*, 187–199. doi:10.1016/j.isatra.2018.11.046 PMID:30545770

Jain, S., Hote, Y. V., Dehuri, P., Mittal, D., & Siddhartha, V. (2020). Experimental validation of fractional order internal model control design on buck boost converter. *Measurement and Control*, 1–19.

Kennedy, J., & Eberhart, R. (1995). Particle swarm optimization, *Proceedings of the IEEE International Conference on Neural Networks*, *4*, 1942-1948. 10.1109/ICNN.1995.488968

Lakshmanprabu, S. K., & Banu, U. S. (2019). Adaptive multiloop IMC based PID controller tuning using bat optimization algorithm for two interacting conical tank process. *International Journal of Advanced Intelligence Paradigm*, *13*(3).

Li, P., & Zhu, G. (2019). IMC based PID control of servo motors with extended state observer. Mechtronics, Elsevier, 62(2019).

Li, Y. L., & Bing, J. Q. (2019). Analytical optimization of IMC-PID design based on performance/robustness trade-off tuning strategy for the modified Smith structure, *Journal of Low Frequency Noise, Vibration, and Active Control. Sage (Atlanta, Ga.)*, 1–16.

Mahapatro, S. R., Subudhi, B., & Ghosh, S. (2019). Design and experimental realization of robust decentralized PI controller for a coupled tank system. *ISA Transactions*, *89*, 158–168. doi:10.1016/j.isatra.2018.12.003 PMID:30594337

Mirjalili, S., & Lewis, A. (2016). The whale optimization algorithm. *Advances in Engineering Software, Elsevier*, *95*, 51–67. doi:10.1016/j.advengsoft.2016.01.008

Mirjalili, S., Mirjalili, S. M., & Lewis, A. (2014). Gray wolf optimizer. *Advances in Engineering Software*, *69*, 46–61. doi:10.1016/j.advengsoft.2013.12.007

Muresan, C. I., Birs, I., & De Keyser, R. (2021). An alternative design approach for fractional order internal model controllers for time delay systems. *Journal of Advanced Research*, *31*, 177–189. doi:10.1016/j.jare.2021.01.004 PMID:34194841

Nath, U. M., Dey, C., & Mudi, R. K. (2019). *Designing of IMC-PID controller for higher-order process based on model reduction method and fractional coefficient filter with real time verification*. Chemical Product and Process Modelling. doi:10.1515/cppm-2019-0089

Patel, H. R., & Shah, V. A. (2018). A framework for fault tolerant control for an interacting and non-interacting level control system using AI, In *Proceedings of the ICINCO*, (180-190).

Pati, A., & Negi, R. (2019). An optimized 2-DOF IMC-PID-based control scheme for real time magnetic levitation system. *International Journal of Automation and Control*, *13*(4), 413–439. doi:10.1504/IJAAC.2019.100468

Podlubny, I. (1994). *Fractional order systems and fractional controllers, Slovak Academy of Sciences, Institute of Experimental Physics,* (1st ed.). Springer.

Rayalla, R., Ambati, S. R., & Uday Bhaskar Babu, U. (2019). *Improved fractional filter IMC-PID controller design for enhanced performance of integrating plus time delay processes, Indian Chemical Engineer*. Taylor & Francis.

Rodrigues, T., Muhammad, A. Q., Zahoor, M., Kamga, C., Kwok, E., & Lee, M. (2019). Analytical design of constraint handling optimal two parameter internal model control for dead-time processes. *Korean Journal of Chemical Engineering*, *36*(3), 356–367. doi:10.100711814-018-0215-5

Sathish Kumar, K., Kirubakaran, V., Radhakrishnan, T. K. (2017). Real time modeling and control of three tank hybrid system. *Chemical Product and Process Modelling*, 13(1).

Saxena, S., & Birada, S. (2022). Fractional-Order IMC controller for high-order system using reduced-order modelling via Big-bang Big-crunch algorithm. *International Journal of Systems Science*, *53*(1), 168–181. doi:10.1080/00207721.2021.1942587

Singla, R., Agarwal, A., Kumar, V., Pachauri, N., & Verma, O. P. (2019). Industrial simulation of PID and modified M-PID controller for coupled tank system. *Advances in Signal Processing and Communications, Springer*, *526*, 421–428. doi:10.1007/978-981-13-2553-3_41

Tasoujian, S., Salavati, S., Franchek, M., & Grigoriadis, K. (2019). Robust IMC-PID and parameter varying control strategies for automated blood pressure regulation. *International Journal of Control and Automation, and Systems, Springer*, *17*(7), 1803–1813. doi:10.100712555-018-0631-7

Tavakoli, S., Sadeghi, J., Griffin, I., & Fleming, P. J. (2018). PI controller tuning for load disturbance rejection using constrained optimization. *International Journal of Dynamics and Control, Springer*, *6*(1), 188–199. doi:10.100740435-016-0290-6

Tepljakov, A., Petlenkov, E., & Belikov, J. (2011). FOMCON: A MATLAB toolbox for fractional order system identification and control. *IJAMCS*, *2*(2), 51–62.

Vadamalu, R. S., & Beidl, C. (2019). Adaptive internal model based harmonic control for torsional vibration reduction. *IEEE Transactions on Industrial Electronics*, *67*(4), 3024–3032. doi:10.1109/TIE.2019.2908579

Vavilala, S. K., & Vinopraba, T. (2018). Design of fractional order controllers for conical tank process, In *Proceedings of the IEEE International Conference ICETIETR*, Ernakulam, India. 10.1109/ICETIETR.2018.8529047

Vinopraba, T., Sivakumaran, N., Narayanan, S., & Radhakrishnan, T. K. (2012). Design of internal model control based fractional order PID controller. *Journal of Control Theory and Applications*, *10*(3), 297–302. doi:10.100711768-012-1044-4

Wang, T., Wang, H., Hu, H., Qing, J., & Wang, C. (2022). Lion swarm based optimization based tuning method for generalized predictive fractional-order PI to control the speed of brushless direct current motor. *IET Electric Power Applications*, *16*(8), 879–895. doi:10.1049/elp2.12201

Wen, S., Hu, X., Zhang, D., Zhang, B., Sheng, M., Lam, H. K., & Zhao, Y. (2020). Fractional order internal model control algorithm based on the force/position control structure of redundant actuation parallel robot. *International Journal of Advanced Robotic Systems*, *17*(1), 1–13. doi:10.1177/1729881419892143

Wu, Z., Yuan, J., Chen, Y., Li, D., Chen, Y. Q. (2019). Fractional order [PI] controller and Smith like predictor design for a class of high order systems, *Systems and Control*.

Yumuk, E., Guzelkaya, M., & Eksin, I. (2019). Analytical fractional PID controller design based on Bode's ideal transfer function plus time delay. *ISA Transactions*, *91*, 196–206. doi:10.1016/j.isatra.2019.01.034 PMID:30738584

ADDITIONAL READING

Conejoro, J. A., Franceschi, J., & Pico-Marco, E. (2022). Fractional Vs ordinary control system: What does the fractional derivative provide? *Mathematics*, *10*(2719), 2719. doi:10.3390/math10152719

Gad, A. G. (2022). Particle swarm optimization algorithm and its applications: A systematic review. *Advances of Computational Methods in Engineering*, *29*, 2531–2561.

Gupta, N., Kaur, M., Gupta, R. (2022). Ant colony optimization based optimal tuning of fractional order (FO) PID controller for controlling the speed of a DC motor. *Journal of Engineering Research.*

Kumar, D., Aryan, P., & Raja, G. L. (2022). Design of novel fractional order internal model controller based Smith predictor for integrating processes with large dead time. *Asia-Pacific Journal of Chemical Engineering*, *17*(1). doi:10.1002/apj.2724

Patel, V., Guha, D., Purwar, S. (2022). Optimized cascade fractional order 3-DOF controller for frequency regulation of a hybrid power system using marine predators algorithm, *International Journal of Numerical Modelling*. .

KEY TERMS AND DEFINITIONS

Integral Performance Indices: Indices like ISE, IAE, ISTE etc. which are used to assess the performance of the controllers.

Offline Optimization: The controller parameters obtained from the optimization are not changed during the controller operation in the plant.

Swarm Intelligence Algorithms: Algorithms which use a swarm of different biological species to optimize a function.

Chapter 9
Identifiability and Detectability of Lyapunov Exponents in Robotics

Nikolay Nikolayevich Karabutov
https://orcid.org/0000-0002-3706-7431
MIREA-Russian Technological University, Moscow, Russia

ABSTRACT

Lyapunov exponents (LE) are one of the most effective tools for analyzing the quality of systems in robotics. They are used to optimize the robot work, evaluate the quality of their work, and solve various tasks in control systems with robots. Calculating the full spectrum of LE is a complex problem. It relates to the identifiability recoverability and detectability of LE. The identifiability recoverability and detectability issues of Lyapunov exponents were not considered. This problem is relevant. The authors propose an approach to verify these characteristics for the linear dynamical system. It bases on the analysis of geometric frameworks (GF) that depends on the structural properties coefficient (SPC). The SPC reflects the change in Lyapunov exponents, and GF guarantees decision-making on the LE type. They obtain (1) conditions of fully detectable LE (these conditions correspond to the determination of an indicators complete set in the robot control system) and (2) s-detectability conditions with level v-no recoverability if the system contains no recoverable lineals.

INTRODUCTION

Lyapunov exponents (LE) widely used to analyze the qualitative behavior of robotic systems and solve various control problems. Usually, the largest Lyapunov exponents (LLE) determines. So, LE (Szaniewski et al., 2015) uses to select the controller structure for the control of an industrial robot. The stability improvement of the prototype bipedal robot (Yunping et al., 2013) is also based on the LLE estimation. LLE is the basis for the stabilization problem solution of a passive bipedal robot. Reddy and Ghoshal (2016) consider the stabilization problem of the robot described by nonlinear differential equations. LE changes analysis guarantees the compensation of chaotic robot movements. Schramm et al. (2020) con-

DOI: 10.4018/978-1-6684-5381-0.ch009

siders improving the quality of robot control based on the apple of neural networks. It shows that models trained on a small sample of data give chaotic or divergent behavior in some areas. Improving the model properties based on the LLE estimation. Influence compensation of vibrations on the control quality of flexible robots considers by Ban et al., (2021). Various approaches based on the frequency approach, LE identification, Poincaré maps analyzed. The effectiveness of the proposed approaches shown. Puzikov et al., (2019) consider the stabilization task of a robotic system (RS) in the neighbourhood of a steady state. The resulting faults RS detective based on the LE estimation. Some issues of LE analysis and application in robotic systems consider by Nasr et al., (2018), Azar and Vaidyanathan, (2014).

Most often, LE estimates base on time series analysis. It assumes that a priori information knows about the system structure. Sokos (2010) represented an overview of LLE computation for systems various classic. An estimating Lyapunov exponents algorithm (Gencay et al., (1992)) proposes an unknown dynamical system. It bases on the method for evaluating multidimensional networks with direct communication.

Several approaches were proposed for the evaluation of LE (see, for example, Sano and Suwada, 1985; Wolf, 1985). In the future, these approaches were changed (see, for example, Rosenstein et al. (1993); Balcerzak et al. (2018). Reconstruction of the systems attractor and time series analysis is the basis of the proposed algorithms. The basis of these algorithms is the Takens theorem (Takens, (1981). Similar ideas were implemented by Rosenstein et al. (1993).

Soloviev et al., (2020) presented an overview of other procedures and algorithms for LE estimation. Glyzin et al., (2005) propose a modification of the Benetti algorithm for calculating LLE. The proposed approach develops methods for reconstructing the system attractor.

Most approaches and methods of evaluating LE do not contain input impacts. As a rule, these algorithms and methods give the LLE estimate. Explain this by the class of systems and the difficulties of the Lyapunov exponents spectrum evaluating.

Note that the study of the qualitative behavior of the system is the main direction of research on LE. It does not affect other issues of identification theory. But the LE estimation and analysis provide the other identification problems solution for the system. The structural identification problem is one of the directions for using LE. Calculating the full spectrum of LE is the complicated problem. It links to the identification, recoverability and detectability of Lyapunov exponents. The problem of the LE identifiability, recoverability and detectability not considered.

The approach to estimating the LE identifiability of a robotic system proposes below. It is the basis for deciding on the LE detectability under a priori uncertainty. Karabutov (2015; 2018) proposes the approach on LE spectrum estimation. The approach analyzes the properties of geometric frameworks describing LE changes dynamics. The case of stationary and periodic RS considers.

PROBLEM STATEMENT

Consider the dynamic system describing the movement of a robot

$$\dot{X} = AX + BU, \quad Y = C^T X, \tag{1}$$

where $X \in \mathbb{R}^m$ is state vector, $U \in \mathbb{R}^k$, $Y \in \mathbb{R}^n$ are input and output, $A \in \mathbb{R}^{m\times m}$, $B \in \mathbb{R}^{m\times k}$, $C \in R^{n\times m}$. Matrix A is the Hurwitz matrix.

The measurement information has the form

$$I_o = \{Y(t), U(t), t \in J = [t_0, t_1]\}. \tag{2}$$

The solution of the system (1) writes as

$$X(t) = X(t_0, u, t) \tag{3}$$

where X is an operator defined by matrices A,B and initial conditions $X(t_0)$.

For $X(t)$, the system (1) solution has the form

$$X(t) = X_g(t) + X_q(t), \tag{4}$$

where $X_q(t)$ is particular solution with $U \in I_o$, $X_g(t)$ is general solution with $U=0$.

Denote the general solution of system (1) as $X_g(X_0,t)$.

Problem: determine estimates for $X_g(t)= X_g(X_q,X_0,t)$ on the set I_o and obtain (i) conditions for identifiability and detectability LE spectrum, (ii) the order of the system (1).

We propose the estimation $X_g(t)= X_g(X_q,X_0,t)$ method (Karabutov, (2015)) based on data processing a set I_o. Therefore, we believe that the evaluation $X_g(t)$ is known.

RECOVERABILITY AND DETECTABILITY OF RS

Consider the basic concepts from the general theory of systems (Kwakernaak et al., 1972). The issues of recoverability and detectability closely interrelated. The LE calculation base on obtaining estimates for system state variables (SV). The possibility of determining SV relates to the problem of system (1) recoverability.

Definition 1. Let $y(t, t_0, X_0, u)$ describe the output change of the system

$$\dot{X}(t) = A(t)X(t) + Bu(t), y(t)=C^T X(t), \tag{5}$$

With $X_0=X(t_0)$. Then the system called *fully recoverable* if for all t_1 exists a moment $t_0 \in (-\infty, t_1)$ such that from

$$y(t,t_0,X_0,u) = y(t,t_0,X_0',u) \tag{6}$$

$\forall u(t)$ and $t_0 \in [t_0,t_1]$ follow $X_0 = X_0'$.

From definition 1, it follows that if the system is fully recoverable and the system output observes until the moment t_1, then the moment $t_0<t_1$ exists such that the system state determines uniquely. Therefore, if $X(t_0)$ is known the value $X(t_1)$ can be determined.

Theorem 1. *System (5) is fully recoverable if and only if such a moment* $t_0 \in (-\infty, t_1)$ *exists for all* t_1, *that from equality*

$$y(t, t_0, X_0, 0) = 0 \tag{7}$$

fellow $X_0=0$.

Definition 2. Let A, C be matrices of dimension $n \times n$ and $l \times n$. Then the pair (A, C) is *fully recoverable* if the system

$$\dot{X}(t) = AX(t), \ \ y(t) = CX(t) \tag{8}$$

is fully recoverable.

There may be a situation when the system state is not fully recoverable. This case complicates the estimation of the system current state.

Let $y=Y$, $u=U$, $y \in \mathbb{R}$, $u \in \mathbb{R}$ in (1).

Definition 3. The space of non-recoverable states PN_n of the system (1) is a *linear subspace* of states X_0 if $y(t, t_0, X_0, 0)=0$ $t \geq t_0$.

Theorem 2. *The non-recoverable states* PN_n *subspace of the system (1) is the null space of the recoverability matrix*

$$D = \begin{pmatrix} C \\ CA \\ CA^2 \\ \vdots \\ CA^{n-1} \end{pmatrix}.$$

The state property belonged to PN_n: the system motion at $u=0$ converges to zero. Therefore, a non-recoverable state component does not give growth of error. A system with this property is detectable.

Definition 4. If the subspace of non-recoverable states belongs to the subspace of stable states, then system (1) is *detectable*.

Definition 5. A pair (A, C) is *detectable* if the system

$$\begin{cases} \dot{X}(t) = A(t)X(t), \\ y(t) = CX(t) \end{cases} \tag{9}$$

is detectable.

Theorem 3. *Any asymptotically stable system (1) is detectable. Any fully recoverable system (1) is detectable.*

Consider the system (8) and a transition matrix $\Phi(t, t_0)$ corresponding to it. System (8) is fully recoverable if and only if

$$M(t,t_1) = \int_t^{t_1} \Phi(\tau,t)C^T(\tau)C(\tau)\Phi(\tau,t_0)d\tau > 0 \ (\forall t_1) \text{ and } (-\infty<t_0<t_1). \tag{10}$$

If

(a) $\alpha_0 I \le M(t_1 - \sigma, t_1) \le \alpha_1 I \quad \forall t_1, (\alpha_0,\alpha_1)>0,$
(b) $\beta_0 I \le \Phi^T(t_1 - \sigma, t_1)M(t_1 - \sigma, t_1)\Phi(t_1 - \sigma, t_1) \le \alpha_1 I \quad \forall t_1, (\alpha_0,\alpha_1,\beta_0)>0,$

then the system (8) is uniformly fully recoverable.

Uniform recoverability guarantees the identification of the state at a given time interval. Uniform full recoverability of system (1) follows from full recoverability.

ESTIMATION

$X_g(t)$

Form the set $\{X_g(t)\} = \{X(t)\} \backslash \{X_q(t)\}$ to evaluate LE. Apply the approach (Karabutov, 2015; 2018). Present I_o as

$$I_o = I_o^q(J_q) \cup I_o^g(J_g), \tag{11}$$

where $J_q \cup J_g = J \subseteq \mathbb{R}$; I_o^q, I_o^g are sets contained information about X_q and X_g.

Statement 1. (Karabutov, 2015). *The model*

$$\hat{X}_q(t) = \hat{A}_q W(t) \quad \forall t \in J_q, \tag{12}$$

applies to the identification of $X_g(t)$ *on the set* I_o^q, *where* $\hat{A}_q \in \mathbb{R}^{2\times 2}$ *is matrix of model parameters,* $W = [uu']^T$.

Determine the evaluate $\widehat{X}_q(t)$ for particular solution of the system (1) on the set I_o^g. Next, find the estimate of the general solution

$$\widehat{X}_g(t) = X(t) - \widehat{X}_q(t) \quad \forall t \in J_g, \tag{13}$$

where $\widehat{X}_g(t) = \left[\hat{y}_g(t)\dot{\hat{y}}_g(t)\right]^T$.

LE. STRUCTURAL PROPERTIES COEFFICIENT

Define LE (Lyapunov, 1992) for the real function $h(t)$ as

$$\chi[h] = \overline{\lim_{t\to\infty}} \frac{\ln|h(t)|}{t}, \tag{14}$$

where $\overline{\lim\limits_{t\to\infty}}$ is upper limit.

For the stationary system (1), LE $\chi_i \left(i = \overline{1,m}\right)$ coincide with the eigenvalues λ_i real part of the matrix A.

Let the general solution $X_g(t)$ $\forall t \in J_g$ estimate known for the system (1) and the system (1) is stable. Therefore, the system (1) is recoverable and discoverable. Apply (14) to $\hat{y}_g(t)$

$$\chi\left[\hat{y}_g\right] = \overline{\lim_{t\to\bar{t}}} \frac{\ln\left|\hat{y}_g(t)\right|}{t}, \tag{15}$$

where $\bar{t} \in J_g$ is the maximum value (upper bound) on the interval $J_g \subset J$.

(15) is the largest (maximum) LE. If the limit (14) exists, then $\chi\left[\hat{y}_g\right]$ is the estimate of the matrix A maximum eigenvalue. Therefore, $\chi\left[\hat{y}_g\right]$ is an indicator of the stability degree for the system (1). If m=2, then we obtain for $\dot{\hat{y}}_g$

$$\chi\left[\dot{\hat{y}}_g\right] = \overline{\lim_{t\to\bar{t}}} \frac{\ln\left|\dot{\hat{y}}_g\right|}{t}. \tag{16}$$

Perron bottom index (Izobov, 2006) define as

$$\eta[h] = \underset{t\to\infty}{\underline{lim}} \frac{\ln|h(t)|}{t}, \tag{17}$$

where $\underset{t\to\infty}{\underline{lim}}$ is the bottom limit.

Karabutov (2009; 2018) proposes using LE to solve identification problems. The approach bases on the analysis of the structural properties coefficient (SPC) (Karabutov, 2015)

$$k_s(t,\rho) = \frac{\rho(\hat{y}_g)}{t} \tag{18}$$

where $\rho(\hat{y}_g) = \rho_g = \ln|\hat{y}_g(t)| \quad \forall t \in \bar{J}_g \subset J_g, \bar{J}_g = [t_0, \bar{t}]$ determine from (15). The approach to estimation LE (see, below) bases on the GF analysis.

FRAMEWORKS FOR ESTIMATE LE

Consider sets

$$I_{k_s} = \left\{k_s\left(t,\rho\left(\hat{y}_g(t)\right)\right), t \in \bar{J}_g\right\},\ I_{k_s^{'}} = \left\{k_s\left(t,\rho\left(\dot{\hat{y}}_g(t)\right)\right), t \in \bar{J}_g\right\}. \tag{19}$$

Define the mapping (framework) $S_{k_{s,\rho}} \to I_{k_s} \times I_{k_s^{'}}$ on $I_{k_s}, I_{k_s^{'}}$. The framework $S_{k_{s,\rho}}$ reflects changes in dynamics of indicators depending on LE. Consider the function on $I_{k_s^{'}}$

$$\Delta k_s^{'}(t) = k_s\left(t,\rho\left(\dot{\hat{y}}_g(t+\tau)\right)\right) - k_s\left(t,\rho\left(\dot{\hat{y}}_g(t)\right)\right), \tag{20}$$

where $\tau>0$.

Form the set

$$I_{\Delta k_s^{'}} = \left\{\Delta k_s\left(t,\rho\left(\dot{\hat{y}}_g(t)\right)\right), t \in \bar{J}_g\right\} \tag{21}$$

and introduce the frameworks

$$SK_{\Delta k'_{s,\rho}} \to I_{k_{s,\rho}} \times I_{\Delta k'_{s,\rho}},\ LSK_{\Delta k'_{s,\rho}} \to I_{k_{s,\rho}} \times B\left(I_{\Delta k'_{s,\rho}}\right) \tag{22}$$

where $B\left(I_{\Delta k'_{s,\rho}}\right) \subset \{-1;1\}$. Define elements of the binary set $B\left(I_{\Delta k'_{s,\rho}}\right)$ as

$$b(t) = \begin{cases} 1, if\ \Delta k'_s(t) \geq 0, \\ -1, if\ \Delta k'_s(t) < 0, \end{cases} \tag{23}$$

where $t \in \bar{J}_g$.

Remark 1. Choose the upper limit boundary in (15) based on changes in $S_{k_{s,\rho}}$.

Remark 2. Determine the choice of function $b(t)$ values range by the convenience of its graphical analysis.

Consider the frameworks $SK_{k^i_{s,\rho}} \to I_{k_s} \times I_{k^i_s}$ and $LSK_{\Delta k'_{s,\rho}}$, where i is the i -th derivative of $\hat{y}_g(t)$,

$$I_{k^i_s} = \left\{ k_s\left(t, \rho\left(\hat{y}^{(i)}_g(t)\right)\right), t \in \bar{J}_g \right\}. \tag{24}$$

LLE $\chi_l\left[\hat{y}_g\right]$ locate on the left side of the framework $SK_{k^i_{s,\rho}}$. All other $\chi_i\left[\hat{y}_g\right]$ coincide with Perron bottom indices $\eta_i\left[\hat{y}_g\right]$. They locate to the right of $\chi_l\left[\hat{y}_g\right]$. Indices $\chi_i\left[\hat{y}_g\right]$ correspond to local minima on $SK_{k^i_{s,\rho}}$. Minima $SK_{k^i_{s,\rho}}$ coincides with the change in the sign of the framework $LSK_{\Delta k'_{s,\rho}}$. Hence, we obtain the set $M^m_{LE} = \left\{\hat{\lambda}_1, \hat{\lambda}_2, \ldots, \hat{\lambda}_m\right\}$ of eigenvalues for the system (1). If the system (1) is fully L -detectable (see below), then we obtain the full spectrum for LE, which corresponds to the full space of lineals $L^m(t)$.

IDENTIFIABILITY AND DETECTABILITY OF LE

Stationary System

Consider the input $u \in PE_\alpha$, where PE_α is the constancy excitation property

$$PE_\alpha\text{: } u^2(t) \geq \alpha \tag{25}$$

fair for $\exists \alpha > 0$ and $\forall t \geq t_0$ at some interval $T>0$. With such an input, the phase portrait of the system is closed, the system (1) is identifiable (Karabutov, 2018), and the model (10) identified.

Definition 6 (Smirnov, 1968). A set $\mathbb{F}$ of elements from L_2 is *lineal* if from the fact that functions $\varphi(p) \in \mathbb{F}$, $\phi(p) \in \mathbb{F}$, we have $c\varphi(p) \in \mathbb{F}$ for any c and $\left(\varphi(p) + \varphi(p)\right) \in \mathbb{F}$.

Definition 7. If Lyapunov exponents set detects based on lineals, then system (1) is L *-detectable.*

We present a modification of theorem 1 (Karabutov, 2015) for the system (1) with simple eigenvalues.

Theorem 4. *Lyapunov exponents are LP -detectable or fully detectable, and system (1) has the order m if the function b(t) changes sign m–1 times on the interval* $\left[t_0,t^*\right]\subset \bar{J}_g \quad \left(t^*\leq \bar{t}\right)$.

Proof of theorem 4. Consider the system (1) with the stable matrix A. Let the roots be simple and ordered as $\lambda_1>\lambda_2>\ldots>\lambda_m$. According to Bylov et al., (1966), the solution of system (1) is the lineal $L^m(t)$, which is part of the linear space $L^m(t)$. Since the system is linear, then (Bylov et al., 1966)

$$0\equiv L^{l_0}\left(t\right)\subset L^{l_1}\left(t\right)\subset\ldots\subset L^{l_n}\left(t\right)\equiv L^m\left(t\right), \tag{26}$$

Call the expression (26) a pyramid L^m, where l_i is the dimension of the lineal $L^{l_i}\left(t\right)$. Denote the difference of two adjacent lineals

$$L_k=L^{l_0}\left(t\right)\subset L^{l_k}\left(t\right)\setminus L^{l_{k-1}}\left(t\right) \tag{27}$$

as pyramid step. The stage weight is $m_k=l_k-l_{k-1}$.

l_i corresponds to the number of solutions that contain the lineal $L^{l_k}\left(t\right)$. The index λ_m belongs to $L^{l_1}\left(t\right)$ at the initial time interval. As t increases, the solution moves to the pyramid stage $L^{l_2}\left(t\right)$. The indicator λ_{m-1} dominates on $L^{l_2}\left(t\right)$. Changing properties of function $b(t)$ reflects this. Next, apply induction along on steps of the pyramid we obtain the entire space $L^m(t)$. ■

Corollary of theorem 4. *If the change number of function b(t) signs is less m–1, then the pyramid* L^m *contains steps that are undetectable.*

Proof of corollary from theorem 4. Let the number of signs shift υ of the function $b(t)$ be $\upsilon<m-1$. Therefore, (26) contains lineals $NL_k(k)$, $k\geq1$. These lineals $NL_k(t)$ belong to L_m implicitly and are undetectable (non-computable). System (1) is stable. Therefore, these sets $\{NL_k\}$ exist "implicitly" (in the sense of computation). Consequently, the observed lineal pyramid L_υ has υ observed steps and $L_\upsilon=L^m\setminus\left\{NL_{k|k>0}\right\}$. ■

Denote $\sigma=m-1-\upsilon$. We receive statement from the corollary to theorem 4.

Definition 8. System (1) is σ *-detectable* with an υ -non-recoverability level if υ lineals are undetectable.

Remark 3. We interpret LP -detectability and recoverability as the possibility of calculating LE.

Remark 4. If the system (1) is LP -detectable, then the set M_{LE}^m of Lyapunov exponents elements correspond to the steps of the pyramid L^m.

Theorem 5. If *the system (1) is stable and recoverable, and the framework* $LSK_{\Delta k'_{s,\rho}}$ *changes sign m–1 times, then (i) Lyapunov exponents are LP -detectable, (ii) the structure* $SK_{k^i_{s,\rho}}$ *contains the set* M^m_{LE}.

The proof of theorem 5 follows from the stability and recoverability of the system (1). It is analogous to the proof of theorem 4.

Corollary of theorem 5. *If the change in signs of the framework* $LSK_{\Delta k'_{s,\rho}}$ *is less m–1, then pyramid* L^m *contains undetectable steps.*

System With Periodic Coefficients

Consider the system

$$\dot{X} = A(t)X + Bu, \;\; y = C^T X, \tag{28}$$

where $X \in \mathbb{R}^m$ is state vector, $u \in \mathbb{R}, y \in \mathbb{R}$ are input and output, $A(t) \in \mathbb{R}^{m\times m}, B \in \mathbb{R}^m, C \in \mathbb{R}^m$.

Assumptions.

A1. $A(t)$ is the continuous Frobenius matrix and

$$\|A(t)\| \leq \alpha_A \tag{29}$$

where $\alpha_A>0$, $\|\bullet\|$ is matrix norm.

A2. $A(t)$ is almost periodic, i.e. from any sequence (Bylov et al., 1966)

$$A_i(t) = A_i(t-\tau_i), \tag{30}$$

we can choose the subsequence, which converges uniformly to the almost periodic matrix $\bar{A}(t)$ for $\forall t \geq t_0$.

A3. $A(t)$ is Hurwitz matrix for almost $\forall t \geq t_0$.

Apply the approach (see the section "Estimation $X_g(t)$") to system (28) and obtain the set $\{X_g(t)\}$. Consider the class of almost periodic functions by Demidovich (1967).

Definition 9. The numerical set $\Xi=\{\xi\}$ is *relatively dense* on the real axis $-\infty<x<\infty$ if number $l>0$ exists such that each segment $a \leq x \leq a+l$ contains at least one element of our set, i.e. we have at any a

$$[a, a+l] \cap \Xi \neq 0. \tag{31}$$

Definition 10. If the inequality

$$\left|f\left(x+T_f\right)-f\left(x\right)\right|\langle\delta,\delta\rangle 0 \tag{32}$$

fulfil for any $x\in(-\infty,\infty)$, then $T=T_f(\delta)$ is *almost the period* (or δ *-almost period,* or δ *-shifting*) of the function $f(x)$ accurate within δ.

Definition 11. If a relatively dense set of almost periods T_f for the function $f(x)$ exists with accuracy within δ, then the function $f(x)$ is *almost periodic in Bohr sense (BF -function)*

$$|f(x+T_f)-f(x)|<\delta \text{ by } x\in(-\infty,\infty). \tag{33}$$

Elements of the set $\{X_g(t)\}$ and, therefore $y_g(t)$, belong to the class of exponential-sinusoidal functions. Hence, the condition (32) may not be fulfilling. Therefore, ensure that $\hat{y}_g(t)$ belongs to the class of BF -functions. Perform these steps (Karabutov, 2018).

Consider some point $t\in R$ and its neighborhood O_t. Determine the average value $\hat{y}_g(t)$ $\forall t\in O_t$

$$\alpha=\bar{\hat{y}}_{g,O_t}=\frac{1}{N_t}\sum_i\hat{y}_i, \tag{34}$$

where N_t is the number of points on O_t, $t_i\in O_t$ is the current coverage of interval O_t with step τ.

Obtain for $\forall t\in O_{t+T_{\hat{y}_g}}$

$$\pi=\bar{\hat{y}}_{g,O_{t+T_{\hat{y}_g}}}=\frac{1}{N_{t+T_{\hat{y}_g}}}\sum_i\hat{y}_i, \tag{35}$$

Definition 12 (Karabutov, 2018). A function $\hat{y}_g(t)\in(0,\infty)$ is α_π *-almost periodic in the Bohr sense* ($BF_{\alpha\pi}$-function) if a relatively dense set of almost periods T_f exists for the function $\hat{y}_g(t)$ with accuracy within δ for any $\delta>0$; i.e. a positive number $l=l(\delta)$ exists such that any segment $[a, a+l]$ contains at least one number T_f, for which

$$\left|\frac{\hat{y}_g\left(t+T_f\right)}{\pi}-\frac{\hat{y}_g\left(t\right)}{\alpha}\right|<\delta \quad \text{by } t\in[0,\infty). \tag{36}$$

Consider the framework $SK_{\Delta k'_{s,\rho}}$ defined on $I_{k_{s,\rho}}\times I_{\Delta k'_{s,\rho}}$ and described by the function $f_{sk}(t)$: $k_s\to\Delta k_s$. $k_s(t)$ is a $BF_{\alpha\pi}$-function. Therefore, $f_{sk}(t)$ contains areas D_{sk} that have sharply varying amplitude.

Definition 13 (Karabutov, 2015). The domain D_{sk} is $\alpha\pi$ *-domain* of the function f_{sk} on the interval $J_{sk}=[t, t+T]$ ($T>0$) if it corresponds to the change $BF_{\alpha\pi}$-function $k_s(t)$ on this interval.

Theorem 6. *(Karabutov, 2018) If the system (28) is stable and contains prime eigenvalues, then frameworks* $SK_{k^i_{s,\rho}}, i=\overline{1,m}$ *contain information about Lyapunov exponents.*

Theorem 7. *If the system (28) is stable and recoverable, and the function* $f_{sk}(t)$ *contains at least m of domain* D_{sk} *on the interval* $\left[t_0,t^*\right]\subset \bar{J}_g$ $\left(t^*\leq \bar{t}\right)$, *then the system (28) has the order m and is LP -identifiable (LP -detectable).*

The proof of theorem 7 follows from theorem 6.

The system (28) is stable and recoverable. According to theorem 7, system (28) is identifiable on the set $\left\{SK_{k^i_{s,\rho}}\right\}$. As shown by Karabutov (2015), the local minima located on the framework $SK_{k^i_{s,\rho}}$ coincides with domains D^i_{sk} of the framework $SK_{\Delta k^i_{s,\rho}}$. As the result of the analysis D^i_{sk}, we obtain a set M_{LE} contained LE estimates of the system (28).

The cardinal number M_{LE} does not match the LE number. M_{LE} characterizes the available set of lineals for the system (28).

Denote the upper bound of the smallest $\eta_i\left[\hat{y}_g\right]$ as κ_m.

Theorem 8. *If (a) assumptions A1-A3 fulfils for the system (28), (b) the system (28) is recoverable, (c) set* $\left\{k^i_{s,\rho}(t)\right\}$ *elements are* $BF_{\alpha\pi}$*-functions, (d) the framework* $SK_{\Delta k^i_{s,\rho}}$ *contains a point at which the sign of the derivative changes, then the indicator* κ_m *is detectable.*

Proof of theorem 8. Since assumptions A1-A3 fulfil, the system is stable and recoverable. According to theorem 6, LE is identifiable (detectable) on the framework $SK_{k^i_{s,\rho}}$. $k^i_{s,\rho,j}(t)\in\left\{k^i_{s,\rho}(t)\right\}$ (j>0) is the $BF_{\alpha\pi}$-function. Therefore, the framework $SK_{k^i_{s,\rho}}$ has local minima the global minimum of which is the upper estimate κ_m for the least $\eta_i\left[\hat{y}_g\right]$. This confirms the detectability $\eta_i\left[\hat{y}_g\right]$. ■

Remark 5. We make the decision based on the analysis of several frameworks $SK_{k^i_{s,\rho}}$ ($i\geq 1$). Therefore, κ_m is the upper bound of $\kappa_{m,i}$.

Remark 6. The approach based on the analysis of frameworks $SK_{k^i_{s,\rho}}$ gives the estimate of the minimal $\eta_i\left[\hat{y}_g\right]$. It is the main advantage of this approach. If the structure $SK_{\Delta k^i_{s,\rho}}$ contains a single sharp change in the value, then it is a sign $\eta_m[\bullet]$. As $|\eta_m[\bullet]| > |\chi_i[\bullet]|$, where $i=\overline{1,m-1}$, then $\eta_m[\bullet]$ is equal to κ_m, and it is an upper bound of the least LE. It is because the corresponding lineal Lm(t) has a minimal definitional domain. The function $\hat{y}_g(t)$ is not $\alpha\pi$ -almost periodic on $L^m(t)$, so its parameters α,π decrease rapidly. Therefore, condition (32) not fulfills. So, the lineal L^m(t) contains only one value that corresponds to κ_m.

Theorem 9. *Let (a) assumptions A1-A3 fulfil for the system (28), (b) the system (28) is recoverable, (c) set* $\left\{k^i_{s,\rho}(t)\right\}$ *elements are* $BF_{\alpha\pi}$*-functions, (d) the structure* $SK_{\Delta k^i_{s,\rho}}$ *contains not fewer m of regions*

D_{sk}^{v} (v≤m), corresponding to local minima on the structure $SK_{k_{s,\rho}^{i}}$. So the set M_{LE} is LP -detectable or fully detectable.

Proof of theorem 9. The system (28) is stable and recoverable, and the structure $SK_{\Delta k_{s,\rho}^{i}}$ contains m regions D_{sk}^{v}. $SK_{\Delta k_{s,\rho}^{i}}$ is a framework obtained from $SK_{k_{s,\rho}^{i}}$. Therefore, local minima correspond to regions D_{sk}^{v} on $SK_{k_{s,\rho}^{i}}$ and reflect the position of the LE. It follows from theorems 7 and 8. Therefore, the set M_{LE} is identifiable or LP -detectable. ■

Remark 7. Since eigenvalues $\lambda_i(t)$ of the matrix A are periodic functions lineals $L^i(t)$ and $L^{i+1}(t)$ can intersect. This case gives an infinite LE spectrum.

The system (28) is non-stationary. Therefore, estimate the area to which the set M_{LE} belongs. Explain this by the power of the set M_{LE} (see remark 7). Note that this area is bounded by the exponent κ_m from below.

Specify the accessible region for M_{LE} and the number determined the mobility of the largest LE. From below, this area limited by the indicator κ_m. Use the time range $\left[0, \bar{t}\right]$ to select desired parameters, where $\bar{t}$ determines according to (15). Select the change χ_1 area based on the analysis of the framework $SK_{\Delta k_{s,\rho}^{i}}$. Since the area $D_{sk}^{1} \subset SK_{\Delta k_{s,\rho}^{i}}$ is an indicator of the occurrence χ_1, then the fragment $V_{k_{s,\rho}^{i}}^{1}$, which changes on the interval $J_{k_{s,\rho}^{i}}^{1}$, corresponds to D_{sk}^{1} on $SK_{k_{s,\rho}^{i}}$. Therefore, $\chi_1 \in J_{k_{s,\rho}^{i}}^{1}$. Then the permissible mobility boundary for χ_1 is (Karabutov, 2018)

$$\chi_1 \le \sup J_{k_{s,\rho}^{i}}^{1}. \tag{37}$$

So, fair

Theorem 10. *If conditions of theorem 9 hold true, then the permissible mobility boundary for χ_1 satisfies condition (37).*

Consider the criterion for evaluating the LP -detectability of the system (28). The choice of adequate indicators is a complex problem under uncertainty. Such indicators (Bylov, (1966)) as strength and reachability are not verifiable under uncertainty. The set M_{LE} and the mobility boundary (37) are results of the LP -detectability. Therefore, use the results of the structures properties analysis. Apply the approach to confirm the detectability of the set M_{LE}. It bases on checking the χ -adequacy of obtained LE estimates (Karabutov, (2018)).

Consider the framework $S_{\hat{y}_g,\hat{\dot{y}}_g}$ described by the function $f_{\hat{y}\hat{\dot{y}}}: \hat{y}_g \to \hat{\dot{y}}_g$ in the space $R_y = \left(\hat{y}_g, \hat{\dot{y}}_g\right)$. The system (28) satisfies A1-A3 assumptions. Therefore, $S_{\hat{y}_g,\hat{\dot{y}}_g}$ contains regions that reflect the $\alpha\pi$ -almost periodic behavior of the system. Next, consider frameworks $SL_{\Delta k_{s,\rho}^{i}}$ and $SL_{k_{s,\rho}^{i}}$ described by functions

$$f_{SLk^i_{s,\rho}} : \hat{y} \to k^i_{s,\rho},\ f_{SL\Delta k^i_{s,\rho}} : \hat{y} \to \Delta k^i_{s,\rho}. \tag{38}$$

Definition 14. Lyapunov exponents χ_i estimates are χ*-adequate* in the space R if definition domains χ_i coincide with domains of the $\alpha\pi$-almost periodicity of the framework $S_{\hat{y}_g,\hat{\dot{y}}_g}$.

Introduce fragments $D^j_{sl} \subset SL_{\Delta k^i_{s,\rho}}$ ($j \geq 1$) in space R_y. Denote the domain of definition D^j_{sl} as dom D^j_{sl}. The following modification of theorem 3 (Karabutov, 2018) is valid.

Theorem 11. *Let (i) system (28) be stable and recoverable, (ii) the set M_{LE} is LP-detectable, (iii) definition domains D^j_{sl} of framework $SL_{\Delta k^i_{s,\rho}}$ coincide with domains of the $\alpha\pi$-almost periodicity $S_{\hat{y}_g,\hat{\dot{y}}_g}$. Then elements estimate of set ML_E are χ-adequate to domains of $\alpha\pi$-almost periodicity $S_{\hat{y}_g,\hat{\dot{y}}_g}$.*

Proof of theorem 11. According to theorem 7, the function $f_{sk}(t)$ contains m of domains D^j_{sk}. Structures $SK_{\Delta k^i_{s,\rho}}$ and $SL_{\Delta k^i_{s,\rho}}$ have identical value ranges. This statement is true for D^j_{sk} and D^j_{sl}. Areas D^j_{sk}, D^j_{sl} determine the LE change. From the equality of values areas D^j_{sk}, D^j_{sl}, obtain the congruence of the domain of definition for D^j_{sk}, D^j_{sl}. If definition domain of the fragment $D^j_{sl} \subset SL_{\Delta k^i_{s,\rho}}$ and the region $\alpha\pi$-almost periodicity $S_{\hat{y},\hat{\dot{y}}}$ coincide, then we obtain, taking into account the dependence between $f_{SLk^i_{s,\rho}}$ and $f_{SL\Delta k^i_{s,\rho}}$, that some subset $\tilde{M}_{LE} \subset M_{LE}$ covers the corresponding $\alpha\pi$-almost periodic region $S_{\hat{y}_g,\hat{\dot{y}}_g}$. Consequently, estimates χ_i are χ-adequate in space $R_y = \left(\hat{y}_g, \hat{\dot{y}}_g\right)$. ■

So, we propose the approach and indicators for the detectability estimation of robotic systems.

Remark 8. If we analyze LE for non-stationary RS, then the $BF_{\alpha\pi}$-function application gives the solution to this problem.

Examples

Examples 1. Consider the system whose phase portrait is shown in Figure 1. We have information set (4). Input is $u(t)=5+2\sin(0.2\pi t)$. As follows from Figure 1, oscillations occur in the system.

Apply model (12) on the interval [5; 40] s and obtain the general solution of the system. The coefficient of determination for the model is 0.95. Synthesize the model (12) and obtain estimates $\hat{y}_g(t)$ and $\hat{\dot{y}}_g(t)$. The model parameters vector for determining $\hat{\dot{y}}_g(t)$ is [-0.17; -0.89; 0.27]^T. Show the system phase portrait in Figure 2. It confirms the presence of periodic changes in the system.

As it follows from the figures, the system is stable and recoverable. Therefore, the system LE must be detectable. We present the results confirming the LP-detectability of Lyapunov exponents in Figure 3. Presented structures $SK_{\Delta k^i_{s,\rho}}$ and $S_{k^1_{s,\rho}}$ have the form (17),

$$k^1_s(t,\rho) = \frac{\rho\left(\hat{\dot{y}}_g(t)\right)}{t}. \tag{39}$$

Figure 1. System phase portrait

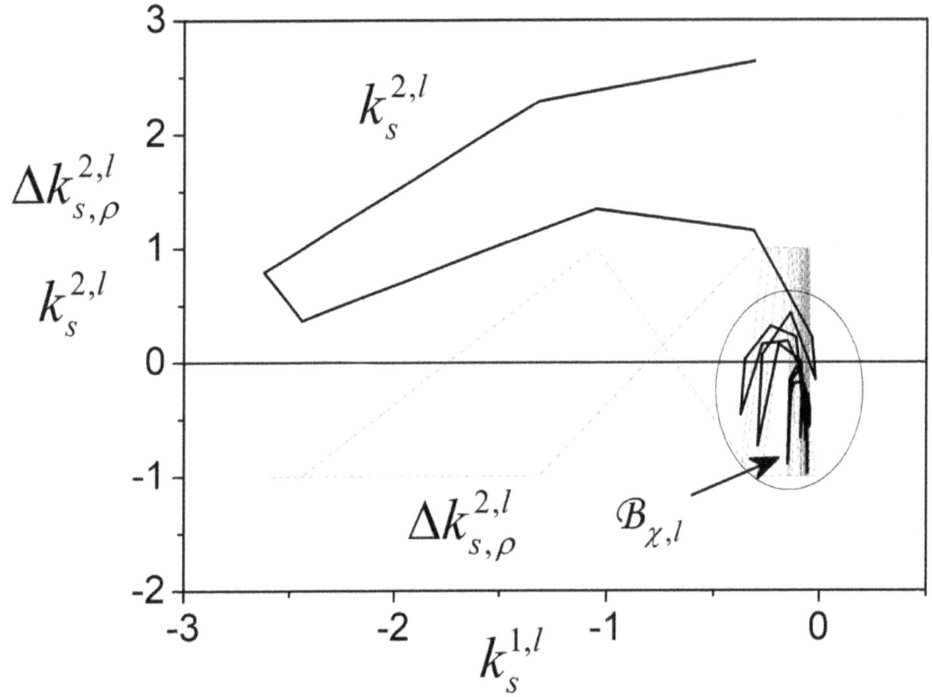

Figure 2. System phase portrait in space $R_y = \left(\hat{y}_g, \hat{\dot{y}}_g\right)$

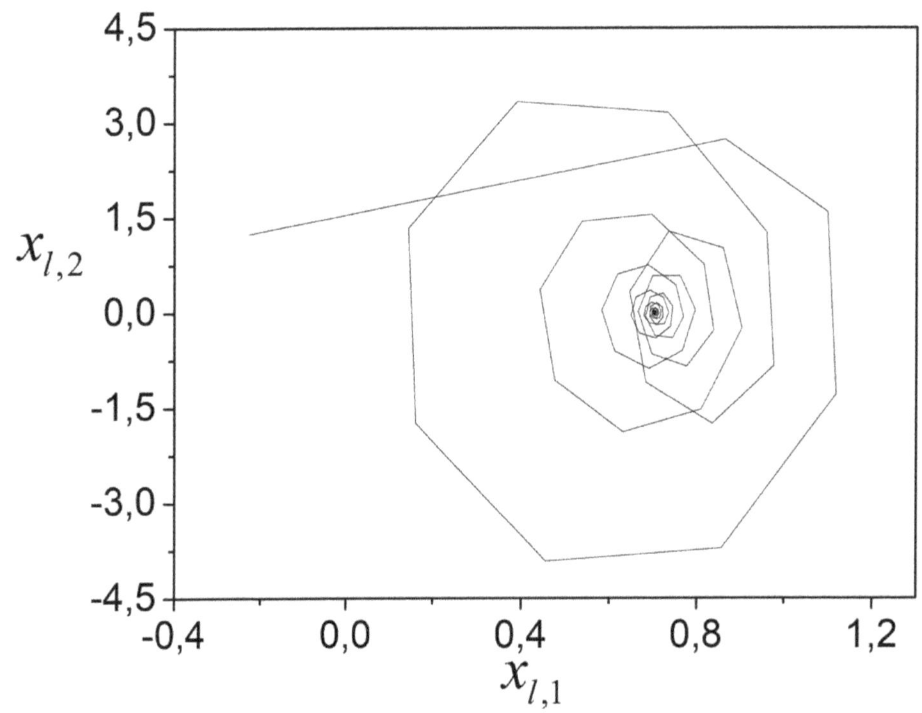

Figure 3. Lyapunov exponents sets

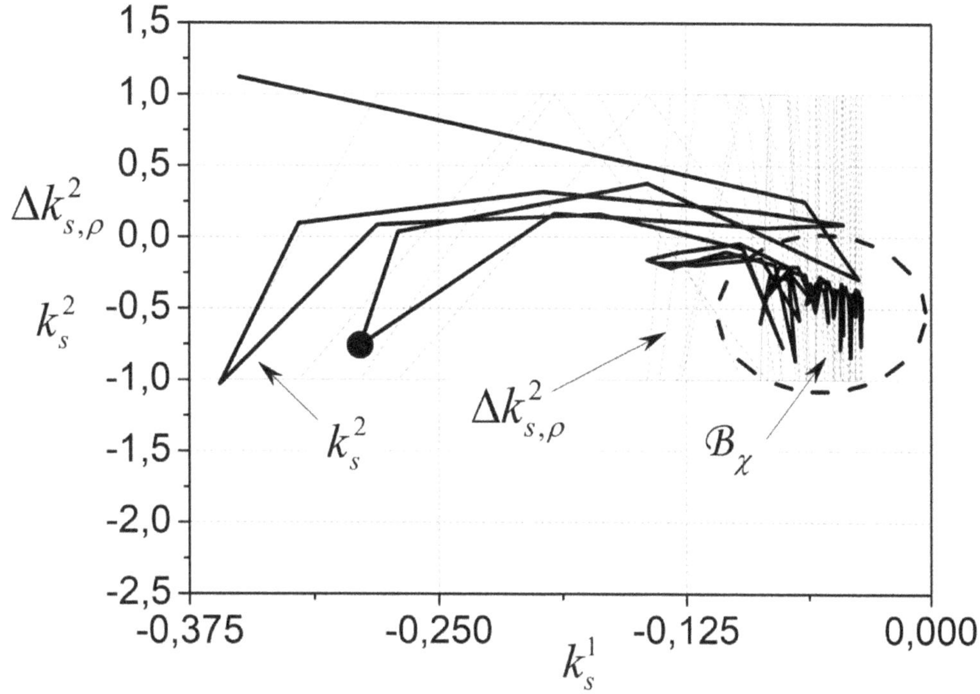

Perform the $SK_{\Delta k'_{s,\rho}}$ analysis and obtain M_{LE}= {-1.8, [-1.21, -0.88]}. Apply theorem 7 and receive the estimate for the system order. It is equal to 2. The upper bound for the least LE is κ_m= -1.8. The permissible mobility boundary of the largest index χ_1 is –0.8. We have another observed set M_{LE}^1 on the interval $k_s \in$[-0.2, -0.13]. This result confirms the conclusion given in remark 7. Figure 4 show the LE χ-adequacy estimation.

Return to the original system (28). We consider a second-order system with a matrix

$$A(t)=\begin{bmatrix} 0 & 1 \\ a_1(t) & a_2(t) \end{bmatrix}, a_1(t)= -3+0.2\sin(0.02\pi t),\ a_2(t)= -4+0.3\sin(0.04\pi t). \quad (40)$$

The eigenvalues of the matrix A: $\lambda_1(t)\in$[-1.325, -0.819], $\lambda_1(t)\in$[-2.37, -3.48]. Simulation results confirm the LE detectability for the system (28).

Examples 2. We present the order identification results for system (1) with matrix and spectrum

$$A=\begin{bmatrix} 0 & 1 & 0 \\ 0 & 0 & 1 \\ -6 & -11 & -6 \end{bmatrix}, \sigma(A)=(-1; -2; -3). \quad (41)$$

in Figure 5.

Figure 4. Results of χ *-adequacy estimation*

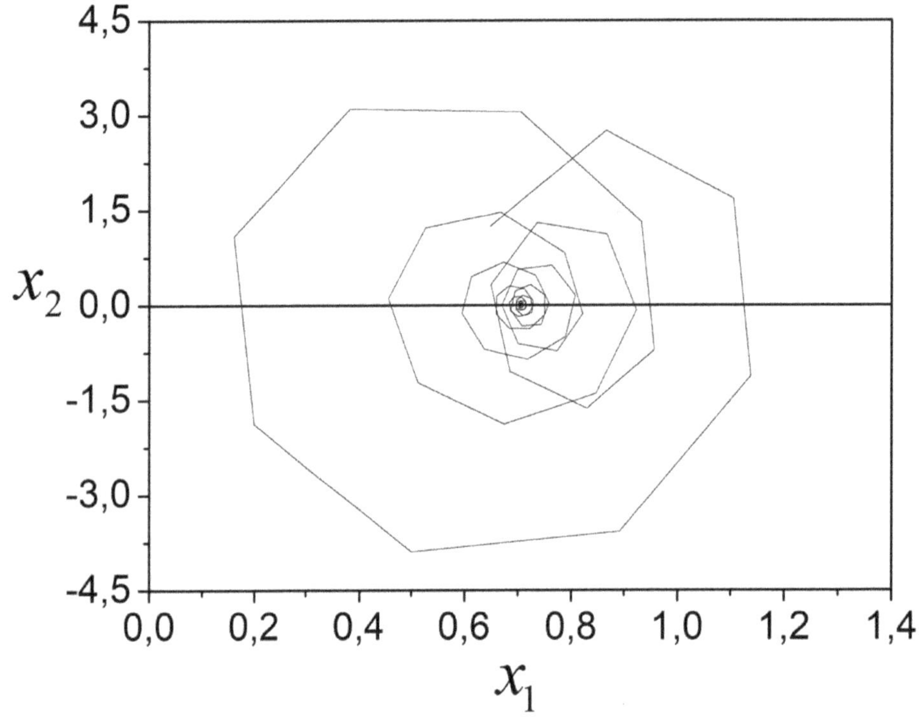

Apply the approach proposed in the "Estimation $X_g(t)$ " section and form sets

$$\left\{\hat{y}_g(t), 0 \le t \le 16s\right\},\ I_{\Delta k_s^i}, I_{k_s}, I_{k_s^i},$$

i=1,2. We present estimation results of the system order and *LP* -detectability LE, applying on theorem 4, in Figure 5, where

$$k_s = k_s(t,\rho),\ k_s^1 = k_s\left(t, \rho\left(\hat{\dot{y}}_g(t)\right)\right),\ k_s^2 = k_s\left(t, \rho\left(\hat{\ddot{y}}_g(t)\right)\right).$$

The numbers 1, 2, 3 indicate LE estimates of eigenvalues λ_i, i=1,2,3 in Figure 5.

Results presented in Figure 5 show that Lyapunov exponents are detectable, and the system is identifiable and recoverable.

Figure 5. Evaluation of order and Lyapunov exponents for third-order system

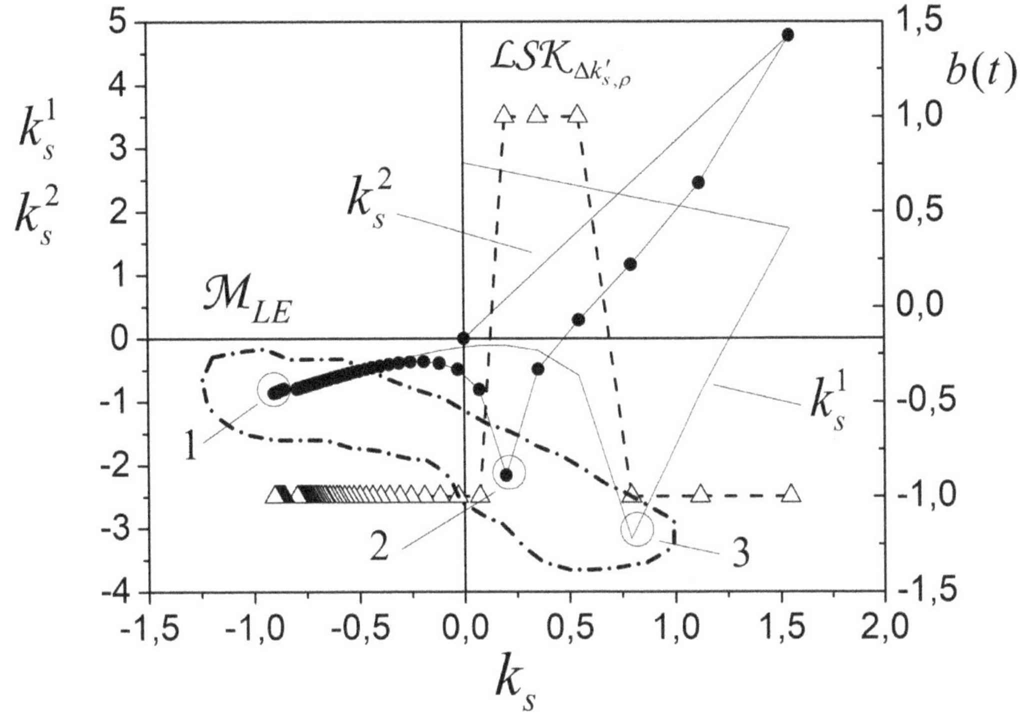

Examples 3. Consider the model of the single-link manipulator robot with a flexible hinge (Veluvolu et al., 2011).

$$\begin{pmatrix} \dot{x}_1 \\ \dot{x}_2 \\ \dot{x}_3 \\ \dot{x}_4 \end{pmatrix} = \begin{pmatrix} 0 & 1 & 0 & 0 \\ -\dfrac{k}{J_m} & -\dfrac{b}{J_m} & \dfrac{k}{J_m} & 0 \\ 0 & 0 & 0 & 1 \\ -\dfrac{k}{J_l} & 0 & -\dfrac{k}{J_l} & 0 \end{pmatrix} \begin{pmatrix} x_1 \\ x_2 \\ x_3 \\ x_4 \end{pmatrix} + \begin{pmatrix} 0 \\ \dfrac{k_\tau}{J_m} u \\ 0 \\ -\dfrac{mgh}{J_l} \sin(x_3) \end{pmatrix}, \begin{pmatrix} y_1 \\ y_2 \end{pmatrix} = \begin{pmatrix} x_1 \\ x_3 \end{pmatrix}, \tag{42}$$

where x_1, x_3 are the angular rotations of the motor and the link; u is motor control torque; J_m, J_l are inertia moments of the motor and the link; k is the elastic constant; b is the viscous friction coefficient; k_τ is the amplifier gain; m is the link mass; h is the center of mass and g is the gravity. The system (42) considers the influence of viscous friction and elastic deformation of the flexible joint. The measurement vector is $Y=[x_1,x_3]^T$. System parameters: J_m=0.037 Kg m^2, J_l=0.093 kg m^2, κ=0.18 Nm/rad, b=0.0083 Nm/rad, m=0.21 kg, g=9.81 m/s^2, h=0.15 m, k_τ=0.18 Nm/v.

Consider the system (42) phase portrait (PP) on channel "x_1-x_2" (Figure 6). Input $u(t)$ does not satisfy the condition of constant excitation. But, despite this, damping oscillations occur in the system. The system is stable and, therefore, recoverable. Structure (phase portrait) the system is open-ended. But internal oscillations are a source for analyzing the system identifiability.

Figure 6. System (42) phase portrait on channel "x_1-x_2"

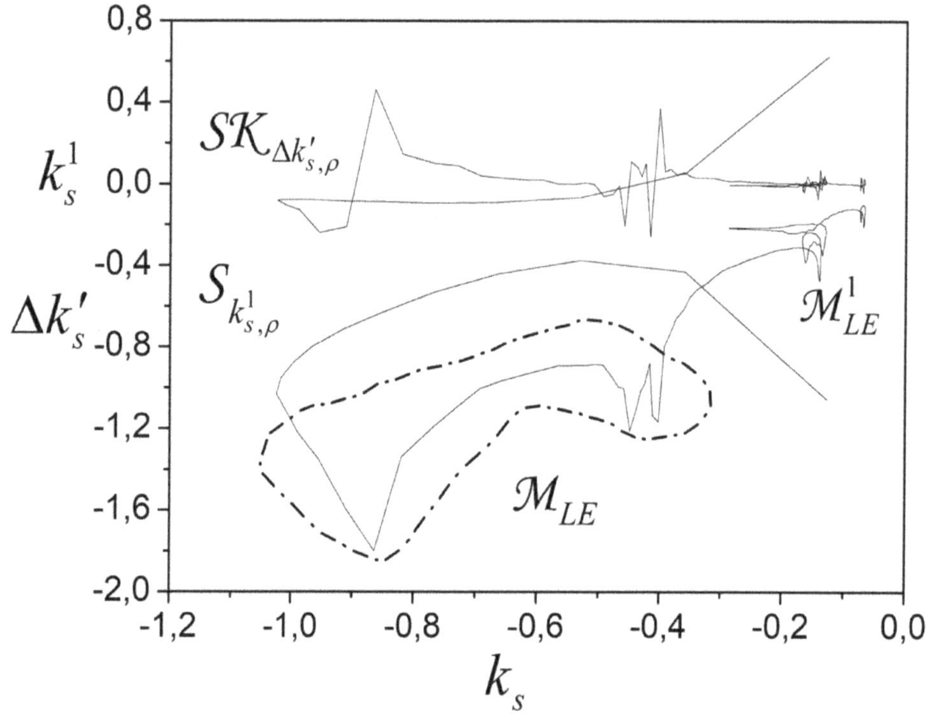

Frameworks $S_{k^1_{s,\rho},k^2_{s,\rho}}$ and $LSK_{\Delta k^2_{s,\rho}}$ presented for evaluating Lyapunov exponents in Figure 7. Here $\hat{y}_1$ is the estimate x_2,

$$k_s^1(t,\rho)=\frac{\rho(y_1(t))}{t},k_s^2(t,\rho)=\frac{\rho(\hat{y}_1(t))}{t},LSK_{\Delta k^2_{s,\rho}}\to I_{k^1_{s,\rho}}\times B\left(I_{\Delta k^2_{s,\rho}}\right). \tag{43}$$

As the system is stable, we consider only negative LE. The oscillations presence reflects by function $b(t)$. $b(t)$ represents by the framework $LSK_{\Delta k^2_{s,\rho}}$ (Figure 7) described by the function $LSK_{\Delta k^2_{k_{s,\rho}}}\to I_{k^2_{k_{s,\rho}}}\times B\left(I_{\Delta k^2_{k_{s,\rho}}}\right)$. System state variables are $BF_{\alpha\pi}$-functions. Apply theorem 8. As Figure 7 follows, the order of the system is 4, and the evaluation of the Perron bottom index is known (it is equal to –0.7). The area to which the set LE $\chi_1[x_2]$ belongs, we denote as B_χ. B_χ (Karabutov, (2018)) corresponds to the presence of a complex root. $\chi_3[x_2]$ locate between $\eta[x_2]$ and area B_χ. $\chi_3[x_2]$ belongs to the region approaching $k_s^2=0$ from below. The lineal corresponding to $\chi_3[x_2]$ is linearly dependent, and, therefore, the system is σ -detectable with level υ=1. The roots spectrum of the system is $\sigma(\lambda)$= [-0.65 0 ±0.8+8.16i]. The system (42) is sensitive to input influences.

Figure 7. Structures for evaluating LE

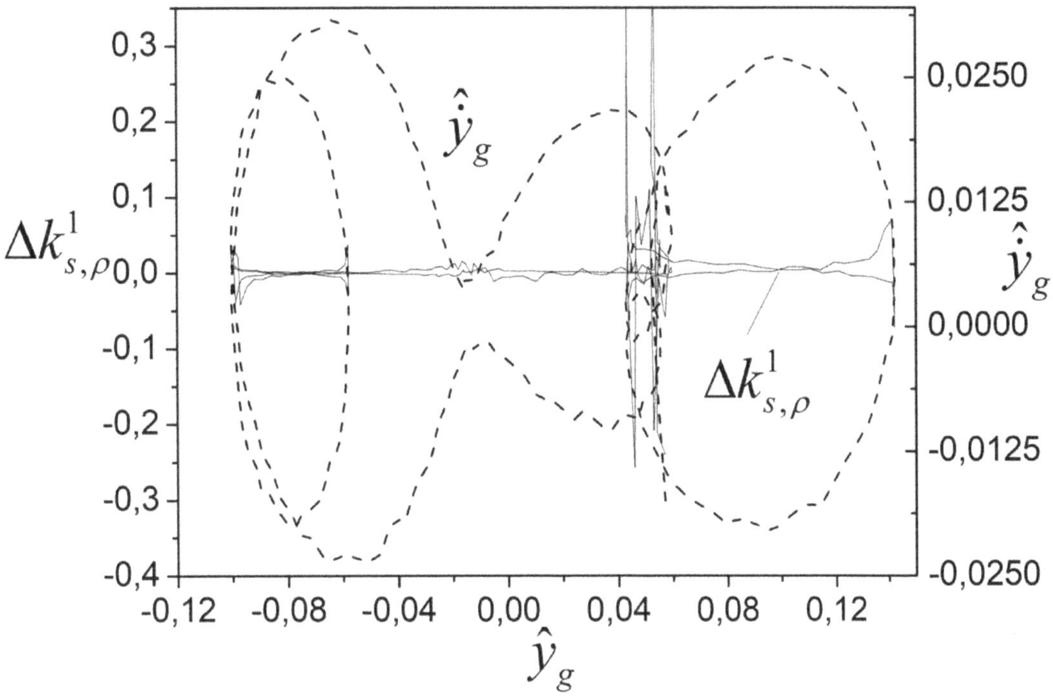

Consider the linearized system for (42). We present estimation results of the LE identifiability and detectability in Figures 8 and 9. Show the system phase portrait in the space $(x_{l,1}, x_{l,2})$ in Figure 8, where $x_{l,1}, x_{l,2}$ are state variables of the linearized system. PP has a similar form in space $(x_{l,3}, x_{l,4})$. The system is stable, and damping oscillations occur in it. Therefore, the system is identifiable, and state variables are $BF_{\alpha\pi}$-functions. Figure 9 shows the frameworks $SL_{k^1_{s,\rho},k^2_{s,\rho}}$ and $LSKL_{\Delta k^2_{s,\rho}}$, confirming the detectability LE:

$$SL_{k^{1,l}_{s,\rho},k^{2,l}_{s,\rho}} \to I_{k^{1,l}_{s,\rho}} \times I_{k^{2,l}_{s,\rho}},\ LSK_{\Delta k^{2,l}_{s,\rho}} \to I_{k^{1}_{s,\rho}} \times B\left(I_{\Delta k^{2l}_{s,\rho}}\right), \tag{44}$$

where

$$SL_{k^{1,l}_{s,\rho},k^{2,l}_{s,\rho}} \to I_{k^{1,l}_{s,\rho}} \times I_{k^{2,l}_{s,\rho}},\ LSK_{\Delta k^{2,l}_{s,\rho}} \to I_{k^{1}_{s,\rho}} \times B\left(I_{\Delta k^{2l}_{s,\rho}}\right), \tag{44}$$

We see (Figure 9) that the linearized system (42) contains two complex indicators $\chi_i[x_{l,2}]$, and the order of the system is 4. So, the linearized system (42) is identifiable, and the set of LE is detectable.

Remark 9. 1. Linearization changes the location of the LE spectrum and affects the stability degree. But the linearization improves the system quality, as the system (42) is from the stability boundary. Note that the system is sensitive to parameter changes. 2. The LE distribution depends on elasticity and viscous friction. So, if elasticity is not considered, then RS (42) will be at the boundary of stability, and low influences cause loss of controllability.

Figure 8. Phase portrait of linearized system (42)

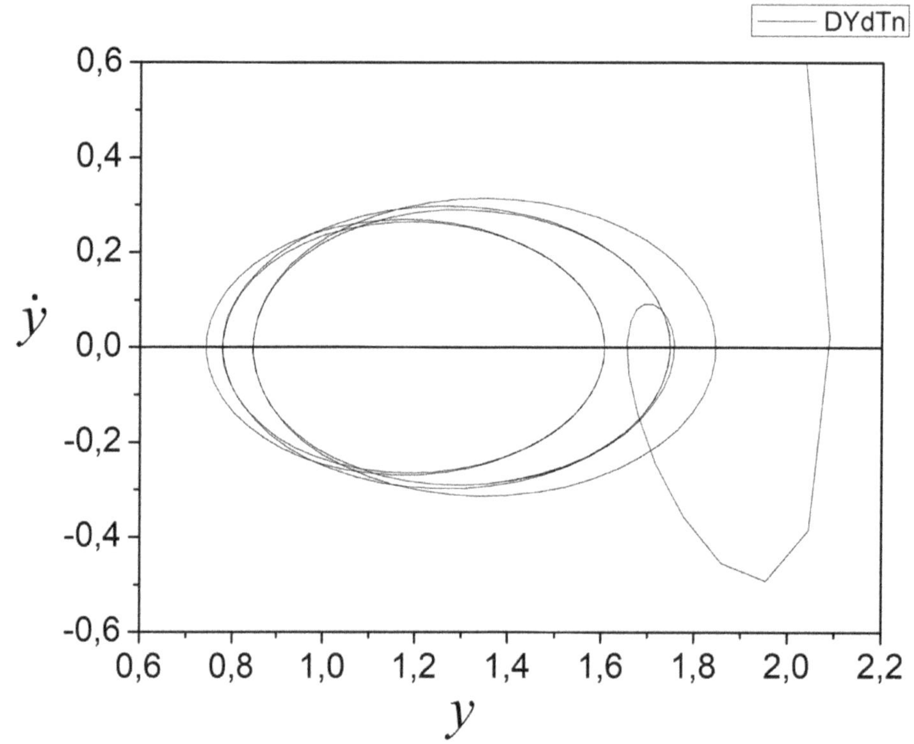

Figure 9. Frameworks for LE estimating of linearized system (42)

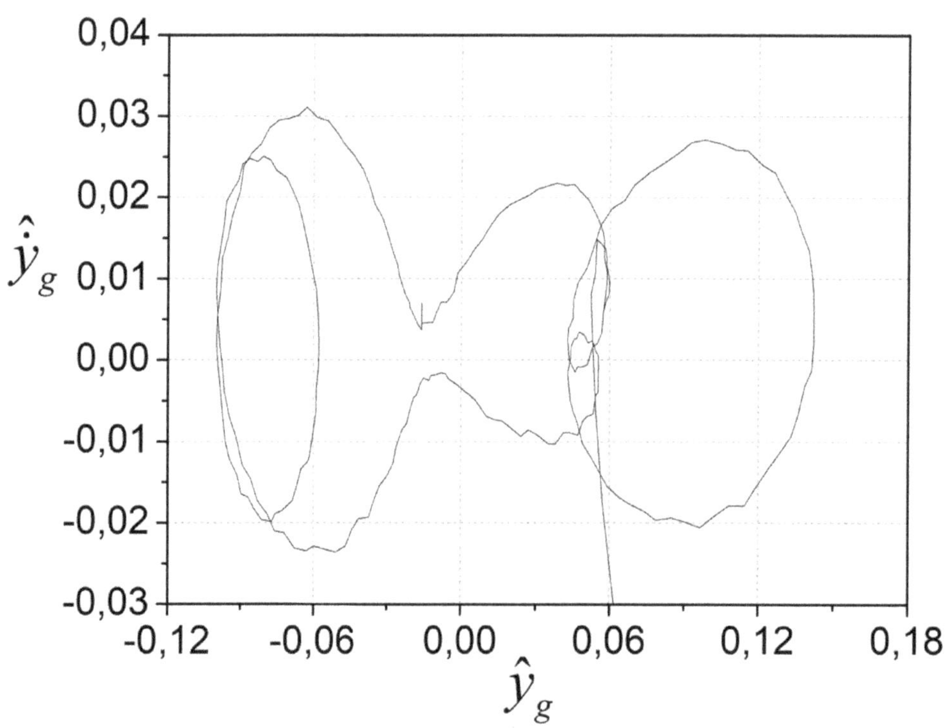

CONCLUSION

The problem of identifiability, recoverability and detectability of Lyapunov exponents consider. We have obtained conditions that guarantee the calculation (detection and identifiability) of the Lyapunov exponents set. They are the result of the analysis structures reflected the change dynamics Lyapunov exponents. We obtain (i) conditions of the σ -detectability with the level of υ -non-recoverability and LP -detectability for Lyapunov exponents of the stationary system; (ii) conditions for the detectability of Lyapunov exponents for the linear robotics system with periodic coefficients. We show that linearization gives to the change of the Lyapunov exponents set for the nonlinear robotics system.

REFERENCES

Azar, A. T., & Vaidyanathan, S. (Eds.). (2014). *Handbook of Research on Advanced Intelligent Control Engineering and Automation*. IGI Global. doi:10.4018/978-1-4666-7248-2

Balcerzak, M., Pikunov, D., & Dabrowski, A. (2018). The fastest, simplified method of Lyapunov exponents spectrum estimation for continuous-time dynamical systems. *Nonlinear Dynamics*, *94*(4), 3053–3065. doi:10.100711071-018-4544-z

Ban, C., Cai, G., & Wei, W. (2021). Dynamic Response and Chaotic Behavior of a Controllable Flexible Robot. Research Square, 1-14. doi:10.21203/rs.3.rs-1116482/v1

Bylov, B. F., Vinograd, R. E., Grobman, D. M., & Nemytsky, V. V. (1966). *Theory of Lyapunov indexes and its application to stability problems*. Nauka.

Demidovich, B. P. (1967). *Lektion on mathematical stability theory*. Nauka.

Gencay, R., & Dechert, W. D. (1992). An algorithm for the n Lyapunov exponents of an n-dimensional unknown dynamical system. *Physica D. Nonlinear Phenomena*, *59*(1-3), 142–157. doi:10.1016/0167-2789(92)90210-E

Glyzin, D. S., Glyzin, S. D., Kolesov, A. Y., & Rozov, N. K. (2005). The Dynamic Renormalization Method for Finding the Maximum Lyapunov Exponent of a Chaotic Attractor. *Differential Equations*, *41*(2), 284–289. doi:10.100710625-005-0159-5

Izobov, N. A. (2006). *Introduction in theory of Lyapunov indexes*. Minsk BGU.

Karabutov, N. (2015). Structural methods of estimation Lyapunov exponents linear dynamic system. *International Journal of Intelligent Systems and Applications*, *7*(10), 1-11.. doi:10.5815/ijisa.2015.10.01

Karabutov, N. (2018). About Lyapunov exponents identification for systems with periodic coefficients. *International Journal of Intelligent Systems and Applications*, *10*(11), 1-10.. doi:10.5815/ijisa.2018.11.01

Karabutov, N. N. (2009). *Structural Identification of Systems. Analysis of Information Structures*. URSS/ Librokom.

Karabutov, N. N. (2018). *Frameworks in identification problems: Construction and analysis*. Lenand.

Kwakernaak, H., & Sivan, R. (1972). Linear optimal control systems. Wilev-Interscience.

Lyapunov, A. M. (1992). *General Problem of the Stability of Motion*. Taylor & Francis. doi:10.1080/00207179208934253

Nasr, S., Abadi, A., Bouallegue, K., & Mekki, H. (2018). Chaos Engineering and Control in Mobile Robotics Applications. *Proceedings of the 15th International Conference on Informatics in Control, Automation and Robotics (ICINCO 2018)*, *2*, 364-371. 10.5220/0006867103640371

Puzyrov, V., Acho, L., Pujol, G., & Rodellar, J. (2019). Fault detection based on Lyapunov exponents estimation for stabilized mechanical systems. *Journal of Theoretical and Applied Mechanics*, *57*(2), 519–531. doi:10.15632/jtam-pl/105471

Reddy, B. S., & Ghosal, A. (2016). Asymptotic Stability and Chaotic Motions in Trajectory Following Feedback Controlled Robots. *Journal of Computational and Nonlinear Dynamics*, *11*(5), 051012. doi:10.1115/1.4032389

Rosenstein, M. T., Collins, J. J., & De Luca, C. J. (1993). A practical method for calculating largest Lyapunov exponents from small data sets. *Physica D. Nonlinear Phenomena*, *65*(1–2), 117–134. doi:10.1016/0167-2789(93)90009-P

Sano, M., & Sawada, Y. (1985). Measurement of the Lyapunov spectrum from a chaotic time series. *Physical Review Letters*, *55*(10), 1082–1085. doi:10.1103/PhysRevLett.55.1082 PMID:10031723

Schramm, L., Sintov, A., & Boularias, A. (2020). *Learning to Transfer Dynamic Models of Underactuated Soft Robotic Hands*. doi:10.1109/ICRA40945.2020.9197300

Skokos, C. (2010). *The Lyapunov Characteristic Exponents and Their Computation*. Lecture Notes in Physics, doi:10.1007/978-3-642-04458-8_2

Smirnov, V. I. (1968). A Course of Higher Mathematics: Vol. 5. *Integration And Functional Analysis*. Nauka.

Soloviev, V., Bielinskyi, A., Serdyuk, O., Solovieva, V., & Semerikov, S. (2020). *Lyapunov exponents as indicators of the stock market crashes*. http://ds.knu.edu.ua/jspui/handle/123456789/3080

Szaniewski, M., Dąbrowski, A., Balcerzak, M., & Zawiasa, P. (2015). Application of the Lyapunov Exponent Based on Current Vibration Control Parameter (CVC) in Control of an Industrial Robot. *Mechanics and Mechanical Engineering*, *19*(1), 51–61.

Takens, F. (1981). *Detecting strange attractors in turbulence*. Springer. doi:10.1007/BFb0091924

Veluvolu, K., & Soh, Y. (2011). Fault reconstruction and state estimation with sliding mode observers for lipschitz non-linear systems. *IET Control Theory & Applications*, *5*(11), 1255–1263. doi:10.1049/iet-cta.2010.0171

Wolf, A., Swift, J. B., Swinney, H. L., & Vastano, J. A. (1985). Determining Lyapunov exponents from a time series. *Physica D. Nonlinear Phenomena*, *16*(3), 285–317. doi:10.1016/0167-2789(85)90011-9

Yunping, L., Lipeng, W., Ping, M., & Kai H. (2013). Stability Analysis of Bipedal Robots Using the Concept of Lyapunov Exponents. In *Robust Control with Engineering Applications*. doi:10.1155/2013/546520

Chapter 10
Issues, Challenges, and Progress of Autonomous Robotic Platoons

Alex Gunagwera
Istanbul Sabahattin Zaim University, Turkey

Aydin Tarik Zengin
https://orcid.org/0000-0002-0860-4509
Istanbul Technical University, Turkey

ABSTRACT

The number of vehicles on the roads worldwide grows annually. With the increase in the population of on-road vehicles comes the increase in traffic-related problems such as emissions, traffic accidents, traffic jams, and increased fuel consumption, to mention but a few. The field of robotics presents the autonomous robotic platooning concept, which presents a very promising future for the transportation industry in general and intelligent transportation industry, in particular. With autonomous platooning, efficient road usage, reduced fuel consumption and emission are possible. This promising field, however, faces issues and challenges barring the deployment of autonomous platoons. In this chapter, the authors introduce the user to the concept of autonomous platooning, present the current state-of-the-art of autonomous robotic platoons and issues that they still face, and ultimately, discuss their progress and future trends.

INTRODUCTION

Owing to urbanization, the tremendous improvement in the automobile industry, and related fields such as communication, the number of vehicles on the roads is growing at a very rapid rate nowadays. More than 50 million cars have been being produced annually since 2010 with a total of approximately 1 billion cars running the streets of the world today, (Worldometer, 2019). With the increase in vehicles throughout the world comes a significant number of challenges and problems. These range from traffic

DOI: 10.4018/978-1-6684-5381-0.ch010

congestion, air pollution, increased fuel consumption, traffic accidents to efficient road usage regardless of the nature of the environment in which we use the vehicles.

The field of autonomous robotic platoons, in general, presents promising solutions to the problems culminating from the increase in the number of vehicles in the world today. The concept of autonomous robotic platoons further presents enhancements to the transportation industry. Autonomous robots may interchangeably be referred to as autonomous vehicles in this literature for more specific scenarios or applications. Autonomous Vehicular Platoon Systems (AVPS), (Su & Ahn, 2016), have attracted the attention of not only researchers or automotive industries but also the attention of governments worldwide. A group of two or more Autonomous Vehicles (AVs) traveling safely at small inter-vehicular gaps, and at high speed together as one entity is referred to as an AVPS. Details on AVPS are provided in subsequent sections. AVPS are the most promising innovation likely to pragmatically either solve or substantially alleviate most of the problems facing the transport industry today, (Chang et al., 1993; Tsugawa et al., 2001). Off-the-shelf solutions like constructing wider roads and flyovers to mitigate traffic congestion are costly, and require extra manual, mechanical and technical labor. In this chapter, the history, vivid definition, progress, and future works concerning AVPS are presented.

The topic of autonomous robots, in general, has gained a lot of attention nowadays from academia, the military, national governmental bodies, to mention but a few. Autonomous vehicles include autonomous land, underwater or aerial vehicles with vast applications ranging from but not limited to robotic academic laboratories, military, industrial, health facilities, and intelligent transport system-oriented applications. Given its cross-disciplinary nature and the significant potential the field of autonomous vehicles possesses, researchers have carried out many studies in the autonomous vehicles field, in general and autonomous vehicle platoons, in particular. Autonomous vehicle platoons, a subset of autonomous vehicles, are not only a promising solution to many of the problems facing the transport department globally but also a great development in general. Extensive research suggesting improvements to these developments, solutions to the problems, novel application ideas, and so much more have been and are being made at a fast rate. However, there is less literature review to summarize the up-to-date state of the art comprehensively. It is with this motivation that this literature review is presented.

Despite the main focus of this study being projected towards autonomous land vehicle platoons, the concepts discussed in this literature have got a cross-disciplinary reach. Thus, the researchers, students, and practicing engineers in related academia and industrial fields such as computer science in general, electrical, and automobile engineering interested in intelligent transport systems, control theory, robotic swarms and autonomous vehicle platoons, in particular, will find them helpful.

Previous related surveys include (Tong, Hussain, Bo, & Maharjan, 2019), whose literature mainly focuses on studies that utilize artificial intelligence to solve problems facing the vehicle to everything bases systems. (Kuutti et al., 2018) review and evaluate the localization techniques used in autonomous vehicles and their applications, whereas (Marina Martinez et al., 2018) go through the driving style recognition studies for intelligent vehicle control. (Sawade & Radusch, 2015) review automated driver assistance systems and further tries to categorize them. (Jia et al., 2015) surveys platoon vehicle cyber-physical systems and their related works. (Kavathekar & Chen, 2011) review literature on vehicle platoons published in the range of 1994 to 2010 inclusive.

The rest of this chapter is organized as follows: The Autonomous Vehicles section covers autonomous vehicles; brief history and levels of vehicle automation. In the Platoon Driving section, the concept of platoon driving is discussed. Communication is tackled in the communication section followed by a discussion on energy consumption in the Energy section. Then the benefits, issues and challenges faced by AVPS plus future works are covered. In the following section, the most recent trends and progress of AVPS are discussed and finally conclude this chapter in the Conclusion section.

REVIEW METHODOLOGY

In this Section, the review criteria used is presented. This chapter provides answers to the questions, such as what autonomous vehicles and autonomous vehicle platoons are. Furthermore, their progress is covered, including the issues they are currently facing, and their future on a global scale. Given that the field of autonomous vehicles is vast, and cross-disciplinary, the chapter commences this literature by clearly and distinctly defining the various terms commonly and, or interchangeably used in the field. Then the chapter provides a comprehensive updated state of the art rooting back since autonomous vehicle platoons were just a dream - by way of a brief history, subsequent works, currently faced issues and challenges and, finally, the future of autonomous vehicle platoons.

Initially, keyword-based searching was performed mainly in the IEEE Xplore Digital Library for literature relevant to the topic with the primary emphasis being put on articles published within the previous decade i.e., between the years, 2010 and 2020. The authors also put relevant citations from the amassed literature into consideration. As a result, the authors needed to, occasionally, use other domain-oriented digital libraries such as ACM Digital Library, Elsevier. In general, we queried the search keywords into scientifically indexed databases e.g., Scopus, and Web of Science (peer-reviewed). On rarer occasions, the authors used google scholar for obtaining appropriate, up-to-date, or omitted literature.

The search terms used are a combination of the following keywords: ”autonomous vehicle platoons”, ”automated vehicle platoons”, ”control”, ”energy efficiency”, ”human interaction”, ”fully autonomous”, ”semi-autonomous vehicle platoons”, ”communication”, ”security”. If a given query yielded too many articles to consider, the authors chose articles according to relevance, publication date, number of citations, and the impact factor of the journal in which they were published.

AUTONOMOUS VEHICLES

Autonomous Vehicles (AVs) are capable of navigating from a given point to a desired destination on their own without the interference of a human driver. These vehicles are equipped with hardware and software that enable them to accomplish this feat. Varying hardware and sensors are used by AVs to perceive, understand the environment about them to be able to make optimal and correct decisions. The sensors used can broadly be categorized in two i.e. proprioceptive sensors which measure the state of the vehicle itself. They measure vehicle states such as speed, battery charge, wheel position to mention but a few. The other category is that of the exteroceptive sensors. These sensors measure the state of the environment in which the vehicle or corresponding robot is. They measure values such as the environment temperature, humidity. Some of the most prominent hardware and sensor technologies employed are outlined below:

- **Cameras:** A camera can be defined as a device for recording films, photographs, visual images, or video signals. The different types of cameras utilized in ADS include.

    - **Omni-direction Cameras:** These cameras have got a 360 angle of view. They provide vision and can replace camera arrays. They can be used for localization, mapping, and navigation, (Gaspar, Winters, & Santos-Victor, 2000).
    - **Monocular Cameras:** These cameras are capable of sensing color. The ability to sense color is of great importance in a variety of AV navigation activities, such as recognizing traffic lights. Monocular cameras can also be used for Simultaneous Localization And Mapping (SLAM), (Xiao Chen, Hu, Zhang, Shi, & Li, 2018).
    - **Neuromorphic Cameras:** These work differently from traditional camera types. They are bio-inspired cameras that, instead of capturing images at a fixed rate, operate by asynchronously measuring the brightness of each pixel and are basically triggered when stimuli occur, (Gallego et al., 2019). The stimulus, in this case being pixel brightness. They are also referred to as event cameras. They report brightness changes as a stream of asynchronous events instead of intensity frames.
- **RADAR:** Originally used as an abbreviation for Radion Detection And Ranging (RADAR), it is a technology used to approximate the range and direction of objects by emitting radio frequency (RF) waves reflected by objects such as ships, vehicles, and airplanes. It was adopted and is now used as a common noun to describe these technologies. RADAR is lightweight technology; they can be fit inside vehicle side-mirrors, (Levanon, 1988).
- **Light Detection and Ranging (LIDAR) or Lidar:** This a rather more recent technology as compared to RADAR. It uses laser light at higher frequencies than RF – RF is 300MHz with about 1m wavelength while UV and infrared is around 200-1000nm) to estimate the range and direction of objects. Hence, Lidar attenuates faster due to the high frequencies but is more accurate, (Reutebuch, Andersen, & McGaughey, 2005).
- **Ultrasonic and proprioceptive sensors:** Ultrasonic sensors serve a similar purpose to that of LIDAR and RADAR, but they use sound waves to achieve their objective. On the other hand, proprioceptive sensors provide useful information such as vehicle speed, orientation, yaw, speed. Some vehicles nowadays are also equipped with Inertial Measurement Units (IMU), wheel encoders, tachometers, altimeters for odometry, positions and velocity estimations, speed and altitude measurements.

The various navigation techniques are discussed in greater detail in the following sections.

BRIEF HISTORY

The idea of Automated Vehicles (AVs) has been a dream for at least 80 years, with significant developments 92 emerging in the 1990s. General Motors sponsored the installation of a model of what they envisioned as a 93 future with automated highways at the New York World's Fair in 1939, (Herman, 2012). By 1960, General 94 Motors had constructed a working prototype, (Rannenberg, 2016). Then followed the so-called Demo '97, 95 in 1991, where the United States Department of Transport (USDOT) was instructed to demonstrate an automated vehicle and highway system in San Diego, California, by 1997. Close headway platooning 97 was demonstrated using buses, cars, and trucks. Other car-makers like Toyota also participated in this 98 event staged for thousands of onlookers and attracted a lot of attention from the media (Bishop, 2005).

Thus, the National Automated Highway System Consortium (NAHSC) proved to the general public that autonomous driving was not just a dream in 1997. In Europe as well, the Eureka PROMETHEUS research project (1987-1995) aiming at remedying the deficiencies of the road traffic system, was started. The Eureka program was funded by EUREKA members states. The program comprised of car manufacturers, research institutes, and universities from those six countries. Currently, EUREKA consists of 41 members, (E-Network, 2013). Qualified teams were invited to participate in the grand challenge. Each group was required to complete a 150-mile route in the Mojave Desert using its autonomous vehicles by the United States Defense Advanced Research Projects Agency (DARPA) in 2004, (Hooper, 2004). However, none of the participants managed to complete the 150 miles and hence the one-million U.S. dollar prize remained unclaimed. In 2007, The DARPA urban challenge (Markoff, 2007), was held. During all the DARPA challenges, no human interference was allowed during the competition. In November 2010, the Hyundai autonomous challenge was held in South Korea, where contestants only had to construct autonomous vehicles capable of following a course and performing missions, (Cerri et al., 2011). An outstanding achievement was in October 2010 when Google announced that their autonomous cars had logged over 140,000 miles – which was a first in robotics research by that time, (Thrun, 2010). This motivated automobile industries to start looking towards the actual production of commercial autonomous vehicles. Meanwhile, research and autonomous vehicle challenges all over the world on similar technology went on steadily. For example, from Vislab (Bertozzi et al., 2010a, 2010b; A. Broggi, Medici, Zani, Coati, & Panciroli, 2012). By 2018, significant progress in the field of Intelligent Transport Systems (ITS), in general, and autonomous vehicular systems AVs, in particular, had been attained. These systems targeted various application areas and functionalities generalized in two main categories: outdoor and indoor applications. Project SARTRE (Safe Road Trains for the Environment) presented by (Robinson, Chan, & Coelingh, 2010) launched in 2009 in Europe was finalized in January 2012 with a successful final phase demonstration of a three-car road semi-autonomous platoon tested in Sweden, (Arturo & Nombela, 2010; SARTRE, 2019). Similarly, Highly Automated Vehicles for Intelligent Transport (HAVE-it), an FP7 (Framework 7 Programme) project launched in Europe in February 2008 was concluded in June 2011, (Hoeger et al., 2008). It mainly focused on areas of that included task repartition between the automated vehicle system and the driver. The final demonstration event of the project included seven vehicles. Research concerning automated driving systems is accelerating quickly, owing to the development in related technical fields such as the Internet of Things (IoT). For example, the AUTOPILOT project, launched in January 2017 to develop services that bring together automated driving and IoT, is now entering a more advanced project stage with six pilot cities and forty-five partners, (Vermesan & Bacquet, 2017).

AUTONOMOUS DRIVING

Engineers in the automobile industry sought to come up with a joint word to describe technology comprising of hardware and software components that collectively aim at executing Dynamic Driving Tasks (DDTs) with little or no dependency on human drivers. But a human driver can assume control of the vehicle at any point as smoothly and seamlessly as possible whenever the need arises. In earlier literature, the word "autonomous", (Bartz, 2009; Beard et al., 2005; Kanade, Thorpe, & Whittaker, 1986; Pomerleau, 1989) was most frequently used, and it is still used in some texts while others referred to the technology as "automated": (Shladover et al., 1991; Smith & Starkey, 1995). Another utilized term is, “driverless” (Szalay, Tettamanti, Esztergár-Kiss, Varga, & Bartolini, 2018), or “self-driving”. Recently field Engineers from the Society of Automotive Engineers (SAE) and the National Highway Traffic Safety Administration (NHTSA) intervened in the United States (U.S.) and all converged on the term “automated” as the most appropriately describing and least misleading term as stipulated in j3016, (SAE, 2013). It is also note-worthy that automated driving should be distinguished from crash or collision avoidance systems, (Vahidi & Eskandarian, 2003). Crash avoidance systems aid the driver when something goes awry, whereas automated driving refers to taking over the entire DDTs. That is, the operational, tactical, and (in some cases) strategic aspects of driving. The operational aspect includes lateral and longitudinal motion control of the vehicle i.e., steering, braking, accelerating, decelerating, and so on. The tactical aspect ranges from responding to and using signals, other real-time events such as being overpassed by cars, determining when and how to change lanes, turn etc. efficiently. Strategic aspects include determining routes from source to destination, waypoints.

VEHICLE AUTOMATION LEVELS

SAE provides a detailed description and level classification of Automation Driving Systems (ADS) depending on the extent to which they automate the DDTs - wholly or partially- in the recently revised J3016 201806, (international, 2016). It categorizes this automation into six levels ranging from level zero (no driving automation) to level 5 (full driving automation). These levels, as described by SAE, are as follows.

- **Level Zero:** No driving automation at this level. The human driver – hereafter also simply referred to as driver - executes the entire DDT whether or not enhanced safety systems are installed. The human driver carries all the DDTs; Sustainable lateral and longitudinal control of the vehicle’s motion, and all Object and Event Detection and Response (OEDR). Whenever and where applicable, the driver is also responsible for any DDT fallback needed.
- **Level One:** At this automation level, either the sustained and Operational Design Domain (ODD)-specific execution of lateral or longitudinal vehicle motion control subtask of the DDT (but not both simultaneously) is carried out by the automation system, and the driver is expected to carry out all other tasks. In short, it is either a mode-specific hands-off or legs-off automation level with limited ODD. It is commonly referred to as Driver Assistance.
- **Level Two:** Also referred to as the Partial Driving Automation. At this level, the automation system can handle both the longitudinal and lateral vehicle motion control subtasks of the DDT. The driver, however, is still expected to keep eyes on the road, do all other DDTs, and supervise the performance of the system. ODDs are limited at this level as well. At all levels above Level two, the ADS performs the entire DDT when and while engaged.
- **Level Three:** The ADS executes the entire DDT, but the driver is expected to respond appropriately to any system notifications and or assume control when required to intervene. The driver also performs the DDT fallback. Hence, it is also known as Conditional Driving Automation.
- **Level Four:** At this level, the ADS performs the entire DDT and DDT fallback. Nothing is expected from the user/driver. Level four is also described as High Driving Automation. All the above-mentioned levels are ODD-specific. They cannot perform outside and, or beyond the specified domain.

- **Level Five:** This is Full Driving Automation. The ADS performs the entire DDT and DDT fallback without expectations from the driver under all road, weather, and reasonable environment conditions.

A visual graph summarizing the levels of automation updated in 2018 is available on the SAE official (International, 2021; Shuttleworth, 2019). ADS are still being researched and further developed all over the world. The most prominent of them all is the Waymo Project (Gibbs, 2017). Some projects have already launched self-driving vehicles to the public in some parts of the world. For instance, Google's Waymo launched in Phoenix in 2018, (Team-Waymo, 2019). SAE international held a Self-Driving Car demonstration in 2018. It was held in partnership with many international companies such as Partners for Automated Vehicle Education (PAVE), Dataspeed inc, National Safety Council (NSC), to mention but a few (Visnic, 2018). Level 4 vehicles were presented to the general public for free usage trials in these demonstrations. Similar tests and events are being held around the world to date. Tesla has also been working on implementing and improving the "full", self-driving feature in its cars since 2016. The company held public tests around April 2019, (Kessler, 2015). Also, CES featured Nvidia's Self-Driving Car in 2017, (Shapiro, 2017).

PLATOON DRIVING

Platoon driving refers to having more than one vehicle linked and traveling as one unit. The linked vehicles are usually under the same longitudinal and, or lateral motion control (Figure 1), which implies, at the very least, level two automation. These vehicles follow one another a given distance apart. They are linked via headway control mechanisms such as LIDAR-based, (Levinson et al., 2011), RADAR-based (Guizzo, 2011), electro-based using Vehicle to Vehicle (V2V) communication systems, (Karnadi, Mo, & Lan, 2007; Tonguz, Wisitpongphan, Bai, Mudalige, & Sadekar, 2007), to mention but a few, or a combination of more than one system. The frontmost vehicle will hereafter be referred to as the Lead Vehicle (LV) and denoted by LV in figures, whereas the rest of the vehicles, referred to as the followers (F), will be designated by the letter F in the figures. Platooning is a very promising technique in Intelligent Transport Systems (ITS) due to the benefits it promises to bring along. These range from road-capacity efficiency, air-drag reduction, hence, fuel consumption reduction, and, thus, air pollution reduction as well. Given the above benefits, researchers, the automobile production industry, and scholars have paid significant attention to vehicular platooning in general and automated vehicular platooning in particular. Automated vehicular platooning further reduces the workload of employees, (Kimura, Matsunaga, Okajima, & Kotaki, 2017), and reduces fatigue resulting from driving long distances, especially during long-distance highway driving. As a result of the incredible amount of attention AVPS have attracted from not only academic scholars, automobile industries but also governmental bodies, a plethora of studies and works to realize and improve vehicle platooning have been made so far. Efforts by the above-mentioned parties interested in AVPS, have yielded techniques and strategies aimed at improving and, ultimately, bringing AVPS closer to deployment. Studies carried out range from Adaptive Cruise Control (ACC), (R. K. Jurgen, 2006; Sivaji & Sailaja, 2013), Cooperative Adaptive Cruise Control(CACC) (Naus, Vugts, Ploeg, Van De Molengraft, & Steinbuch, 2010; Shladover, Nowakowski, Lu, & Ferlis, 2015), String stability, (D.V.A.H.G. Swaroop, 1997; Swaroop, 1996), inter-vehicular communication, general safety to security. Cruise Control refers to the proper maintenance of speed set by the vehicle-driver without requiring further intervention by the driver. The Cruise Control mechanism handles this by controlling the throttle, accelerator-decelerator, pedal, and brake linkage. ACC is an improved form of Cruise Control where the vehicle maintains a given gap between the preceding vehicle and itself, hereafter referred to as the leader in this Section, by automatically accelerating or decelerating as required. The vehicle uses sensors such as RADAR to measure the headway distance between the leader and itself and then carry out the appropriate operation; accelerate or decelerate as described by (Sivaji & Sailaja, 2013). CACC is an extension of ACC where communication is enabled. Connected and Automated Vehicles (CAVs) cooperatively perform DDT tasks, (Shladover et al., 2015; Wang et al., 2018). Vehicles communicate so as to control speed and successfully follow each other. They achieve this by sharing their parameters autonomously through Vehicle to Vehicle (V2V) or, Vehicle to Infrastructure (V2I) communication without necessarily requiring central management, (Van Arem, Van Driel, & Visser, 2006; Wischhof, Ebner, & Rohling, 2005). As defined by SAE, both ACC and CACC are under level one automation. One of the primary objectives of CACC is the vehicle following objective. However, it is of paramount importance that the effect of disturbances throughout the string as vehicle index increases does not amplify, and the string remains stable. String stability, (D.V.A.H.G. Swaroop, 1997; Öncü, 2014; Swaroop, 1996) is, given that the initial errors in vehicle-spacing in a platoon are bounded, all spacing errors between vehicles can assuredly be uniformly bounded in time by string stability. The detailed, comprehensive design of a string stable CACC system was presented by (Naus et al., 2010) accompanied with corresponding experiments. (Anayor, Gao, & Odekunle, 2018) proposed a CACC-based solution in which kinematics information were shared by vehicles using V2V communication. Their solution maintained steady velocity, keeping constant headway distance between vehicles, and also ensuring string stability. The solution they proposed targeted mixed traffic comprising of both human-driven vehicles and autonomous vehicles. (Z. Huang, Chu, Wu, & He, 2019) used hybrid automata to enable each vehicle in the platoon to plan its path. They argued that vehicle cooperative driving is essentially a hybrid system combining discrete events such as platoon maneuvering operations like merging and splitting. Therefore, they present a novel control algorithm coupling Artificial Intelligence Potential Field (APF) approach and a Model Predictive Control (MPC) to realize both motion control and path planning in a synchronized manner. Secondly, using hybrid automata, they defined a model that enables cruise control of a single vehicle and platooning of many vehicles. The effectiveness of their method was verified via simulations of typical traffic scenarios. Vehicle Platooning has got a wide range of application environments and areas ranging from military, mobile robots, agriculture, underwater, industrial, indoors, such as hospitals, care homes, airports, to mention but a few. A virtual platoon control system using Augmented Reality to assist welfare vehicle users, especially the elderly, in narrow environments and indoors such as corridors and elevators was developed and tested by (Kimura et al., 2017). The vehicle users are provided with a head mount display (HMD) that displays the virtual vehicle projection of the real-world to the user. (Y. Zhao & Ogai, 2011) suggested a way to drive safely at high speeds and reduce carbon dioxide emissions and also improve the flexibility and efficiency of transport systems. The Quality of Service (QoS) of Cellular Vehicle to Everything (C-V2X) and radio coverage issues for multiple vehicle strings on a highway(high-density vehicle platooning) was covered by (Qi & MacH, 2019). They further proposed a mechanism for dynamic platooning while considering the change in coverage conditions. (Jesús Mena-Oreja, Gozalvez, & Sepulcre, 2019) analyzed the effect of safe gaps, maximum platoon length on traffic flow under mixed traffic scenarios. The authors claim that the configuration of these parameters has got a significant impact on efficient traffic flow under mixed traffic scenarios. Additionally, emphasize that these parameters determine if and how platooning maneuvers are executed in mixed traffic where human driven vehicles coexist with AVs.

Automation technology is expected to enhance the Transport System and, ultimately, people's ways of living. Meticulous work has been done to make autonomous platoon driving not only a reality but also to improve the autonomous platoon moving systems. (Deng, 2016) proposed a general simulation framework to enhance the analysis and modeling of Heavy-Duty Vehicle platooning. The model was made in consideration of the possible acceleration patterns and the longitudinal driving behaviors of HDV platoons. Autonomous HDV platooning on highways improves not only energy efficiency, but also road usage efficiency, and relieves drivers' fatigue from driving for long hours, (Deng, 2016; Fabricio & Liang, 2013; Viscelli, 2018). (Deng, 2016) further quantified the impacts of HDV platooning on traffic flow and fuel efficiency concluding that increasing HDV platooning in traffic yielded improvements in traffic efficiency while preserving traffic safety for passenger vehicles. Studies to control and enhance communication for reliable platooning, (B. Liu et al., 2017) and facilitate autonomous platoon formation for Vehicular Ad-hoc Network (VANET) enabled vehicles, (Su & Ahn, 2016), were also performed. (Karlsson et al., 2015) comprehensively analyzed the effect spatial diversity imposes on V2V communication during the various platoon operations; merging, splitting, cruising, and formation of the platoon. They drew conclusions about which scenarios/operations can benefit most from vertical spatial diversity and which ones do not. (Harfouch, Yuan, & Baldi, 2017) proposed a model-reference adaptive control that augmented the conventional CACC with an adaptive control term that compensated for every vehicle's driveline dynamics thereby enabling adaptive control of heterogeneous platoons. Their proposal addressed the conventional CACC strategies that assume homogeneous vehicles. A homograph-based visual servo control approach was presented by (Cao & Liu, 2017), who transformed the platoon control problem into a trajectory tracking problem presented by (C. Chang & Yuan, 2018). Their approach only required that the image had to be taken in the ideal position of the leader. A virtual robot would then be then generated based on the leader and the homograph. The homograph matrix entries were utilized to construct the velocity estimator for the leader and control law to drive the follower in the leader-follower platoon system. Two-wheeled robot simulations were used to verify their approach. Chang and Yuan presented a combined trajectory tracking and adaptive cruise control for lateral and longitudinal control of a platoon comprising of mini vehicles all in a decoupled controller. They concluded, based on their experimental results, that vehicle platooning could achieve string stability while driving at low velocity along a trajectory. A practical scenario considering platoons and individual vehicles should be considered when designing autonomous vehicle systems. Understanding the relationship between beaconing performance and the requirements of the control mechanism is also of paramount importance. (B. Liu et al., 2017) proposed a joint control-communication mechanism between the platoon and individual vehicles with both periodic beacon message and event-based safety message delivery. They performed beacon safety message checks and enabled interrupts of event-based messages to be sent and delivered on the same network. (R. H. Huang, Chang, Tsai, & Liang, 2018) argued that CACC experiences the butterfly effect while exhibiting asynchronous brakes that easily lead to backward shock waves and is difficult to remove. They, thereby, proposed a mobile edge computing-based vehicular cloud of CACC as a plausible solution to tackle this problem. (Zou & Li, 2019) investigated the control of event-driven connected vehicular platoons with mixed time-varying delay. They developed an event-driven information interaction method for connected vehicular platoon systems, using the Lyapunov stability criterion and demonstrated the effectiveness of the proposed method using numerical simulations and experiment results with experimental electric cars.

Figure 1. Platoon Driving - longitudinal/lateral control

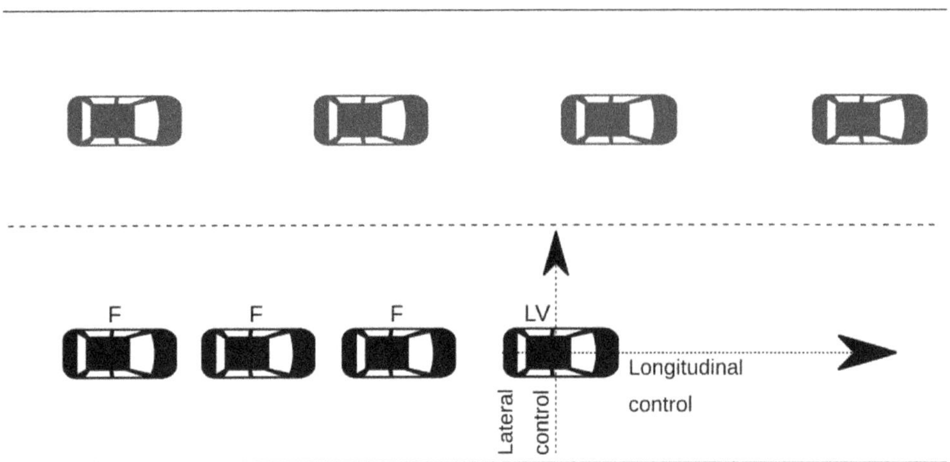

HUMAN-VEHICLE INTERACTION AND COOPERATIVE PLATOON DRIVING

In this Section, the authors discuss the interaction between humans and autonomous vehicles and autonomous vehicle platooning. People's views about these ideas and the steps taken by responsible parties to show the general public and government bodies the benefits of these ideas. Promising as the automated vehicle platooning idea may be, concerns about how various categories of humans; drivers, passengers, pedestrians, workers - will interact with these vehicles are inevitable. AVs have been involved in traffic accidents, some of which were fatal, (Boudette, 2016). An average of 9.1 AV accidents per million miles driven, while regular vehicles face 4.1 crashes per million miles, (Miller, 2015). Due to such accidents involving AVs and the current statistics, news, and reports, the general public is still wary of using or buying AVs. People will start to trust them once they become four or five times safer, (Stewart, 2018b). Furthermore, the transition to fully autonomous vehicle driving platoons will be a gradual process. So, the issue of mixed traffic platooning needs to be addressed as well. Nonetheless, meticulous research about AVs and AV platooning is underway. The global market for level 4 and 5 AVs alone is expected to reach 60 billion dollars by 2030, (Wagner, 2019b) and about 36 billion U.S. dollars for semi-autonomous vehicles by 2025, (Wagner, 2019a). Engineers, AV manufacturing companies, and academicians are working hard to investigate and improve the quality of AVPS. For example, during tests, most companies make sure two safety drivers are in the AV; one to watch the road while the other to monitor the vehicle system responses. Meanwhile, literature on the subject has been published for AVPS. (Zheng et al., 2014) presented an emergency-avoidance braking system for AVPS. They carried out a detailed experimental study for the emergence braking of automated truck platoons using a driving simulator. In addition, they presented a modification on the braking capability of the trucks in a platoon for safety control. They tested various scenarios, such as, abrupt braking of the preceding vehicle without warning. They achieved this by studying the behavior of human drivers. Their study primarily aims at avoiding rear-end crashes during emergence decelerations or braking by the preceding trucks. A great deal of platoon driving systems that directly interact with human users have been examined, as well. Platoon driving systems for automobiles and robots that assume straight roads with good visibility, highway truck platoon driving systems, and indoor platoon driving systems for vehicles that operate in relatively narrower environments, (Kimura et al., 2017; Sugano, Okajima, & Matsunaga, 2015; Sugano, Okajima, Samura, & Matsunaga, 2016; S. Zhao, Zhang, Wu, Ogai, & Tateno, 2015). Such narrow environments include corridors, airport halls with smaller trajectory paths and in more complicated buildings such as industries, hospitals, airports, etc. (Sugano et al., 2015) addressed the control of robotic platoons in narrow environments by proposing a robust novel system for controlling platoon driving along wheel track. The proposed system contained an inter-vehicle distance control with the Model Error Compensator(MEC) and steering control with waypoints along the wheel track of the preceding vehicle. They evaluated the effectiveness of their proposed approach through driving experiments conducted in a narrow corridor corner. Authors claimed that small electric vehicle platoons minimize the labor required from workers and improve the lives of the patients who use welfare vehicles, especially in hospitals. (Sugano et al., 2016) proposed a controller consisting of longitudinal and lateral control using MEC with the adaptive target tracking while considering curvature of the wheel track of the preceding vehicle. A novel truck driving human-machine interface system for automatic platooning of trucks was built by (Zheng et al., 2013). The constructed interface is a truck-driving-simulator for the automatic platooning of trucks that evaluates drivers' behavior in different driving conditions. They developed three types of HMI systems: numeric, and graphics. Ten professional truck drivers aided in the experimental evaluation of the proposed simulator system.

Autonomous Vehicle Platoon Maneuvers: Merging and Splitting

Efficient, smooth, and safe execution of maneuver operations such as splitting/leaving, joining/merging (Figure 2), longitudinal, and lateral (Figure 1) operations of vehicles in a platoon is of paramount importance. Executing these operations in a platoon, however, is not an easy feat. Traffic is mixed nowadays, including cars handled by human drivers. Actions of such vehicles should be put into consideration when designing autonomous platoon systems. Efficient platoon operations increase optimal road usage, reduce fuel consumption, carbon dioxide emissions, road congestion, among others. Thus, it is imperative that the safe and efficient execution of these operations should be guaranteed. In this section, the authors present selected studies concerning platoon maneuvers and their challenges. A trajectory generation approach was proposed to enable the merging and splitting of vehicles in a platoon in urban environments, (Valdês, Espinosa, Awawdeh, & Pastor, 2006). A solution was set for the vehicles that wish to split from or join(merge) the platoon by using a reference trajectory defined by the leader and pre-defined locations of park areas outside the leader-defined course. (Amoozadeh, Deng, Chuah, Zhang, & Ghosal, 2015) presented a multiagent solution to the platoon linear configuration and control problem. Each vehicle was handled as an agent and its behavior was specified by a model hence enabling the computation of the vehicle's speed and distance to preceding vehicle. Ultimately, platoon evolution through merging and splitting was implemented based on the Multiagent System (MAS) architecture. According to their approach, a vehicle willing to join the platoon simply had to follow the last member of the platoon. The vehicle joining the platoon controls the insertion process using Newton's force model. On the other hand, any vehicle could split from the platoon. Information such as weights and index sent to it would, instead, be sent to the preceding and following vehicles. Stability of the platoon in their solution merged as a global result of individual vehicle interactions. Controllers have also been applied varyingly to facilitate the merging and splitting maneuvers of platoons, (Amoozadeh et al., 2015; Dasgupta, Raghuraman, Teja, & Dauwels, 2017). A novel PID controller was suggested to realize the merging and splitting of platoons. MATLAB and VISSIM micro-simulator (Group, 2015) were used to verify the efficacy of their controller

(Dasgupta et al., 2017). Merging and splitting for VANET-enabled vehicles in autonomous platoons was explored and explained in detail by (Su & Ahn, 2016). It was achieved via V2V communication. The fact that full automation needs to be introduced gradually necessitates the investigation of the impact of non-automated vehicles on both automated vehicle platoon maneuver operations and traffic flow in general. PERMIT (Jesus Mena-Oreja & Gozalvez, 2018) aimed at fulfilling the above-stated objective by presenting a SUMO based simulator that implemented all maneuvers; join, merge, split, and leading while putting into account mixed traffic scenarios. Maneuvers were implemented using Finite State Machines (FSM). (Z. Huang et al., 2019) used hybrid automata with explicit rules to develop models and implement merge and split platoon operations as events following the single-vehicle cruise along with multi-vehicle platooning. They carried out numerical simulation tests to ensure the safety and viability of their approach. (Peng, Ziran, Kanok, & Matthew, 2017) synthetically analyzed the impact of disturbances when vehicles leave or join the CACC system range. Furthermore, they optimized energy-consumption while opening or closing gaps during platoon maneuvers. Different simulation scenarios they performed in MATLAB reflected 51% – 77% saving increment in consumed energy during intra platoon gap closing and opening.

Figure 2. Merging and Splitting Maneuvers of AVPS

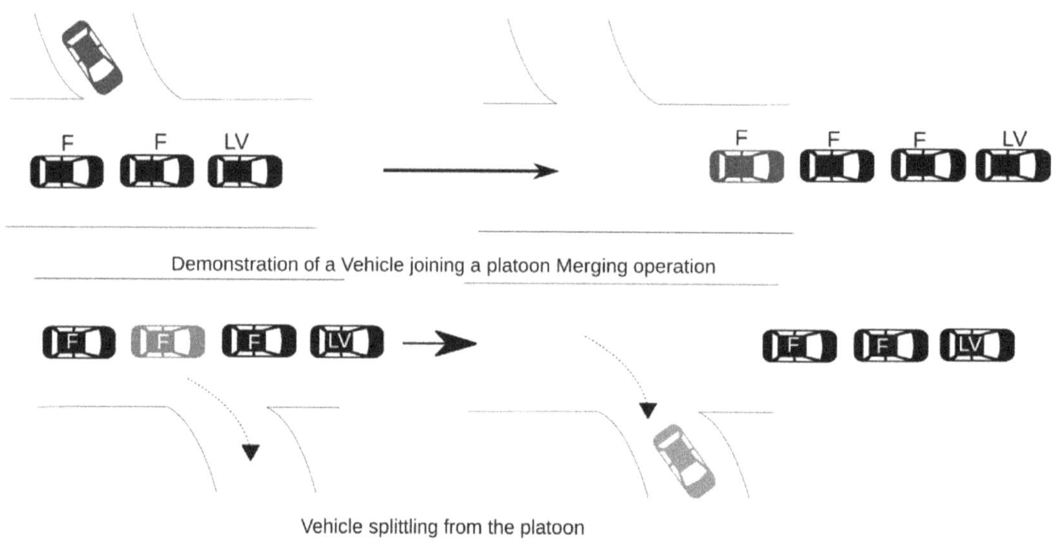

Platoon Navigation Systems

Orientation, location, and speed of a mobile system are the parameters that a navigation system tries to estimate as precisely and reliably as possible. The Global Navigation Satellite Systems (GNSS) (mann-Wellenhof, Bernhard and Lichtenegger, Herbert and Wasle, 2007) including Global Positioning System (GPS) (Alkan, Karaman, & Şahin, 2005; Ananda, Bernstein, Cunningham, Feess, & Stroud, 1990; Patent No. 5,225,842, 1993), Galileo, etc., and sensor technology Micro-Electromechanical Systems (MEMS) play a significant part in not only the construction but also the improvement of navigations systems. Mobile system(vehicle, robot, planes, and humans) navigation and guiding systems are comprised of various technologies ranging from designing and planning of paths, tracing, detecting, avoiding obstacles and collisions, mapping to localization mention but a few. Navigation systems are also necessary for autonomous vehicles and autonomous vehicle platoons – especially where the Lead Vehicle (LV) is also automated – this, by no means, reduces the necessity of navigation systems in semi-automated vehicle platoons. Navigation systems may be centralized or decentralized, (Freddi, Longhi, & Monteriù, 2015; Hallé, Laumonier, & Chaib-draa, 2004; Stilwell & Bishop, 2001; Zaher, Madeleine El and Gechter, Franck and Hajjar, Mohammad and Gruer, 2016) as well. Navigation system structures can be broadly generalized into two categories; Multi-Unit Connected systems and Single-Unit Connected systems. These can be implemented using the modular format or the end-to-end format. The type of navigation system structure is greatly dependent on the kind of platooning system under consideration. Vehicles can be equipped with the tools that facilitate easy localization and navigation information acquisition. These include GPS sensors and cameras. In some architectures, information necessary for vehicles to make their decisions can be provided by road infrastructure embedded in autonomous-vehicle-dedicated lanes – in the so-called infrastructure assisted concept. Providing information further facilitates inter-vehicular communication during merging, splitting, entry, and exit platoon maneuvers. A great deal of literature features the various technologies used to implement and achieve the functionalities that facilitate navigation of different types of AVs and AV platoons. Systems in which these functionalities are implemented in modular format divide the system into various components that perform separate tasks. Such components may include object detection components, mapping and positioning, path, and general strategic planning. The components link the actuator outputs to sensor inputs, (McAllister et al., 2017). In such systems, sensor data is passed to the object detection and positioning components which is finally followed by decision making (Alberto Broggi et al., 2013; Levinson et al., 2011; Maddern, Pascoe, Linegar, & Newman, 2017; Urmson, Chris and Anhalt, Joshua and Bagnell, Drew and Baker et al., 2008; Ziegler et al., 2014). These components can be mounted or installed on one vehicle or different vehicles. The second approach is the vehicle-independent one where the automated vehicle carries all the necessary components for successful DDT operations at all times – referred to as Ego-only systems. Such a vehicle may or may not depend on other assistance infrastructure or vehicles. Most single AV systems fall under this category e.g., (Team-Waymo, 2019; Ziegler et al., 2014). Finally, the end-to-end systems generate motion directly from the sensor inputs, (C. Chen, Seff, Kornhauser, & Xiao, 2015; Kendall et al., 2019; LeCun, Muller, Ben, Cosatto, & Flepp, 2005). End-to-end systems can be implemented using learning techniques such as;

- Deep Reinforcement learning: Here, the vehicle learns the optimum way of driving and requires online intervention, (El Sallab, Abdou, Perot, & Yogamani, 2017),
- Neuro-evolution: this technique uses evolutionary algorithms, (Fortin, De Rainville, Gardner, Parizeau, & Gagńe, 2012) to generate artificial neural networks (ANN), (A. K. Jain, Mao, & Mohiuddin, 1996). This variant does not require backpropagation, (Koutník, Cuccu, Schmidhuber, & Gomez, 2013).
- Supervised Deep Learning: In this technique, data is collected – say from an expert human driver and the system is trained using the collected data, (Bojarski et al., 2016; Cheng, Cheng, Meng, & Zhang, 2019). It can be executed offline.

End-to-end deep learning is a very powerful technique currently, owing to the abundant data availability and improvement in computing and computer vision areas. (Wayve, 2019) recently published that their car and algorithm can drive on never-seen-before roads in urban areas equipped with just a camera and basic Satellite Navigation (sat-nav) route-map after twenty hours of training, (Cowden & Singh, 2013). The training is done using end-to-end machine learning. The AV units also need to cater to the planning and decision-making part. Leaders, in addition to local planning, should have access to global plan information; how the automated vehicle/platoon is to navigate from a given point to the target destination. In AV platoons, all vehicles do not necessarily need to be able to make a global plan. Local plans and decisions (such as when to change lanes, detect other vehicles and objects to avoid collisions) suffice.

ADS then employ different techniques to detect, recognize or track the various objects detected. Information acquired by the onboard or unit sensors can also be used to aid the planning, mapping, figure out the driving style of nearby vehicles or decision-making systems. Numerous studies for object detection of 2D, 3D objects or both, (Andreopoulos & Tsotsos, 2013; Xiaozhi Chen et al., 2015; Xiaozhi Chen, Ma, Wan, Li, & Xia, 2017; L. Liu et al., 2019; Redmon, Divvala, Girshick, & Farhadi, 2016; Song & Xiao, 2014) and object tracking (He, Takeuchi, Ninomiya, & Kato, 2016; Held, Thrun, & Savarese, 2016; Hwang, Kim, Choi, Lee, & Kweon, 2016) were performed. Recently machine-learning, in general, and deep learning, in particular, approaches, and algorithms have displayed auspicious performance. In ADS being able to detect or track objects may prove insufficient for practical autonomous dynamic driving. Hence, further studies to assess situational risk and uncertainty assessments, (Bahram, Hubmann, Lawitzky, Aeberhard, & Wollherr, 2016; Gal, 2016) and decision making are carried out to enable autonomous vehicles mimic, and gradually, outperform experienced human drivers. These studies are essential for optimal fuel consumption, road-capacity efficiency, and general safety. AVs are enabled to predict the future behavior of surrounding vehicles hence being able to make optimal decisions. Driving style recognition is of significant importance as well. Examples include; Google's self-driving car's collision with the bus was because the ADS assumed the bus driver was going to yield during the lane-change (Davies, 2016). The term driving style has no established definition. However, it can be defined with respect to either fuel utilization (Corti, Ongini, Tanelli, & Savaresi, 2013; R. Jurgen, 2011; Manzoni, Corti, De Luca, & Savaresi, 2010) or aggressiveness (Doshi & Trivedi, 2010; Karginova, Byttner, & Svensson, 2012; Marina Martinez et al., 2018). (Marina Martinez et al., 2018) presented a detailed survey about driving style recognition. Artificial Neural Networks(ANN), deep learning, and reinforcement learning are among the most famous techniques used to train ADS systems.

Mapping and Localization

The process of determining one's position relative to a reference frame is referred to as localization (Kuutti et al., 2018). Precise localization is essential to ADS, in particular, and all mobile robots, in general because the AVs are required and expected to be accurately positioned in the right lane at all times, on a local scale (Ort, Paull, & Rus, 2018). On a global scale, precise localization is a condition for successful navigation. Magnusson suggests a three-dimensional normal-distribution-based technique that can be used in mobile robot localization, (Magnusson, 2009). Techniques used for ADS localization and mapping range from Apriori map-based, Simultaneous Localization And Mapping (SLAM), Sensor fusion - Usually using a combination of GPS and Inertial Measurement Units (IMU) to Dead reckoning – the technique of calculating an object's current position with reference from a previously determined position and utilizing the available orientation and speed over a given period of time rather than using physical environment structures such as landmarks or physical map features. - A priori map-based localization relies on matchmaking. Online readings of the mobile unit are compared to the information on the a priori map, and the best match provides the localization of the mobile unit, (Burgard, Wolfram and Brock, Oliver and Stachniss, 2008). There are two main subdivisions of this technique; Landmark Search and Point cloud matching. Landmark searching: This involves the use of landmarks such as road signs, poles road markers. It is computationally less expensive than point cloud. Additionally, it is sufficiently robust provided landmarks are available in abundance, (Hata & Wolf, 2014). Point Cloud matching: With this method, the corresponding mobile unit (robot or vehicle) scans a point cloud. The scanned point cloud is then rotated about its axis iteratively and compared to a relatively larger a priori point cloud. Localization is achieved that way, (Takeuchi & Tsubouchi, 2006). SLAM, as implied by its name, enables localizing the robot and making the map online at the same time. Hence, it does not require a priori information, (Bailey & Durrant-Whyte, 2006; Chang et al., 2007; Eliazar & Parr, 2003). Sensor fusion i.e., dead-reckoning combined with GPS, works by correcting accumulated errors of dead-reckoning in intervals with absolute position reading provided by GPS, since dead-reckoning may give rise to error accumulation, (Kao, 1991; Omar, Yanzhong, Bo, & Gul, 2016). Different types of SLAM exist today, these include Visual SLAM (vSLAM) where a localization system uses 3D vision is used for localization and mapping in cases where the environment or location sensors are inexistent. Acoustic SLAM (aSLAM) is another variation of the SLAM problem. In this methodology, a sound-based map is generated. This technique relies greatly on the principles of sound localization and therefore requires that the robotic or vehicular system in question should be equipped with a microphone array. The microphone array is used to better estimate the Direction of Arrival (DoA) of sound from the sources which are in turn used to construct a sound map. Various SLAM implementation techniques exist in literature with the most popular being the Extended Kalman Filter (EKF SLAM) based methodology. This technique is majorly feature based. It uses the likelihood approach to associate data. GraphSLAM is a graph-based SLAM implementation technique that relies on the information matrices established from data interdependencies.

Local and Global Path Planning and Decision Making

Ultimately, the ADS of AVs have to navigate from the point of origin to a designated final destination. In this Section, a review of how this goal is achieved is presented. Global route planning has been an area of interest for quite a long time now, and tremendous progress has been achieved – accurate navigation and outstanding performance have been nothing out of the ordinary for more than a decade. The problem can easily be formulated as a simple shortest pathfinding problem applied to a directed graph. Local planning refers to how best this can be achieved while avoiding obstacles and collisions in the mobile unit's locale. Techniques utilized by the state-of-the-art include.

- Graph search approaches that include algorithms like the A*, (Duchon et al., 2014), Dijkstra, (Zhang & Zhao, 2014). These methods output a discrete path from origin to the required destination.
- Sampling-based approaches such as the Rapidly-exploring Random Tree (RRT), (Haitao, Qingbao, & Shoujiang, 2007) and the Randomized Potential Planner(RPP), (LaValle & Kuffner, 2001) aim at building the connectivity of the Configuration Space – the so-called C-Space – by randomly sampling it, (Salzman, 2019).

COMMUNICATION

AV platoons are expected to improve transportation efficiency and safety. In fact, according to the USDOT, deployment of AV technology would culminate in the reduction of non-impaired accidents by at least 80%, (Belcher, Merlis, McNew, & Wright, 2018) - more remarkable than any safety measure currently available. Even if they are 50% wrong, the result is still significant enough and worth implementing. Reliable communication amongst vehicles in AV platoons is essential. Vehicles in the platoon should be able to seamlessly share information such as acceleration, deceleration, speed, and obstacles detected, especially by the LV, (K. S. Chang et al., 1991). Robust V2V and V2X communication improve AV platooning, road capacity usage, and energy efficiency as AVs can move at reduced inter-vehicular gaps safely. Meticulous studies have been carried out to enable and enhance communication in AV platoons. In contrast, V2V communication is not a condition for the introduction and usage of AVs, it is a condition for AV platooning. USDOT concluded that 94% of road accidents are caused by human error. It then issued a Notice of Proposed Rulemaking (NPRM) that instructs new commercial vehicles sold after 2023 to have connected vehicle technology. Vehicles sold in the United States in 2023 will be required to have Dedicated Short-Range Communications (DSRC) based V2V connectivity if approved. Many countries worldwide are taking serious steps to facilitate AV platoons. For instance, China committed to both DSRC and C-V2X as well and planned to deploy C-V2X on 80% of its roads and in 80% of its vehicles by 2020. Hence, research has been conducted related to the implementation and improvement of inter-vehicular communication. (Mei, Zheng, Zhao, Lei, & Wang, 2018) performed a joint radio resource allocation optimization in LTE-V2V network for the cooperative awareness messages (CAM) periodically exchanged in vehicle platoons. A Null-Space-Based Behavioral control (NSB) approach for multi-robots was proposed by (Marino, Parker, Antonelli, Caccavale, & Chiaverini, 2009) to aid the tackling of the multi-robot border patrolling problem. A prevalent framework that has facilitated numerous AV platoon related works is VANET, (Singh & Agrawal, 2014). It is an active research field that has attracted a lot of attention and has laid the foundation for technologies such as Intelligent Transport Systems (ITS). (Kaur & Meenakshi, 2018) analyzed two geographic routing protocols; Anchor based Street and Traffic-Aware Routing (A-Star) and Greedy Perimeter Stateless Routing (GPSR) protocols. They claim that simulation of VANETs on real map scenarios provides accurate results. (Guo & Wen, 2016) focused on vehicular platoon control in VANETs subject to limited capacity and random packet dropouts. They introduced binary sequences and derived a closed-form methodology for platoon control. A novel V2V and prediction based platooning system targeting the reduction of inter-vehicular gaps while maintaining stability is introduced by (Zhang et al., 2017). For more information about VANETs, (Singh & Agrawal, 2014) review the routing protocols in VANET. It is, however, noteworthy that the most critical requirements in VANET are protocols and systems are low latency and high reliability. IEEE 802.11 is the leading used standard in VANETs. The pioneer standard being the IEEE based on carrier Sense Multiple Access and collision detection (IEEE 802.11 CSMA/CD), (M. Jain & Saxena, 2017). Some of the popular protocols include:

- IEEE 802.11a uses 5 GHz, with a theoretical maximum throughput of 54Mbps but only provides 24Mbps.
- IEEE 802.11b, the most popular, uses 2.4GHz with a theoretical maximum throughput of 11Mbps but in practice only providing 7.5Mbps.

- A new protocol, IEEE 802.11p, was developed specifically for VANET. It is suitable for high-speed vehicle communication. Its range is within 5.850Ghz – 5.925GHz in North America. DSRC also uses 802.11p at 5.9GHz,

(Chaurasia, Verma, & Tomar, 2011) discussed the various attack types on VANET and analyzed the security issues facing VANET, whereas (Kakkavas, Casta, Stirling-gallacher, & Nossek, 2018) investigated the performance of multi array 5G V2V positioning of vehicles having many antennae. (Shen, Zhang, Yang, Ensor, & Etwork, 2018) presented platooning a system that does not require vision sensors. It only relies on V2V communication. They further evaluated the effects of communication reliability and latency on the performance of the system of vehicles using the 5G V2X hardware prototypes and 802.11 as the communication protocol. (Janssen, Zwijnenberg, Blankers, & de Kruijff, 2015) explored and evaluated the performance of sensor sharing networks considering the message rate and sharing range.

ENERGY EFFICIENCY

Well-organized platoon control has advantages in terms of fuel consumption, safety, emissions, and energy saving. This Section focuses on works that aim at energy-saving. (Van De Hoef, Johansson, & Dimarogonas, 2018) proposed a fuel-efficient coordination scheme for large truck platoons putting into consideration the departure time, arrival deadlines, start location, and destination of the vehicles provided by higher level planning. They demonstrate, via simulations, that their approach leads to large fuel savings. The Platoon Route Optimization Picking up Automated Vehicles (PROPAV) problem is tackled in (Hadded et al., 2018). They proposed a linear programming-based solution of minimizing the collection duration, number of platoons and the total energy required by the platoon leader throughout the whole operation. While considering travel time, road inclinations and fuel consumption, (Nemeth et al., 2012) suggested the design of the velocity of the platoon based on the H-infinity method. (Blum & Eskandarian, 2004) formulated a centralized optimal solution to the problem of fuel-efficient platoon splitting and merging. The proposed algorithm accounted for speed-dependent fuel consumption, and platooning decisions. They took the shortest paths of each vehicle.

BENEFITS, ISSUES AND CHALLENGES FACED BY AVPS

In this section, the authors present the benefits of AVPS followed by the issues and challenges they face. A summary of the benefits and major factors that bar the deployment of AVPS is also additionally provided in Table 1.

Benefits

AV platooning systems have got wide-reaching benefits. The beneficiaries range from automobile manufacturing industries, the military, underwater research engineers, welfare vehicle manufacturers to heavy-duty vehicle managers. The benefits of AV platooning can generally be categorized into two; business-oriented and societal-oriented.

- Business-oriented benefits include;
    - Optimal asset utilization: Platooning reduces road congestion, thereby reducing total vehicle idle time
    - Labor costs. Platooning leads to driver efficiency optimization and allows resting times which is more pronounced in Automated Highway Systems (AHS), especially for truck platooning where the drivers need to travel long journeys. It also reduces the workload of employees in welfare institutions, (Kimura et al., 2017; Sugano et al., 2015; 2016).
    - Fuel consumption: A test involving loaded vehicles at 65000 lbs running up to 70mph recorded 5.3% less fuel consumption by the lead vehicle (LV) truck, whereas the trailing truck saved 9.7% of the fuel, (Brandt, Frame, Warden, & Antonio, 2010).
- Societal Benefits
    - Road efficiency. (Janssen et al., 2015) calculated that two trucks in a platoon with an inter vehicular gap of 0.3 seconds or, equivalently, at 9m inter-vehicle gap moving at 60mph reduced the length of those trucks by 46% from 82m to 44m. Hence almost halving the total space occupied by the trucks.
    - Emission Reduction: Platooning enables vehicles to travel at closer gaps, thereby reducing aerodynamic drag hence reducing the amount of fuel consumed. Platooning also reduces traffic congestion leading to further reduction in fuel consumption.
    - Traffic accident reduction: The USDOT concluded that 94% of crashes are due to human error and that removing the human factor from the equation would reduce crashes by about 80%, (Belcher et al., 2018). Furthermore, it would lead to a reduction of accidents caused by human behavior such as driving under the influence of alcohol, over speeding and distractions more so on highways, and, among long-distance traveling truck drivers.
    - Time saving: Time spent while driving can be utilized to accomplish other tasks. One can, for instance, listen to music, eat food, watch tv or engage in relaxation activities as shown in Figure 3.
    - Improve people's quality of life: Transportation of the elderly and physically impaired will be a more comfortable experience. Furthermore, platooning reduces the employees' workload while improving the elderly people's quality of life, (Chang et al., 2007; United Nations D. o. E. a. SA, 2017).

Figure 3. Platoon Driving: Individuals in Follower(F) Vehicles Can Engage in Other Activities

Issues, Challenges, and Future Works Pertaining to AVPS

Vehicle automation and platooning certainly reduce fuel consumption, emissions and decrease long journey-related driving fatigue. However, AVPs face many issues and challenges ranging from security, global standardization to acceptance from the public, to mention but a few. The issue of hackers taking over control of the ADS and potentially causing accidents is non-dismissible. Since some V2V communication systems use Wi-Fi, they are susceptible to attacks from hackers who may take over control of the system and use it to cause accidents (Blum & Eskandarian, 2004). Automating vehicles certainly does not bring new cybersecurity vulnerabilities to the system, but it does bring about higher security risks, because as the levels of automation increase, the driver's role becomes greatly decreased or even non-existent. Studies to address this issue are in progress e.g., (Marques, Casimiro, & Calha, 2009) suggested the use of private keys, where only authenticated messages from vehicle sensors are utilized for communication. It is essential that fail-safe operations be put in place in case attacks such as denial of service attacks are successful. Furthermore, the Society of Automotive Engineers issued a recommended cybersecurity practice in J3061 (Committee & others, 2016).

The other issue is the incorporation of self-driving car and AVP system technologies into the currently existent vehicle infrastructure. This is especially applicable to the V2X design models where vehicles need to interact with special infrastructure to operate smoothly. The costs of incorporating such technologies need to be meticulously analyzed. Furthermore, whether or not it is economically and socially better to design dedicated infrastructure requires further research. Safety requirements and expectations are other issues facing the autonomous vehicle industry.

Following the news and articles about accidents related to AVs; for example, Google's self-driving car's accident, (Miller, 2015), Uber self-driving car's fatal accident, (Claybrook & Kildare, 2018; News, 2019) and Tesla's self-driving car's accident, (Stewart, 2018a), very few people are willing to trust and take relaxing rides in self-driving cars and ADS yet. According to Pew Research center's 2017 survey, (CENTER, 2017), 56% of Americans would not take a relaxing ride in a self-driving car. Therefore, until the rates of accidents improve, and safety becomes more guaranteed, very few people will be willing to use AVs. Conversely, over-cautious systems make a negative impression on the general public. Safety and security are generally used together and, in most cases, used interchangeably. It, however, is vital to point out that safety refers to the smooth and proper working of the system without causing any harm, and security generally refers to the ability to withstand or bar any sort of attacks.

ADS behavior in inevitable accident scenarios is another ethical challenge, (Bonnefon, Shariff, & Rahwan, 2016). This is also referred to as the "Trolley problem". The problem is posed to question how AVs make their decisions given a social dilemma. For instance, in an inevitable accident scenario, will the autonomous vehicle steer of road, and hit one person to save five people in a traveling vehicle? Cases as this are referred to as the Trolley Problem (Bonnefon et al., 2016; de Melo, Marsella, & Gratch, 2021). It is one of the issues facing the deployment of AVPS today. Various experiments and studies have been performed to resolve this issue using mathematical and cultural approaches (Otsuka, 2008; Thomson, 1976). Much as some of the experiments performed and ideologies presented in literature in relation to the Trolley Problem and autonomous robot decision making maybe presented to present awareness or try to create an understanding of how robots may behave given such scenarios, it should be noted that robots, in general, and vehicles, in particular, do not have a conscience of their own. The decision making that they do is dictated by the algorithms they run, and this will be directly dependent on whoever programmed them. Given the fact that what is right or wrong in any given situation will vary from one individual to another, a consistent behavior of robots in such dilemmas is hard to guarantee.

Some AVs greatly rely on heavily detailed maps (digital or not) for accurate navigation. It is crucial that these maps provide up-to-the-minute dynamic information about sensitive topics such as lane closures, weather conditions work zones and so on and so forth. Delays in the communication of such information should also be kept as low as possible. As far as higher applications of AVs go, (Hedrick, Chen, & Mahal, 2001) analyzed the effects of delayed communication on string stability. Delays in wireless communication systems are inevitable due to transmission time and the loss of packets. Additionally, processing this data also requires time. However, the total delays should be kept to a minimum and as low as possible.

The costs and requirements of sustaining and maintaining AV technologies are another area of concern. The level of expertise required from the general public so that every individual can comfortably benefit from these technologies needs to be specified in preparation of minor breakdowns or unforeseen system incapacitations. The other issue is whether or not and to what extent the general public and drivers would need to be trained in order to comfortably operate AV technologies. The cost of the training, if deemed necessary, should also be investigated. Drivers need to be educated about the various operating modes, protocols, and functions of the systems. Drivers should be prepared to handle various irregularities in ADS performance irregularities. Furthermore, these trainings should be carried out with utmost consideration. Very little work has been done to educate the mass about using AVs. It is a potential area for future works and needs attention. Driver and user experience are another issue. AVPs are capable of traveling at shorter inter-vehicular gaps. The relationship between driver comfort to the inter-vehicular space should definitely be studied. Furthermore, for level three and above, whether or not driver vigilance is affected due to the experience being monotonous should be investigated as well. This is because the drivers are less engaged in driver activities at these levels.

Additionally, there's the issue of job loss. Whether some drivers will lose their jobs or simply the structure of the driving career is to change is another hanging issue, (Viscelli, 2018). In the U.S. alone, about 2.1 million people work as truck drivers –including delivery trucks and about 294,000 truck driver jobs may be at risk. Currently, delivery trucks still need the drivers, but once full automation is achieved, whether or not the necessity of these drivers will remain is unknown. Drivers may still be needed to handle other tasks such as document management, truck loading, and off-loading.

Designing AVP systems that are robust throughout the varying weather and road conditions is a great challenge. Adverse weather conditions negatively affect the performance of autonomous robotic and vehicular systems(Bijelic, Gruber, & Ritter, 2018; Walambe, Marathe, Kotecha, & Ghinea, 2021; Zang et al., 2019). These conditions range from hail, heavy fog, rain to snow. These conditions have a negative impact on the behavior and performance of autonomous vehicular systems mainly because they affect the operation of the sensors the autonomous entities use for data acquisition and decision making. (Zang et al., 2019) presents a more detailed and systematic review of how the major sensors such as LIDAR, RADAR and like, used by autonomous robotic systems are affected by adverse weather conditions. AV manufacturers should clarify the requirements under which their products work and do not work. However, AVPs that work throughout all conditions should be the primary goal. The enforcement of AV technologies both by the governments and manufacturers is another issue. Some regions in the U.S. have already legalized the operation of self-driving cars (Laukkonen, 2019). However, self-driving cars cannot operate in developing or under-developed countries yet. How long it will take is still a hanging question. However, IHS believes global autonomous car sales will reach 21 million by 2035, (Korosec, 2016). AVs are designed to eliminate human error, thereby reducing road accidents. This poses a new challenge for the insurance company. Who is responsible for the mistakes that occur in AVs – the insurance company, the driver, or the car manufacturer? Higher levels of automation, by their definition, remove the driver from the above list. It is, therefore, necessary to clarify whether the automobile manufacturers provide insurance for their products or third-party insurance companies do. The conditions under which the insurance is taken should also be stated. According to (Institute, 2018), the insurance companies will be able to determine whether the accidents result in higher percentages of product liability claims as most of the blame will be put on the manufacturers or suppliers for what went wrong instead of driver behavior. Either way, the structure of the insurance policies as we know it will change.

Another significant challenge is the standardization and interoperability of AVP technologies. Standardizing the communication and data interchange technologies to be utilized in AVs is necessary - the type of radio systems to use, frequency to be used by AVs. Standardizing such technologies becomes crucial for platoon systems. Additionally, it is essential to provide critical vehicle operational parameters such as vehicle size, braking status, location, velocity, to mention but a few. For platooning status, the vehicle with better braking and accelerating assets should be a follower, for instance.

Table 1. Provides a summarized outline of the discussed benefits and issues faced by AVPs.

Benefits	Issues and Challenges
Reduction in road congestion and increases road efficiency	Cybersecurity vulnerabilities
Labor cost reduction	Level of knowledge required by the public to benefit from AVPS.
Minimal Fuel Consumption	Where required, setting up special infrastructure essential for AVPS operation.
Emission reduction	Having the public trust AVPS
Traffic accident reduction	Adverse weather conditions
Saves Time	Communication delays amongst AVPS members
Improves people's quality of life, for instance, the disabled and elderly.	Standardization of communication equipment.
	Ethical concerns regarding AVPS (e,g, the "Trolley problem").

CURRENT TRENDS AND PROGRESS

Culminating from the benefits and promising solutions to transport problems, AVs and AVPS have obtained significant focus recently. Immense investigations are being carried out in the field today, yielding encouraging results. One of the significant emerging trends is the provision of Mobility as a Service (MaaS). According to (Global, 2019), sharing a ride costs less than owning a vehicle, especially under 1000km mileage annually. (Ross, 2019), the head of the automotive and transportation practice, anticipates that there will be 700 million connected cars and about 90 million autonomous cars by 2030, which implies improvement and growth in both market and manufacturing of AVs and AVPS. Investments have indeed been made in the autonomous vehicle industry field with the leading autonomous vehicle producers, including Waymo, GM Cruise, ARGO AI, Aptiv, and Hyundai (2019). Aptiv and Hyundai announced in September 2019 that they were to form a joint autonomous driving venture to provide MaaS, with each company owning 50% of the shares, Group (2019). Waymo, having started as Google's self-driving car, already launched its services in quite several states, including Phoenix, in 2018. GM Cruise was planning to start taxi services in 2019, Ross (2019), but they rolled back that target date, delay their commercial launch, and extend their tests (White, 2019). ARGO AI, on the other hand, started in 2016 as a company with 3 employees building Self-Driving software systems. The company is not planning to make self-driving cars itself. Instead, they planned on specializing in producing the software and technology while partnering with automakers. It received a one-billion-dollar investment from Ford in 2017. They reached 500 employees in 2019 and have partnerships with universities such as Carnegie and companies such as Volkswagen AG. Aptiv is a global technology running more than 100 autonomous vehicles in multiple countries globally in various environments. Ford invested in ARGO AI and started its launches for commercial self-driving fleets in 2021 as they planned in (Team, 2019).

CONCLUSION

In this literature, the authors have reviewed both AVs and AVPS. The authors provide a brief history, progress, and state-of-the-art of these concepts while vividly clarifying their meanings and applications. Benefits, issues, and challenges facing autonomous vehicle platooning have also been discussed. Recent theoretical and industrial progress in the related fields has also been presented. The authors further make a few future work suggestions. Autonomous vehicle platooning has attracted the attention of researchers and automobile industries worldwide. A great deal of work has been carried out and is still underway. Autonomous vehicles have already been launched in some parts of the world, e.g., Waymo launched, in Phoenix in 2018. As far as platooning is concerned, practical successful public demonstrations have also been registered; for example, SARTRE's demonstration in 2012. In comparison to two decades ago, significant practical progress has been made following the plethora of academic and theoretical research that has been carried out. Furthermore, despite improvement as far as safety is concerned, compared to the earlier stages of autonomous cars, more than 50% of the people in the U.S say they would not ride in autonomous vehicles. So more excellent safety guarantees need to be put in place. Issues such as unfavorable road and weather conditions persist in many parts of the world. Thus, more robust systems are required. Global standardization of the technologies used in ADS needs to be put in place as well. The structure of insurance should be clearly defined following the deployment of commercial autonomous vehicles. Whether and to what extent the general public should be educated about autonomous vehicles needs work and research. Therefore, despite the promising progress made as far as AVs and AVPS are concerned, they are not ready yet for full deployment to the general public in most parts of the world currently.

REFERENCES

Alkan, R. M., Karaman, H., & Şahin, M. (2005). GPS, GALILEO and GLONASS satellite navigation systems & GPS modernization. *RAST 2005 - Proceedings of 2nd International Conference on Recent Advances in Space Technologies,* 390–394. 10.1109/RAST.2005.1512598

Amoozadeh, M., Deng, H., Chuah, C. N., Zhang, H. M., & Ghosal, D. (2015). Platoon management with cooperative adaptive cruise control enabled by VANET. *Vehicular Communications*, *2*(2), 110–123. doi:10.1016/j.vehcom.2015.03.004

Ananda, M. P., Bernstein, H., Cunningham, K., Feess, A., & Stroud, G. (1990). Global Positioning System (GPS). *IEEE Symposium on Position Location and Navigation. A Decade of Excellence in the Navigation Sciences.*, 497–508. 10.1109/PLANS.1990.66220

Anayor, C., Gao, W., & Odekunle, A. (2018). Cooperative Adaptive Cruise Control of A Mixture of Human-driven and Autonomous Vehicles. *Conference Proceedings - IEEE SOUTHEASTCON,* 1–3. 10.1109/SECON.2018.8479268

Andreopoulos, A., & Tsotsos, J. K. (2013). 50 Years of object recognition: Directions forward. *Computer Vision and Image Understanding*, *117*(8), 827–891. doi:10.1016/j.cviu.2013.04.005

Arturo, D., & Nombela, M. (2010). Sartre: Safe road trains for the environment. *Conference on Personal Rapid Transit PRT@LHR*, 2–3.

Bahram, M., Hubmann, C., Lawitzky, A., Aeberhard, M., & Wollherr, D. (2016). A Combined Model- and Learning-Based Framework for Interaction-Aware Maneuver Prediction. *IEEE Transactions on Intelligent Transportation Systems*, *17*(6), 1538–1550. doi:10.1109/TITS.2015.2506642

Bailey, T., & Durrant-Whyte, H. (2006). Simultaneous localization and mapping (SLAM): Part II. *IEEE Robotics & Automation Magazine*, *13*(3), 108–117. doi:10.1109/MRA.2006.1678144

Bartz, D. (2009, November 16). *Autonomous Cars Will Make Us Safer*. Retrieved from https://www.wired.com/2009/11/autonomous-cars/

Beard, R., Kingston, D., Quigley, M., Snyder, D., Christiansen, R., Johnson, W., … Goodrich, M. A. (2005). Autonomous vehicle technologies for small fixed-wing UAVs. *Journal of Aerospace Computing, Information and Communication*, 92–108. doi:10.2514/1.8371

Belcher, S., Merlis, E., McNew, J., & Wright, M. (2018). *Roadmap To Vehicle Connectivity*. Academic Press.

Bertozzi, M., Bombini, L., Broggi, A., Buzzoni, M., Cardarelli, E., Cattani, S., ... Zani, P. (2010a). The VisLab Intercontinental Autonomous challenge: 13,000 km, 3 months,··· no driver. *17th ITS World Congress*, 1–11.

Bertozzi, M., Bombini, L., Broggi, A., Buzzoni, M., Cardarelli, E., Cattani, S., ... Zani, P. (2010b). The VisLab Intercontinental Autonomous challenge: 13,000 km. *17th ITS World Congress*, 1–11.

Bijelic, M., Gruber, T., & Ritter, W. (2018). Benchmarking image sensors under adverse weather conditions for autonomous driving. *2018 IEEE Intelligent Vehicles Symposium (IV)*, 1773–1779. 10.1109/IVS.2018.8500659

Bishop, R. (2005). *Intelligent Vehicle Technology and Trends*. Artech House.

Blum, J., & Eskandarian, A. (2004). The threat of intelligent collisions. *IT Professional*, *6*(1), 24–29. doi:10.1109/MITP.2004.1265539

Bojarski, M., Del Testa, D., Dworakowski, D., Firner, B., Flepp, B., Goyal, P., . . . Zieba, K. (2016). *End to End Learning for Self-Driving Cars*. Retrieved from https://arxiv.org/abs/1604.07316

Bonnefon, J.-F., Shariff, A., & Rahwan, I. (2016). The social dilemma of autonomous vehicles. *Science*, *352*(6293), 1573–1576. doi:10.1126cience.aaf2654 PMID:27339987

Boudette, N. E. (2016). Autopilot cited in death of Chinese tesla driver. *The New York Times*. Retrieved from https://www.nytimes.com/2016/09/15/business/fatal-tesla-crash-in-china-involved-autopilot-government-tv-says.html

Brandt, A. C., Frame, E. A., Warden, R. W., & Antonio, S. (2010). *SAE J1321 testing using M1083A1 FMTVS interim report.* Force Projection Technologies.

Broggi, A., Buzzoni, M., Debattisti, S., Grisleri, P., Laghi, M. C., Medici, P., & Versari, P. (2013). Extensive tests of autonomous driving technologies. *IEEE Transactions on Intelligent Transportation Systems*, *14*(3), 1403–1415. doi:10.1109/TITS.2013.2262331

Broggi, A., Medici, P., Zani, P., Coati, A., & Panciroli, M. (2012). Autonomous vehicles control in the VisLab Intercontinental Autonomous Challenge. *Annual Reviews in Control*, *36*(1), 161–171. doi:10.1016/j.arcontrol.2012.03.012

Brown, A. K., & Sturza, M. A. (1993). *Patent No. 5,225,842*. United STATES PATENT.

Burgard, W., Brock, O., & Stachniss, C. (2008). Map-based precision vehicle localization in urban environments. *Robotics: Science and Systems, 3*, 121–128. Retrieved from https://ieeexplore.ieee.org/document/6280129

Cao, Y., & Liu, S. (2017). Visual servo control for wheeled robot platooning based on homography. *Proceedings of 2017 IEEE 6th Data Driven Control and Learning Systems Conference, DDCLS 2017*, 628–632. 10.1109/DDCLS.2017.8068145

Cerri, P., Soprani, G., Zani, P., Choi, J., Lee, J., Kim, D., ... Broggi, A. (2011). Computer vision at the Hyundai Autonomous Challenge. *IEEE Conference on Intelligent Transportation Systems, Proceedings, ITSC*, 777–783. 10.1109/ITSC.2011.6082859

Chang, C., & Yuan, Z. (2018). Combined longitudinal and lateral control of vehicle platoons. *2017 International Conference on Computer Systems, Electronics and Control, ICCSEC 2017*, 848–852. 10.1109/ICCSEC.2017.8446824

Chang, H. J., Lee, C. S. G., Lu, Y. H., & Hu, Y. C. (2007). P-SLAM: Simultaneous localization and mapping with environmental-structure prediction. *IEEE Transactions on Robotics*, *23*(2), 281–293. doi:10.1109/TRO.2007.892230

Chang, K. S., Karl Hedrick, J., Zhang, W. B., Varaiya, P., Tomizuka, M., & Shladover, S. E. (1993). Automated highway system experiments in the path program. *Journal of Intelligent Transport Systems*, *1*(1), 63–87.

Chang, K. S., Li, W., Devlin, P., Shaikhbahai, A., Varaiya, P., & Hedrick, J. K. (1991). Experimentation with a vehicle platoon control system. *Vehicle Navigation and Information Systems Conference*, 1117–1124. 10.4271/912868

Chaurasia, B. K., Verma, S., & Tomar, G. S. (2011). Attacks on anonymity in VANET. *Proceedings - 2011 International Conference on Computational Intelligence and Communication Systems, CICN 2011*, 217–221. 10.1109/CICN.2011.43

Chen, C., Seff, A., Kornhauser, A., & Xiao, J. (2015). Deepdriving: Learning affordance for direct perception in autonomous driving. *Proceedings of the IEEE International Conference on Computer Vision*, 2722–2730. 10.1109/ICCV.2015.312

Chen, X., Kundu, K., Zhu, Y., Berneshawi, A., Ma, H., Fidler, S., & Urtasun, R. (2015). 3D object proposals for accurate object class detection. Advances in Neural Information Processing Systems, 424–432.

Chen, X., Ma, H., Wan, J., Li, B., & Xia, T. (2017). Multi-view 3d object detection network for autonomous driving. *Proceedings of the IEEE Conference on Computer Vision and Pattern Recognition*, 1907–1915.

Chen, X., Hu, W., Zhang, L., Shi, Z., & Li, M. (2018). Integration of low-cost GNSS and monocular cameras for simultaneous localization and mapping. *Sensors (Switzerland)*, *18*(7), 1–18. doi:10.339018072193 PMID:29986515

Cheng, J., Cheng, H., Meng, M. Q. H., & Zhang, H. (2019). Autonomous Navigation by Mobile Robots in Human Environments: A Survey. *2018 IEEE International Conference on Robotics and Biomimetics, ROBIO 2018*, 1981–1986. 10.1109/ROBIO.2018.8665075

Claybrook, J., & Kildare, S. (2018). Autonomous vehicles: No driver… no regulation? *Science*, *361*(6397), 36–37. doi:10.1126cience.aau2715 PMID:29976817

Committee, S. (2016). *Sae j3061-cybersecurity guidebook for cyber-physical automotive systems.* SAE-Society of Automotive Engineers.

Corti, A., Ongini, C., Tanelli, M., & Savaresi, S. M. (2013). Quantitative driving style estimation for energy-oriented applications in road vehicles. *Proceedings - 2013 IEEE International Conference on Systems, Man, and Cybernetics, SMC 2013*, 3710–3715. 10.1109/SMC.2013.632

Cowden, S., & Singh, G. (2013). Sat-Nav education: A means to an end or an end to meaning. *Acts of Knowing: Critical Pedagogy in, against and beyond the University*, 41–59.

Dasgupta, S., Raghuraman, V., Teja, T. N., & Dauwels, J. (2017). Merging and Splitting maneuver of platoons by means of a novel PID controller. *2017 IEEE Symposium Series on Computational Intelligence (SSCI)*, 1–8. Retrieved from https://ieeexplore.ieee.org/stamp/stamp.jsp?tp=&arnumber=8280871

Davies, A. (2016, February 23). Google's self-driving car caused its first crash. *Wired*. Retrieved from https://www.wired.com/2016/02/googles-self-driving-car-may-caused-first-crash/

de Melo, C. M., Marsella, S., & Gratch, J. (2021). Risk of Injury in Moral Dilemmas With Autonomous Vehicles. *Frontiers in Robotics and AI*, *7*, 572529. Advance online publication. doi:10.3389/frobt.2020.572529 PMID:34212006

Deloitte Global. (2019). *2019 Deloitte Global Automotive Consumer Study | Advanced vehicle technologies and multimodal transportation*. Retrieved from https://www2.deloitte.com/content/dam/Deloitte/us/Documents/manufacturing/us-global-automotive-consumer-study-2019.pdf

Deng, Q. (2016). A General Simulation Framework for Modeling and Analysis of Heavy-Duty Vehicle Platooning. *IEEE Transactions on Intelligent Transportation Systems*, *17*(11), 3252–3262. doi:10.1109/TITS.2016.2548502

Doshi, A., & Trivedi, M. M. (2010). Examining the impact of driving style on the predictability and responsiveness of the driver: Real-world and simulator analysis. *IEEE Intelligent Vehicles Symposium, Proceedings*, 232–237. 10.1109/IVS.2010.5547969

Duchon, F., Babinec, A., Kajan, M., Beno, P., Florek, M., Fico, T., & Jurišica, L. (2014). Path planning with modified A star algorithm for a mobile robot. *Procedia Engineering*, *96*, 59–69. doi:10.1016/j.proeng.2014.12.098

E-Network. (2013). *Programme for a European traffic system with highest efficiency and unprecedented safety*. Retrieved September 20, 2009, from https://www.eurekanetwork.org/project/id/45

El Sallab, A., Abdou, M., Perot, E., & Yogamani, S. (2017). Deep reinforcement learning framework for autonomous driving. *IS and T International Symposium on Electronic Imaging Science and Technology*, 70–76. 10.2352/ISSN.2470-1173.2017.19.AVM-023

Eliazar, A., & Parr, R. (2003). DP-SLAM: Fast, robust simultaneous localization and mapping without predetermined landmarks. *IJCAI International Joint Conference on Artificial Intelligence*, 1135–1142.

Fabricio, B., & Liang, Z. (2013). Fuzzy community structure detection by particle competition and cooperation. *Soft Computing*, *17*(4), 659–673. doi:10.100700500-012-0924-3

Fortin, F. A., De Rainville, F. M., Gardner, M. A., Parizeau, M., & Gagńe, C. (2012). DEAP: Evolutionary algorithms made easy. *Journal of Machine Learning Research*, *13*, 2171–2175.

Freddi, A., Longhi, S., & Monteriù, A. (2015). Nonlinear decentralized model predictive control for unmanned vehicles moving in formation. *Information Technology and Control*, *44*(1), 89–97. doi:10.5755/j01.itc.44.1.7219

Gal, Y. (2016). *Uncertainty in Deep Learning*. University of Cambridge. Retrieved from http://mlg.eng.cam.ac.uk/yarin/blog_2248.html

Gallego, G., Delbruck, T., Orchard, G., Bartolozzi, C., Taba, B., Censi, A., . . . Scaramuzza, D. (2019). *Event-based Vision: A Survey*. https://arxiv.org/abs/1904.08405

Gaspar, J., Winters, N., & Santos-Victor, J. (2000). Vision-based navigation and environmental representations with an omnidirectional camera. *IEEE Transactions on Robotics and Automation*, *16*(6), 890–898. doi:10.1109/70.897802

Gibbs, S. (2017). Google sibling waymo launches fully autonomous ride-hailing service. *The Guardian*, *7*.

Guizzo, E. (2011). How Google's self-driving car works. *IEEE Spectrum Online*, *18*(7), 1132–1141.

Guo, G., & Wen, S. (2016). Communication Scheduling and Control of a Platoon of Vehicles in VANETs. *IEEE Transactions on Intelligent Transportation Systems*, *17*(6), 1551–1563. doi:10.1109/TITS.2015.2505407

Hadded, M., Lasgouttes, J. M., Nashashibi, F., & Xydias, I. (2018). Platoon Route Optimization for Picking up Automated Vehicles in an Urban Network. *IEEE Conference on Intelligent Transportation Systems, Proceedings, ITSC*, 1987–1992. 10.1109/ITSC.2018.8569809

Haitao, G., Qingbao, Z., & Shoujiang, X. (2007). Rapid-exploring random tree algorithm for path planning of robot based on grid method. *Journal of Nanjing Normal University*, *2*(14).

Hallé, S., Laumonier, J., & Chaib-draa, B. (2004). A decentralized approach to collaborative driving coordination. *IEEE Conference on Intelligent Transportation Systems, Proceedings, ITSC*, 453–458. 10.1109/ITSC.2004.1398942

Harfouch, Y. A., Yuan, S., & Baldi, S. (2017). Adaptive control of interconnected networked systems with application to heterogeneous platooning. *IEEE International Conference on Control and Automation, ICCA*, 212–217. 10.1109/ICCA.2017.8003062

Hata, A., & Wolf, D. (2014). Road marking detection using LIDAR reflective intensity data and its application to vehicle localization. *2014 17th IEEE International Conference on Intelligent Transportation Systems, ITSC 2014*, 584–589. 10.1109/ITSC.2014.6957753

He, M., Takeuchi, E., Ninomiya, Y., & Kato, S. (2016). Precise and efficient model-based vehicle tracking method using Rao-Blackwellized and scaling series particle filters. *IEEE International Conference on Intelligent Robots and Systems,* 117–124. 10.1109/IROS.2016.7759043

Hedrick, J., Chen, Y., & Mahal, S. (2001). *Optimized vehicle control/communication interaction in an automated highway system.* Retrieved from https://escholarship.org/uc/item/8p23v0q4

Held, D., Thrun, S., & Savarese, S. (2016). Learning to track at 100 FPS with deep regression networks. Lecture Notes in Computer Science, 9905, 749–765. doi:10.1007/978-3-319-46448-0_45

Herman, A. (2012). *Freedom's forge: how American business produced victory in World War II.* Random House.

Hoeger, R., Amditis, A., Kunert, M., Hoess, A., Flemish, F., Krueger, H.-P., ... Beutner, A. (2008). Highly automated vehicles for intelligent transport: Haveit approach. *ITS World Congress.*

Hooper, J. (2004). *From Darpa Grand Challenge 2004 DARPA's Debacle in the Desert.* Retrieved September 20, 2009, from Popular Science website: https://www.popsci.com/scitech/article/2004-06/darpa-grand-challenge-2004darpas-debacle-desert/

Huang, R. H., Chang, B. J., Tsai, Y. L., & Liang, Y. H. (2018). Mobile Edge Computing-Based Vehicular Cloud of Cooperative Adaptive Driving for Platooning Autonomous Self Driving. *Proceedings - 2017 IEEE 7th International Symposium on Cloud and Service Computing, SC2 2017,* 32–39. 10.1109/SC2.2017.13

Huang, Z., Chu, D., Wu, C., & He, Y. (2019). Path Planning and Cooperative Control for Automated Vehicle Platoon Using Hybrid Automata. *IEEE Transactions on Intelligent Transportation Systems*, *20*(3), 959–974. doi:10.1109/TITS.2018.2841967

Hwang, S., Kim, N., Choi, Y., Lee, S., & Kweon, I. S. (2016). Fast multiple objects detection and tracking fusing color camera and 3D LIDAR for intelligent vehicles. *2016 13th International Conference on Ubiquitous Robots and Ambient Intelligence, URAI 2016*, 234–239. 10.1109/URAI.2016.7625744

Institute, I. I. (2018, July 30). *Background on: Self-driving cars and insurance.* Retrieved October 29, 2019, from Insurance Information Institute website: https://www.iii.org/article/background-on-self-driving-cars-and-insurance

International, S. A. E. (2016). *Taxonomy and definitions for terms related to driving automation systems for on-road motor vehicles*. SAE International (J3016).

International, S. A. E. (2021, May). *SAE levels of Driving Automation™ refined for clarity and international audience*. SAE International. Retrieved from https://www.sae.org/blog/sae-j3016-update

Jain, A. K., Mao, J., & Mohiuddin, K. M. (1996). Artificial neural networks: A tutorial. *Computer*, *29*(3), 31–44. https://doi.org/10.1109/2.485891

Jain, M., & Saxena, R. (2017). Overview of VANET: Requirements and its routing protocols. *2017 International Conference on Communication and Signal Processing (ICCSP)*, 1957–1961.

Janssen, R., Zwijnenberg, H., Blankers, I., & de Kruijff, J. (2015). *Truck platooning. Driving The Future of Transportation*. TNO.

Jia, D., Lu, K., Wang, J., Zhang, X., & Shen, X. (2015). A survey on platoon-based vehicular cyber-physical systems. *IEEE Communications Surveys & Tutorials, 18*(1), 263–284.

Jurgen, R. (2011). Design and Analysis of an Adaptive Real-Time Advisory System for Improving Real World Fuel Economy in a Hybrid Electric Vehicle (2010-01-0835). In *Electric and Hybrid-Electric Vehicles* (pp. 35–45). Overviews and Viewpoints. Retrieved from https://ieeexplore.ieee.org/document/8522869

Jurgen, R. K. (2006). *Adaptive cruise control*. Academic Press.

Kakkavas, A., Casta, M. H., Stirling-gallacher, R. A., & Nossek, J. A. (2018). Multi-Array 5G V2V Relative Positioning: Performance Bounds. *2018 IEEE Global Communications Conference (GLOBECOM)*, 206–212. Retrieved from https://arxiv.org/abs/1808.07373

Kanade, T., Thorpe, C., & Whittaker, W. (1986). Autonomous land vehicle project at CMU. *CSC '86 Proceedings of the 1986 ACM Fourteenth Annual Conference on Computer Science*, 71–80. Retrieved from https://dl.acm.org/citation.cfm?id=325197

Kao, W. W. (1991). Integration of GPS and dead-reckoning navigation systems. *Proceedings - Society of Automotive Engineers, 2*(2), 635–643. doi:10.1109/vnis.1991.205808

Karginova, N., Byttner, S., & Svensson, M. (2012). *Data-driven methods for classification of driving styles in buses*. SAE Technical Papers. doi:10.4271/2012-01-0744

Karlsson, K., & Larsson, M. (2015). On the effect of vertical spatial diversity on V2V communication for three different platooning scenarios. *2015 9th European Conference on Antennas and Propagation (EuCAP), 5*(2), 1–5.

Karnadi, F. K., Mo, Z. H., & Lan, K. C. (2007). Rapid generation of realistic mobility models for VANET. *IEEE Wireless Communications and Networking Conference, WCNC*, 2508–2513. doi:10.1109/WCNC.2007.467

Kaur, H., & Meenakshi. (2018). Analysis of VANET geographic routing protocols on real city map. *RTEICT 2017 - 2nd IEEE International Conference on Recent Trends in Electronics, Information and Communication Technology Proceedings*, 895–899. doi:10.1109/RTEICT.2017.8256727

Kavathekar, P., & Chen, Y. (2011). Vehicle platooning: A brief survey and categorization. *International Design Engineering Technical Conferences and Computers and Information in Engineering Conference, 54808*, 829–845.

Kendall, A., Hawke, J., Janz, D., Mazur, P., Reda, D., & Allen, J. M. … Shah, A. (2019). Learning to drive in a day. *Proceedings - IEEE International Conference on Robotics and Automation,* 8248–8254. doi:10.1109/ICRA.2019.8793742

Kessler, A. M. (2015). *Elon musk says self-driving tesla cars will be in the us by summer. The New York Times*. Retrieved from https://www.nytimes.com/2015/03/20/business/elon-musk-says-self-driving-tesla-cars-will-be-in-the-us-by-summer.html?_r=0

Kimura, R., Matsunaga, N., Okajima, H., & Kotaki, G. (2017). Design of virtual platoon control system using augmented reality to assist welfare vehicle users. *International Conference on Control, Automation and Systems,* 330–335. doi:10.23919/ICCAS.2017.8204460

Korosec, K. (2016, June 7). Autonomous Car Sales Will Hit 21 Million by 2035, IHS Says. *Fortune*. Retrieved from https://fortune.com/2016/06/07/autonomous-car-sales-ihs/

Koutník, J., Cuccu, G., Schmidhuber, J., & Gomez, F. (2013). Evolving large-scale neural networks for vision-based reinforcement learning. *GECCO 2013 - Proceedings of the 2013 Genetic and Evolutionary Computation Conference*, 1061–1068. doi:10.1145/2463372.2463509

Kuutti, S., Fallah, S., Katsaros, K., Dianati, M., Mccullough, F., & Mouzakitis, A. (2018). A Survey of the State-of-the-Art Localization Techniques and Their Potentials for Autonomous Vehicle Applications. *IEEE Internet of Things Journal, 5*(2), 829–846. doi:10.1109/JIOT.2018.2812300

Laukkonen, J. (2019, November 17). Are Self-Driving Cars Legal in Your State? *LifeWire*. Retrieved from https://www.lifewire.com/are-self-driving-cars-legal-4587765

LaValle, S. M., & Kuffner, J. J. (2001). Randomized kinodynamic planning. *The International Journal of Robotics Research*, *20*(5), 378–400. https://doi.org/10.1177/02783640122067453

LeCun, Y., Muller, U., Ben, J., Cosatto, E., & Flepp, B. (2005). Off-road obstacle avoidance through end-to-end learning. *Advances in Neural Information Processing Systems*, 739–746.

Levanon, N. (1988). *Radar principles*. Retrieved from https://ui.adsabs.harvard.edu/abs/1988wi...book.....L

Levinson, J., Askeland, J., Becker, J., Dolson, J., Held, D., Kammel, S., … Thrun, S. (2011). Towards fully autonomous driving: Systems and algorithms. *IEEE Intelligent Vehicles Symposium Proceedings*, (4), 163–168. doi:10.1109/IVS.2011.5940562

Liu, B., Jia, D., Lu, K., Ngoduy, D., Wang, J., & Wu, L. (2017). A joint control-communication design for reliable vehicle platooning in hybrid traffic. *IEEE Transactions on Vehicular Technology*, *66*(10), 9394–9409. https://doi.org/10.1109/TVT.2017.2702650

Liu, L., Ouyang, W., Wang, X., Fieguth, P., Chen, J., Liu, X., & Pietikäinen, M. (2019). Deep Learning for Generic Object Detection: A Survey. *International Journal of Computer Vision*. doi:10.1007/s11263-019-01247-4

Maddern, W., Pascoe, G., Linegar, C., & Newman, P. (2017). 1 year, 1000 km: The Oxford RobotCar dataset. *The International Journal of Robotics Research*, *36*(1), 3–15.

Magnusson, M. (2009). *The Three-Dimensional Normal-Distributions Transform – an Efficient Representation for Registration, Surface Analysis, and Loop Detection*. Örebro University. Retrieved from http://www.aass.oru.se/Research/Learning/publications/2009/Magnusson_2009-Doctoral_Thesis-3D_NDT.pdf

Mann-Wellenhof, B., Lichtenegger, H., & Wasle, E. (2007). *GNSS-global navigation satellite systems: GPS, GLONASS, Galileo, and more*. Springer Science and Business Media.

Manzoni, V., Corti, A., De Luca, P., & Savaresi, S. M. (2010). Driving style estimation via inertial measurements. *IEEE Conference on Intelligent Transportation Systems Proceedings,* 777–782. doi:10.1109/ITSC.2010.5625113

Marina Martinez, C., Heucke, M., Wang, F. Y., Gao, B., & Cao, D. (2018). Driving Style Recognition for Intelligent Vehicle Control and Advanced Driver Assistance: A Survey. *IEEE Transactions on Intelligent Transportation Systems*, *19*(3), 666–676. https://doi.org/10.1109/TITS.2017.2706978

Marino, A., Parker, L. E., Antonelli, G., Caccavale, F., & Chiaverini, S. (2009). A fault-tolerant modular control approach to multi-robot perimeter patrol. *2009 IEEE International Conference on Robotics and Biomimetics, ROBIO 2009*, 735–740. doi:10.1109/ROBIO.2009.5420581

Markoff, J. (2007). *Crashes and traffic jams in military test of robotic vehicles*. Retrieved from https://www.nytimes.com/2007/11/05/technology/05robot.html

Marques, L., Casimiro, A., & Calha, M. (2009). Design and development of a proof-of-concept platooning application using the HIDENETS architecture. *Proceedings of the International Conference on Dependable Systems and Networks*, 223–228. doi:10.1109/DSN.2009.5270334

McAllister, R., Gal, Y., Kendall, A., Van Der Wilk, M., Shah, A., Cipolla, R., & Weller, A. (2017). Concrete problems for autonomous vehicle safety: Advantages of Bayesian deep learning. *International Joint Conference on Artificial Intelligence*, 4745–4753. doi:10.24963/ijcai.2017/661

Mei, J., Zheng, K., Zhao, L., Lei, L., & Wang, X. (2018). Joint radio resource allocation and control for vehicle platooning in lte-v2v network. *IEEE Transactions on Vehicular Technology*, *67*(12), 12218–12230. https://doi.org/10.1109/TVT.2018.2874722

Mena-Oreja, J., & Gozalvez, J. (2018). PERMIT - A SUMO Simulator for Platooning Maneuvers in Mixed Traffic Scenarios. *IEEE Conference on Intelligent Transportation Systems Proceedings,* 3445–3450. doi:10.1109/ITSC.2018.8569496

Mena-Oreja, J., Gozalvez, J., & Sepulcre, M. (2019). Effect of the Configuration of Platooning Maneuvers on the Traffic Flow under Mixed Traffic Scenarios. *IEEE Vehicular Networking Conference*. doi:10.1109/VNC.2018.8628381

Miller, B. (2015, October 30). First Driverless Car Crash Study: Autonomous Vehicles Crash More, but Injuries Are Less Serious. *Government Technology*. Retrieved from https://www.govtech.com/fs/First-Driverless-Car-Crash-Study-Autonomous-Vehicles-Crash-More-Injuries-Are-Less-Serious.html

Naus, G., Vugts, R., Ploeg, J., Van De Molengraft, R., & Steinbuch, M. (2010). Cooperative adaptive cruise control, design and experiments. *Proceedings of the 2010 American Control Conference,* (1), 6145–6150. doi:10.1109/acc.2010.5531596

Nemeth, B., Csikos, A., Varga, I., & Gaspar, P. (2012). Road inclinations and emissions in platoon control via multi-criteria optimization. *2012 20th Mediterranean Conference on Control and Automation, MED 2012 - Conference Proceedings*, 1524–1529. doi:10.1109/MED.2012.6265855

News, B. (2019, November 6). *Uber in fatal crash had safety flaws say US investigators*. Retrieved from https://www.bbc.com/news/business-50312340

Omar, H. M., Yanzhong, Z., Bo, Z., & Gul, H. U. (2016). Integration of GPS and dead reckoning navigation system using moving horizon estimation. In Proceedings of 2016 IEEE Information Technology, Networking, Electronic and Automation Control Conference (pp. 553–556). ITNEC. https://doi.org/10.1109/ITNEC.2016.7560421.

Öncü, S. (2014). *String stability of interconnected vehicles : network- aware modelling, analysis and experiments*. doi:10.6100/IR762788

Ort, T., Paull, L., & Rus, D. (2018). Autonomous Vehicle Navigation in Rural Environments Without Detailed Prior Maps. *Proceedings - IEEE International Conference on Robotics and Automation*, 2040–2047. doi:10.1109/ICRA.2018.8460519

Otsuka, M. (2008). Double effect, triple effect and the trolley problem: Squaring the circle in looping cases. *Utilitas*, *20*(1), 92–110.

Peng, H., Ziran, W., Kanok, B., & Matthew, B. (2017). Intra-platoon Vehicle Sequence Optimization for Eco-Cooperative Adaptive Cruise Control. *2017 IEEE 20th International Conference on Intelligent Transportation Systems (ITSC)*, 1–6.

Pew Research Center. (2017). *Many Americans would be hesitant to use various automation technologies*. Retrieved from https://www.pewresearch.org/internet/2017/10/04/automation-in-everyday-life/pi_2017-10-04_automation_0-02/

Pomerleau, D. a. (1989). ALVINN: An autonomous land vehicle in a neural network (Technical Report CMU-CS-89-107). *Advances in Neural Information Processing Systems, 1*, 305–313. Retrieved from https://dl.acm.org/citation.cfm?id=89851.89891

PTV Group. (2015). *PTV Vissim*. Retrieved from http://vision-traffic.ptvgroup.com/en-uk/training-support/support/ptv-vissim/

Qi, Y., & Mac, H. T. (2019). QoS and Coverage Aware Dynamic High Density Vehicle Platooning (HDVP). *IEEE Vehicular Technology Conference,* 1–5. doi:10.1109/VTCFall.2018.8690882

Rannenberg, K. (2016). *Autonomous Driving: Technical, Legal and Social Aspects*. doi:10.1007/978-3-662-48847-8_24

Redmon, J., Divvala, S., Girshick, R., & Farhadi, A. (2016). You only look once: Unified, real-time object detection. *Proceedings of the IEEE Conference on Computer Vision and Pattern Recognition*, 779–788.

Reutebuch, S. E., Andersen, H.-E., & McGaughey, R. J. (2005). Light detection and ranging (LIDAR): An emerging tool for multiple resource inventory. *Journal of Forestry*, *103*(6), 286–292.

Robinson, T., Chan, E., & Coelingh, E. (2010). An Introduction To The SARTRE Platooning Programme. *17th World Congress on Intelligent Transport Systems*, 1–11.

Ross, P. E. (2019). Taxis without drivers-or steering wheels. *IEEE Spectrum*, *56*(1), 33–34. https://doi.org/10.1109/MSPEC.2019.8594791

SAE. (2013). Definitions for Terms Related to On-Road Motor Vehicle Automated Driving Systems-J3016. Society of Automotive Engineers: On-Road Automated Vehicle Standards Committee; SAE Pub. Inc.

Salzman, O. (2019). Sampling-based robot motion planning. *Communications of the ACM*, *62*(10), 54–63.

SARTRE. (2019). *Project SARTRE (Safe Road Trains for the Environment)*. Road Traffic Technology. Retrieved from https://www.roadtraffic-technology.com/projects/the-sartre-project/

Sawade, O., & Radusch, I. (2015). Survey and classification of cooperative automated driver assistance systems. *2015 IEEE 82nd Vehicular Technology Conference (VTC2015-Fall)*, 1–5.

Shapiro, D. (2017). *Self-Driving Car Demo Like No Other: The Road to an Unforgettable Cup of Coffee*. Retrieved October 14, 2019, from Nvidia Blog website: https://blogs.nvidia.com/blog/2017/01/04/bb8-ces/

Shen, Z., Zhang, X., Yang, D., Ensor, I. I. S., & Etwork, S. H. N. (2018). Performance Analysis of Extended Sensor Sharing in Vehicular Ad Hoc Networks. *2018 International Symposium on Antennas and Propagation (ISAP)*, 1–2.

Shladover, S. E., Desoer, C. A., Hedrick, J. K., Tomizuka, M., Walrand, J., Zhang, W., Bin, … McKeown, N. (1991). Automatic Vehicle Control Developments in the PATH Program. *IEEE Transactions on Vehicular Technology*, *40*(1), 114–130. doi:10.1109/25.69979

Shladover, S. E., Nowakowski, C., Lu, X. Y., & Ferlis, R. (2015). Cooperative adaptive cruise control: Definitions and operating concepts. *Transportation Research Record*, *2489*(November), 145–152. doi:10.3141/2489-17

Shuttleworth, J. (2019). *SAE Standards News: J3016 automated-driving graphic update*. SAE International.

Singh, S., & Agrawal, S. (2014). VANET routing protocols: Issues and challenges. *2014 Recent Advances in Engineering and Computational Sciences (RAECS)*, 1–5.

Sivaji, V. V., & Sailaja, M. (2013). Adaptive cruise control systems for vehicle modeling using stop and go manoeuvres. *International Journal of Engineering Research and Applications*, *3*(4), 2453–2456. https://citeseerx.ist.psu.edu/viewdoc/download?doi=10.1.1.433.9908&rep=rep1&type=pdf

Smith, D. E., & Starkey, J. M. (1995). Effects of Model Complexity on the Performance of Automated Vehicle Steering Controllers: Model Development, Validation and Comparison. *Vehicle System Dynamics*, *24*(2), 163–181. https://doi.org/10.1080/00423119508969086

Song, S., & Xiao, J. (2014). Sliding shapes for 3d object detection in depth images. *European Conference on Computer Vision*, 634–651.

Stewart, J. (2018a, March 30). Tesla's Autopilot Was Involved in Another Deadly Car Crash. *Transportation, Wired*. Retrieved from https://www.wired.com/story/tesla-autopilot-self-driving-crash-california/

Stewart, J. (2018b, October 10). Why people keep rear-ending self-driving cars. *Wired*. Retrieved from https://www.wired.com/story/self-driving-car-crashes-rear-endings-why-charts-statistics/

Stilwell, D. J., & Bishop, B. E. (2001). *Decentralized Guidance, Navigation, and Control for Platoons of Cooperating UUVs*. Academic Press.

Su, D., & Ahn, S. (2016). Autonomous platoon formation for VANET-enabled vehicles. *2016 International Conference on Information and Communication Technology Convergence, ICTC 2016*, 247–250. doi:10.1109/ICTC.2016.7763478

Sugano, T., Okajima, H., & Matsunaga, N. (2015). Robust and precise platoon driving control of welfare vehicles along wheel track by using model error compensator. *IECON 2015 - 41st Annual Conference of the IEEE Industrial Electronics Society*, 256–261. doi:10.1109/IECON.2015.7392108

Sugano, T., Okajima, H., Samura, S., & Matsunaga, N. (2016). Narrow space driving of welfare vehicles using robust platoon control with adaptive way point tracking. *2016 55th Annual Conference of the Society of Instrument and Control Engineers of Japan, SICE 2016*, 1436–1441. doi:10.1109/SICE.2016.7749175

Swaroop, D. (1996). String Stability of Interconnected Systems - Automatic Control, IEEE Transactions on. *IEEE Transactions on Automatic Control*, *41*(3), 349–357.

Swaroop, D. (1997). *String Stability Of Interconnected Systems: An Application To Platooning In Automated Highway Systems*. UC Berkeley. Retrieved from https://escholarship.org/uc/item/86z6h1b1

Szalay, Z., Tettamanti, T., Esztergár-Kiss, D., Varga, I., & Bartolini, C. (2018). Development of a test track for driverless cars: Vehicle design, track configuration, and liability considerations. *Periodica Polytechnica Transportation Engineering*, *46*(1), 29–35. doi:10.3311/PPtr.10753

Takeuchi, E., & Tsubouchi, T. (2006). A 3-D scan matching using improved 3-D normal distributions transform for mobile robotic mapping. *IEEE International Conference on Intelligent Robots and Systems*, 3068–3073. doi:10.1109/IROS.2006.282246

Team, A. A. I. (2019). *ARGO AI*. Retrieved from https://www.argo.ai/

Team-Waymo. (2019). *WAYMO*. Retrieved from https://waymo.com/mission/

Thomson, J. J. (1976). Killing, letting die, and the trolley problem. *The Monist*, *59*(2), 204–217.

Thrun, S. (2010). *What we're driving at*. Retrieved October 11, 2019, from The Official Google Blog 26 website: https://googleblog.blogspot.com/2010/10/what-were-driving-at.html

Tong, W., Hussain, A., Bo, W. X., & Maharjan, S. (2019). Artificial intelligence for vehicle-to-everything: A survey. *IEEE Access: Practical Innovations, Open Solutions*, *7*, 10823–10843.

Tonguz, O., Wisitpongphan, N., Bai, F., Mudalige, P., & Sadekar, V. (2007). Broadcasting in VANET. *2007 Mobile Networking for Vehicular Environments, MOVE*, 7–12. doi:10.1109/MOVE.2007.4300825

Tsugawa, S., Kato, S., Tokuda, K., Matsui, T., & Fujii, H. (2001). A cooperative driving system with automated vehicles and inter-vehicle communications in Demo 2000. *IEEE Conference on Intelligent Transportation Systems, Proceedings, ITSC*, 918–923. doi:10.1109/itsc.2001.948783

United Nations. (2017). *World Population Prospects: The 2017 Revision, Key Findings and Advance Tables"*. The Department of Economic and Social Affairs: UN.

Urmson, C., Anhalt, J., Bagnell, D., Baker, C., Bittner, R. C., Dolan, M., Duggins, J., Galatali, & Geyer, T. (2008). Autonomous driving in urban environments: Boss and the urban challenge. *Journal of Field Robotics, 25*, 425–466. doi:10.1002/rob

Vahidi, A., & Eskandarian, A. (2003). Research advances in intelligent collision avoidance and adaptive cruise control. *IEEE Transactions on Intelligent Transportation Systems*, *4*(3), 132–153. https://doi.org/10.1109/TITS.2003.821294

Valdês, F., Espinosa, F., Awawdeh, A. B. M. H., & Pastor, J. (2006). Routes for splitting and merging maneuvers of platoon followers in Urban environments. *IEEE International Conference on Emerging Technologies and Factory Automation, ETFA*, 1153–1156. doi:10.1109/ETFA.2006.355236

Van Arem, B., Van Driel, C. J. G., & Visser, R. (2006). The impact of cooperative adaptive cruise control on traffic-flow characteristics. *IEEE Transactions on Intelligent Transportation Systems*, *7*(4), 429–436. https://doi.org/10.1109/TITS.2006.884615

Van De Hoef, S., Johansson, K. H., & Dimarogonas, D. V. (2018). Fuel-Efficient en Route Formation of Truck Platoons. *IEEE Transactions on Intelligent Transportation Systems*, *19*(1), 102–112. https://doi.org/10.1109/TITS.2017.2700021

Vermesan, O., & Bacquet, J. (2017). *Cognitive Hyperconnected Digital Transformation*. doi:10.13052/rp-9788793609105

Viscelli, S. (2018). *Autonomous Trucks and the Future of the American Trucker*. Retrieved from https://laborcenter.berkeley.edu/driverless/

Visnic, B. (2018, May 10). SAE brings autonomy to the public in new demonstration program. *SAE International*. Retrieved from https://www.sae.org/news/2018/05/sae-demo-days-in-florida

Wagner, I. (2019a). *Autonomous vehicle technology - Statistics and Facts*. Statista. Retrieved from https://www.statista.com/topics/3573/autonomous-vehicle-technology/

Wagner, I. (2019b). *Projected size of the global autonomous car market in 2018 and 2030 (in billion U.S. dollars)*. Retrieved October 15, 2019, from Statista website: https://www.statista.com/statistics/428692/projected-size-of-global-autonomous-vehicle-market-by-vehicle-type/

Walambe, R., Marathe, A., Kotecha, K., & Ghinea, G. (2021). Lightweight object detection ensemble framework for autonomous vehicles in challenging weather conditions. *Computational Intelligence and Neuroscience*.

Wang, Z., Wu, G., Member, S. S., Barth, M. J., Wu, G., Member, S. S., & Barth, M. J. (2018). *A Review on Cooperative Adaptive Cruise Control (CACC) Systems*. Retrieved from https://arxiv.org/ftp/arxiv/papers/1809/1809.02867.pdf

Wayve, T. (2019). *Learning to Drive like a Human.* Retrieved October 12, 2019, from https://wayve.ai/blog/driving-like-human

White, J. (2019, July 24). *GM Cruise to delay commercial launch of self-driving cars beyond 2019.* Retrieved from https://www.reuters.com/article/us-gm-cruise/gm-cruise-to-delay-commercial-launch-of-self-driving-cars-to-beyond-2019-idUSKCN1UJ1NA

Wischhof, L., Ebner, A., & Rohling, H. (2005). Information dissemination in multihop inter-vehicle networks. *IEEE Conference on Intelligent Transportation Systems Proceedings, 6*(1), 90–101. https://doi.org/10.1109/ITSC.2002.1041301

Worldometer. (2019). *Annual Global Car Production.* Retrieved from https://www.worldometers.info/cars/

Zaher, M. E., Gechter, F., Hajjar, M., & Gruer, P. (2016). An interaction model for a local approach to vehicle platoons. *International Journal of Vehicle Autonomous Systems, 13*(2), 91–113. Retrieved from https://www.inderscienceonline.com/doi/pdf/10.1504/IJVAS.2016.078760

Zang, S., Ding, M., Smith, D., Tyler, P., Rakotoarivelo, T., & Kaafar, M. A. (2019). The impact of adverse weather conditions on autonomous vehicles: How rain, snow, fog, and hail affect the performance of a self-driving car. *IEEE Vehicular Technology Magazine, 14*(2), 103–111.

Zhang, C., Zang, Y., Calvo, J. A. L., & Mathar, R. (2017). A novel V2V assisted platooning system: Control scheme and MAC layer designs. *2017 IEEE 28th Annual International Symposium on Personal, Indoor, and Mobile Radio Communications (PIMRC)*, 1–7.

Zhang, Z., & Zhao, Z. (2014). A multiple mobile robots path planning algorithm based on a-star and dijkstra algorithm. *International Journal of Smart Home, 8*(3), 75–86. https://doi.org/10.14257/ijsh.2014.8.3.07

Zhao, S., Zhang, T., Wu, N., Ogai, H., & Tateno, S. (2015). Vehicle to vehicle communication and platooning for EV with wireless sensor network. *2015 54th Annual Conference of the Society of Instrument and Control Engineers of Japan, SICE 2015*, 1435–1440. doi:10.1109/SICE.2015.7285493

Zhao, Y., & Ogai, H. (2011). Development of a platooning control algorithm based on RoboCar. *SICE Annual Conference 2011*, 352–355.

Zheng, R., Nakano, K., Kato, S., Ogitsu, T., Yamabe, S., Aoki, K., & Suda, Y. (2013). Human-machine interface system for simulation-based automatic platooning of trucks. *16th International IEEE Conference on Intelligent Transportation Systems (ITSC 2013)*, 535–539.

Zheng, R., Nakano, K., Yamabe, S., Aki, M., Nakamura, H., & Suda, Y. (2014). Study on emergency-avoidance braking for the automatic platooning of trucks. *IEEE Transactions on Intelligent Transportation Systems, 15*(4), 1748–1757. https://doi.org/10.1109/TITS.2014.2307160

Ziegler, J., Bender, P., Schreiber, M., Lategahn, H., Strauss, T., Stiller, C., … Zeeb, E. (2014). Making bertha drive-an autonomous journey on a historic route. *IEEE Intelligent Transportation Systems Magazine, 6*(2), 8–20. doi:10.1109/MITS.2014.2306552

Zou, C., & Li, H. (2019). Event-Driven Connected Vehicular Platoon Control with Mixed Time-Varying Delay. *IEEE Access.* doi:10.1109/ACCESS.2019.2927522

Chapter 11
Study on Enhancing the Lifetime of the Pneumatic Cylinder in Automatic Assembly Line by Parametric Accelerated Life Testing

Seongwoo Woo
Ethiopian Technical University, Ethiopia

Dennis O'Neal
Baylor University, USA

Yimer Mohammed Hassen
Ethiopian Technical University, Ethiopia

Gezae Mebrahtu
Ethiopian Technical University, Ethiopia

ABSTRACT

This investigation shows the implementation of parametric accelerated life testing (ALT) as established. It involves (1) a product BX lifetime that X% of a system population is unsuccessful with ALT plan, (2) fatigue design, (3) ALTs with modifications, and (4) discernment as to whether the design(s) fulfills the objective BX lifetime. As a case study, pneumatic cylinder was investigated. The pneumatic cylinder was unsuccessful in a manufacturing line. To reproduce it, parametric ALT was carried out. At the first ALT, the metal seal made of nickel-iron alloy partially cracked and chipped with metal sound. As an action plan, material of the seal was altered from metal to silicone rubber. At the second ALT, due to seal hardening and wear, the piston seal leaked gas. The failure of the silicone seals in the laboratory tests were alike to those returned from the marketplace. The seal material was changed from silicone rubber to (thermoset) polyurethane. For the third ALT, there were no issues. Finally, the lifetime of the pneumatic cylinder was manifested to have a B1 life of 10 years.

DOI: 10.4018/978-1-6684-5381-0.ch011

INTRODUCTION

Due to aggressive circumstances in the marketplace, system operated by a machinery should be devised to have better functioning and big reliability. As the critical attributes of a system that has technical imperfections are assessed in the developing phase, they are quickly integrated into a product and transported to the end-user. Without either unrestricted test or obvious comprehension of how new design can be utilized by the customer, system release with design imperfections may affect the company's brand in a negative way (Magaziner, 1989).

A pneumatic air cylinder is a part which shall be employed in an automatic assembly. It shall thus be employed for a tool-changing implement for a machine tool. To stop an air cylinder from being unsuccessful in the market in its expected life, a producer might substantiate the reliability of a new cylinder linked with proper ISO guidelines and/or carry out required reliability tests before the system is launched.

The field disaster such as Space Shuttle Challenger in the United States' space project was a lethal mishap that happened on 28 January 1986. It was brought about by the unsuccessfulness of O-ring seals used in the booster junction that was wrongly devised to sustain the unusually cold climate that existed at the period of launch. Consequently, as the Challenger travelled through its flying and exploded, all team staffs were dead. The result of this disaster was a stopping of shuttle launching till rectifications were set in the O-ring design. To secure a system isn't unsuccessful in the market, defective parts should be assessed and altered by utilizing a systematic testing method such as ALT, which shall generate reliability quantitative (RQ) statements before the product is released into the field (Woo et al., 2021).

Material imperfections, such as extraordinarily tiny voids, thin surface, contacts, etc. when subjected to repetitive loads, shall start to be unsuccessful because of mechanical failure such as fatigue. It is the main origin of failure in metallic parts, describing approximately 80–95% of all structural failures (McMillan et al., 1982). Metal fatigue shows in the configuration of cracks that emerge in regions where stress shall condense, such as grooves, sharp-edged, holes, etc. It influences the reliability of systems such as aircrafts, automobiles, refrigerator, construction machine, and atomic plants. Especially, a fatigue testing covers as follows: (1) fluctuating stresses, (2) repetitive stresses, (3) reversed stresses, and (4) random stress cycles. Fatigue can thus be influenced by the cyclic stress range, average stress or stress ratio, R ($=\sigma_{min}/\sigma_{max}$), which is elucidated as the correlation of the least cyclic stress to the greatest cyclic stress (Duga et al. in Campbell, 2008). For interval forms, the heights at both the greatest (high side) and the least (low side) are important. As utilizing an accelerated loading, parametric ALT shall be employed to recognize the structural imperfections such as a stress raiser.

The ALT integrated with the reliability diagram representation was explored as an alternative procedure (Modarres et al., 2016). It covered a test scheme for the product, recognizing to use fatigue failure, and employing a sample size expression, elevated loading, etc. Elsayed (2012) grouped statistics, physics/statistics, and physics/test-based frameworks for investigation. Hahn and Meeker (2004) proposed some feasible procedures to set out an ALT. Conducting an elevated test (McPherson, 1989; 2010) entails some basic idea such as the BX lifetime for the testing strategy, a streamlined life-stress prototype, sample size expression, and fracture mechanics (Anderson, 2017) because unsuccessfulness can instantly take place because of weak parts in the structure. However, present test skills (Braco et al., 2021) cannot replicate the design imperfections in a multi-module structure because it assesses limited component samples & testing time.

To attain a structural robust design operated by machine, designers have utilized established skills such as mechanics of materials and fracture mechanics (Weingart & Stephen, 2007). They have employed quantum mechanics to help pinpoint failures that are arisen from micro-void coalescence (MVC) in numerous metallic alloys and engineering plastics. To realize the fatigue origin of a system, a life-stress prototype shall be employed with typical designs – statics/dynamics, machine design, etc. – and connected skills to help recognize the unsuccessfulness of electronic components due to matter imperfections or tiny cracks when the components are subjected to repeated (mechanical) stresses. Finite element methods (FEMs) cannot recognize such failure sources (Reddy, 2021). As an alternative procedure, there is structural health monitoring (SHM) which allows for the catch of the unsuccessfulness sources (Raghavan & Cesnik, 2007). However, it is not easy to attain test statistics for multi-module system because massive tests shall have been involved, which shall be too overpriced due to sample size & testing time.

To manifest the success of recognizing and modifying the structural imperfections of a product operated by machine, an ALT shall be employed as a structured reliability procedure which produces the RQ statements such as assigned mission time. It includes: (1) a BX life on the parametric ALT strategy, (2) a load analysis, (3) parametric ALTs with the design modifications, and (4) a judgement of whether the system design(s) achieves the aimed BX life. The deduction of the sample size expression, quantum-based time-to-failure, and BX life is supplied. To verify the effectiveness of ALT outcomes in actual lifetime and compare the existing design with last ones, it might be required to observe the release of the brand-new design in the field to make sure it extended the objective life and failure rate. An air cylinder in a machine tool subjected to repetitive pressure loads is utilized to explain these systematic methods.

PARAMETRIC ALT FOR PRODUCT OPERATED BY MACHINE

Concept of BX Life Target

In product operated by machine, power shall be utilized to mechanical advantage to supply a task which necessitate forces and motion by properly employing mechanisms. When systems are subjected to repetitive loading, they shall set an aimed life (or design) for the anticipated stresses in the structure. An instance is a machine tool which can be utilized in a machine metal. One of the apparatuses is the air cylinder. It also covers some subsystems (or modules) — changing device, tool (or pellet) spindle/drive unit, tilting index table, hydropower unit, cooler unit, turret head, CNC controller, etc. The life of the machine tool is determined by the unsuccessfulness of the changed module such as the air cylinder that has structural imperfections (Figure 1).

To conduct a parametric ALT, the BX lifetime such as "Bearing Life", L_B, might be defined as a metric of the system life. The BX lifetime measure arose from the roller and ball bearing company but has grown to be a system life standard employed at various fields today. The ALT shall do whether product satisfies the field requirements for the life & the accumulative failure rate. BX life therefore is a passed time for which X% of a group of parts under consideration is in vain. A "BX life Y years" is a proper representation of defining the system life. If a system is a B20 life 10 years, it signifies that 20% of a collection of items under consideration shall fail in the ten years of working.

For instance, the necessitated product life in a machine tool relies on roughly 2000 parts, incorporating the air cylinder. If air cylinder has structural imperfections, it might affect the possible lifetime of the tool device. If its life is targeted to be a B20 life 10 years, the life of each unit might be devised to be a B1 life 10 years because the machine tool is made up to 20 units and every unit has 100 parts.

*Figure 1. System life determined by new subsystem (**a**) breakdown of machine center; (**b**) system life L_B.*

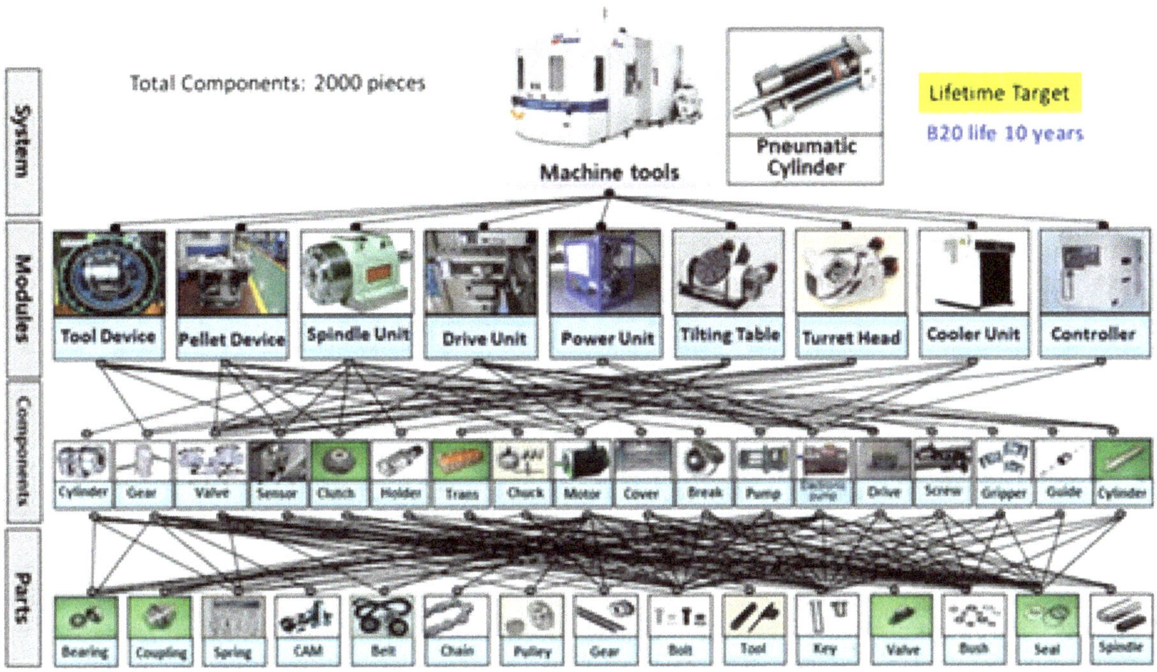

Putting a Whole Parametric ALT Strategy

The reliability of a product operated by machinery shall be assessed as the capability necessitated to run under expressed surround/operational circumstances for a defined period of time (IEEE Standard Glossary of Software Engineering Terminology, 2002).

If T is a random variable meaning the time-to-failure in Figure 2, the reliability, the portion yet outliving at time t, could be expressed as:

$$R(t) = P(T > t) \tag{1}$$

The accumulative distribution function (CDF), $F(t)$, is $F(t)=1-R(t)$. The failure rate of product can be stated as:

$$\lambda(t) = \frac{f(t)}{R(t)} = \frac{dF(t)/dt}{R(t)} = \frac{(1-R(t))'}{Rt} = \frac{-R'(t)}{R(t)} \tag{2}$$

where λ is the failure rate, f is the density function.

If Equation (2) finds the integral till $T = L_B$, the $X\%$ accumulative failure, $F(L_B)$, shall be attained as:

$$A(=X) = \lambda \cdot L_B = \int_0^{L_B} \lambda(t) \cdot dt = -\ln R(L_B) = -\ln(1-F) \cong F(L_B) \tag{3}$$

Figure 2. Reliability R(t) and accumulative distribution F(t)

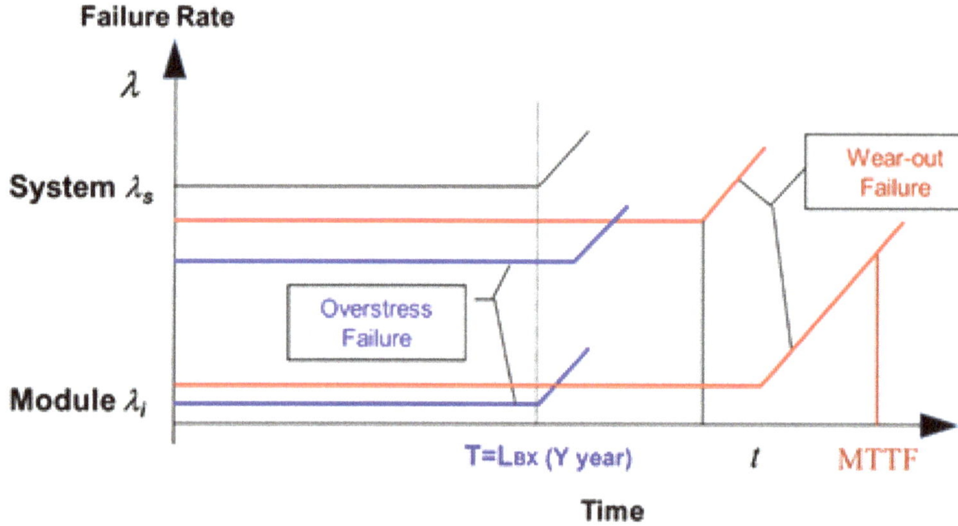

When system reliability is expressed as utilizing the conventional "bathtub curve", there is made up to three zones: (1) in the first area, during the premature system lifetime, the failure rate will be some decrease, (2) in the second area, during its medium life, the failure rate has a comparatively constant, and (3) in the third area, the failure rate has an increasing till the life of the system is fulfilled.

Presuming that T_I shall be the hour of the 1st failure in the second area, reliability $R(t)$ shall be attained:

$$R(t)=P(T_1>t)=P(\text{no failure}\quad \text{in}\ (0,t])=\frac{(m)^0 e^{-m}}{0!}=e^{-m}=e^{-\lambda t} \tag{4}$$

When the design of a system operated by machine is enhanced, the failure rate in the market will drop. On the other hand, the system life increases. For such circumstances, the reliability of system operated by machine shall be expressed as the product of life L_B and failure rate λ:

$$R(L_B)=1-F(L_B)=e^{-\lambda L_B}\cong 1-\lambda L_B \tag{5}$$

This relation will be valid below roughly 20% of the accumulative failure (Kreyszig, 2011).

Tool changing in a machine necessitates a mechanical action: (1) The machinery chooses the switched tool out; and (2) it thus changes the demanded tool. The statistics for the unsuccessful systems in the market were main for comprehending and classifying the use modes of end-users and helping to find structural imperfections which is required to be modified in the system. Based on the field statistics, the possible cause(s) of the air cylinder returned in the field was pinpointed. That is, as targeting the life, L_B, the system operated by machine might thus be altered by recognizing the problematic materials and redesigning them by utilizing an ALT (Figure 3).

Figure 3. Intended task of tool changing device incorporating air cylinder (instance).

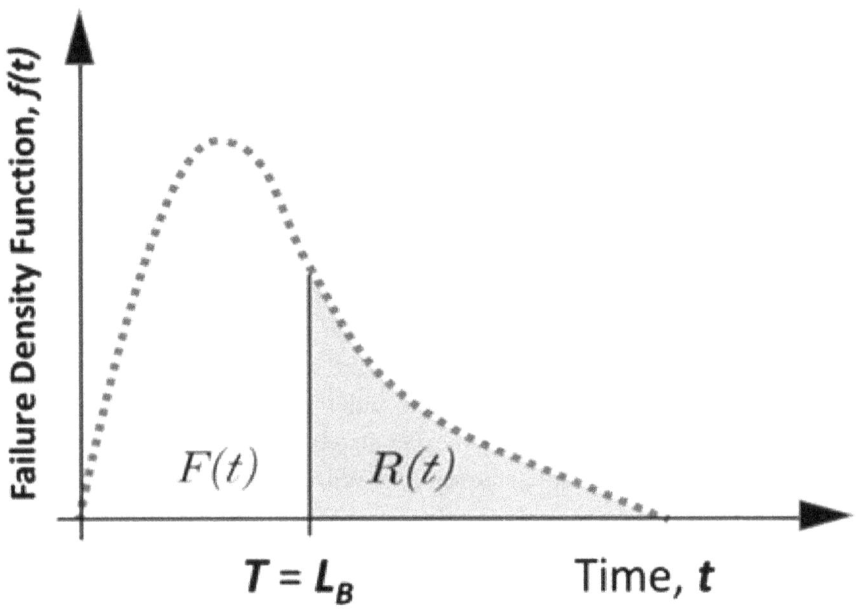

To put the aimed life of a machine tool by an ALT that is composed of nine subsystems (or modules), seen in Figure 1a, there are the subsequent subsystems: (1) modified subsystem, (2) new subsystem, and (3) alike subsystem to the previous design entrenched on an ask in the market. The air cylinder in a tool device investigated is an altered subsystem. It was initially modified to aim on the air cylinder and become the machine tool more aggressive in the market. However, the altered air cylinder was the structural imperfections which was required to be redesigned due to field failures.

Table 1. Entire ALT strategy of modules operated by a machine tool.

Modules	Field Data		Expected Reliability				Aimed Reliability	
	Failure Rate Per Year, λ (%/Year)	BX Life, L_B (Year)	Failure Rate Per Year, λ (%/Year)			BX Life, L_B (Year)	Failure Rate Per Year, λ (%/Year)	BX Life, L_B (Year)
A	0.24	4.2	New	×5	1.20	0.83	0.10	10(BX = 1.0)
B	0.31	3.2	Altered	×2	0.62	1.61	0.10	10(BX = 1.0)
C	0.15	6.7	Altered	×2	0.30	3.33	0.10	10(BX = 1.0)
D	0.35	2.9	Same	×1	0.35	2.9	0.10	10(BX = 1.0)
E	0.30	3.3	Same	×1	0.30	3.33	0.10	10(BX = 1.0)
Others (F/G/H/I)	0.50	10.0	Same	×1	0.50	10.0	0.50	10(BX = 5.0)
Product	1.9	2.9	-	-	3.27	0.83	1.00	10(BX = 10)

The new subsystem A from the market statistics in Table 1 was the failure rate of 0.24% per year and a B1 life 4.2 years. Based on it, the life of the tool device incorporating air cylinder was the anticipated B1 life 0.83 years because they were the failure rate – 1.2% per year. To meet end-user needs, a new aimed life for the system operated by machine such as a pneumatic cylinder in a tool change device was set to be a B1 life 10 years.

Time-to-Failure and Sample Size Expression

Mechanical product transfers power from one location to another by implementing a proper mechanism, such as an air cylinder. When the pressure difference between the two rooms in a pneumatic cylinder push (or pull), the piston rod expands (or draws back). In the process, it guides a piston by linear path move to the required end in an air cylinder. Cylinder parts are subjected to repetitive stresses because of loading/unloading relied on end-masses that are diverse with the implementations. If there is a structural imperfection that has a deficient strength (or stiffness) when the loading is employed, the system operated by machine shall suddenly fail before its aimed life. With the failure, the problematic system will necessitate proper service (Figure 4).

Figure 4. Air cylinder use in the market.

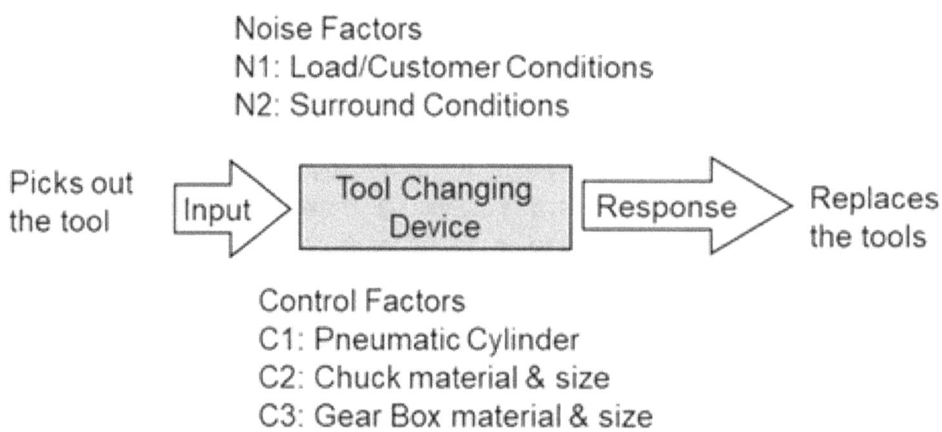

Fatigue in metal shall define the unexpected failure of system parts which produce unsuccessfulness in their life. It is correlated to the amount of stress cycles experienced by a part and the quantity of stress applied on the structure. Fatigue develops if there are some stress raisers such as a notch (or inadequate material). There is also some relationship between metal part's ultimate tensile strength and rigidity and its capacity to withstand fatigue loading. That is, the higher the tensile strength and hardness, the more chance it shall fatigue if subjected to loftier oscillating loading.

Fatigue due to structural imperfections shall be attributed by the subsequent factors: (1) the stress due to loading in the structure and (2) the kind of matters (or form) used. In pinpointing the product unsuccessfulness by an ALT, a designer might thus structure parts with optimal forms and desired material to expand the anticipated lifetime of the system. Finally, the product shall thus sustain enough repetitive loading or cycles to fulfil its aimed life (Figure 5).

Figure 5. Fatigue unsuccessfulness on the system produced by repetitive impact loads and structural imperfections.

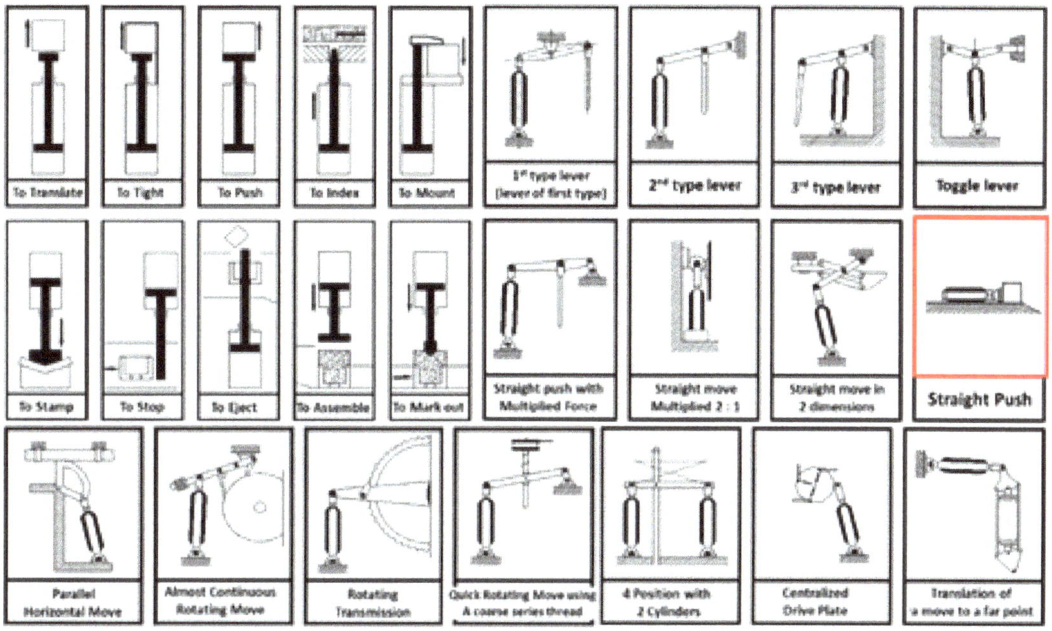

The chief issue for reliability tests is to decide how rapidly the probable failure shall be recognized. That is why structured ALT is required. To do it, it is required to accomplish a simple failure representation and decide the accurate numerical quantity for the prototype. A life-stress (LS) prototype shall be suggested, which holds stresses and reaction quantities. It shall describe some failures such as mechanical fatigue. It shall not only happen due to stresses in a component but also due to the material imperfections such as tiny cracks on the surface of a part.

Figure 6. Possible exchange in a semi-conduct such as silicon when electric-magnetic field is applied.

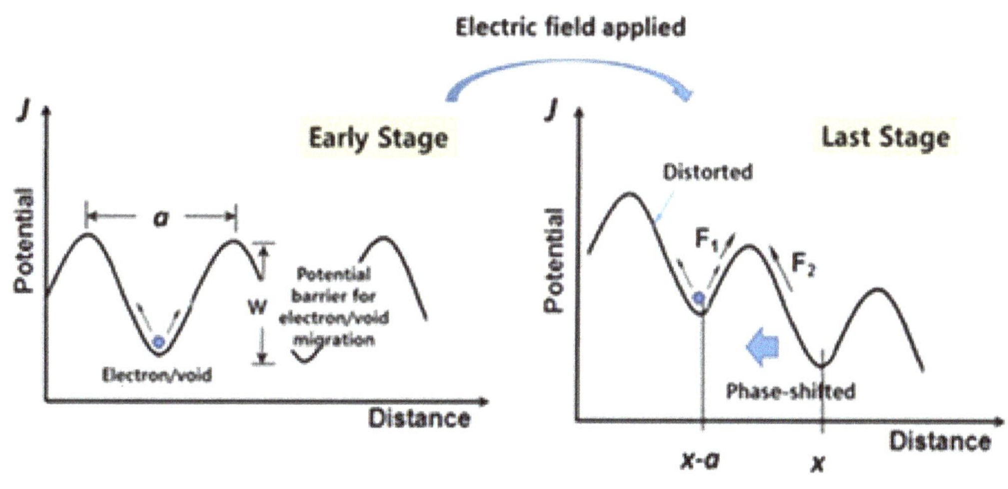

Fatigue starts from matter imperfections – electron/void – which are generated on a microscopic or macro range. From such conceptual viewpoint, it shall be described as transport procedures – the diffusion of shallow level dopants of silicon.

First of all, consider non-interacting electrons moving (freely) in a constant potential wall which we take to be zero. Time independent Schrodinger's equation in one dimension can be expressed as follows:

$$-\frac{h^2}{8\pi^2 m}\frac{d^2\psi_n(x)}{dx^2} = E_n\psi_n \tag{6}$$

where ψn is wave function, En is (electron) energy, h is Planck constant, m is electron mass.

Boundary conditions are: 1) ψ_n finite inside the metal but decaying exponentially. That is, $\psi_n \to 0$ as $x \to \infty$, 2) $\psi_n = 0$ at walls. So, we can find the solution as follows:

$$\psi_n(x) = \sqrt{\frac{2}{a}} sin\left(\frac{n}{a}\right)x;\ E_n = \frac{n^2 h^2}{8ma^2} n > 0 \tag{7}$$

where $\psi_n(x+a)=\psi_n(x)$, a is (periodic) distance, n is principal quantum number.

When an operating force such as an electro-magnetic force, ξ, is applied, the material impurities – voids, related to electronics move, are easily drifted because the walls of junction energy are lowered and distorted/phase-shifted (Figure 6).

Linear transport procedure is expressed by the common configuration:

$$J = LX \tag{8}$$

where J is a flux vector, X is described as a driving force, L is a transport coefficient.

Transport processes are naturally dissipative, which is happening in numerous systems, shall become the system balance when total transport becomes to a stop.

For example, it shall implement the transport process for solid-state diffusion of impurities in semiconduct such as silicon. It includes electro-migration-induced voiding, growth of chloride ions, and catching of electrons or holes. Solid-state diffusion of impurities in a semiconductor such as silicon J shall be abridged as (Grove, 1967):

$$J = \left[aC(x-a)\right]\cdot \exp\left[-\frac{q}{kT}\left(w - \frac{1}{2}a\xi\right)\right]\cdot v$$

$$= -\left[a^2 ve^{-qw/kT}\right]\cdot \cosh\frac{qa\xi}{2kT}\frac{\partial C}{\partial x} + \left[2ave^{-qw/kT}\right]C\sinh\frac{qa\xi}{2kT}$$

$$= \Phi(x,t,T)\sinh(a\xi)\exp\left(-\frac{Q}{kT}\right)$$

$$= B\sin h(a\xi)\exp\left(-\frac{Q}{kT}\right) \tag{9}$$

where C is the concentration, q is the electric energy, ν is the frequency, B is a constant, a is the mean-time between atoms, ξ is the exerted energy field, k is the Boltzmann's constant, T is the temperature, and Q is the energy.

The reaction procedure, which relies on speed, shall be abridged as:

$$K = K^{+} - K^{-} = a\frac{kT}{h}e^{-\frac{\Delta E - aS}{kT}} - a\frac{kT}{h}e^{-\frac{\Delta E + aS}{kT}} = B\sinh(aS)\exp\left(-\frac{\Delta E}{kT}\right) \tag{10}$$

where K is the reaction rate, S is the (chemical) consequence, T is the temperature, E is the activation energy, and Δ is the difference.

The junction, J, from Equations (9) and (10) shall be depicted as:

$$J = B\sinh(aS)\exp\left(-\frac{E_a}{kT}\right) \tag{11}$$

If Equation (11) puts an inversed function, the life-stress (LS) protype could be illustrated as:

$$TF = A\left[\sinh(aS)\right]^{-1}\exp\left(\frac{\boldsymbol{E_a}}{\boldsymbol{kT}}\right) \tag{12}$$

The sine hyperbolic expression $[\sinh(aS)]^{-1}$ in Equation (12) shall be uttered as (Figure 7):

1. $(S)^{-1}$ at the beginning has a little linear outcome;
2. $(S)^{-n}$ has what is judged as a meddle outcome;
3. $(e^{aS})^{-1}$ on end is huge.

A parametric ALT is generally performed in the meddle range, and Equation (12) shall be expressed as:

$$TF = A(S)^{-n}\exp\left(\frac{E_a}{kT}\right) \tag{13}$$

Because the stress in a system operated by machine is not easy to calculate in an parametric ALT, Equation (13) shall be redefined. As the power due to some energy disciplines in Equation (13) is defined as the multiplication of flows and effort, stresses can start from effort in a system (Table 2) (Karnopp et al., 2012).

Stress is a matter amount. It can define the (internal) forces which adjoins sub-atomics of a continuum matter exerted on each other. For a product operated by machine, because stress starts from effort, Equation (13) could be redefined as:

$$TF = A\left(S\right)^{-n} \exp\left(\frac{E_a}{kT}\right) = B\left(e\right)^{-\lambda} \exp\left(\frac{E_a}{kT}\right) \tag{14}$$

where A and B are constants.

Figure 7. Sine hyperbolic stress expression versus S-N curve from a viewpoint of stress scope.

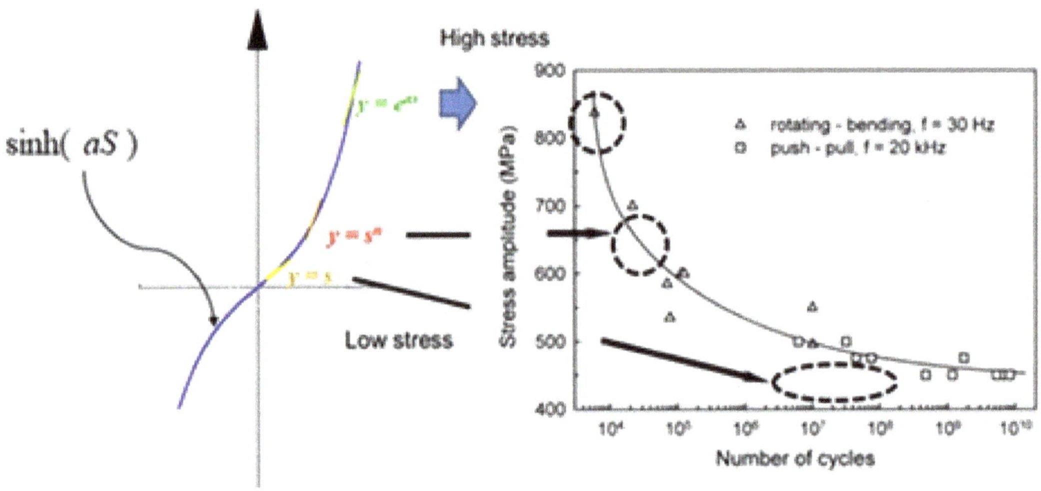

Table 2. System power which is defined as effort and flow.

Structure	Power, $e(t) \times f(t)$	Effort, $e(t)$	Flow, $f(t)$
Translation	$F \times V$	Force, $F(t)$	Velocity, $V(t)$
Rotation	$T \times \omega$	Torque, $\tau(t)$	Angular velocity, $\omega(t)$
Pump/compressor	$\Delta P \times Q$	Pressure difference, $\Delta P(t)$	Volume flow rate, $Q(t)$
Electrical	$V \times i$	Voltage, $V(t)$	Current, $i(t)$
Magnetic	$e_m \times \varphi$	Magnetic force, $e_m(t)$	Magnetic flux, $\varphi(t)$

To get the elevated factor, which shall influence the strength assessment of fatigue in a system, accelerated factor (AF) shall be expressed as the proportion between the raised stress quantities and usual functioning circumstances. It shall be altered to combine with the effort concept:

$$AF = \left(\frac{S_1}{S_0}\right)^n \left[\frac{E_a}{k}\left(\frac{1}{T_0} - \frac{1}{T_1}\right)\right] = \left(\frac{e_1}{e_0}\right)^\lambda \left[\frac{E_a}{k}\left(\frac{1}{T_0} - \frac{1}{T_1}\right)\right] \tag{15}$$

Because elevated testing is carried out at room temp. for air cylinders, Equation (13) shall be restated as:

$$AF = \left(\frac{S_1}{S_0}\right)^n = \left(\frac{e_1}{e_0}\right)^\lambda \tag{16}$$

To attain the assigned mission time of ALTs from the aimed BX life on the test strategy in Table 1, the sample size expression united with equation (15) shall be obtained (Woo et al., 2019).

$$n \geq (r+1) \times \frac{1}{x} \times \left(\frac{L_B^*}{AF \cdot h_a}\right)^\beta + r \tag{17}$$

where the sample size expression shall be expressed as n ~ (failure numbers + 1)·(1/accumulative failure)·((aimed life/(planned test cycles)) ^ $\beta + r$.

Equations (17) also shall be expressed as (Wasserman, 2003). If the objective life of a product operated by machinery, such as the air cylinder, is assured to be B1 life 10 years, the assigned mission time shall be obtained for samples. Through parametric ALTs, the structural imperfections of product shall be recognized to fulfil the aimed life.

Case Study: Enhancing Life of a Newly Designed Air Cylinder in an Automatic Line

Engineer employs pneumatic cylinder because it is silent, cleaned, and do not need huge spaces for fluid storing place. The air cylinder is a mechanical system by relatively simple adapting mechanism and can be utilized in a variety of mechanical system. As manufacturer utilizes tools in a line, engineer can easily alter the tools by just pushing them such as the red-colored device in Figure 4. Air cylinders shall be utilized to help carry devices operated by machine. An air cylinder mounted in a tool device should be devised with an expected life.

To convey it along a path to the needed location, some structural parts in the air cylinder are necessitated. They are made up to a rod cap, head cap, piston, etc. By the anticipated end-user circumstances, the air cylinder in a machine tool was subjected to repeated loading for usual working. There were structural imperfections in a new air cylinder which will shorten its anticipated life. However, it was unknown which components in the pneumatic air cylinder were unsuccessful. To recognize and alter them, it is essential to apply the structured reliability method – parametric ALT (Figure 8).

Based on the market statistics, as an air cylinder was subjected to repetitive loading in its expected life for common enduser use, the cracked (or harden and worn) piston seal created the air cylinder to accidentally lose contents and stop working. The field statistics about the problematic air cylinders were critical for comprehending the use patterns of end-users and helping to decide design imperfections – wrong-selected material rubber or polyurethane – which should be altered in the structure. According to the market statistics, it was required to replicate the unsuccessfulness of the problematic air cylinder. If the aimed life were put, it shall recognize which part(s) in the air cylinder might be altered to increase its life by ALT.

As market statistics were exclusively appraised, the troublesome products accidentally appeared to lose contents near the piston. As the air cylinder was subjected to repetitive loads, it encountered the following presumed failure processes: (1) continually heating of the piston seal; (2) hardening and wear of the piston seal; and (3) being over Minimum Operation Pressure (MOP) and Stroke Time (ST). The problematic pneumatic air cylinders returned from the market had structural imperfections – inappropriate seal (metal) material.

Figure 8. Air cylinder in a line.

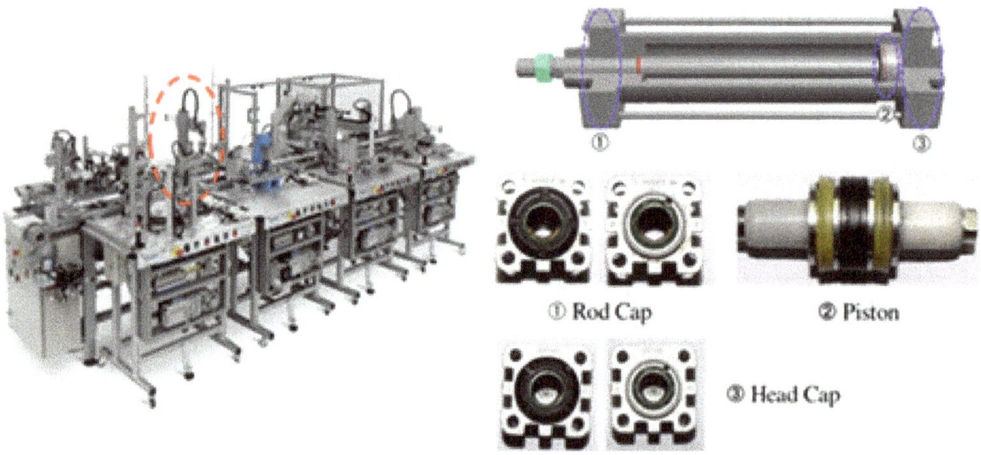

Because the factory line, incorporating the air cylinder, had to be stopped, producers shall have the problematic product replaced. To prevent the field issue, it was important to replicate the potential failure of the air cylinder in laboratory circumstances in a chamber mounted for apparatus. Initially, to supply the toleration of loading of the air cylinder, seal material such as Fe-36Ni Invar Alloy holding 36% Ni was utilized because of its fine sealing characteristics. Due to their rigidity and comparatively tall oxidation resistance at excessive temperatures, nickel alloys have been utilized in various fields – gas turbines, rocket engines, boat propeller, etc. It is too uncomplicated to work and shall be extracted into wire. As the fatigue of nickel alloys in the present design of an air cylinder was assessed by ALT, design modifications shall be attained for fulfilling the aimed life (Figure 9).

The pressure of the air cylinder because of loads shall be expressed as:

$$\Delta P = \Delta F_{int} / A \tag{18}$$

where ΔF_{int} = force difference in an air cylinder due to loading.

Under elevated circumstances, the life-stress prototype (LS model) from Equation (14) shall be defined:

$$TF = A(S)^{-n} \exp\left(\frac{E_a}{kT}\right) = B(\Delta P)^{-\lambda} \exp\left(\frac{E_a}{kT}\right) \tag{19}$$

where A and B is quantity, k is Boltzman's quantity, E_a is the activation energy, T is the temperature, n is the quotient, and λ is the accumulative damage quantity.

As ALT is carried out at room conditions, Equation (15) shall be defined as

$$AF = \left(\frac{S_1}{S_0}\right)^n = \left(\frac{\Delta P_1}{\Delta P_0}\right)^\lambda \tag{20}$$

where S_1 (or ΔP_1) is stress due to pressure difference at raised circumstances, and S_0 (or ΔP_0) is stress due to pressure difference at normal circumstances.

Figure 9. Unsuccessful seal material in the market.

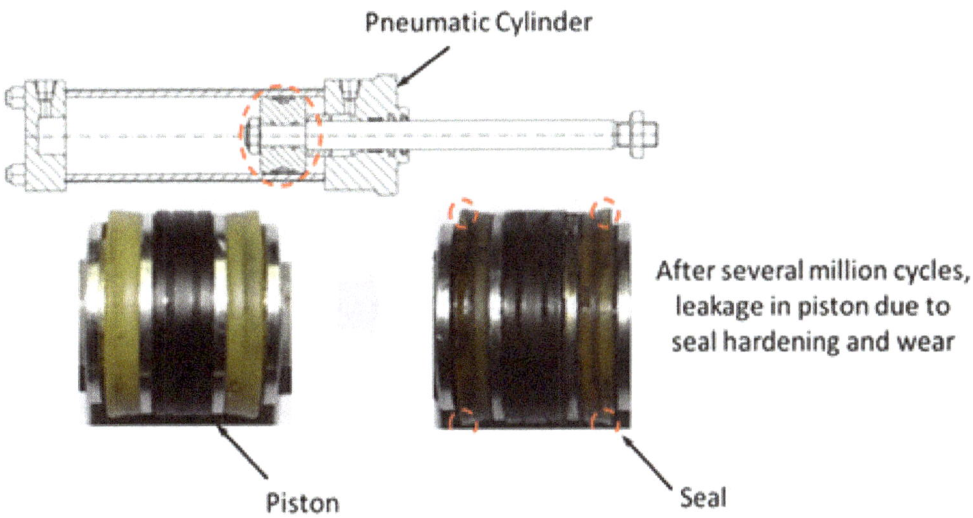

For an air cylinder in a tool device, the normal working temperatures varying from 0 °C to 43 °C with humidity varying from 0% to 95%. Vibration circumstances presumed for functioning the air cylinder spanned from 0.2 to 0.24g of acceleration. As changing a tool was anticipated to happen a mean of ten to thirty times per day. Under the most severely circumstances, the air cylinder functioning for ten years shall roughly experience 109,500 cycles.

To decide the stress quantity for parametric ALT, according to the permissible use limit of air cylinder in bench-marked statistics which were attained from competitive manufacturers, we employed the step-stress life testing that might assess the life at regular usage circumstance for some raised loading (or load masses) (Tang, 2003). As the diverse stress quantity due to end-load mass was altered, we shall see the failure time of the air cylinder at particular stress quantities such as 0.8MPa, 1.2MPa, and 1.4MPa which depended on end-load masses in some areas. For this ALT, the pressure difference was eventually accelerated to have 0.8 MPa with a cumulative damage quantity, λ, of 2. The whole AF calculated in Equation (20) was 1.6.

The assigned mission time for the parametric ALTs determined from Equation (17) were 200,000 cycles for ten samples if the shape parameter, β, was assumed to have 2.0. This ALT was devised to make sure an aimed life — B1 life 10 years — with a roughly 60% level of confidence that it shall be unsuccessful less than once for 200,000 cycles.

Figure 10. Apparatus for parametric ALTs.

RESULTS AND DISCUSSION

To put the stress quantity by the step-stress lifetime testing, we observed the unsuccessful cycles at the subsequent stress – 0.8 MPa, 1.2 MPa, and 1.4 MPa which shall be attained from end-mass. For 0.8 MPa, the air cylinder made sound at 6000 cycles and 10,000 cycles. For 1.2 MPa, the air cylinder made sound at 4,000 cycles and 8,000 cycles. And, for 1.4 MPa, the air cylinder made sound at 1000 cycles. So, we determined the stress quantity as 0.8MPa for ALT because it had a relatively better data-linearity, contrasted to other stress quantities.

Originally, when 0.8 MPa as the accelerated level in a air cylinder was pressurized, the air cylinder (n = 10) made sound at 6000 cycles, 10,500 cycles, and 11,000 cycles which had a 70 dB (greatest) and caused moderate-to-severe hearing losing. As disassembling three troublesome specimens, one specimen was a partly cracked sample and two specimens were chipped due to the repetitive loads. The failed sections were investigated by utilizing scanning electron microscopy (SEM). The place of the multiplicative crack starting and the microscopic growing at the correct origin of the crack were recognized. We therefore thought that repetitive pressure loading was the very possible root of unsuccessfulness for the partially cracked seal in air cylinders. In other word, as pressure loading was repetitively exerted to the seal along the axial orientation of the air cylinder, it might result in such partly cracked and chipped specimens. It was concluded that this unsuccessfulness originated from an inadequate seal matter – Fe-36Ni Invar Alloy. Though Fe-36Ni Invar Alloy had fine sealing characteristics, it was finally unsuccessful under repetitive pressure loads. The seal matter such as Fe-36Ni Invar Alloy in the air cylinder was altered to a silicone rubber (C1) (Figure 11).

Figure 11. Structural problems of air cylinder in 1st ALT

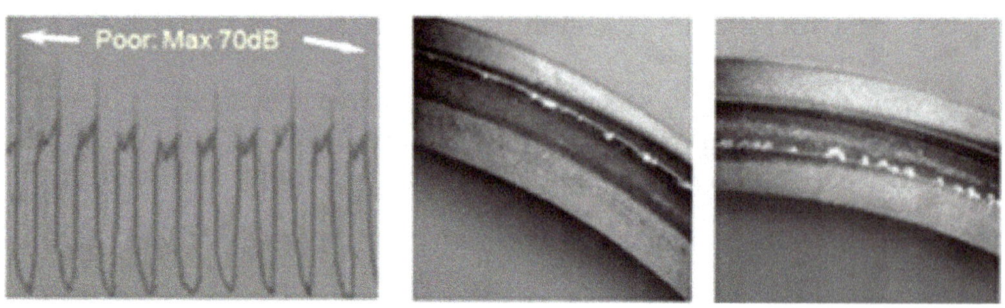

Various cylinders (n = 10) in the 2nd ALT was unsuccessful at 50,000 cycles (one specimen), 100,000 cycles (two specimens), 110,000 cycles (one specimen), and 115,000 cycles (one specimen). The failure mode of the specimens was over the lowest operation pressure and piston stroke hour. When the troublesome air cylinders were taking apart, the hardening and wear of piston seals were like those of specimen failed from the field. The failure process occurred: (1) repeated operating stress, (2) friction heat, (3) lube vaporization, (4) elevating seal hardening, and (5) elevated seal wear. The forms and locations of the failed specimens attained from the 2nd ALT and the market (or usual circumstance) were alike (Figure 12).

Figure 12. Failed parts in market and ALT

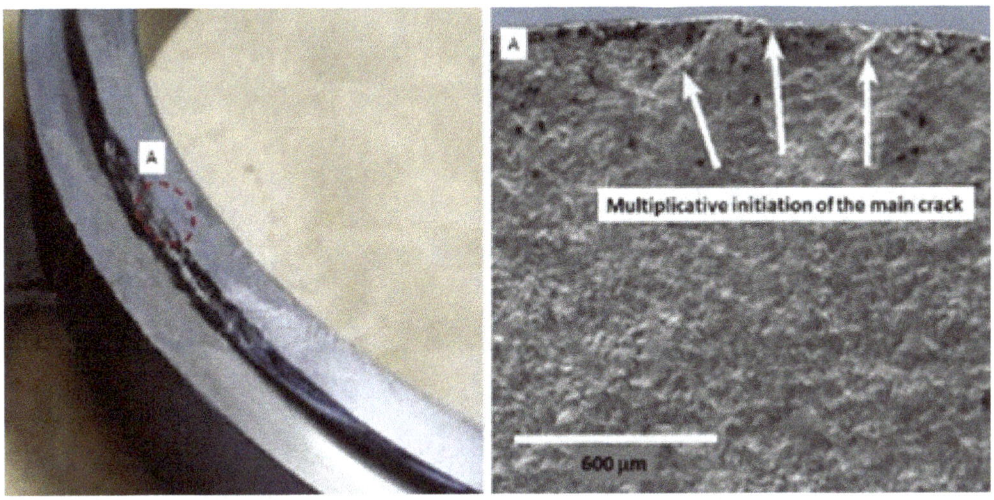

The failure of 2nd ALT and field also appeared an alike form on a Weibull chart (Figure 13). Because the statistics for the 2nd ALT and field had alike gradients on the chart, we could expect that each load circumstance of the 2nd ALT and that from the market over the system life were alike. It could be expected that the test specimens in the room would be similarly unsuccessful to those in the market. The shape parameter, β, from the Weibull plot was affirmed to be 1.97, contrasted with the approximated value of 2.0. Based on both test results (Figure 12 & 14) and result of the Weibull chart (Figure 13), this ALT was accomplishing a desired result. because it pinpointed the structural imperfections that were accountable for the field unsuccessfulness.

To attain the unsuccessfulness of the air cylinder at an accelerated circumstance of 1.2 MPa and 23 °C, parametric ALT was performed. As a consequence, the air cylinder was unsuccessful in 5 out of 6 samples: roughly 30,000 cycles (one specimen), 70,000 cycles (two specimens), and 80,000 cycles (two specimens). The possible failure pattern of the specimens was over the minimum operation pressure and stroke time piston. As the failed cylinders were taking apart, we found that there was hardening and wear of the piston seal and the rod cap impeded by a slurry. The repetitive loads of the air cylinder created the wear of the piston seal that produced the slurry and its flow into the port. The slurry thus gathered at the port and congested it.

Figure 13. Field statistics and outcomes of 2nd ALT on Weibull plot.

Figure 14. Congested rod cap in the ALT.

As investigating the failed product in the 2nd ALT, we knew that the hardening and wear of piston seal in the air cylinder originated from an inadequate matter. For this ALT, it was a silicone rubber. Though silicone rubber was fine sealing characteristics, it was yet unsafe to malfunction under repetitive loads. The seal matter such as silicone rubber, C2, was altered to a (thermoset) polyurethane with excellent erosion and wear resistance that supplied improved lifetime.

In the 3rd ALT, there were no structural issues in the air cylinder till it passed 220,000 cycles. We inferred that the structural alternations obtained from the 1st and 2nd ALTs were practical in fulfilling the aimed life —B1 life 10 years – of product.

CONCLUSION

To lengthen the life of a new product such as an air cylinder utilized in a machine tool, we utilized a reliability methodology that included a quantum-based life-stress prototype by transport procedure and a sample size expression. It included the following: (1) the BX life generated the ALT strategy, (2) parametric ALTs with structural alternations, and (3) determination if the design attained the desired number of assigned mission cycles. The failure and redesign of a pneumatic cylinder were used as a test case for the methodology.

- In the 1st ALT, the cylinder (n = 10) made sound around 6,000 cycles, 10,500 cycles, and 11,000 cycles. As taking apart three troublesome samples, we found they are one partly cracked specimen and two chipped specimens due to heavy load. The seal matter – Fe-36Ni Invar Alloy – in an air cylinder was altered to silicone rubber.
- In the 2nd ALT, we caught the hardening and wear of piston seal like those of market specimens. It happens: (1) repeated working stress, (2) friction heat in seal, (3) lube vaporization in seal, and (4) elevating seal hardening & wear. We modified the seal matter such as silicone rubber, C2, to (thermoset) polyurethane.
- During the 3rd ALT, there were no issues discovered. The air cylinder systems might attain the aimed life — B1 life 10 years. By inspecting troublesome market systems and conducting ALTs with structural alternations, it could enhance the life of the pneumatic cylinder.

This structured method has been applied successfully to other systems operated by machine and, as demonstrated here, was effective in improving the reliability of the pneumatic cylinder. It should be applicable to other mechanical systems. Designers need to understand why multi-module system is unsuccessful in their life. If there are structural imperfections in the newly designed module structures and these are subjected to repetitive loading during its operation, the system can be unsuccessful before its anticipated life. Engineers might need to pinpoint the (dynamic) loading of the system operated by machine so that the elevated testing expressed as the proportion between greatest stress in contrast to least stress shall be performed till the required mission cycles (reliable quantitative specification) are obtained from the sample size equation. In the meantime, ALT shall be employed to pinpoint and alter the structural imperfections of a system operated by machine.

REFERENCES

Anderson, T. L. (2017). *Fracture Mechanics—Fundamentals and Applications* (3rd ed.). CRC. doi:10.1201/9781315370293

Braco, R., Prates, P., Costa, J. D. M., & Berto, F. (2018). New methodology of fatigue life evaluation for multiaxially loaded notched components based on two uniaxial strain-controlled tests. *International Journal of Fatigue*, *111*, 308–320. doi:10.1016/j.ijfatigue.2018.02.027

Duga, J. J., Fisher, W. H., Buxaum, R. W., Rosenfield, A. R., Buhr, A. R., & Honton, E. J. (2008). Fatigue. In F. C. Campbell (Ed.), *Elements of Metallurgy and Engineering Alloys*. ASM International.

Elsayed, E. A. (2012). *Reliability Engineering*. John Wiley & Sons.

Grove, A. (1967). Physics and Technology of Semiconductor Device (1st ed.). Wiley International Edition.

Hahn, G. J., & Meeker, W. Q. (2004). *How to Plan an Accelerated Life Test (E-Book)*. ASQ Quality Press.

IEEE Standard Glossary of Software Engineering Terminology. (2002). *IEEE STD 610.12-1990. Standards Coordinating Committee of the Computer Society of IEEE.* Available online: https://ieeexplore.ieee.org/document/159342

Karnopp, D. C., Margolis, D. L., & Rosenberg, R. C. (2012). *System Dynamics: Modeling, Simulation, and Control of Mechatronic Systems* (6th ed.). John Wiley & Sons. doi:10.1002/9781118152812

Kreyszig, E. (2011). *Advanced Engineering Mathematics* (10th ed.). John Wiley and Son.

Magaziner, I. C., & Patinkin, M. (1989). Cold competition: GE wages the refrigerator war. *Harvard Business Review*, *89*, 114–124.

McMillan, S. C. (1982). *The Economic Effects of Fracture in the United States*. Final Report.

McPherson, J. (1989). Accelerated testing. In Electronic Materials Handbook Volume 1: Packaging. ASM International Publishing.

McPherson, J. (2010). *Reliability Physics and Engineering: Time-to-Failure Modeling*. Springer. doi:10.1007/978-1-4419-6348-2

Modarres, M., Kaminskiy, M., & Krivtsov, V. (2016). *Reliability Engineering and Risk Analysis: A Practical Guide* (3rd ed.). CRC Press. doi:10.1201/9781315382425

NBS Special Publication 647-2. (n.d.). Battelle Laboratories.

Raghavan, A., & Cesnik, C. E. (2007). Review of guided-wave structural health monitoring. *Shock & Vibration Digest*, *39*(2), 91–114. doi:10.1177/0583102406075428

Reddy, J. N. (2021). *An Introduction to Nonlinear Finite Element Method with Applications to Heat Transfer, Fluid Mechanics, and Solid Mechanics* (2nd ed.). Oxford Press.

Tang, L. C. (2003). Multiple-steps Step-stress Accelerated Life Test. In *Handbook of Reliability Engineering* (pp. 441–455). Springer. doi:10.1007/1-85233-841-5_24

Wasserman, G. (2003). *Reliability Verification, Testing, and Analysis in Engineering Design*. Marcel Dekker.

Weingart, R.G., & Stephen, P. (2007, Aug.). Timoshenko: Father of Engineering Mechanics in the U.S. *Structure Magazine*.

Woo, S., & O'Neal, D. (2021). Reliability Design of Mechanical Systems Such as Compressor Subjected to Repetitive Stresses. *Metals*, *11*(8), 1261. doi:10.3390/met11081261

Woo, S., Pecht, M., & O'Neal, D. (2019). Reliability design and case study of the domestic compressor subjected to repetitive internal stresses. *Reliability Engineering & System Safety*, *193*, 106604. doi:10.1016/j.ress.2019.106604

Chapter 12
Narrow Gap Welding Principle

Duong Vu
Duy Tan University, Vietnam

ABSTRACT

The effectiveness of the welding process depends on some attributes, such as costs, the production time, and the quality of welding joints. This matter is more vital when dealing with the thick plates of steel for special structures in industry. To provide the quality when joining thick plates, especially with narrow gap and small chamfer angle, it is necessary to design the progressive process focusing on the main parameters, including the arc power, the translational velocity, and the rotational velocity of the tip. To evaluate the stability and the reliability of this technique, it must also perform the microstructure analysis in different zones of welding joint. In this chapter, previous publications related to this topic were gathered and analyzed. Particularly, several findings in recent studies for improving this progressive technology in the manufacturing area were discussed in detail.

INTRODUCTION

Narrow gap welding has found wide applications in many industries such as nuclear, aerospace, shipbuilding, petroleum and automotive industries, especially in fabrication of heavy construction from thick steel plate. The high productivity can be achieved using narrow gap welding (NGW). In welding the thick steel plate, there is the challenge with shrinkage of construction due to additional time of welding and residual stress. The one pass conventional large heat input welding or low heat input multi-layer welding by Gas Metal Arc Welding (GMAW) are not a reasonable solution in this context. In shipbuilding and container industry for transportation of long-distance cargo, it is necessary to design construction reaching 100 mm in thickness of high-strength steel. From the other aspect, the added time of welding makes the production cost higher. Due to these conditions, the development of the new improved welding technology is prevailed (Sumi et al., 2015; Uchino et al., 2020; Murayama et al., 2015).

DOI: 10.4018/978-1-6684-5381-0.ch012

DEFINITIONS

The formulation of this phenomenon is defined by the American Welding Society (AWS) as follows: a variation of a multiple pass welding process with filler metal, using a small root opening and V-groove with a small groove angle". Another significant feature yields a weld joint with a high ratio of depth to width.

Narrow Gap Welding (NGW) refers to narrow groove weld between two thick plates, applying to butt welding Type I or X groove. Normally, butt weld is considered NGW when the chamfer angle is typically in ranges 2-20º. Accordingly, the thickness of plates can reach 50 – 100 mm (see Figure 1, 2, 3). This requires less weld metal and less heat input in comparison with traditional welding technologies.

Figure 1. Narrow Gap Welding Joint

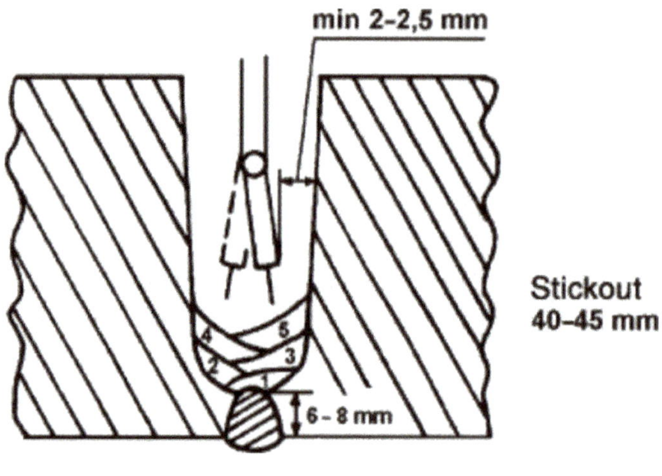

Figure 2. Narrow- groove welding

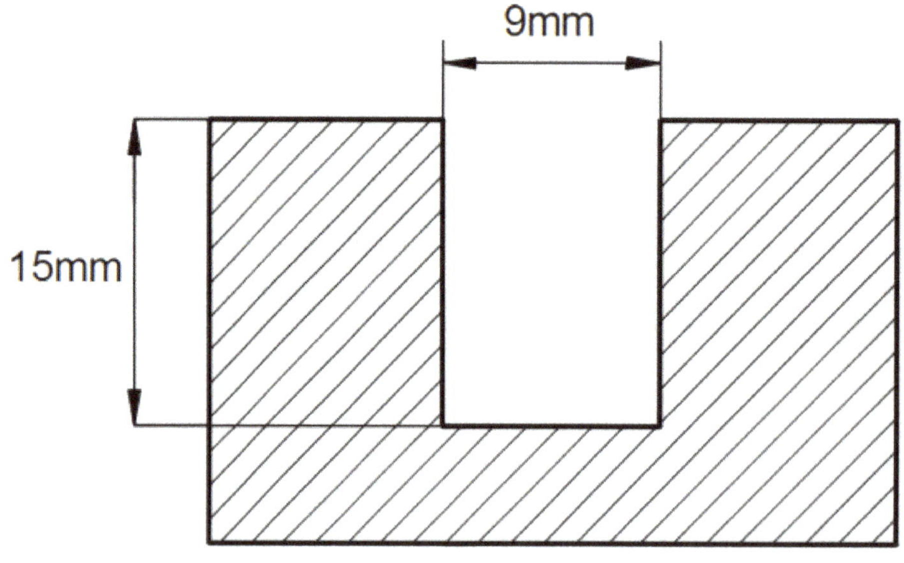

Concerning the groove width in welding joint, various interpretations have been given depending on the progression of welding equipment, but the nominal value is in range: 9-18 mm (Jia et al., 2018).

Figure 3. U-shaped groove for narrow-groove butt- joint welding

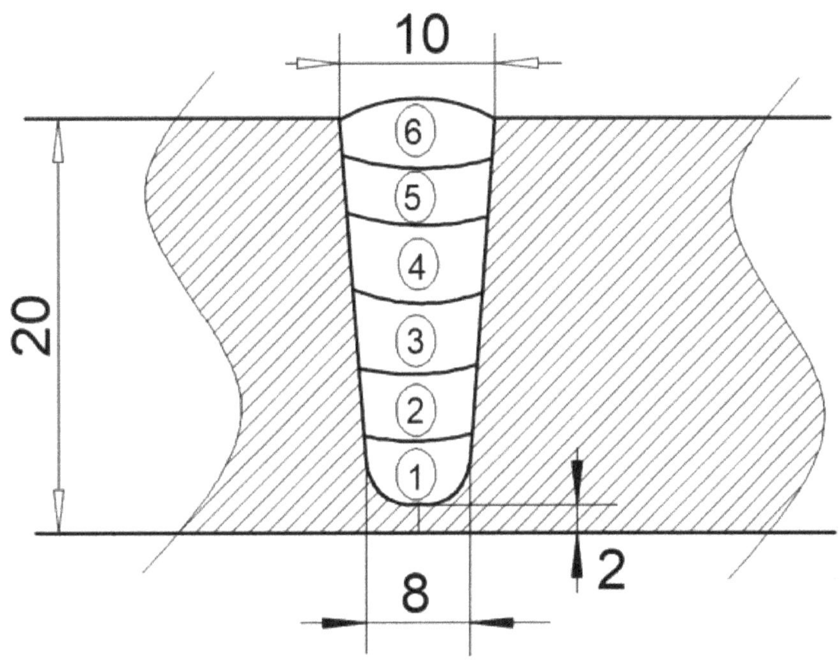

ADVANTAGE AND DISADVANTAGE

Advantage

- The process provides better saving of consumable and welding time. Excellent weldproperties.
- Low distortion and residual stress due to the joint preparation is almost parallel- side and less heat input.
- Reduced labor costs.

Disadvantage

- The risk of sidewall infusion.
- The poor joint accessibility limits the removal of any failure when detected.
- J-groove preparation is more expensive in comparison with other common types of groove design.

PROGRESSIVE PROCESSES

Since of many advantages, the NGW techniques are well- established for: gas metal arc welding (GMAW), submerged arc welding (SAW), gas tungsten arc welding (GTAW), Hybrid Laser Welding (HLW).

Narrow Gap Gas Metal Arc Welding

The traditional Gas Metal Arc Welding uses the CO_2 or mixed gas: Ar - 15% CO_2 as the shielding gas. This mixed gas eliminates the amount of spatter generated CO_2 gas. The narrow gap welding could become unstable due to accumulation of spatter on the welding tip and groove sidewall. This tends to weld defect as infusion or porosity. Too high gas flow or improper rate of CO_2 can generate porosity. This defect impacts on the mechanical properties of welding joint.

To resolve this problem, Japanese research (Sumi et al., 2015) developed a welding wire containing a very small amount of rare earth metal and direct current electrode negative method, where welding wire forms the negative terminal. This innovation helps to obtain a deep penetration and significant reduction in spatter creation. The core value of this process consists of forming a stable conical arc and making a fine-continuous spray transfer of molten droplets to the welding pool (see Figure 4).

Figure 4. The new solution for narrow gap GMAW

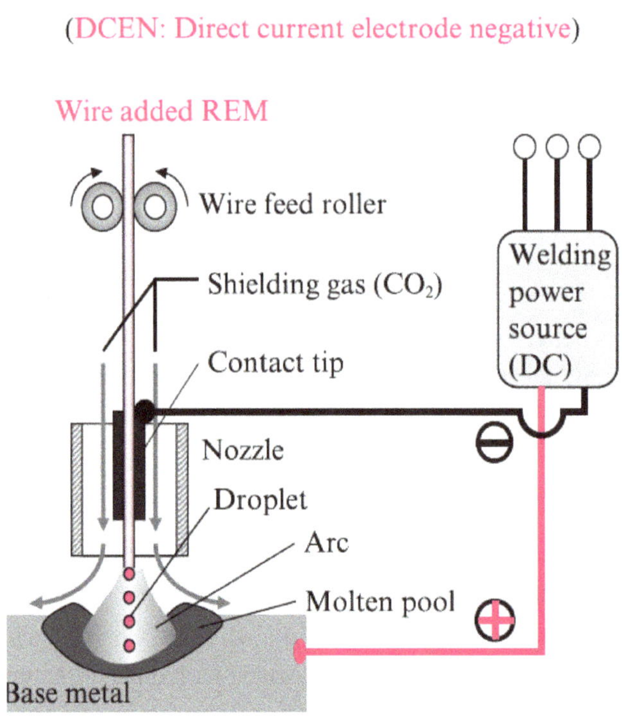

For the extra thick plates, they proposed the new design technique using curved electrode tip with two-pass one - layer (see Figure 5). The curved electrode tip allows to expand melting area in the groove sidewall because the arc strike the groove corner and sidewall.

Figure 5. Illustration of two-pass one - layer technique

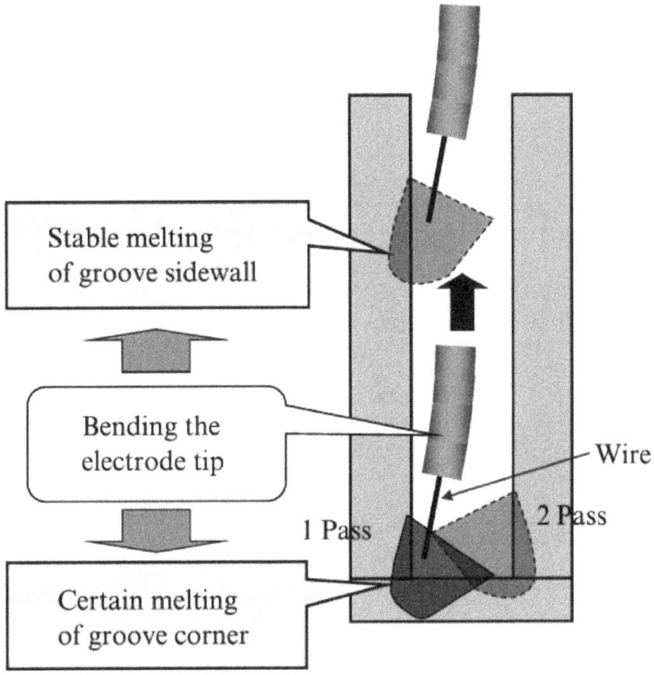

One of the disadvantages in the narrow gap GMAW is exceeding side wall penetration caused by the arc wandering during welding. Various solutions were developed to overcome these drawbacks, such as rotating arc, swing arc and wavelike wire system. But it is necessary to find new solution to decrease the cost of manufacturing since these systems are expensive and require highly skilled staff in the operation. That is why the cold wire gas metal arc welding (CW-GMAW) was developed. The principle of this method consists of the standard gas metal arc welding (GMAW) with an extra cold wire (non-charged) fed into the arc welding- welding pool zone. (See Figure 6).

Narrow Gap Gas Tungsten Arc Welding

Narrow gap TIG welding suits for welding of workpieces having thickness up to 30 mm with hot or cold wire.

In spite of many advantages, the only swing amplitude of the welding arc is limited by a narrow gap. This is the reason why it is inadequate in heating the groove corner and sidewall. It is noted that there is a lack of penetration at the weld bead. Some solutions are developed to improve in this context. Based on the effect of a swing arc system in narrow gap gas metal arc welding, the authors continued in deep the optimization of welding parameter experimentally (Wang et al,2016). It is found that with raising swing frequency, overage penetrated depth into sidewall grows noticeably.

Figure 6. Cold wire GMAW system

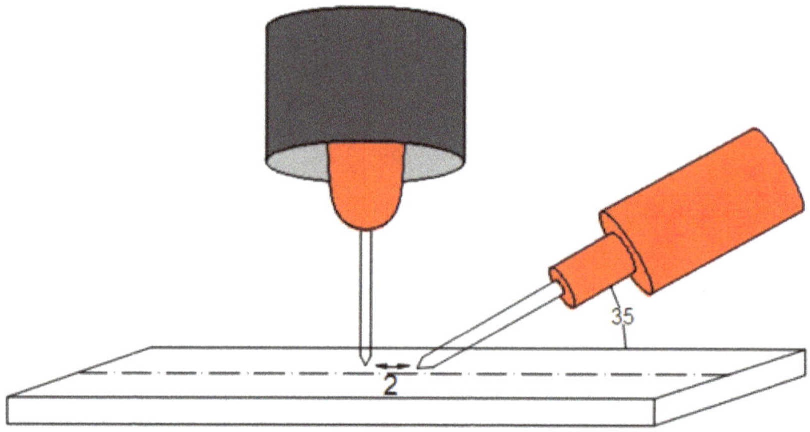

Figure 7. Swing arc narrow gap welding system

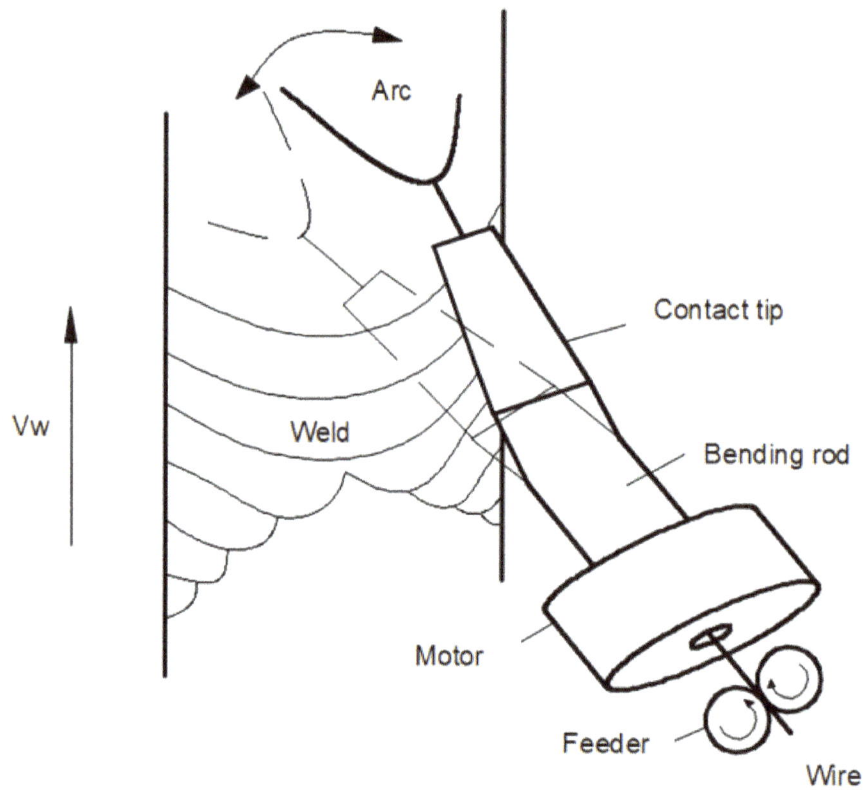

Regarding side wall penetration, one new method suggests combining tungsten inert gas welding (TIG) and metal inert gas welding (MIG) welding under the term hybrid TIG-MIG, which improves the heat input in comparison to using only one of them separately (Huang et al,2019). Beside the swing arc, one new development related to the rotating arc method, which helps to increase not only the penetration but also to remove the slag at the interlayers (Guo et al., 2019). In this work, the relationship between the defect and the welding parameters is established. The valuable outcome from this study is the optimum welding parameters of rotating arc narrow gap horizontal welding. To underline the role of the stirring effect of the rotating arc, its impact is investigated on the molten pool (Wei et al,2020). In comparison with the conventional GTAW process, the mechanical force and inertial centrifugal force are main driving forces to make the flow pattern of the molten pool changing. Under the combination of tandem technology to the narrow gap welding, it is expected a uniform fusion weld if the appropriate welding parameters obtained. However, it should note that this welding equipment is complex.

To overcome these disadvantages, the new direction of research focuses on the extra magnetic field and result reveals the effective impact in regulating the arc welding process. This system is easy operating and have the low cost (Sharma et al., 2019; Jian and Wu, 2020). The magnetic field, applied to the arc welding, many physical properties such as current density, heat flux and fluid flow will be changed, results in the critical influence on the weld formation.

Narrow Gap Submerged Arc Welding

The conventional submerged arc welding (SAW) using the V and U- groove is adopted for many years to join the thick plates. But with the increasing wall thickness the narrow gap SAW gradually replaced and recognized as standard welding technology today. It is difficult to show any modifications have been necessary for introduction narrow gap SAW to replace a conventional SAW due to only the edge angle needed to be reduced. The weld root buildup is keeping unchanged. It is useful to apply this approach when quality requirements demand the modification of edge angle without application of expensive and time consumption additional welding procedure qualification test (WPQT).

Normally SAW is used for welding structure with plate thicknesses up to 70 mm and one-side welding with one run, one pass. When the thickness is over 70 mm, welding is conducting by shielded metal arc welding (SMAW), multi-pass welding by SAW. To eliminate the disadvantage of conventional welding procedure for heavy thick plates, this problem was solved by the combining the two processes. The high - speed rotating arc welding process is used for square shape gap groove in the root part of the groove (Iwata et al., 2009). At the surface of the single V-groove, the welding is performed by SAW (see Figure 8).

Regarding the groove angle, its value can approach the range: 1° to 3 °, in extreme case it reaches zero (0 °). The ultra-narrow gap welding has been studied. The ultra-narrow gap welding yielded the deposition rate but there is the significant risk of undercuts and lack of fusion. That is why the ultra-narrow gap SAW is hardly applied. There is the study on the relationship between lack of fusion and the weld shape. This correlation can be defined by the heat input to groove and the distance squared between the heat input and the corner of the groove bottom (Abe et al., 2021).

Laser Narrow Gap Welding of Dissimilar Metal

The problem for narrow gap welding of dissimilar metal is set fort hardly in the production of nuclear reactors when deal with thick section low alloyed and stainless steel, especially in pressurized water reactor. Some study focused on dissimilar metal welding in the cooling circuit, where austenitic steel pipe connecting to forge low alloy pressure vessel nozzles. To overcome this challenge, one sophisticated process was dedicated – Laser Narrow Gap Welding of Dissimilar Metal. The main core of this method is as follows (see Figure 9).

Figure 8. Combined Groove Detail

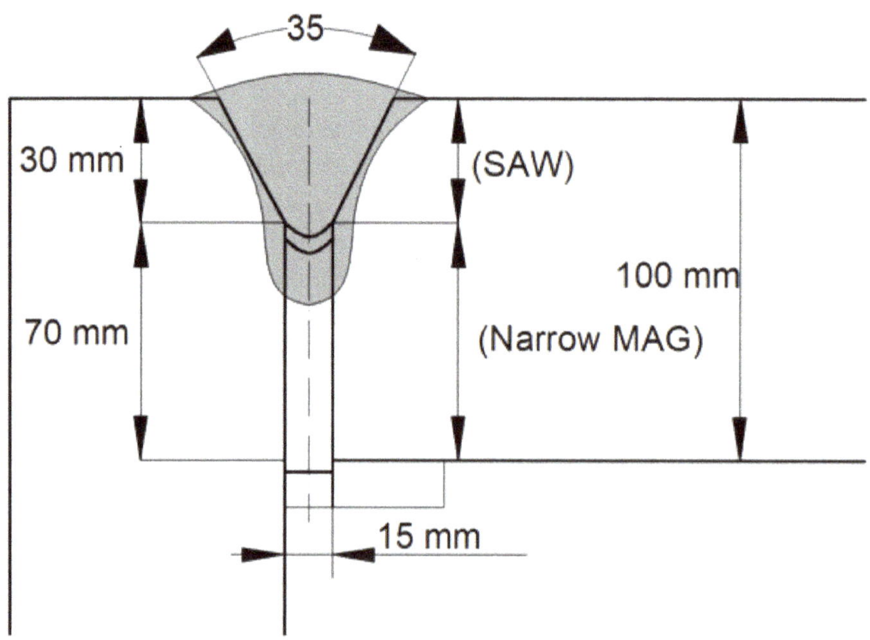

Figure 9. Principle of Laser Narrow Gap Welding

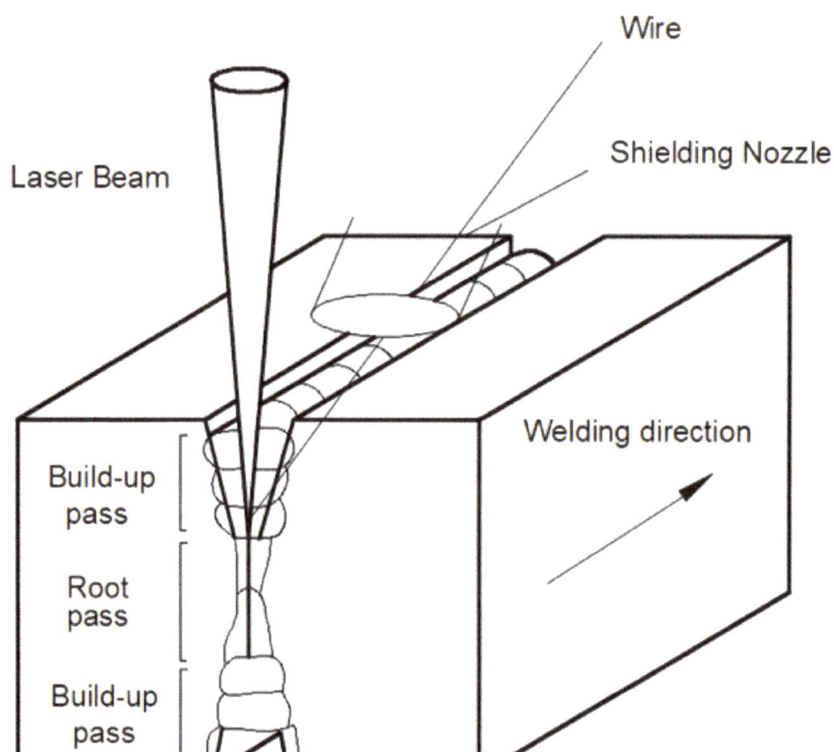

This novel welding process has attractive interest upon the invention of high beam quality laser source such as fiber lasers. For successful implementation of this process, some aspects are investigated carefully, namely

- Design a welding equipment and welding setup adopting the process parameters in welding thick section up to required thickness.
- Analysis the distortion and welding flaws during welding process, using the modern methods of microscopy.
- Study the mechanical properties of weld metal in welds and heat-affected zone, focusing on the analysis of hardness, tensile strength, impact energy and fatigue strength due to temperature change cycles.

The typical groove detail in this process is shown in Figure 10. The author of this work (Vaisto, 2019) proposed the following welding parameters:

Table 1. Parameter for dissimilar metal weld

Laser power	2.8 – 3.5	[kW]
Welding speed	0.08	[m/min]
Welding wire feed rate	1.0	[m/min]
Argon flow rate	Up to 30	[l/min]

Figure 10. The V-groove for laser narrow gap welding of dissimilar metal

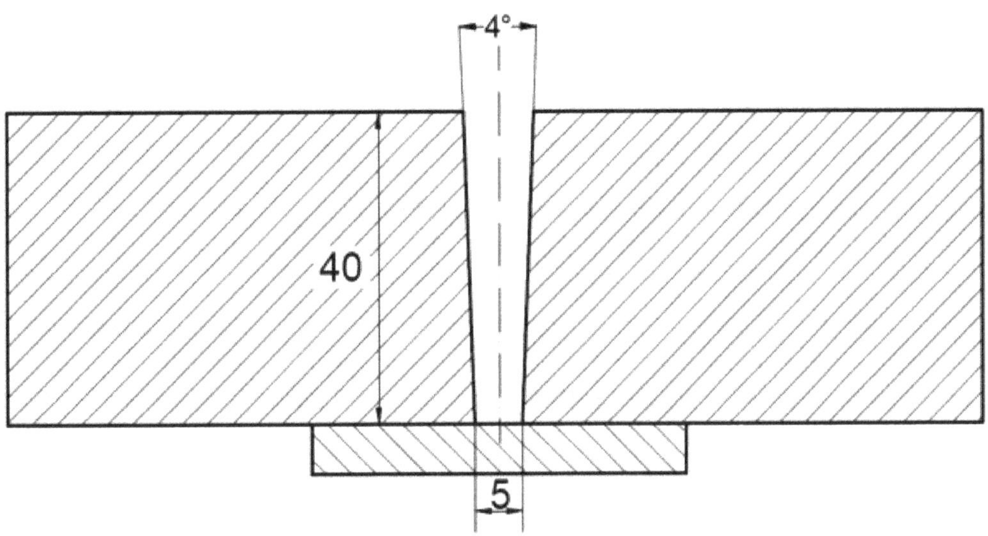

METHOD OF EVALUATION QUALITY

In general, for the evaluation the quality of narrow gap welding, all aspects of quality can be qualified using standards from ASME IX (ASME, Boiler and Pressure Vessel Code Section IX, *Welding General Requirements*, 2019), namely as follows:

Radiographic Analysis According to the ASME IX

The scope of this standard concern the welding, brazing and fusing. The nomenclature of test according to ASME IX consists of (but not limited): groove weld, face and root bends- transverse /longitudinal, plate macro, hardness and chemical analysis specimens, test flow position, tension-pipe, short-term hydrostatic test specimen, welding variable procedure specification, groove weld tension test.

Macro-Graphical Analysis Based on Different Etching Processes According to ASME IX

Macro-graphical analysis is applied to reveal the structure of metallic specimen in large scale. The metallic specimen depending on the material of the substrate, undergo the etching procedure in different appropriate chemical solutions. This test is frequently used for investigation of weld structure, definition of weld penetration, porosity, crack in weld metal and heat-affected zone. This test also helps to quantitatively evaluate the phase composition in alloy structure.

Residual Stress Analysis was Performed Using the Contour Method (Traore et al., 2014).

The contour method is a strong instrument for measurement and analysis of residual stress state in material. This method had the background of the theory of elasticity. The principle of superposition assumes that stress relaxation is pure elastic during the cutting. But if the plasticity occurs during cutting, this can reflect errors in the contour results.

Impact Toughness Analysis According to ASTM E 23 (ASTM, E23- 16b) and ISO 148-1: 2016 (ISO, BS EN ISO 148-1: 2016)

Impact test is a basic method to determine the toughness of material. The core of this test is the calculation of energy absorbed during fracture. Based on the test results one can be useful in assessing the material suitability for a specific application. According to ASTM impact test regulation, testing can be performed in temperature range: - 452° F to + 500° F. The Charpy test is mainly applied for the metallic materials, while the Izod test is frequently used for the plastic material.

ULTRASONIC TESTING

Ultrasonic testing of Narrow Gap Welding has some specific requirements due to the space and location limit. It is suggested so-called mock-up scheme and block according to ASME VIII (Zhou et al., 2016).

The innovative technique had been developed. The core nature of this invention is application of conventional ultrasonic dual element focusing for the inspection of thick-walled Austenitic Stainless Steel. The authors in (Zhou et al., 2016) certified successfully the qualification and approved that the new ultrasonic technology with specific designed scanner illustrates unique evidence in the mock-up and specimen with artificial flaw.

CONCLUSION

The study and development of the narrow gap welding is the prospective way to resolve the challenge existing in many important industries such as heavy machinery, nuclear, shipbuilding, weapon manufacturing and petrochemical ones.

Along with the opening of new limits for the groove design, it is noted that the development of new materials, special welding equipment and the qualification techniques are considered the significant problems, requiring the effort from the scientific and technology community.

The narrow gap welding is the promising area where not only researchers but also investor and manufacturer can explore more profitably in the long-term target.

REFERENCES

Abe, Y., Fujimoto, T., Nakatani, M., Shigeta, M., & Tanaka, M. (2021). Study on Proper Welding Condition for Ultranarrow Gap Submerged Arc Welding. *Welding International*, *35*(7-9), 369–381. doi:10.1080/09507116.2021.1980298

ASME. (2019). *Boiler and Pressure Vessel Code*. Welding General Requirements.

Guo, N., Wang, M., Guo, W., Yu, J., & Feng, J. (2014). Study on forming mechanism of appearance defect in rotating arc narrow gap horizontal GMAW. *International Journal of Advanced Manufacturing Technology*, *2014*(75), 15–20. doi:10.100700170-014-6127-6

Huang, J., Chen, H., He, J., Yu, S., Pan, W., & Fan, D. (2019). Narrow gap applications of swing TIG-MIG hybrid weldings. *Journal of Materials Processing Technology*, *271*, 609–614. doi:10.1016/j.jmatprotec.2019.04.043

ISO. (2016). *BS EN ISO 148-1: 2016- Metallic Materials*. Charpy Pendulum Impact Test. Test Method.

Iwata, S., Murayama, M., & Kojima, Y. (2009). Application of narrow gap welding process with high speed rotating arc to box column joints of heavy thick plates. *Jef Technical Report*, *14*(14), 16–21.

Jian, X., & Wu, H. (2020). Influence of the longitudinal magnetic field on the formation of the bead in narrow gap gas tungsten arc welding. *Metals*, *10*(10), 1351. doi:10.3390/met10101351

Murayama, M., Oazamoto, D., & Ooe, K. (2015). Narrow gap gas metal arc (GMA) welding technologies. *JFE Technol. Rep*, *20*, 147–153.

Jia, C., Yan, Q., Wei, B., & Wu, C. (2018). Rotating-Tungsten Narrow-Groove GTAW for Thick Plates. *Welding Journal*.

Sharma, P., Chattopadhyaya, S., & Singh, N. K. (2019). A review on magnetically supported gas metal arc welding process for magnesium alloys. *Materials Research Express*, *6*(8), 082002. doi:10.1088/2053-1591/ab1e67

Sumi, H., Kataoka, T., & Kitani, Y. (2015). Application of narrow gap welding process with "J-STARTM welding" to shipbuilding and construction. *JFE Tech. Rep*, *20*, 112–117.

Traoré, Y., Hosseinzadeh, F., & Bouchard, P. J. (2014). Plasticity in the contour method of residual stress measurement. *Advanced Materials Research*, *996*, 337–342. doi:10.4028/www.scientific.net/AMR.996.337

Uchino, I., Uemura, T., & Gotoh, K. (2020). A study on adopting λ-shape groove for laser-arc hybrid welding to construct thick plate butt welded joints. *Welding International*, *34*(7-9), 357–371. doi:10.1080/09507116.2021.1936931

Vaisto, T. T. (2019). Thesis submitted to the University of Manchester for degree of Doctor of Philosophy in the Faculty of Science and Engineering.

Wang, J., Zhu, J., Zhang, C., Xu, G., & Li, W. (2016). Effect of arc swing parameters on narrow gap vertical GMA weld formation. *ISIJ International*, *56*(5), 844–850. doi:10.2355/isijinternational.ISIJINT-2015-660

Wei, B., Jia, C., Wu, W., Fang, C., & Wu, C. (2020). Stirring effect of the rotating arc on the molten pool during non-axisymmetric tungsten NG-GTAW. *Journal of Materials Processing Technology*, *285*, 116769. doi:10.1016/j.jmatprotec.2020.116769

Zhou, L., Li, M., Chen, H., Ma, G., & Wang, W. (2016). Research on ultrasonic inspection of the narrow gap welds of the main coolant line (MCL) in Chinese evolutionary pressurized reactor (CEPR) nuclear power plant. In *19th World Conference on Non-Destructive Testing* (pp. 13-17). Academic Press.

Chapter 13
Deep Learning and Analysis of Cardiovascular Imaging

Salvador Castillo-Rivera
https://orcid.org/0000-0002-6500-3136
Universidad Carlos III de Madrid, Spain

Ana María Castillo-Rivera
University of Jaén, Spain

ABSTRACT

Artificial intelligence in the healthcare field has gotten the attention of professionals and researchers because the automated analysis of medical images can enhance diagnosis and therapy considerably. Deep learning (DL) puts forwards tools to deal with this kind of problem. Data science is presumable to address significant changes in the study of cardiovascular imaging allowing change appreciably in the next years encouraged by this learning. For medical professionals, it is fundamental to keep track of the development to make sure that it can have a significant impact on clinical practice. DL in medical image analysis is an interdisciplinary area that requires the support of medical experience and computer techniques. Nevertheless, the optimal clinical application of DL needs tricky formulation of problems, selection of the most proper algorithms and data, and balanced interpretation of outcomes. A current challenge comes from echocardiography. The need for human interpretation has restricted echocardiography's capability for precision medicine.

INTRODUCTION

In 1950, Alan Turing set out the idea of using computers to simulate intelligent behaviour. John McCarthy in 1956 put the term Artificial Intelligence (AI) as "the science and engineering of making intelligent machines" (Kaul et al., 2020).

AI presents different subfields as Machine Learning (ML), Deep Learning (DL), and computer vision (Kaul et al., 2020). It covers medical fields for differing purposes (Becker, 2019). AI in medicine has made progress over the past five decades. The developments of ML and DL have increased the chances

DOI: 10.4018/978-1-6684-5381-0.ch013

for personalized medicine (Kaul et al., 2020). Helping the medical field in diagnosing diseases, patient treatment, reducing errors, etc (Giovanni & Olivier, 2020). In addition, it can enhance diagnostic rigour, efficiency, disease, therapeutic monitoring and overall patient clinical results (Kaul et al., 2020). In the field of medicine, ML can be put in an application to clinical decision-making to individualize patient attention instead of following a static algorithm (Kaul et al., 2020).

Holzinger et al. (2019) have pointed out that medical practice understandable AI is required for different motives such as education, research and clinical decision-making. Modern AI methods can aid medical professionals. They must possess the pieces of knowledge to understand and trail the machine's decision. Nowadays, the current computers have enabled the advances in medical AI applications have risen due to the considerably improved computing power and the extensive amount of digital data available. Their applications may manage the enormous amount of data generated in medicine and discover patterns that would be contrarily concealed (Secinaro et al., 2021).

AI points to changing medical applications; however, many practical implementations are in early stages, needing further investigation and development. Furthermore, healthcare care can improve if medical professionals get used to these advances (Amisha, 2019).

According to Becker (2019), AI and its likely implications have begun to be contemplated more seriously in the education of future physicians. In fact, some studies have reported that most medical students concur that AI will make better certain medical areas and it should be incorporated in medical training.

Some determined settings in clinical practice profit from the application of AI, such as the diagnosis of disease based on histopathological examination or medical imaging, among others (Giovanni & Olivier, 2020).

Some current applications of AI in cardiology are the early detection of atrial fibrillation and the prediction of the risk of cardiovascular diseases such as acute coronary syndrome and heart failure (Giovanni & Olivier, 2020).

ML has progressed to DL, and it is made up of algorithms to generate Artificial Neural Networks (ANNs) that can acquire knowledge of making decisions on their own. Computer vision is the procedure by which a computer obtains information and knowledge from a set of images or videos (Kaul et al., 2020).

Medical applications are a huge challenge for AI, ML and DL. Medical decisions must cope with data sets in arbitrarily high-dimensional spaces. These are often affected by uncertainty, noise, inaccurate and missing data, etc. A dare of future medicine is modelling the complexity of patients to fit medical decisions and therapies. Until the visual analysis of the data, tricky situations, especially in the fusion, integration and mapping of heterogeneous data must be coped with. AI in the setting of medicine must consider that different data may derive an appropriate result. Thus, medical professionals must have a chance to understand how and why a decision has been made about the corresponding machine. Methods and models are required to establish the machine decision-making process and to derive and understand the learning and knowledge mining process (Holzinger et al., 2019).

This chapter is focused on DL and cardiovascular image analysis in research and clinics. First, it is described deep neural networks used in this field such as Recurrent Neural Networks (RNNs), Convolutional Neural Networks (CNNs), Fully Convolutional Networks (FCNs), Generative Adversarial Networks (GANs) and Stacked Autoencoders (SAEs). Second, it is introduced DL applications in different areas in cardiovascular image analysis, such as image classification, object detection, image segmentation. Third, the principal imaging modes that have been used in cardiology have been studied. Fourth, it is dealt with the role of DL in echocardiography. Finally, the conclusions discuss open challenges and future research.

BACKGROUND

AI is a subfield of Computer Science that tackles the use of computers to simulate the cognitive functions of humans and accomplish tasks that revolve around algorithms in an intelligent mode. ML is a subset of AI and brings out the learning feature of intelligence. The ML approach needs a set of properties to be directly measured from the data, see Figure 1 (Ahuja, 2019).

Figure 1. Diagram of AI and its subsets (Dorado-Díaz et al. 2019).

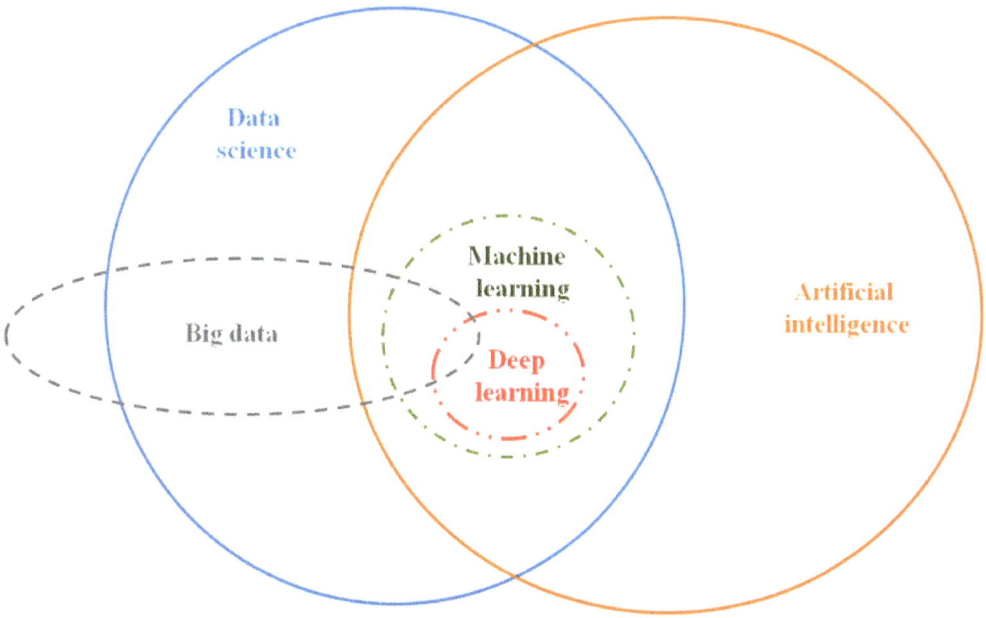

AI can diminish cost and upgrade image acquisition, analysis and decision-making. The problems with efficiency, timing and failed diagnoses occur at all phases of the imaging processing chain. The leading areas of AI for imaging will concern diagnostic support and image interpretation, among others. Diagnostic aid will be supplied by measurements and image segmentation, both automated. Likely the accuracy of cardiovascular imaging together with "big data" from the electronic health register and pathology can lead to improvement in featuring of the disease as well as personalised therapy (Dey, 2019).

The concept of ML can be defined as the algorithm that evaluates large amounts of data, acquire knowledge of it, and responds to it. These algorithms presented reduced success in getting optimal AI yield and processing complex data. As a result, it was put forward to make ANN more complex by raising the number of hidden layers and neurons. The need for implementing more robust algorithms along with the exponential growth in data sets and more sizeable availability of graphics processing units derived in DL algorithms (Akay & Hess, 2019).

The weights in neural networks are overall set up randomly, and their corresponding values require to be adjusted, in this way, the networks derive a suitable output. This process is often called "training" the neural network. Deep Neural Networks (DNNs) can be trained using supervised learning, i.e., during training, the neural network may access both the input as well as the reference. In the supervised context of imaging analysis, neural networks are optimized by giving a network with example input images, the output is predicted and the error is figured out. Once these steps are done, it is computed the contribution of every coefficient in this error. Then, all coefficients of the neural network are updated to reduce the error. This is replayed numerous times until the error does not remarkably decrease anymore (Litjens et al., 2019).

An ANN is a computational programme composed of interconnected nodes that simulate the brain's neuronal functions (Akay & Hess, 2019). It is made up of a large number of artificial neurons. A single artificial neuron may hold multiple inputs that are mixed into a single output. Each input is multiplied by a coefficient, and all of them multiplied are summed. In many cases, an additional coefficient is included in this summation which is called the bias. Then, the resulting amount is processed by a nonlinear function. The power of ANNs consist of combining a lot of neurons in layers, highly nonlinear connections between the input and the obtained output can be modelled, which is required to get accurate predictions (Litjens et al., 2019).

The connections between the nodes are made stronger or debilitated as the learning process advances. Learning algorithms make use of statistics and analytics to allow computers to upgrade the performance of a set task using data without getting particular instructions. They operate through supervised and unsupervised learning to balance and identify inputs (Akay & Hess, 2019). Supervised learning works with categorized data that are put into the input layer, which generates an output. Supervised ANNs are suitable when the result of the feature vector is known. ANN structure has to be chosen and fitted to reach the best performance. There are essential parameters that need to be fixed on each application as the number of layers, nodes for each layer and the activation functions for nodes in each one of the layers. Unsupervised learning uses an identical method as ANNs and can be taken into account as a less complex version of a DL network. The unsupervised ANN requires labelling and grouping, the unprocessed data without previous knowledge of patterns in the input item. It can drive to less accurate outcomes. The output cannot be worked out because the number of classes is unknown, and it has to be determined by the ANN obtained results (Akay & Hess, 2019). To derive optimal properties, DL uses multiple layers of neurons to learn different levels of abstraction (Gandhi et al., 2018).

A DNN is an ANN that consists of more layers allowing the improvement of predictions from data. A remarkable advantage of DNNs is that their yield constantly improves as the set of the training dataset raises. It has become a new subfield of ML called deep learning, which uses algorithms such as DNN and CNNs. The concept is that a neural network can work as the feature extractor instead of only acting as a classifier in the conventional ML. CNNs are mainly applied to image segmentation and classification (Ahuja, 2019); this network uses the same structure that an ANN. However, the first one has more hidden layers between the input and output layers. The output values are successively calculated within the network's hidden layers, with each input vector amounting to the previous layer's output vector to derive a weighted sum. DNNs have been extensively used in biomedical imaging, segmentation, brain decoding, etc (Akay & Hess, 2019).

RNNs and CNNs are the most widespread DNNs. A CNN, shown in Figure 2, is set up of convolution, non-linear and pooling layers. The first type of layer derives local weighted sums at every layer by estimating filters that are recurrently applied across the total data set to upgrade training efficiency. The second type of layer raises the characteristic maps' non-linear properties. The third type of layer runs sub-sampling of non-overlapping regions in feature maps to allow the CNN to add local features to detect complex characteristics. These networks process numerous sets of data in particular two-dimensional images. RNNs are appropriate for sequential information processing, especially audio recordings and natural language processing (Akay & Hess, 2019).

Figure 2. A CNN to identify objects in an image.

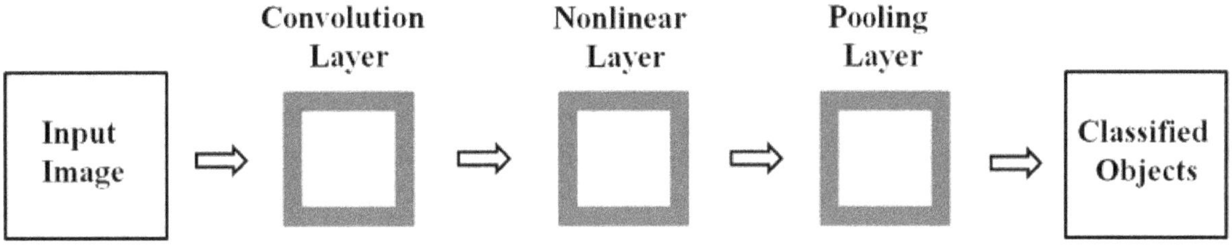

Recurrent Neural Networks

An RNN is a type of network for sequential data processing. It uses hidden states for memorizing the patterns of the data. An RNN can be viewed as a deep feed-forward neural network where all the layers possess the same parameters. The backpropagation in the time domain is employed to train these networks. RNNs that are implemented for sequential data cannot be directly used in two-dimensional images, because there are no explicit time sequences in image data. This issue is coped with through multidimensional RNNs, which work out data with more than one spatiotemporal dimension. This is obtained by replacing the standard single recurrent connections with multidimensional connections so that the network can derive a suitable performance of the surrounding context, Xing et al. (2021).

An RNN considers the first request of a sequence, achieves a prediction and later on takes its befitting output in combination with the succeeding item of the sequence for forthcoming predictions (Litjens et al., 2019).

Convolutional Neural Networks

A CNN is a feed-forward neural network that works convolutional processes in one or more layers of the network. These networks use local connections and allow using of weights for the convolutional layers. A CNN is generally made up of a set of pile-up convolution-pooling layers succeeded by fully connected layers. Regarding input images, convolutional layers work out output feature maps with different convolutional kernels using nonlinear activation functions and one kernel per feature map. Pooling layers sum up the responses within a particular neighbouring area and diminish the dimension of feature maps. Some CNN architectures make us stride convolutional layers, which enable parameter learning. To learn higher levels of feature summary is used fully connected layers which are more particular to determining tasks, wherein connected layers are joined to all units in the previous layer. A CNN may be trained using a suitable loss function in the last fully connected layer, using the backpropagation algorithm for regression and classification, among others. For classification, a cross-entropy loss is generally used, while for regression is frequently employed a mean squared error. Due to deep CNN training often needing a large amount of data, these could not be available in some sets of medical imaging. It is common to drive data rise for model training such as image rotation, colour shifting, etc. The dropout technique is commonly used to train these networks to avoid model overfitting, Xing et al. (2021).

Fully Convolutional Networks

A variant of CNNs is a fully convolutional network (FCN). It is a deep network made up of no fully connected layers. A CNN can be transformed into an FCN by substituting a fully connected layer with a particular convolutional layer. It has a kernel size the same as the dimension of the feature map introduced into the layers. Thereby, FCNs do not need fixed-size image input, which is computationally expensive sizeable images, and they are employed in CNNs such as dense prediction. FCNs can carry out pixel-to-pixel learning for straight pixel-wise prediction. It may be done by putting in one or multiple transposed or up sampling convolutional layers. It is especially suitable for applications such as image segmentation. Based on FCNs, U-Net is a popular deep network. It is an encoder-decoder U-shaped deep architecture and it is used in medical imaging, inclusive pathological image analysis. The encoder is composed of stacked convolution-pooling layers, and it learns feature representations from input images. The decoder is made up of a set of up sampling and convolutional layers. These are symmetrically mapping and they are learned visual models back to the input space for pixel-wise estimation. One relevant feature of U-Net is inserting long-range skip connections between the encoder and decoder. Thus, high-resolution data in the encoder can be employed for more accurate object localization. U-Net can generate a prediction map that presents the same size as the input image by implementing padding or cropping to the output feature maps in the decoder, allowing U-Net a fast inference for arbitrary-sized images, being this relevant for the field of digital pathology, Xing et al. (2021).

Generative Adversarial Networks

A GAN is an unsupervised learning model that simultaneously trains two deep neural networks: a generator and a discriminator. The generator learns to get training data and derivate fake samples. They keep the same statistics as the training data. The discriminator learns to discern fake samples from real data. These two networks are integrated into a two-player minimax game for model training. The applications of GANs and their variants have expanded in the medical imaging field and have arisen in digital pathology, Xing et al. (2021).

Stacked Autoencoders

An autoencoder is an unsupervised learning neural network. It is made up of an input layer, a hidden layer, and an output layer. The output layer commonly has a smaller number of units than the input layer, and it aims to reconstruct the input from the hidden layer as well as to learn a compressed representation of input data. It is placed on the network constraints, for example, restraining the number of hidden to obtain suitable features of input data. Autoencoders may be stacked to set up a deep, multilayer network that can model complex data structures by using every hidden layer to a level of feature abstraction. A stacked autoencoder can be trained without supervision in a layer-wise manner. It can be carried out by training one layer at a time, wherein the feature representations are used from lower layers as the input of higher layers. Once unsupervised training is done, the stacked autoencoder can be mixed with an additional layer. It can be taken conjointly and fine-tuned with backpropagation in a supervised process for discriminatory learning tasks, Xing et al. (2021).

ROLE OF DEEP LEARNING IN DIAGNOSTIC SUPPORT

DL methods simulate human learning by using multiple layers of CNNs. These methods have been applied to image segmentation using, for example, Marginal Space Learning which manages parameter evaluation in a series of approaches of object orientation, position and anisotropy.

Most of the diagnostic help for cardiovascular imaging takes on the segmentation of anatomical systems such as valves, ventricles, etc, as well as the accurate measurements of different parameters that cover perfusion defect and ejection fraction, among others.

Segmentation is the method of content extraction that receipts as an input a medical image, volume, or sequence of both of them to derive linked shapes, such as two-dimensions (2D) contours or three-dimensions (3D) meshes. Cardiovascular image segmentation is a developing sector in which DL has mainly shown considerable yield improvements.

Multiple segmentation methods are based on key points that attach the contour to particular and well-defined anatomical. One of the most current methods for robust localization of anatomical key points relies on multiscale deep reinforcement learning. This method is trained not only to determine the target anatomical key point but also how to find this point by following and learning an optimal navigation path across the volume. Once particular key points in the data are known several segmentation methods can be used to outline the heart and vascular structures. One of them is based on a database of annotated echocardiography images of the left ventricle. The method has been expanded to 3D by formulating a learning procedure in marginal spaces of restricted dimensionality and using as a classifier a probabilistic boosting tree. Furthermore, DL methods put together with level sets have been implemented for the segmentation of the left ventricle from cardiac magnetic resonance images. Segmentation is often followed by a temporal tracking algorithm that goes together with the myocardial border across frames to assess myocardial motion.

Modelling and segmentation of the heart valves have caused an interest due to the apparition of minimally invasive valve implant methods, which need valve modelling and acute dimensioning of the implants. The four heart valves and the pulmonary trunk have been modelled and segmented from cardiac computed tomography by using marginal space learning. This method has employed a constrained multilinear form model to simulate Spatio-temporal variations of the heart valves. DL has also been used to carry out the automated quantification of epicardial adipose tissue to gauge its volume and density from calcium-scoring computed tomography.

There are two frequent imaging procedures for cardiac examination how are M-mode and Doppler echocardiography. The first one presents high temporal and spatial resolution along one scan line and it is effective in capturing fine motion patterns. The second is mainly used to obtain a velocity-time image for evaluating stenosis and valvular regurgitation. ML can be utilized to automate a set of measurements associated with M-mode and Doppler.

DL has been applied to plenty of facets of cardiac image analysis. Magnetic Resonance Imaging (MRI) is the most explored technique and left ventricle segmentation is the most frequent task. The set of applications is thoroughly varied: tracking, segmentation, slice classification, image quality assessment, coronary centreline tracking, etc (Litjens et al., 2017). Automated cardiac segmentation from MRI datasets is a fundamental item in the proper diagnosis and management of cardiac pathologies. According to Tran (2017) through the application of a deep fully CNN can be set out to approach the problem of automated left and right ventricle segmentation.

DEEP LEARNING IN CARDIOVASCULAR IMAGING

In the field of medicine, ML and DL can establish connections between different pathologies. However, many medical professionals are unknown how they work. Medical imaging has a significant role in the diagnosis and it is the first clinical-stage for additional invasive diagnostic testing and treatments, which can expose a patient to unnecessary risk (Gandhi et al., 2018).

DL algorithms simultaneously learn significant features and the prediction model from the input image to the derived outcome. It is frequently called end-to-end learning (Litjens et al., 2019).

In unsupervised learning, the reference is not affordable during training, and the input data are solely available. In this case, the data are named unlabeled data. This setting is less widespread in medical imaging. The main motive is that medical image data are extensively available; however, the required labels are less straightforward to get, and generally need high expert input. Unsupervised learning enables the leveraging of this set of unlabeled data. GAN is an unsupervised deep learning algorithm. These are commonly used for image improvement or generation. GANs are composed of two networks that have opposing aims. One network takes action as a generator and the other acts as a discriminator evaluating if the obtained image is feasible. It enables optimization of the generator without requiring accurate labels (Litjens et al., 2019).

The main imaging modes that are being used in cardiology cover Echocardiography, Magnetic Resonance Imaging, Optical Coherence Tomography, Intravascular Ultrasound, Computerized Tomography, Fundus Photography (Bizopoulos & Koutsouris, 2019).

Echocardiography

Echocardiography is an imaging mode that employed ultrasound waves to represent the heart area. DL is used in echocardiography mostly to tackle quality score assessment and left ventricular segmentation, among others. Deep Belief Networks (DBNs) have been employed for left ventricular segmentation in echocardiography. These networks are probabilistic generative models that are made up of multiple layers of stochastic, latent variables, (Hinton, 2009; Ranzato et al., 2008). Echocardiography has also been used for classification purposes. For example, CNN has enabled the implementation of a residual 3D CNN for ejection fraction rating from transthoracic echocardiogram images. Quality score assessment has been dealt with by using echocardiography (Bizopoulos & Koutsouris, 2019).

Magnetic Resonance Imaging

Magnetic Resonance Imaging (MRI) is built on the interaction of a system of atomic nuclei and an external magnetic field which derives an internal image of the studied object (Bizopoulos & Koutsouris, 2019). DL applications in MRI have typically focused on segmentation, concretely; cardiac structures in short axis reconstructions for evaluation of cardiac function. DL segmentation methods are on par with human specialist performance. Notwithstanding the combination of conventional techniques with

DL has been the paths of research, most of the algorithms have entirely been based on CNNs (Litjens et al., 2019). The leading applications of this method include Right and Left Ventricles and entire heart segmentation. CNNs have been used for left ventricular segmentation using MRI. Left Ventricular / Right Ventricular segmentation has been studied through several datasets and CNNs as well as Fully Convolutional Network. In addition, some methods based on regression models have allowed the setting up of a regression segmentation framework to trace the boundaries of both ventricles. Full heart segmentation has been studied using CNN and 3D fully convolutional networks (Bizopoulos & Koutsouris, 2019).

Optical Coherence Tomography

Optical Coherence Tomography (OCT) is an intravascular imaging technique that obtains cross-sectional images of arteries with high resolution. It is also able to reproduce quantitative measurements of the coronary geometry. Linear-regression CNN and linear_Support Vector Machine (SVM) have allowed the study of OCT (Bizopoulos & Koutsouris, 2019). SVM uses supervised learning models with associated learning algorithms to study data for classification and regression analysis, Khandoker et al. (2009), Son et al. (2010). DL for cardiovascular ultrasound presents applications in identification, rating, segmentation, tracking, etc. Some works also include direct pathology classification from echocardiography (Litjens et al., 2019). Intravascular Ultrasound (IU) makes use of ultra-miniaturized transducers assembled on adapted intracoronary catheters to generate radial anatomical imaging of intracoronary calcification and plaque creation. CNN and GAN have been mainly employed in the study of IU (Bizopoulos & Koutsouris, 2019).

Computerized Tomography

Computerized Tomography (CT) is a non-invasive technique for the diagnosis of obstructive artery disease (Bizopoulos & Koutsouris, 2019). DL has been used for various applications in CT as coronary artery calcium score evaluation, localization and segmentation of cardiac zones (Bizopoulos & Koutsouris, 2019). Most of them have focused on the identification and segmentation of anatomical structures for several clinical aims in both non-contrast as well as coronary CT angiography. Due to CT images being commonly obtained with a low radiation dosage and how image quality may be influenced by cardiac motion, DL has aided in the reduction of noise and deficiencies (Litjens et al., 2019). CNN has been mainly used in CT (Bizopoulos & Koutsouris, 2019).

Fundus Photography

Fundus photography allows the evaluation of retinopathy progress in patients wherein the image intensity depicts the quantity of reflected light of a determined waveband. The non-invasive character of the fundus and its current use to estimate cardiovascular risk predictors makes it an encouraging modality of higher functionality in the field of cardiology. Vessel segmentation in fundus imaging has mainly been studied using CNNs while a fitting approach in some cases has been the use of the conditional random field as a postprocessing stage for vessel segmentation refining (Bizopoulos & Koutsouris, 2019).

According to the work that needs solving, determining neural network architecture can be selected, and it can be optimally adapted for that particular task. The architecture points out the number of layers and neurons in the neural network and how these are connected. The most widespread architecture in image analysis is CNN, which significantly reduces the number of coefficients in the network by sharing weights at each position in the image. CNNs are made up of both convolutional and pooling layers. The last ones do not hold coefficients or weights but decrease the sample of the data, generally by a ratio of two. It diminishes the resolution of the image but raises the field of view of successive layers, helping CNNs to include further contextual information. The structure of CNNs can be fitted to carry out image segmentation. FCNs get an image as input and straight forecast an image-sized segmentation; for example, in the context of cardiovascular imaging can be applied to left or right ventricle segmentation. RNNs are also employed in medical imaging. These are oftentimes used for sequential data, such as electrocardiography or cine magnetic resonance imaging, among others (Litjens et al., 2019).

DEEP LEARNING IN ECHOCARDIOGRAPHIC IMAGING

The echocardiogram is maybe the most commonly used tool in the field of cardiology. It is employed mostly for its capability in the early diagnosis and treatment of heart diseases. Among other characteristics, it is an inexpensive and non-invasive technique that can accurately display the pressure gradient of heart lesions. It is regarded to be safe because uses sound waves rather than radiation. The echocardiogram performs an essential role in the diagnosis of cardiac diseases and utilizes standard two-dimensional, three-dimensional as well as a Doppler ultrasound to generate images of the heart (Wahlang et al., 2021).

The first item to understand computer-assisted echocardiographic is establishing if computers can learn to identify the corresponding views (Madani et al., 2018). Clinical echocardiography has evolved from the two dimensions of visual appearance to semi-quantitative assessment. However, this last one is limited by human error in acquisition and interpretation (Gandhi et al., 2018). An example of DL via supervised learning using CNN to identify echocardiographic views consists of the input is a standard echocardiographic image, and the output is the probabilities of the feasible outcomes, which are used to derive the most likely view. Abstraction of multiple layers through the process of DL enables the computer to detect which view the image was taken from. CNN can learn the features from data in the training phase, namely, the echocardiographic image data set. It can be expanded to multiple views given an adequate amount of data. The method presents some restrictions as a long time for analysis and the complexity of learning. It can be appropriate for applications for image recognition. However, there is some uncertainty in the analysis of big data for predicting clinical results and prognosis because of the lack of capacity to explain the features of the learning process (Gandhi et al., 2018).

AI has raised the reproducibility of measurements and increased efficiency and workflow. Echocardiography is a fundamental evaluation and quantification of left ventricular size and function. There are different methods and techniques for tracing biplane disc summation to estimate ejection fraction and left ventricular volumes (Gandhi et al., 2018). DL can likely figure out complex relationships and patterns when is applied to vast image repositories integrating the whole of the properties of the image, which can show hidden connections related to the natural history and prognosis of different cardiac pathologies. DL has been utilized to derive properties representation from image data directly. Cohorts of patients with cardiac amyloidosis and patients with hypertrophic cardiomyopathy with varying patterns of left ventricular thickening have been used for training two separate CNNs for disease detection. The corresponding models have been able to identify cardiac amyloidosis and hypertrophic cardiomyopathy with high precision (Gandhi et al., 2018). DL automatically evaluates the outcomes from the image without the requirement to predetermine specific imaging features. The deep layers of the CNN can obtain determined low-level information from the early image and incorporate these to set up higher-order structural information. It allows the detection of complex items from the image. This method may be suitable for studying echocardiographic findings of several heart diseases (Kusunose et al., 2020).

Echocardiographic images incorporate some video clips for evaluation of the complex cardiac structure and need standard views for diagnosing cardiovascular diseases. Determination of the view is the first stage in studying an echocardiogram. Nevertheless, echocardiographic data are normally labelled partially and exist discrepancies in image features in the dataset. It is needed a proper identification of individual echocardiographic views to work out an achievable algorithm for the evaluation of cardiovascular diseases (Kusunose et al., 2020).

According to Kusunose et al. (2020), the classification of views can be achieved by a CNN algorithm. It was designed and trained three CNN models to identify five standard echocardiographic views. The first one employed as input the averaged image over time. The input image was two dimensions with a size of 120 x 120 pixels. The second one used ten selected images of the same size. These were trained separately and the averaged probability was handled in the prediction phase. Other models such as "ImageNet" were utilized for testing the performance (Kusunose et al., 2020).

Two main findings can be highlighted in Kusunose et al. (2020), the first was the rating of videos, the analysis of them can be a cumbersome task that requires numerous trivial labours, such as item tracking and frame to frame colour variation. A proper view classification may lead to this process being most methodological and cost-effective, through a reduction of coding and training time. Second, it was obtained a suitable accuracy of the echocardiogram prediction of 0.5% error rate from the database for left ventricular ejection fraction. The study showed the good running of a CNN algorithm and that this algorithm can be employed to accomplish the echocardiographic prediction model for ejection fraction. The model consisted of five layers with 64 to 128 nodes that derived the best performance. The optimal cutting size of the input images was also monitored by comparing the pixel sizes of 100, 120, and 200 in the corresponding space. Regarding these tests, it was determined to use a pixel size of 120 in the classification. The authors claimed that the model was fitted for application to echocardiographic images. However, they pointed out that the number of patients was limited in the study and DL algorithms may need thousands of images in some cases. To sum up, the outcomes support in principle that CNN may be especially effective in the view classification. It should be considered larger numbers of patients for measuring the efficacy of automatic diagnosis systems in a clinical stage (Kusunose et al., 2020). For the study of valvular heart diseases as regurgitation, the echocardiogram has become a remarkable option. The main types of regurgitation are mitral, aortic and tricuspid. It can be diagnosed from abnormal flow patterns using colour imaging (Wahlang et al., 2021).

According to the different types of regurgitation, the images may be classified as normal or abnormal. Classification has a relevant role in the estimation of an area or region holding abnormalities for the diagnosis of any disease. For this purpose, can be used Long Short-Term Memory (LSTM), Variational Autoencoder plus CNN along with Support Vector Machine (SVM). The two-dimensional static echocardiogram images and three-dimensional static doppler images can be classified into two classes, that is, normal or abnormal. Video graphic echocardiogram images can be also classified as normal or abnormal using these methodologies (Wahlang et al., 2021). Long Short-Term Memory (LSTM) is an RNN based model and has feedback connections that can process single data points such as images (Hochreiter & Schmidhuber, 1997). This technique has been found appropriate for videos and can aid in predicting the next frame in a video and is implemented in numerous areas of videography. Autoencoder is a generative model based upon the CNN concept, which enables to reconstruct of its output employing encodings.

SVM is an extended supervised learning method for the classification of medical imaging. This method is successful for high-dimensional spaces and regression (Wahlang et al., 2021).

Once the pre-processing of the images is carried out, they can be saved into two types of files called Comma-Separated Values (CSV) files, for the testing and training phase. A CSV file is a text file that employs a comma to separate values. The images are worked out using various methodologies such as LSTM/Variational Autoencoder - CNN/SVM for validation purposes (Wahlang et al., 2021). To accomplish the classification of video graphics images, it is required to apply the spatiotemporal DL i.e., to pull out each frame and work on each of them. Each frame is allocated a class in the training and validation stage. Frames are cut off and resized. Afterwards, the network (LSTM and Variational Autoencoder - CNN) is fed by the training set for classification (Wahlang et al., 2021).

The application of CNN to echocardiography can get the classification of major transthoracic echocardiogram views at the expert level. The set of training data can be a large range of physiological and clinical variations, which can prove applicability to real information. Madani et al. (2018) have proposed a model that used some of the same characteristics in echocardiograms that human specialists use to make their decisions (Madani et al., 2018).

Ultrasound technology is used in echocardiography to obtain high spatial, temporal resolution images of the heart and its surrounding structures. It has become the most frequent imaging technique in cardiovascular medicine. CNNs have shown that DL applied to echocardiography on a large dataset can evaluate cardiac function, recognize local cardiac structures and prognosticate systemic phenotypes that change cardiovascular risk. It is a relevant advance because; all of them are not easily identifiable to human specialists (Ghorbani et al. 2020).

According to Ghorbani et al. (2020), the DL model, EchoNet, has been able to identify the presence of left atrial enlargement, left ventricular hypertrophy, ventricular volumes and ejection fraction, etc. These outcomes can make more efficient some tasks in the clinical workflow because they supply initial interpretation, which helps in the prediction of phenotypes cumbersome for human evaluation.

DL based methods supply a successful and effective procedure to segment determined organs or tissues as the left ventricle and coronary vessels, among others. There is a large set of methods designed for ventricle segmentation in magnetic resonance and ultrasound domains, mainly. The aim of ventricle segmentation is to outline the endocardium and epicardium of the left ventricle and/or right ventricle. These segmentation maps are relevant for obtaining clinical indices, such as left and right ventricular end-diastolic volumes, left and right ventricular end-systolic volumes, and ejection fraction. These segmentation maps are fundamental for three-dimensional shape analysis, three-dimensional plus time-motion study and survival prediction (Chen et al., 2020).

CONCLUSION

Medicine needs new models that allow understanding complex data. It is currently carrying out relevant advances in medical imaging analysis and automated clinical decision-making, among others. It establishes challenging research problems in the sectors of robotics and computer vision, mainly. DL has shown its capability of tackling this kind of problem.

There is a singular area of medicine where AI is having an effect as clinical decision-making and disease diagnosis. This chapter provides an overview of the applications of DL in the field of cardiovascular images to tackle the challenges that have been successfully carried out and the next ones to come, for example, the role of DL in echocardiography. Advances in DL have enhanced cardiovascular image analysis in clinical practise and clinical research. The main deep neural networks employed for medical image analysis have been dealt with, i.e., Recurrent Neural Networks, Convolutional Neural Networks, Fully Convolutional Networks, Generative Adversarial Networks and Stacked Autoencoders.

The role of DL in diagnostic support has been tackled; most of the diagnostic help for cardiovascular imaging makes use of the segmentation of anatomical systems. Segmentation is a method that receipts as an input a medical image, volume, or sequence of both to them to get linked shapes as 2D contours or 3D meshes. Cardiovascular image segmentation is a developing sector, and DL is playing a considerable role in setting improvements. The leading imaging modes that have been used in cardiology have been studied as Echocardiography, Magnetic Resonance Imaging, Optical Coherence Tomography, Intravascular Ultrasound, Computerized Tomography, and Fundus Photography.

DL can likely calculate complex relationships and patterns when applied to extensive image repositories, which can integrate all properties of the image, allowing the prognosis of different cardiac pathologies. DL has been utilized to get properties representation from image data directly. The echocardiogram is employed for its capability in the early diagnosis and treatment of heart diseases. The echocardiogram has a fundamental role in the diagnosis of cardiac diseases. It uses standard two-dimensional, three-dimensional and Doppler ultrasound to obtain images of the heart. DL model for echocardiography has also been studied. Echocardiography has been developed from the two dimensions of visual appearance to semi-quantitative evaluation; however, this one is limited by human error. In summary, the chapter has presented an outlook of DL and the analysis of cardiovascular images to shed light on the present issues as well as the upcoming challenges in this field.

ACKNOWLEDGMENT

This work was partially supported by Comunidad de Madrid (Spain) under the Multiannual Agreement with UC3M in the line of Excellence of University Professors (EPUC3M23), in the context of the 5th.\ Regional Programme of Research and Technological Innovation (PRICIT).

REFERENCES

Ahuja, A. S. (2019). The impact of artificial intelligence in medicine on the future role of the physician. *PeerJ*, *7*, e7702. doi:10.7717/peerj.7702 PMID:31592346

Amisha, M., Malik, P., Pathania, M., & Rathaur, V. K. (2019). Overview of artificial intelligence in medicine. *Journal of Family Medicine and Primary Care*, *8*(7), 2328–2331. doi:10.4103/jfmpc.jfmpc_440_19 PMID:31463251

Becker, A. (2019). Artificial intelligence in medicine: What is it doing for us today? *Health Policy and Technology*, *8*(2), 198–205. doi:10.1016/j.hlpt.2019.03.004

Bizopoulos, P., & Koutsouris, D. (2019). Deep Learning in Cardiology. *IEEE Reviews in Biomedical Engineering*, *12*, 168–193. doi:10.1109/RBME.2018.2885714 PMID:30530339

Chen, C., Qin, C., Qiu, H., Tarroni, G., Duan, J., Bai, W., & Rueckert, D. (2020). Deep Learning for Cardiac Image Segmentation: A Review. *Frontiers in Cardiovascular Medicine*, *7*, 25. doi:10.3389/fcvm.2020.00025 PMID:32195270

Dey, D., Slomka, P. J., Leeson, P., Comaniciu, D., Shrestha, S., Sengupta, P. P., & Marwick, T. H. (2019). Artificial Intelligence in Cardiovascular Imaging: JACC State-of-the-Art Review. *Journal of the American College of Cardiology*, *73*(11), 1317–1335. doi:10.1016/j.jacc.2018.12.054 PMID:30898208

Dorado-Díaz, P. I., Sampedro-Gómez, J., Vicente-Palacios, V., & Sánchez, P. L. (2019). Applications of artificial intelligence in cardiology. The future is already here. *Revista Espanola de Cardiologia*, *72*(12), 1065–1075. PMID:31611150

Gandhi, S., Mosleh, W., Shen, J., & Chow, C.-M. (2018). Automation, machine learning, and artificial intelligence in echocardiography: A brave new world. *Echocardiography (Mount Kisco, N.Y.)*, *35*(9), 1402–1418. doi:10.1111/echo.14086 PMID:29974498

Ghorbani, A., Ouyang, D., & Abid, A. (2020). Deep learning interpretation of echocardiograms. npj. *Digital Medicine*, *3*, 10. PMID:31993508

Giovanni, B., & Olivier, L. M. (2020). Artificial Intelligence in Medicine: Today and Tomorrow. *Frontiers in Medicine*, *7*(27), 1–6. PMID:32118012

Hinton, G. E. (2009). Deep belief networks. *Scholarpedia.*, *4*(5), 5947. doi:10.4249cholarpedia.5947

Hochreiter, S., & Schmidhuber, J. (1997). Long short-term memory. *Neural Computation*, *9*(8), 1735–1780. doi:10.1162/neco.1997.9.8.1735 PMID:9377276

Holzinger, A., Langs, G., Denk, H., Zatloukal, K., & Müller, H. (2019). Causability and explainability of artificial intelligence in medicine. *WIREs Data Mining Knowl Discov.*, *9*(4), e1312. doi:10.1002/widm.1312 PMID:32089788

Kaul, V., Enslin, S., & Gross, S. A. (2020). History of artificial intelligence in medicine. *Gastrointestinal Endoscopy*, *92*(4), 807–812. doi:10.1016/j.gie.2020.06.040 PMID:32565184

Khandoker, A. H., Palaniswami, M., & Karmakar, C. K. (2009). Support vector machines for automated recognition of obstructive sleep apnea syndrome from ECG recordings. *IEEE Transactions on Information Technology in Biomedicine*, *13*(1), 37–48. doi:10.1109/TITB.2008.2004495 PMID:19129022

Kusunose, K., Haga, A., Inoue, M., Fukuda, D., Yamada, H., & Sata, M. (2020). Clinically Feasible and Accurate View Classification of Echocardiographic Images Using Deep Learning. *Biomolecules*, *10*(5), 665. doi:10.3390/biom10050665 PMID:32344829

Litjens, G., Ciompi, F., Wolterink, J. M., de Vos, B. D., Leiner, T., Teuwen, J., & Išgum, I. (2019). State-of-the-Art Deep Learning in Cardiovascular Image Analysis. *JACC: Cardiovascular Imaging*, *12*(8 Pt 1), 1549–1565. doi:10.1016/j.jcmg.2019.06.009 PMID:31395244

Litjens, G., Kooi, T., Bejnordi, B. E., Setio, A. A. A., Ciompi, F., Ghafoorian, M., van der Laak, J. A. W. M., van Ginneken, B., & Sánchez, C. I. (2017). A survey on deep learning in medical image analysis. *Medical Image Analysis*, *42*, 60–88. doi:10.1016/j.media.2017.07.005 PMID:28778026

Madani, A., Arnaout, R., & Mofrad, M. (2018). Fast and accurate view classification of echocardiograms using deep learning. npj. *Digital Medicine*, *1*, 6. PMID:30828647

Ranzato, M., Boureau, Y., & LeCun, Y. (2008). Sparse feature learning for deep belief networks. Proceedings of Advances in Neural Information Processing Systems, 1185-1192.

Secinaro, S., Calandra, D., Secinaro, A., Muthurangu, V., & Biancone, P. (2021). The role of artificial intelligence in healthcare: A structured literature review. *BMC Medical Informatics and Decision Making*, *21*(1), 125. doi:10.118612911-021-01488-9 PMID:33836752

Son, Y. J., Kim, H. G., Kim, E. H., Choi, S., & Lee, S. K. (2010). Application of support vector machine for prediction of medication adherence in heart failure patients. *Healthcare Informatics Research*, *16*(4), 253–259. doi:10.4258/hir.2010.16.4.253 PMID:21818444

Tran, P. V. (2017). *A fully convolutional neural network for cardiac segmentation in short-axis MRI.* arXiv:1604.00494.

Wahlang, I., Maji, A. K., Saha, G., Chakrabarti, P., Jasinski, M., Leonowicz, Z., & Jasinska, E. (2021). Deep Learning Methods for Classification of Certain Abnormalities in Echocardiography. *Electronics (Basel)*, *10*(4), 495. doi:10.3390/electronics10040495

Chapter 14
Applications of Robotics in Gynecological Surgery

Ciro Comparetto
https://orcid.org/0000-0002-2985-4478
Department of Obstetrics and Gynecology, San Paolo Private Clinic, Italy

Franco Borruto
https://orcid.org/0000-0003-0667-8851
Independent Researcher, Monaco

ABSTRACT

Robotic-assisted surgery is a branch of engineering that develops robotic machines that allow the operator to perform surgery by maneuvering, from a distance, a robot that is not completely autonomous but capable of performing controlled maneuvers. It is a technique that has recently come into use, even in selected centers, and represents a further step in the field of minimally invasive surgery (MIS). It has basically the same indications, but, at the moment, it is reserved for selected patients. Compared to traditional video-assisted surgery, it has some important differences. The surgeon is physically distant from the operating field and sits at a console, equipped with a monitor, from which, through a complex system, he controls the movement of the robotic arms. To these are fixed the various surgical tools—forceps, scissors, dissectors—which a team present at the operating table introduces into the cavity where the surgery is performed.

INTRODUCTION

In the field of medicine, the branch that has made the most progress in the last 20 years is surgery. In truth, surgical techniques have undergone a complete revolution as a result of scientific study and the introduction of new technology in the field. Today, endoscopic approaches, such as laparoscopy, robotic surgery, and hysteroscopy, are the gold standard for treating numerous illnesses. Robotic surgery, also known as computer-assisted telemanipulation, is a type of operation performed using sophisticated sur-

DOI: 10.4018/978-1-6684-5381-0.ch014

gical platforms that are capable of replicating, by way of miniaturization, the movements of the human hand within the human body, or in any event in the surgical field (Palep, 2009).

The most recent advancement in technical advancements used in surgery is currently robotic surgery. Contrary to popular belief, robotic surgery is not a procedure carried out by a machine. It is a minimally invasive laparoscopic procedure that needs to be overseen and carried out by qualified medical professionals. Both traditional laparoscopic surgery and robotic surgery, in which a robot precisely mimics the movements of the surgeon, are safe treatment options; however, only highly specialized surgeons with adequate training, essential skills, and extensive experience in minimally invasive laparoscopic and robotic surgery should perform either procedure. As a result, the robot in the operating room is unable to make decisions or act on its own (Gültekin et al., 2021).

Robotic surgery is used by surgeons to precisely help during surgery. Many people believe that this kind of surgery is less intrusive because the incisions are smaller. Through robotic surgery, the surgeon is able to control the robot's arms, which are attached to high-precision tools. This device improves the precision and safety of surgery because it offers the surgeon the ability to perform movements that the human hand alone would not be able to do. Obviously, a great experience and skill of the physician who uses it is always indispensable at the base of robotic surgery (Ng & Tam, 2014; Guerin et al., 2022).

MINIMALLY INVASIVE SURGERY

A group of procedures known as minimally invasive surgery (MIS) allows for the performance of surgical procedures through incisions that are only a few centimeters (cm) deep, thereby reducing the trauma associated with the treatment. Compared to traditional ("open") surgery, which involves a cut of several cm to expose the part to be treated, minimally invasive technique allows to reach the area on which to intervene with tools such as cameras, lights, and scalpels, which pass through a few cm. The last 40 years have seen the development of minimally invasive procedures. Although the research and study date back to the 19th century, the first surgery was in 1987, when a French physician, Dr. Phillipe Mouret, completed the first cholecystectomy with laparoscopic technique (Al-Akash et al., 2009).

This procedure was developed starting from the knowledge acquired in diagnostics applied above all to gynecology. Diagnostic laparoscopy, already used in the past specially to analyze the abdominal cavity, has evolved to become one of the most used surgical techniques. In fact, laparoscopy, endoscopy, and robotic surgery are included in MIS (Cho et al., 2010).

It is possible to apply this type of surgery, depending on the pathology, to all medical specialties that require the execution of interventions, such as cardiac surgery, urology, gynecology, orthopedics, etc. With MIS, today, a whole series of interventions can be made: in addition to observation and diagnosis, removal of anatomic parts and tumors, restoring heart and lung function, inserting prostheses (knee and hip), making transplants (such as kidney), etc. Among the pathologies that currently see a greater use of the minimally invasive procedure are (Ulmer, 2010):

- gallbladder stones and appendicitis;
- colorectal, stomach, and liver cancer;
- pathologies that involve the removal of the spleen;
- Crohn's disease and ulcerative colitis;
- obesity;

- abdominal disorders: adhesions, gastroesophageal reflux, hernia (inguinal, umbilical, hiatal, etc.); and
- numerous gynecological pathologies.

Compared to traditional surgery, MIS, with its small incisions, reduces trauma, therefore the time of hospitalization and rehabilitation are shorter. There is also a lower risk of infection and a better aesthetic result (Marra et al., 2019).

When undergoing a minimally invasive procedure, the first thing that catches the eye is the size of the incisions: the small holes are about 1-2 cm in diameter, which is the space needed to introduce surgical instruments and an endoscopic probe. This is equipped with a camera and light that can transmit the images it takes inside the body to a monitor in the operating room. The surgeon then proceeds to move the instruments. Preparation for MIS is no different from that required for traditional surgery (Cohen & Einarsson, 2011; Lim & Kang, 2017). Certainly, the times of hospitalization and rehabilitation, compared to those of traditional surgery, even if it depends a lot on the treated pathology and the state of the patient, are definitely reduced in those who undergo this type of procedure (Koo, 2018).

Laparoscopic Surgery

Laparoscopic surgery is a minimally invasive surgical procedure, therefore it only requires small incisions. It can be used to diagnose some disorders as well as treat them. Depending on the purpose, the number of incisions may vary because, in the operative one, surgical instruments must also be introduced. The name derives from the instrument, "laparoscope", which is introduced during the surgical procedure in the area of intervention, for example in the abdomen. The laparoscope is a small, 5–10 mm-diameter tube that has a fiber optic camera that sends images to a monitor and a lighting system. This allows the surgeon to position himself correctly inside the abdomen (or pelvis) and carry out the procedure. Laparoscopic surgery has many advantages. With laparoscopy, there is a quicker postoperative recovery, a lower risk of complications, less intraoperative bleeding, and a less noticeable cosmetic and psychological impact even if the surgical hours are similar to those of standard surgery and may require general anesthesia (Lee et al., 2016; Picerno et al., 2017). Numerous abdominal and pelvic operations can be performed laparoscopically.

Diagnostic Laparoscopy

To make a diagnosis, non-invasive techniques such nuclear magnetic resonance imaging (MRI) and/or ultrasonography are typically preferred. However, when these instruments are not able to give clear results, it may be necessary to resort, in extreme cases, to diagnostic laparoscopy which, despite being minimally invasive, is still a surgical procedure. It is usually used in the diagnosis of gynecological pathologies:

- pelvic inflammatory disease (PID), i.e., acute or chronic inflammation, affecting the woman's reproductive organs and adjacent structures (usually due to infections with *Chlamydia trachomatis*, *Mycoplasma hominis*, and *Neisseria gonorrhoeae*);
- endometriosis (presence of endometrial tissue outside its natural location, i.e., the uterus);
- infertility and ectopic pregnancy, i.e., outside the uterus (extrauterine) or in an inadequate location of the uterus (intrauterine);

- ovarian cysts; and
- uterine fibroids (benign tumors of the uterus).

Other pathologies that can be investigated are:

- cryptorchidism (failure of one or both testicles to descend from the abdominal cavity to the scrotum);
- appendicitis (inflammation of a small section of the large intestine); and
- abdominal and/or pelvic pain for no apparent reason.

Finally, tumors of the abdominal/pelvic organs are also diagnosed: liver, pancreas, kidneys, ovaries, bile duct, and gallbladder. In the case of cancer, diagnostic laparoscopy offers the possibility of taking a small sample of cells from the diseased organ to be subsequently analyzed in the laboratory (biopsy). Furthermore, if the surgeon finds a problem that can (or must) be solved immediately, diagnostic laparoscopy can become, during the same session, also operative (therapeutic) (Weickert et al., 2005).

Operative Laparoscopy

During the procedure, the surgeon may remove the inflamed appendix (appendectomy), the gallbladder (cholecystectomy), a severely inflamed section of the intestine that does not recover with other therapies (e.g., Crohn's disease), and portions of fatty tissue, for example to reduce body weight. A tumor, an embryo in the case of an ectopic pregnancy, a fibroid, or even the uterus (hysterectomy) can also be removed. With this operation, it is also possible to perform a hernial surgery (typical in inguinal hernia), stop a hemorrhage caused by a gastric ulcer, etc. (Cuschieri, 1994; Nezhat et al., 2011)

Endoscopy

Endoscopy refers to a technique of direct observation of the internal part of an organ or cavity, using an optical instrument (a tube equipped with lighting and connected to a camera and a monitor). In this instance, however, the area is accessed by the abdominal wall or through one of the body's natural orifices, as in hysteroscopy and laparoscopy. Endoscopy, like laparoscopy, is also a technique that can be used both for diagnosis and observation, and for treatment and surgery (endoscopic surgery). It is widely used in gynecology, to treat conditions that can also be addressed with laparoscopy, which however involves abdominal incisions (Bailey et al., 2015; Thubert et al., 2016).

Endoscopy is widely used in otolaryngology, where endoscopic sinus surgery is very common to treat injuries or problems inside the nose, which prevent normal breathing, affect the sense of smell, or cause pain in the face and head. In the context of general surgery and the digestive system, endoscopy or laparoscopy may be indicated for obesity surgery (bariatric surgery). In the field of neurosurgery, in an emergency, this procedure can be useful in treating a cerebral hemorrhage or a cerebrovascular accident (stroke) (Montori, 1999).

The main advantage of endoscopic surgery is that it is less invasive than other surgical techniques and allows for a shorter postoperative period. Generally, the procedures are performed in day-hospital (DH) and the patient, after a few hours, can go home and resume daily activities, unless otherwise indicated by the physician.

Robotic Surgery

Robotic surgery is the development of laparoscopic surgery: an advancement in MIS techniques is represented by robotic surgery, which consists of a series of techniques put into practice by a specialist surgeon, through a high-precision robotic system, which allows to better see the area to be treated and to perform movements with surgical instruments, otherwise impossible for a human being, as we will see in detail below (Liu et al., 2014).

Robotic surgery is performed through the use of robots. It is a very advanced technique, which allows the surgeon to perform operations in a very precise way. The surgical instrument is fixed to the robot's arms and the specialist manipulates it from a computer station. To carry out the procedure, the robot mimics the movements of the surgeon's hands. In robotic surgery, the surgeon controls the motions of the robot, which is outfitted with all the instruments required to complete the procedure. The specialist practices small incisions to introduce the instruments into the body and the endoscope, a thin tube equipped with a camera. As a result, the operating room's 3D visuals are visible to the surgeon compared to the two-dimensional (2D) view and gives freedom of movement in all directions of space, which allows to perform delicate techniques more precisely (Rardin, 2014).

Theoretically, the surgeon could also operate from another place but, currently, he is usually in the operating room, not standing in front of the patient, but near, in front of a console, that is a command station from which he controls all the instruments. The peculiarity is the 3D view of the operating field. He is practically immersed in the operative reality because he does not have a 2D vision screen (as in laparoscopy), but 3D, as it happens in virtual reality, with very high precision (Moawad et al., 2019). The technology also makes it possible to superimpose the diagnostic examination on the operating field and simultaneously have more elements, more useful information, in the intraoperative phase, to intervene in a more precise, targeted, and effective way (Lawrie et al., 2019).

Compared to traditional surgery, robotics allows to make smaller and more precise incisions. With this technique, therefore, it is possible to perform through holes of 1-2 cm, surgical interventions that in the classic technique require wider cuts. This results in less trauma for the patient, less intense pain after surgery, less risk of infections and bleeding, and consequently a shorter recovery time and a quicker return to normal life, as well. The hospitalization time, in the case of colorectal surgery, for example, can be reduced by as much as 40% due to the lower risk of complications, especially in older and more fragile patients due to the presence of other pathological conditions (Kenngott et al., 2012). Compared to laparoscopy, which has rigid instrumentation, robotics allows movements with 7 degrees (°) of freedom and miniaturized (Gala et al., 2014; Marchand et al., 2021). As with any surgery, there are risks due to general anesthesia and infections, although, being less invasive, they are lower than other techniques (McLarney & Rose, 2011; Aceto et al.,2019).

This type of surgery can be carried out in various procedures involving hip and knee arthroplasty, but also kidney transplantation. Robotics is also successfully used for interventions on various internal organs. The technique, in fact, is used for very high-precision interventions especially at the level of structures such as the colorectum, stomach, liver, and pancreas, where laparoscopy has great difficulty: in the repair of heart valves, removal of tumors in the liver, lungs, stomach, colon, and pancreas, creation of a gastric bypass, in gynecology in the treatment of endometriosis, hysterectomy, closure of the Fallopian tubes, and in urology, in tumors of the bladder and prostate (radical prostatectomy) (Falcone & Steiner, 2002; Liu et al., 2012).

HISTORICAL OVERVIEW

The Czech author Karel Čapek's science fiction (sci-fi) book "R.U.R." (Rossumovi univerzáln roboti, Rossum's Universal Robots), a three-act utopian drama published in 1920 and staged at the National Theater in Prague on January 25th, 1921, is credited with coining the word "robot". In this work, the term robot appears for the first time, invented deriving it from the Czech word "robota" ("work") to indicate replicating androids, designed to free man from slavery and physical fatigue (Čapek, 1920).

In the most frequent meaning, "robotics" are defined machines capable of autonomously performing even complex functions, which involve active interaction with the environment. The adjective "robotic", although extremely widespread, is therefore to be considered improper in the case of surgery, since it can hint at an operational and decision-making autonomy that, at least at the moment, does not exist. In fact, it is always the surgeon, through the master manipulators, who controls all the movements of the instruments that are performed during the surgery.

The history of robotic surgery begins in the late 1950s, when the Pentagon developed the first robotic surgical device, called "Robotic-Assisted MicroSurgery" (RAMS). This was designed to perform microsurgery in war situations and was carried out by a partnership between the National Aeronautics and Space Administration (NASA) Jet Propulsion Laboratory (JPL) of Pasadena and Micro-Dexterity System Incorporation (Inc.) to treat war wounded from remote locations; in fact, thanks to two extremely precise mechanical arms piloted by the surgeon, it was thought to be able to perform operations hundreds of kilometers (km) away. The war experiments, which were unsuccessful due to the challenges of signal transmission, nevertheless marked the beginning of a new era in the field of surgery: the robotic era, in which the surgeon performs surgery for the first time in the history of medicine while seated at a console rather than using his own hands (Satava, 2003).

In 1985, the first surgical robot was the Programmable Universal Manipulation Arm (PUMA) 560, which was used to perform neurosurgical biopsies (Theodore et al., 2018). The other systems, born later, were used to perform various types of services, of which the robots used in orthopedics for knee and hip surgery, through the Robodoc and the Acrobat, are very well known, both developed during the '90s (Jacofsky & Allen, 2016). Since then, many steps forward have been made and, arriving at the last decade, we must certainly include (and take as examples) the two devices that for a certain period have been competing for the world market, especially the North American one.

The manufacturing companies, both from the United States (US), Computer Motion (Goleta, CA) and Intuitive Surgical Inc. (Mountain View, CA), are the most successful, having produced and marketed numerous examples of their robots. The first produced the Aesop and Zeus systems, which were endorsed in 1997 and 2001, respectively, by the Food and Drug Administration (FDA). The second, founded in 1995 in California, was the Intuitive Surgical Inc., which first patented the Stanford Research Institute (SRI) Green Telepresence Surgery System, used for the first time in Europe in 1998 by Alain Carpentier in cardiac surgery, later renamed in the "DaVinci" system, in honor of the Italian Renaissance genius who already in 1400 imagined an automatic machine driven by pulleys and wires and who also included anatomy among his fields of interest (Pugin et al., 2011).

The DaVinci robot received the approval of the FDA in 2000 and has gradually acquired more and more slices of the market, coming to have a monopoly in the sale of surgical robots, especially after the acquisition of the rights of Computer Motion, thus also obtaining the rights of Aesop and Zeus robots. On September 9th, 2001, Jacques Marescaux performed a cholecystectomy in transatlantic telesurgery for the first time, effectively marking the beginning of a new era. For this operation, the Zeus system was used; the patient was in Strasbourg, 6200 kilometers distant, while the operating surgeon was in New York. In 2003, the fusion of Computer Motion and Intuitive Surgical gave rise to a single company, Intuitive (George et al., 2018).

Since its introduction, thousands of surgeries have successfully utilized the DaVinci surgical system; its safety, efficacy, and superiority of clinical results are proven in hundreds of scientific papers (Han & Advincula, 2019). There are about 100 in Italy, but not all of them work at full capacity. Currently, the most employed are about 80 and are found mainly in the North. Italy is one of the countries in which there are more devices, but to amortize the costs it would be necessary to distribute them more rationally on the territory. However, the sector is under development. The biggest obstacle to the spread of these techniques is the lack of trained surgeons. Yet, Italy was the leader because some pioneers, such as Pier Cristoforo Giulianotti, who currently works in America and who is among the world's leading experts, come from there (Hockstein et al., 2007; Giulianotti, 2020).

FEATURES, aDVANTAGES, AND DISADVANTAGES

Compared to traditional surgery, robotic surgery has numerous advantages for both the patient and the physician. All the operations that can be performed with the laparoscopic technique can be carried out with the aid of the robot, thus avoiding the cut on the abdomen and respecting the body integrity of the patient. With robotic surgery, operating times have been reduced (reduced anesthesia and less physical stress for the patient) (Hottenrott et al., 2020; Ardon & Hernandez, 2022).

Furthermore, carbon dioxide (CO_2) pressures in the abdomen are reduced and therefore even patients with particular problems can be operated on. About 90% of patients, even those undergoing complicated anticancer operations, go home the same day as the operation. Leading surgical and cancer institutions have conducted considerable research on the drive towards same-day discharge, which has been continuously growing. Patients can recover more quickly with their families in the comfort of their homes than in hospitals, which has proven to be safe and extremely pleasant (Nahas et al., 2016; Sanabria et al., 2020).

Robotic operations often have recovery times that are half as long as conventional techniques, allowing patients to resume their regular routines much sooner and lessening the financial load on them to return to work.Compared to open treatments, MIS results in significantly reduced long-term adhesion and scar tissue formation (Weinberg et al., 2011). The use of the robot is especially beneficial for pelvic surgery, therefore in gynecology (recto-vaginal septal and intestinal endometriosis) or in surgery for obese patients, where the bowel restricts the field of action (Scheib et al., 2014; Menderes et al., 2019).

The surgeon can more clearly see nerves, arteries, and some ligament structures thanks to the robot's 3D vision. The robotic platform allows to obtain very high levels of precision and accuracy of the surgical gesture. With the robot, which gives the surgical gesture a precision that cannot be compared with other techniques, it is possible to overcome the limits linked to the difficulty of treating pathologies in anatomic locations difficult to reach with laparoscopy. In fact, the use of the robot provides the surgeon with greater safety to treat anatomic areas that are difficult to access, facilitates a wide and clear view of the inside of the patient, and ensures stable and tremor-free handling of surgical instruments (forceps, scissors, etc.) and the surgeon's hands moving with the greatest amount of precision. These qualities enable the surgeon to perform with greater precision and accuracy than those possible up to now with traditional laparoscopy, especially in maintaining the correct innervation, avoiding injuries during surgery. This translates into undeniable benefits for the patient: concurrently, in fact, the robotic surgical system ensures patients all the benefits of the traditional minimally invasive technique, which are linked to a higher quality of life and cosmetic outcome because there are small skin scars present. With all these benefits, the future of surgery, and of the mixture of robotics and medicine, seems clear and marked. This technology is used for an increasing number of interventions. But the surgeon's experience weighs heavily in the balance between advantages and disadvantages (Falcone & Goldberg, 2003; Gastrich et al., 2011).

The following are a few characteristics and benefits of the robotic surgical system:

- expansion of the operator's capabilities: the command of instruments with simple movements of the wrist allows the surgeon to overcome the limits of "open" and laparoscopic surgical techniques by using a minimally invasive technique for a variety of intricate procedures;
- greater speed in procedures, decreased invasiveness of interventions, and increased precision in configuring the accuracy of the surgical movement;
- improved user interface, which ensures integration between the robotic system and other equipment in the operating room; the simplicity of the touch-screen interface of the viewing trolley monitor allows staff to speed up the preparation of the operating room for the various interventions, with a reduction in time;
- the possibility to see the operating field in high definition (HD) 3D with perfect view of the anatomic landmarks and to use natural movements, similar to those of the hands and arms, unlike laparoscopic surgery;
- the presence of four mechanical arms that support the endoscopic instruments and the camera without "shaking" and without "getting tired";
- the use of robotic arms linked to the control panel prevents natural hand tremors and increases precision;
- the extreme maneuverability of the endoscopic instruments with 7° of freedom of movement and possibility of movements and rotations superior to the human hand;
- enhanced comfort during operation because to the console's ergonomics, without sacrificing surgical accuracy;
- reduction of the laparoscopic fulcrum effect, which enables the surgeon to move more normally, typical of open surgery;
- extreme ease of access to "difficult" anatomies;
- detailed view of the cleavage planes;
- considerably increased precision in the demolitive and reconstructive procedures;
- greater ease in performing complex surgical maneuvers;
- greater oncological radicality;
- greater preservation of healthy tissues and organ functions;
- maintenance of correct innervation ("nerve-sparing" surgery);
- small incisions with reduction of tissue trauma;
- less bleeding and therefore less need for transfusions;
- reduction of hospital stays and postoperative pain;
- reduction of recovery times;
- faster recovery in carrying out daily activities and work with significant savings from an economic point of view;
- reduced risk of postoperative infections;

- greater safety for the patient; and
- excellent aesthetic results.

On the other side, robotic surgery has drawbacks, which focus on four areas:

- cost of the system: first of all, the purchase cost of the machine must be considered which, operating under a world monopoly system, costs between 2 and 2.5 million euros (€), but there are additional expenses: the instrumentation, which require spare parts, maintenance (an average of 150 thousand € per year), and so on; an average of 2,000-3,000 € per intervention can be reached; it has been calculated that, to amortize them, a robot would have to carry out at least 500 interventions per year; then, for supporting the system itself; since the tools are partially reusable, the costs are substantial (use limited for some to 10 interventions and for others to 20); however, it is foreseeable that costs could undergo a significant reduction (the patent will expire shortly), similarly to what has been observed in other high technology sectors, with the increasing diffusion of the method in the near future (Swan & Advincula, 2011; Tiwari, 2014);
- in order to operate with the robot, very high skills of the surgeon and room staff are required, to be acquired through specific training; fortunately, the training of dedicated personnel is generally included in the purchase cost of the machine (Lee et al., 2011; Foote & Valea, 2016);
- robot set-up and its logistics: problems during system setting, from the point of view of the errors that are made and the time it takes (initially everything is linked to the lack of experience and the problems tend to decrease as manual skills of the operating room team increase); due to the distance between the first operator and the rest of the team, communication issues between the operating surgeon and the rest of the surgical team can arise in the operating room; and
- safety: using robotic equipment and systems can make surgeons feel too secure, and they can be misled from the point of view of perception, as computer and robotic processes produce circumstances that are apparently similar to old practices yet significantly different from them; a first example of the difference consists, in fact, in the lack of tactile feedback, not yet present and most likely being studied (Varghese et al., 2019; Han & Advincula, 2021).

INDICATIONS

In recent years, traditional MIS (laparoscopic-thoracoscopic) has been combined with robotic MIS. Robotic surgery has become a viable option for numerous surgical operations over the past ten years, advancing the area of MIS. Its rate of diffusion has varied based on a number of variables, such as the cost-effectiveness ratio and the learning curve associated with the various operations (Schreuder et al., 2012; Xie, 2015).

Robotic surgery is applied to several types of specialties, including:

- urological surgery (prostatectomy, nephrectomy, pyeloplasty, cystectomy, varicocele, ureterectomy, testicular resection, etc.);
- gynecological surgery (hysterectomy, myomectomy, repair of vaginal prolapse, salpingectomy, tubal anastomosis, sacral colpopexy, dermoid cyst, etc.);

- cardiovascular and thoracic surgery (lobectomy, coronary anastomosis, thrombectomy, pericardial window, pneumonectomy, pacemaker implant, mediastinal resection, etc.) (Rueda et al., 2018);
- general surgery (gastric bypass, gastrectomy, colon resection, gastroplasty, appendectomy, adrenalectomy, intestinal resection, etc.);
- endocrine surgery (adrenalectomy, thymectomy, trans-axillary thyroidectomy, etc.);
- pediatric surgery (Pelizzo et al., 2019; Meinzer et al., 2020); and
- otolaryngological surgery (trans-oral surgery is used to treat certain disorders through the mouth).

Some techniques, such robotic prostatectomy, have had a very quick spread. About 80,000 robotic prostatectomies were were out in the US in 2009. Excellent oncological and functional outcomes following robotic prostatectomy have led to full approval of robotic surgery in the field of urology for further procedures such partial nephrectomies, cystectomies, and pyeloplasty (Ravivarapu et al., 2021).

Gynecology is the second most popular area for robotic surgery. Robotic surgery currently would seem to represent the most valid and least invasive therapeutic option in carrying out complex interventions such as, for example, the radical uterine resection and the systematic lymph node removal both at the pelvic and the abdominal level, the suspension of the uterus and of the vagina, and the radical excision of deep endometriosis (Zakashansky et al., 2008; Andres et al., 2022). This method has gained popularity among women because to its extremely pleasing aesthetic outcomes and among surgeons due to its ability to circumvent the challenges of performing traditional laparoscopic surgery in the small pelvic area.

Although the adoption of robotics in general surgery has been slower and more gradual, the robotic platform is remarkably advantageous in a number of complex operations (liver and pancreas surgeries, as well as colorectal and stomach surgeries) that are carried out by a small percentage of general surgeons. The learning curve for such complicated procedures is very steep and time-consuming. But according to the surgical outcomes thus far, there have been less difficulties with esophageal, pancreatic, hepatic, and colorectal surgery, and the robot now enables a wider range of minimally invasive treatments for a larger number of physicians. In example, there is evidence that robotic rectal cancer treatment has better results in preserving urinary and sexual function than conventional laparoscopic and open surgery. As a result, robotic colorectal surgery is growing in popularity. Additionally, compared to laparoscopic surgery, the rate of conversion to open surgery occurs far less frequently – pretty much never – during the actual procedure.

Other fields in which robotic surgery is applied are endocrine surgery (adrenalectomy, thymectomy, and trans-axillary thyroidectomy), and trans-oral resections of mucous lesions of the upper respiratory tract in the otolaryngology field. Robot use was first utilized in cardiac surgery, which subsequently encountered some difficulties, due to the fact that some procedures have mostly been replaced by minimally invasive per-cutaneous treatment, while multiple by-passes are technically extremely difficult, especially as regards the access to the back of the heart.

Since 2000, more than a million robotic treatments have been carried out globally.

TECHNICAL-OPERATIVE ASPECTS

In conventional laparoscopic surgery, the patient has a few tiny incisions made by the physician. While the assistant operates the equipment at the patient's bed, the surgeon penetrates those incisions to carry out the procedure using a camera and small tools. Additionally, these straight-stick tools can only move up or down. The same is true for robotic surgery. Little incisions are made by the surgeon. But instead of the surgeon holding the instruments at the patient's bedside, a robotic platform does so. The surgeon also uses hand controls to operate the tools while standing close to a console. In contrast to laparoscopic surgery, the instruments move in all directions. The surgeon now has better access to work. As a result, surgeons can benefit from executing intricate surgeries with higher accuracy (Magrina, 2007).

The operating room's surgical console serves as the robot's command center. The surgeon observes the surgical field, while comfortably seated at the console, through a visual system and controls a camera that has been introduced through an 8 mm tube (trocar) through a little incision in the abdomen of the patient to be operated on. Thanks to the camera, equipped with a HD viewer, which provides processed images in 3D, adding a realistic depth to the vision of the operating field, the surgeon gets a clear, magnified and HD 3D view of the area on the display. The actual robot is the surgical trolley, which is naturally situated at the operating table. It has four mechanical arms (Bouquet de Joliniere, 2016).

The surgical assistants who are specialists in this procedure are present at the operating table to help with the intervention phases when the robotic tools are introduced into the belly and secured to the surgical trolley's arms using safety devices. The equipment, in continuous evolution, is extremely versatile in its movements: its articulation allows 7 movements on its own axis, with 90° angles which, in particular cases, facilitate the achievement of narrow and deep anatomic spaces. The operator controls the instruments and directs the vision with the hands using two joysticks and using the pedals with the feet; with the movement of his hands, he gives movements to the instruments. The movements of the operator's wrist and hands are transmitted with precision and without tremors to the endoscopic micro-instruments inserted in the abdomen of the patient. The system enables the surgeon to practice minimally invasive surgery since the movements of the robotic instruments are articulated like those of the hand. The robot mimics the movements of the surgeon's hand using its articulated grippers and robotic arms in real time and with absolute fidelity and precision (Fanfani et al., 2016).

The HD 3D camera's ability to use fluorescence offers countless benefits for the operator's field of vision as well as for better identifying anatomic features. It also makes it possible to find the sentinel lymph node using the dye indocyanine green (ICG). Specifically, the 3D/ICG and 4K vision systems are visible on 55° 4K/3D screens and can be managed remotely from a touchscreen with preset scenarios and video stream in full 4K Over IP. For planned interventions in the gynecological oncology field, the new infrared technology in partial white light allows the identification of the ICG lymphatic tracer with the aim of identifying and selectively removing the sentinel lymph node. This approach makes it possible to significantly reduce surgical morbidity and improve the ability to stratify patients by risk groups (Rocha et al., 2016; Mach et al., 2020).

Robotic surgery requires that each maneuver be performed under the direct control of the surgeon, which cannot be performed otherwise; recurrent safety checks prohibit the robotic arms or tools from moving independently. The surgeon uses two handpieces, positioned under the display, which perform 360° movements, i.e., in full rotation, and with which he can control the endoscope and robotic instruments, maintaining his wrists and hands parallel to his eyes. The surgical instruments are moved simultaneously and precisely by the surgeon's hand, wrist, and fingers thanks to the technology. Other surgical equipment is connected to the motorized arms such as forceps, scissors, dissectors (whose exclusive design ensures optimal dexterity), which the team present at the operating table introduces in the surgery site. The pedals at the base of the console allow the control of the auxiliary surgical instruments (electrocautery, ultrasound, etc.) and include a clutch mechanism that allows the handpieces to be re-positioned in line with respect to the robotic arms for greater ergonomics and oculomotor coordination (Alkatout, 2016).

The versatility of the movements of robotic tools is the most important advantage. The versatility of the instruments, in fact, allows to reach narrow and deep anatomic spaces and this feature is particularly advantageous in pelvic surgery and in gynecology. The second fundamental feature of the robot is the 3D vision. The image that the surgeon sees on the console screen is enlarged, in HD and 3D, thanks to the two tiny cameras mounted on a single robotic (optical) instrument that is inserted into the operating room using a laparoscopic trocar; the computer then captures and reconstructs the image, giving a flawless representation. The 3D vision helps the surgeon in the various surgical times to better visualize nerves, vessels, and some ligament structures.

The minor physical stress of the surgeon should not be overlooked, since the operator is comfortably seated at the console and without fatigue can perform long operations that are generally tiring with the classic laparoscopic technique, such as oncological operations (Cardenas-Trowers et al., 2018; Das et al., 2019). When performing robotic surgery, the surgeon's hands at the manipulators align with the instrument tips in the stereo viewer. This is so that the natural alignment between eyes, hands, and instruments can be obtained. It is possible also to switch between on-screen and multi-image displays. The normal alignment between hands, eyes, and instruments, as occurs in traditional surgery, in robotic surgery is guaranteed by the machine whose stereo viewer also allows the display of full screen or that of multiple images, as well as other scans such as that of the ultrasound or of the electrocardiogram (ECG) (Mettler et al., 2011). The surgeon's hands move the console masters (or joysticks) that induce movements on the robotic instruments; after an adequate software filtering, these movements are optimized by decreasing the tremor, maintaining a correct ratio between the movement of the masters and the robotic instrumentation itself. In robotic surgery, the physiological tremor of the operator's hands and involuntary movements are thus minimized.

The instruments connected to the robotic arms (several types of graspers, forceps, scissors, a clip applicator, and an ultrasound dissector) are interchangeable according to the needs of the surgeon and the operation being performed. The console is situated inside the operating room itself for logistical reasons, and a fiber optic connection connects it to the robotic trolley. Longer distances could theoretically be covered by such a connection (different hospital, city, or country). The operating theaters, in fact, are also equipped with advanced videoconferencing systems that allow to put the room and the surgeon in contact with other centers, managing video streams even in full HD 3D mode and at the same time recording and documenting images and videos of the clinical case. In addition, each event can also be broadcasted within the hospital in Web Live Streaming mode for online only broadcasts with the possibility of creating on-demand multimedia libraries of surgical procedures and educational material for participants and trainees. In these cases, the only limitation is the very high cost of the connection and the interval between visual transmission and reaction, albeit minimal, of the surgeon's movement (Harding & Williams, 2016; Steil et al., 2019).

A microphone system incorporated into the console that can transmit audio to speakers in the operating room is also an option for verbal communication between the surgeon and the surgical team at the operating table. The second operator aids the surgeon at the console in the sterile area and is given full priority in maintaining patient safety. Regardless of issues that develop during the robotic surgery procedure, the surgical team at the operating table can unplug the robot and carry out the rest of the operation laparoscopically or using a standard "open" technique. The surgical procedure's technical difficulties and the surgeon's experience affect the conversion rate. It varies between 1% for robotic prostatic surgery and 6-10% for robotic pancreatic surgery, which is particularly complex. It is highly unusual for conversion to occur as a result of robotic equipment malfunctions. Other than those associated with the laparoscopic method, the robot has no special morbidities (Reza et al., 2010).

The instruments are in continuous technological evolution to better meet the needs of surgeons in the various disciplines and for the various types of interventions. The evolution of nanotechnology and mechanics will ensure that current robotic surgery is only the first step towards an ever less invasive and precise surgery. However, it must be borne in mind that the robot does not replace the surgeon. Patients frequently believe that a robot operates freely due to artificial intelligence (AI). This is the furthest thing from the truth. From the console, the surgeon directs every action. He views the robotic platform as a constant aide that keeps the instruments in place and directly transmits the surgeon's hand movements. It can be defined as the evolutionary formula of laparoscopy, but it is certainly much more. In addition to giving the surgeon stereoscopic vision thanks to the use of a two-channel optical video-endoscope and a double three charge-coupled devices (3CCD) high-resolution camera, the technology enables continuous articulation of all hand and wrist movements within the patient. In comparison to conventional surgery, this real-time magnification device gives the physician a phenomenal image quality. More than 1000 photos are delivered every second by the system, and each image is processed to remove background noise. Additionally, it enables the filtering of the surgeon's hands' tremors and scaling of the action of the instruments. The operating room instrumentation may be moved precisely and easily thanks to the movement actuation module. The robot is "moved" and controlled by a surgeon who uses the system to improve vision, which becomes 3D, and the hand which is more precise and freer than that of the human wrist. So, the robot does nothing autonomously, the choice of procedures and steps to be performed is exclusively the surgeon who guides it. The robot enhances in the realization what the surgeon thinks and decides to do. The human limit of precision, speed and vision is bridged but the strategic decision of the intervention remains entirely with the physician who knows the patient in detail and modulates what is necessary to do for his pathology. This combination leads to the realization of a more precise, faster, less invasive, and more complete surgery (Hamet & Tremblay, 2017; Moglia et al., 2021).

The DaVinci™ Robotic Surgical System

The DaVinci™ Robotic Surgical System (Intuitive Surgical Corporation, Sunnyvale, CA, US) is touted as the most cutting-edge platform for robotic surgery and as the ideal tool for highly complex MIS since it allows for the circumvention of traditional surgery's restrictions (Advincula & Song, 2007; Quemener et al., 2012).

The DaVinci robot is an integrated system consisting of three main components (Advincula & Wang, 2009; Sinha et al., 2015):

- the operating room's control panel for the robot is called the surgical console; the console is outside the sterile environment, and the surgeon is seated at it and operates by means of two manipulators (similar to joysticks) that perform 360° movements and pedals that guide the instrumentation; through the viewer the surgeon sees, via the 3D endoscope monitor, everything that the camera frames in the operating field;
- the surgical (patient) trolley: it is the real robot and it is located at the operating table; it supports the robot's arms, which physically perform the operation; the surgical instruments are mounted on the robot arms, and there are assistants at the operating table to help with the interventional stages; the robots' arms are fitted with EndoWrist instruments, which have a wrist that can spin

virtually 360°; the instruments have a 7° range of motion, which is far more than a human wrist's; the main feature of EndoWrist instruments is the wrist which guarantees the free movement that is impossible for the hand of man; and

- the viewing trolley: it contains the central processing unit; it consists of a monitor, an electrosurgical unit, and shelves on which to place other instruments; it is also equipped with a full HD video system; the endoscope records the operation field and sends it back to the head of the HD stereo camera, which has a 60° field of view; together with the Intuitive Surgical endoscopes, the vision system allows to enlarge the operating area by 6-10 times; the following are the components of the viewing cart:
    - electronic system (core): it processes the video image;
    - endoscope controller: it contains a high-intensity light source to illuminate the surgical site and the electronics for processing the image detected by the endoscope;
    - video processor: it receives and processes the video input from the endoscope and sends it through the system electronics to the touch-screen and 3D viewer;
    - endoscopes: the DaVinci HD Vision System uses oblique or straight tip endoscopes; the guide light and endoscope communication signals are fixed to the endoscope at the beginning of the operation; the endoscope cable connects directly to the viewing trolley to provide communication and illumination; the heat from the optical fibers helps minimize fogging of the endoscope lenses; the processing of the image coming from the left and right channels allows 3D viewing on the console; the DaVinci HD Vision System is compatible with dedicated 30° and 0° 12 mm diameter endoscopes; the endoscope acquires a 3D HD video of the surgical site; the 3D video is processed by the system electronics in the viewing trolley and displayed on the 3D viewer of the surgical console and on the touchscreen of the viewing trolley;
    - touchscreen with a series of controls for the endoscope and video configurations;
    - oxygen (O_2) bottle holder;
    - auxiliary shelves: movable shelves for supplemental surgical instruments that are optional, like CO_2 insufflators; and
    - electrosurgical unit: integrated unit for the activation of robotic and laparoscopic mono-polar and bipolar instruments.

The surgeon operates the robot arms that are attached to the endoscopic equipment through tiny incisions while physically separated from the operating field and situated at a station with monitors and controls. During the surgery, the position of the robotic arms is not random but, based on the type of operation, is optimized through a laser pointer; their movements are fluid, jerky, and intuitive. The operating field is 3D-projected, with still and very high-resolution images. The surgeon at the console uses the robot to mimic his actions, but the robot's ability to scale (de-multiply) movements and control natural hand tremors allows for more precise surgery. Additionally, the surgeon can detect the smallest anatomic components, which are challenging to perceive with the unaided eye, thanks to the expanded and 3D image of the operating field. The system, performing autonomously more than 1 million safety checks per second, offers maximum safety and reliability during operation. Surgeons and operating room staff are kept up to date on the system state and the patient's health via audio-video feed-backs. A huge touch-screen monitor with the possibility of making freehand drawings on the projection of the operating field facilitates communication within the team (Tsuda et al., 2015; Azadi et al., 2021).

Although the DaVinci robot in Italy may still seem like a novelty, DaVinci technology is not brand new: it was born in the military field. The DaVinci robot was born in Silicon Valley (US) and has been designed, developed, and marketed by Intuitive Surgical Inc. since 1999. The first intervention took place in Germany in 2000 (Palep, 2009). The first DaVinci robotic system (IS1200) called Standard was introduced on the market in 1998 and in Italy in 1999, while in 2006, the DaVinci S system (IS2000) was produced (Elliott et al., 2006). The DaVinci Si HD (IS3000) version was introduced to the international and domestic market in 2009. The DaVinci Si system has the possibility of placing an additional console side by side to accommodate a second surgeon (for example a trainee surgeon) (Nezhat & Lakhi, 2016). In 2014, Intuitive Surgical launched the DaVinci Xi HD (IS4000) model, the fourth generation of the system, the robot's most recent iteration. In the final quarter of 2013, almost 3000 DaVinci systems of different configurations had been installed worldwide (Patel & Hemal, 2018).

The DaVinci robotic system, now in its fourth generation, testifies the most widespread and advanced expression of this technology in the world. Twenty years after the introduction of the DaVinci surgical system, this powerful MIS tool is now a standard of excellence in many surgical fields. The DaVinci robot is in fact the most advanced platform in the world for MIS. Its technical characteristics allow the robot to find different applications, from urology to gynecology, from thoracic surgery to general surgery.

Robotic surgery in Italy has recently passed the decade of activity. From complex instruments limited only to certain types of interventions, such as the first robot-surgeon Zeus, we have come to the development of a new extremely precise and adaptable technique (Marescaux & Rubino, 2003). The innovative machinery, pearl of MIS currently used is now present in 54 hospitals throughout the country. Italy is second only to the US in the use of this equipment. The interest in this new star of technology applied to medicine is increasing year after year, thanks above all to the wide use that can be made of it. In Italy, in 2011 approximately 7,000 interventions were carried out, with an average annual increase of approximately 40%. About 30% of these involved gynecological operations (Santoro & Pansadoro, 2013). More than 11,000 procedures were performed in 2014, a 20% increase over 2013. In 2017, 17,462 interventions with the robot were carried out, +16% compared to 2016. In 2018, 20.450 operations using the DaVinci surgical system were carried out in Italy, 18% more than in 2017: a datum that indicates growing confidence with the system. Today, 4400 are installed worldwide and after the US, with 2862 robots, there is Europe, with 742 (Fong et al., 2018).

The data speak for themselves: a growth that is actually reducing the use of laparoscopy, now overtaken by this technology. Of course, focusing on these technologies means focusing on training: to use the DaVinci needs about 30-40 training sessions. The numbers allow to affirm that robotic surgery also in Italy is reaping important clinical benefits. Despite the high cost of this technology, about 3 million euros, the map of Italy is becoming more and more homogeneous: with the installation of the 100th DaVinci robot at the Catania Polyclinic, Italy joined France as a leading nation in this field. There are 111 DaVinci surgical platforms in Italy and they are now in all regions, with a clear prevalence in the North: Lombardy has more than all, 22, then Tuscany with 13, and Veneto with 12. Rarer in the South, with Basilicata and Calabria having one each. But this is the trend now. Around 5,000 are instead present in the world; in many areas, today, it is considered more effective and safer than traditional surgery for the precision that the system allows, the low impact on the patient (MIS), and the reduction of postoperative recovery times. In Italy, this type of surgery is used above all in the urological field (almost 12 thousand, covering about 68% of the activity), while in the world the gynecological use prevails. In Italy, 68% of robotic surgery operations are carried out by urologists because robotic surgery was born for radical prostatectomy. The robot is able to amplify the image 10 times, so a nerve becomes almost like a pen.

This allows to safeguard sexual functions and therefore quality of life. AB Medica of Lainate, Milan (which distributes the DaVinci platform in Italy and provides training and assistance), with continuous training plans and remote assistance, and the heads of the departments where the DaVinci is installed, work on this front: robotic surgery is now an essential gold standard (Medical Advisory Secretariat, 2010; Rardin, 2011).

The surgeon still has complete control over the game in this so-called "master-slave" setup because, sitting at the console, he controls the various tools: pliers, scissors, etc. The concept of surgery in general has changed: more and more the operating room resembles the control room of an airplane and is completely different from the concept of scalpel, needle, and thread that we had previously. Everything is more technological (Frick & Falcone, 2009; Levinson et al., 2013).

The "DaVinci Xi" model is the latest and most advanced version of DaVinci technology. Compared to the previous version (Si model), the new robot operates in the opposite manner. The position of the arms is determined by a computer as they drop from above according to anatomic patterns chosen by the surgeon, in relation to the various interventions. The parallel work of the arms reduces the conflict between tools and allows them to move in opposite directions to the usual direction of operation. This innovation allows to operate on different anatomic sites, even opposite ones (upper and lower abdomen) with a single position of the robot. Another interesting feature is the new "Grab and Move" arm handling system, which allows them to be handled with a great reduction in operating times. Added to these innovations is the fact that the new Xi is equipped with all the options available to the Si, including fluorescence vision, robotic staplers, and other hemostasis tools (American College of Obstetricians and Gynecologists' Committee on Gynecologic Practice & The Society of Gynecologic Surgeons, 2020; Lee et al., 2021).

The cost of the Da Vinci does not seem to prevent its sustainability, at least for the centers that carry out a certain number of operations per year (about 200): the hospitalization times of patients operated on with the robot collapse and also those who, for example, are operated on for a prostatectomy after 2 days can already be discharged, allowing not only a benefit in terms of health but also a saving in terms of beds. Robotics is destined to remove more and more space from open surgery and laparoscopic surgery, the future of surgery goes towards robots (Iavazzo et al., 2014; Korsholm et al., 2018).

APPLICATIONS IN GYNECOLOGICAL SURGERY

The surgical approach in gynecology is now diversified and multiple:

- conventional laparotomic surgery;
- vaginal surgery;
- traditional laparoscopic surgery; and
- robotic surgery.

The choice of which "surgical path" to undertake must take into account the patient, who is always placed "at the center". Robotic gynecological surgery (or robotic-assisted surgery in gynecology), represents only a small percentage of general robotic surgery, since 75% of robotic-assisted interventions are the prerogative of urology. However, among the latest frontiers acquired in the gynecological field, gynecological robotic surgery represents the novelty that has aroused the most interest in recent years (Kim et al., 2008; Liu et al., 2014).

Robotic gynecological surgery has made it possible to merge the best of traditional laparotomy surgery (simple, precise, and safe movements, but with large abdominal incisions to access the pelvic cavity) with the advantages of conventional laparoscopic surgery, where the surgical field is accessed through small incisions. These aspects are particularly important and crucial in gynecological surgery, where it is necessary to work in the pelvic cavity, in close contact with important structures such as the rectum, ureters, blood vessels, and the bladder. Robotic gynecological surgery, as a result, is the most innovative and sophisticated technique available today and represents a valid alternative to both traditional "open" techniques and standard laparoscopy (Visco & Advincula, 2008; Truong et al., 2016).

Robotic gynecological surgery is indicated for many benign pathologies (uterine fibroids, genital prolapse, etc.), but it is with the most complex and delicate gynecological procedures, such as the radical treatment of deep endometriosis or in oncological surgery, both for the treatment of uterine body and cervical tumors, which expresses all its advantages and potential. Potentially all the interventions achievable in laparoscopy are convertible into robotics as well as many of those traditionally performed in laparotomy. However, there are some that, more than others, benefit from the robotic approach. These are all the procedures that involve a meticulous and delicate reconstructive time, such as pelvic floor restructuring procedures. In addition to these, also the "cut" demolition operations woman by woman (tailored surgery) in cases of deep endometriosis of the female pelvis's anterior or posterior portion. However, there is still no clarity as to when it is better to use one or the other method: given the costs of robotic surgery, it can be deduced that it should be used only when the cheapest laparoscopy is not suitable for intervention (Holloway et al., 2009; Moon et al., 2020).

A US study compared the outcomes of the two methods on benign gynecological pathologies based on studies of the last 10 years. Reason for inclusion was having reported mortality within 30 days of surgery. 21 studies responded to study requests and were rated with a variety of scales. The results indicated that mortality associated with minimally invasive gynecological surgical techniques is generally low and there are no distinctions between laparoscopy and robotic surgery. (Behbehani et al., 2020). Robotic surgery in gynecology is hence less invasive laparoscopic surgery with the assistance of a robot, which is particularly appropriate for performing highly complex operations. In fact, as we have seen, the versatility of the movements of the robotic instruments, which allows to reach narrow and deep anatomic spaces, is a particularly advantageous feature in pelvic surgery, therefore in gynecology. In the early '90s, minimally invasive laparoscopic surgery began to outline the first steps of what, after a decade, would become the standard of gynecological surgery. This new therapeutic approach could be the surgical standard of this century for many pathologies. The evolution of laparoscopic surgery is certainly robotic surgery, which, in addition to the advantages of MIS, adds instruments that try to overcome the limits of the tools offered by laparoscopic surgery. The intervention with the robot avoids cutting the abdomen even for gynecological oncology operations. In addition, it must be remembered that surgery is always carried out by the surgeon and not by the machine (Wexner et al., 2009).

The DaVinci Robot and, in particular the Xi model, represents the most advanced robotic system used in Europe dedicated to MIS applied to gynecology. Since the introduction of the DaVinci robot in surgery about 10 years ago, robotic surgery in gynecology today represents about 30% of all interventions performed with this robotic technology. With the latest robot update, interventions made with the DaVinci have numerous additional benefits. Robotic surgery has made it possible to take a step forward compared to laparoscopic surgery, which is still used in some types of surgery, by expanding the possibility of operating with extreme precision even oncological pathologies in anatomic locations that are difficult to reach, extending to interventions previously carried out with traditional surgery, the benefits of minimally invasiveness (Oehler, 2009).

The two-channel optical video-endoscope and dual high-resolution cameras used by the DaVinci robotic system give the surgeon stereoscopic vision. This allows the surgeon to have a magnified view in real time of the operating field, with a significantly increased and better image quality compared to traditional surgery, while the robot adjusts the way the instruments move and eliminates the physiological hand tremor of the surgeon., making precise and fluid movement of the instrumentation over the area of the intervention. Furthermore, as we will say after, thanks to the recent update of the DaVinci robot, the procedures can be carried out through a single, tiny incision made at the umbilical level which, if on the one hand improves the aesthetic result, on the other allows to reduce postoperative pain even more. The possibility of operating with a small umbilical incision of about 2 cm in the abdominal cavity through which the tools move, giving the surgeon more room to manipulate, is not only an aesthetic fact, but in any case is always important for the psychological impact it can have on the patient. It is important also in the reduction of postoperative pain, which is due to the fact that only one access is made, instead of 4 or 5 as in the previous version of the robot, and are reduced therefore also the traumas of the tissues and the muscular wall, i.e., no extensive scar on the abdomen, reduced operating time, less anesthesia, reduced hospitalization time, and faster recovery after surgery (Ayala-Yáñez et al., 2013).

Unlike other surgical disciplines, the scientific literature does not record better outcomes for the robot than for most traditional techniques. The advantages found so far relate, on the one hand, to the benefit for patients of having a single scar in an inconspicuous position and, on the other, to ergonomic results. For the human hand, operation from a single access is tiring and unnatural, there is a problem of co-axiality because often the tools are used with a movement along two axes converging on a point. Through a single-entry hole, the “hand” of the robot can better manage the different useful tools (Desai et al., 2014).

Robotic surgery in gynecology can be applied in the following cases (Wimberger & Schindelhauer, 2014):

- oncological surgery:
  - treatment of cervical cancer;
  - treatment of ovarian cancer;
  - treatment of endometrial cancer and uterine sarcomas; and
  - treatment of pelvic recurrences of genital neoplasms;
- hysterectomy (with or without adnexectomy);
- treatment of uterine myomas (myomectomy);
- tubal surgery;
- treatment of endometriosis:
  - treatment of endometriotic nodules of the posterior compartment: rectum, sigma, parameters, uterosacral ligaments;
  - treatment of endometriotic nodules of the anterior compartment: vesicouterine fold; and
  - removal of endometriotic ovarian cysts; and
- pelvic floor surgery: surgical repair of pelvic static and pelvic organ prolapses (POP) and/or urinary incontinence:
  - multi-compartment prolapses involving bladder, uterus, and rectum (promontofixation)

- recurrence of vaginal vault prolapses after hysterectomy (Dubuisson lateral suspension or promontofixation) (Dubuisson et a., 2016);
- uni-compartmental genital prolapses (Dubuisson lateral suspension);
- central suspension (Bob Shull) (Shull et al., 2000); and
- supra-cervical robotic hysterectomy with promontosacropexy with synthetic mesh.

All of the patients with the aforementioned disorders can benefit from robotic surgery are patients in which laparoscopy or vaginal surgery are less effective. It is especially indicated in obese patients (Burke et al., 2011; Yurashevich & Habib, 2021). Any patient thinking about gynecological surgery should consult with her dependable surgeon to go over the procedure's potential outcomes. Patients should have a thorough discussion with their doctor about whether a robotic operation is appropriate for their condition. For instance, a lot of women may receive the information that they are not good candidates for robotic surgery because of a history of several prior operations, whereas in reality, they may be able to undergo MIS in a very safe manner. Asking about the doctor's experience with minimally invasive procedures is crucial, as is finding out what to anticipate after surgery, particularly with regard to limitations and return-to-work schedules. These inquiries will aid in preparing patients for the healing process (Thomas et al., 2020; Martinello & Loshak, 2020).

Oncological Surgery

In Italy, the fields of application of robotic surgery are "essentially" in gynecological oncology (Cho & Nezhat, 2009; Mattson & Bender, 2020):

- total hysterectomy with adnexectomy for endometrial or ovarian cancer;
- cervical cancer treated with a radical hysterectomy along with pelvic and lumbo-aortic lymphadenectomy;
- rectosigmoid and pelvic wall and diaphragm recurrency resections; and
- omentectomy and appendectomy.

In the oncological field, procedures for ovarian, cervical, and endometrial cancer can be carried out by robotic systems (Dornhöfer & Höckel, 2008; Lin et al., 2009; Iglesias & Ramirez, 2011; Lucidi et al., 2017; Liu et al., 2021; Poddar & Maheshwari, 2021). The removal of metastatic tumors during extremely difficult ovarian or endometrial cancer operations can benefit from robotic surgery (Boggess, 2007; Magrina & Magtibay, 2020). These procedures were frequently carried out in the past via an "open approach" or a sizable abdominal incision. Pathology generally managed with traditional surgery (cut on the abdomen) today, also thanks to robotic surgery, can be performed with 3-4 small 8 mm incisions on the abdomen. In the last two decades, several series of published case histories have highlighted the feasibility and safety in terms of care of the traditional laparoscopic procedure compared to the laparotomy route. With this technique, in selected cases, patients with cervical cancer, endometrial cancer, uterine sarcomas, and pelvic recurrences of genital neoplasms can be operated on (Ramirez et al., 2008; Shafer & Boggess, 2008; Blache et al., 2018; Cianci et al., 2021; Courcier et al., 2021; La Russa et al., 2021). The pelvic and lumbo-aortic lymph nodes can also be removed (Wisner et al., 2015, 2017).

With the 3D view of the robot, the nerves that are present around the uterus are better visualized: avoiding injuring them during hysterectomy preserving their functionality prevents the patient from postoperative urinary problems, such as urinary retention or urinary incontinence ("nerve-sparing" technique). In this way, the integrity of bladder and intestinal function is guaranteed (Neto et al., 2015; Hiramatsu, 2021).

The predicted blood losses and length of the minimum hospital stay are in favor of the robotic treatment, and the mean operative time using the robotic technique is lower than the laparoscopic one (about 3 days) (Nezhat, 2008; Clair & Tewari, 2020).

Hysterectomy

Robotic hysterectomy is presently the most often done operation worldwide due to the increasing usage of robots in gynecological surgery with DaVinci robotic technology, together with radical prostatectomy. Robotic hysterectomy is one of the surgeries that has recorded the greatest increase in recent years (Advincula, 2006; Lungu et al., 2021).

In hysteroadnexectomy, the different phases of the operation are the same followed also in the case of an intervention of this type carried out in laparotomy or laparoscopy. It is therefore not the operating technique that adapts to the robot, but vice versa. Resection of the uterus's circular ligament is done initially, which allows access to the retro-peritoneal organs, in particular the ureter, to make it safe. The uterine artery is then identified, which is closed with a surgical clip at the origin; then, the influndibulum-pelvic ligament (which contains the vessels directed to the ovary) is identified, isolated from the ureter, and then coagulated and sectioned. The operator moves to work in the anterior compartment where, to cause the bladder to slip downhill, away from the uterus, he separates the vesico-uterine septum; then, the artery and vein of the uterus are coagulated and dissected at the level of the body of the uterus. These same operations are repeated on the right side. At this point, the uterus is isolated from all its "hooks". A vaginal circular incision, below the cervix, is then made with electrified scissors; at this point, the Fallopian tubes, ovaries, and uterus are extracted vaginally. The surgery is completed by making a continuous stitch suture of the vaginal stump (Kho et al., 2007; Thurston et al., 2019).

Numerous scientific works testify the advantages of robotic hysterectomy over traditional surgeries and laparoscopy. In general, however, among the most relevant advantages of gynecological surgery with the DaVinci robot is the reduction of hospitalization days (which in the case of hysterectomy are reduced to about 3), the reduction of blood losses and post-surgical complications, and, in the case of cancer surgery, a more precise removal of the lymph nodes. To date, 3 and 4 arms are used to perform robotic radical hysterectomy without or with lymphadenectomy, respectively, with a hospital stay of about 3 days (Fanning et al., 2009; Whiteside et al., 2019). It should also be considered that fatigue and anti-ergonomic postures, to which the surgeon may be forced during an operation performed with other techniques, can constitute an additional risk factor for any operator, even the most qualified and expert, after many hours of major surgery. This could be an important element in an assessment of clinical risk management (Rosenblatt et al., 2013; Catanzarite et al., 2018).

Myomectomy

Myomectomy is the removal of uterine fibroids. The uterine fibroid, under surgical indication, is removed with laparoscopic technique or, in selected cases, with robotic surgery. The surgery is called "myomectomy" and requires, through laparoscopic access with 4 entry routes, that the myoma is enucleated and then removed. Laparoscopic surgery, in addition to avoiding the scar on the abdomen, seems to reduce the possibility of adhesions, an important factor especially in childbearing age in women who are looking for children (Falcone & Parker, 2013; Closon & Tulandi, 2016). The average hospital stay is 2 days (Stein & Ascher-Walsh, 2009; Marín-Buck et al., 2021).

Laparoscopic removal of uterine fibroids reached its peak in the '90s, although laparotomic myomectomy is still widespread. At the basis of the surgical choice to still perform laparotomic myomectomy today there are, on the one hand, the quality of the suture and, on the other, the experience of the gynecologist surgeon whose technique is difficult to learn. It has now been demonstrated that the laparoscopic suture has the same characteristics as that performed by the laparotomic route; supporting data refer mainly to the conduct of pregnancy and the risk of uterus rupture, which appear to be superimposable between the two surgical accesses. The suture with the robot is easy to perform and, therefore, even faster and this reduces the operating time. The quality of the suture is identical to that performed laparotomically. Thanks to the robotic needle-holders and the versatility of their movements, it is possible to perform sutures of the uterus after the removal of the myoma even in multiple layers with speed, reducing surgical times and therefore the bleeding of the uterine muscle (Agdi & Tulandi, 2010; Gingold et al., 2018).

Tubal Surgery

The first surgery with the robot was performed on Fallopian tubes, one of the most complex surgical acts, such as their reopening after tying them for voluntary sterilization. The indication to proceed with robotic surgery for this operation lies precisely in the quality of the sutures, which can be performed easily thanks to robotic tools. In fact, the anastomosis of the Fallopian tubes involves a series of knots with extremely thin absorbable threads (absorbable thread 6-0 and 8-0) which are used to reconstruct the muscular layer and the Fallopian tube's serosa. Furthermore, 3D vision is a valid support for this surgery, which is extremely precise (microsurgery). A first experience, reported by the literature, gave a 50% success in terms of pregnancies in the absence of ectopic pregnancies. By virtue of the qualities that the robot expresses in this intervention, if there are the indications to perform an attempt at tubal anastomosis, this must certainly be made with the robotic technique (Rodgers et al., 2007; van Seeters et al., 2017).

Endometriosis

Endometriosis of the recto-vaginal septum is currently addressed with laparoscopic surgery (Eltabbakh & Bower, 2008; Bafort et al., 2020). The gynecologist surgeon who performs the procedure with the robot certainly finds advantages, such as being able to perform complex movements in confined spaces. Removal of bladder endometriosis is another important surgical indication to be addressed with this technique (Nezhat et al., 2017). Another intervention that can be carried out with robotic surgery is endometriosis-related rectum excision. Robotic surgical procedure is well established by general surgeons for rectal cancer. For rectal endometriosis, robotic surgery has the advantage for the surgeon of refining the procedure in such confined spaces. It should be tried to avoid the cutaneous ostomy and the average hospital stay is 5 days, dictated only by the greater control due to the patient, but the woman is already independent from the 1st-2nd day (Ferrero et al., 2021; Hur & Falcone, 2021).

Pelvic Floor Surgery

The standard of care for genital prolapse is vaginal surgical treatment (Ross & Preston, 2009; Chesson & Hallner, 2013). The surgical treatment of the recurrence of genital prolapses, when the first surgery has removed the uterus, finds in the abdominal path the best way with the placement of a mesh (promontosacropexy) (Mourtzinos & Raz, 2006; Mueller, 2013). The positioning of a mesh, to treat prolapse relapse, which has always been performed with the cut on the abdomen, took advantage in the '90s of the evolution of the procedure by laparoscopy, with increased operating times, but with the benefits of MIS (McDermott & Hale, 2009; Raju & Linder, 2021).

Avoiding laparotomy in patients in which a permanent mesh is placed significantly reduces the incidence of peritoneal infections and eliminates the risk of hernia (hernia of the abdomen due to cutting). Promontosacropexy involves anchoring a mesh to the vagina (posterior wall, vaginal vault, and anterior wall) and subsequent attachment to the promontory of the sacrum. The number of sutures to be performed is variable (about 8-12) and the operating time in the laparoscopic procedure is determined precisely by the number of sutures (Gadonneix et al., 2005). It is clear that the ease with which the sutures are performed with robotic surgery allows to significantly reduce these times, making the operating time similar, if not less, to laparotomy surgery.

In the treatment of high-grade prolapses, robotic surgery allows to better position the support mesh of the vaginal vault at the promontory of the sacrum. The intervention is very precise, vascular structures and nervous integrity are preserved, avoiding repercussions such as constipation or urinary incontinence, which can be a consequence of the classic surgical approach. The first advantage of this robotic-assisted approach, compared to the laparoscopic route, is the lower hospitalization and the lower rate of conversions in laparotomy and complications, and compared to the vaginal route ("gold standard" of POP pelvic reconstructive surgery) a control from above and in selected cases a faster recovery. The days of hospitalization are reduced by an average of 3 days and the intervention with the robot is shorter than the laparoscopic route and therefore less anesthesia for the patient is required (Morling et al., 2017). Even among women with a uterine prolapse who want to keep their uterus, it is possible to undergo surgery with the robotic technique (isthmopromontosacropexy) with the shown advantages: shorter hospitalization, fewer complications, less invasive interventions, faster recovery with a peaceful resumption of work, and... much less anxiety (Schachar & Matthews, 2020; Pushkar et al., 2021).

SINGLE-ACCESS ROBOTIC SURGERY

In recent years, thanks to technological development and the growing interest in MIS, laparoscopic surgery with single umbilical access has developed. The use of specific and highly technological tools, in fact, allows for a single 2 cm incision at the level of the umbilicus to perform abdominal surgery; all the surgical instruments necessary for the operation are introduced into the abdomen, therefore without the addition of scars on the skin and in ever greater safety. The benefits pertain to aesthetics, postoperative pain reduction, and issues linked to trocar positioning (Jung et al., 2009; Giovannopoulou et al., 2021).

The most suitable gynecological interventions with this type of approach are for benign pathologies of the ovaries and tubes, but also interventions on the uterus such as myomectomies and hysterectomies. So, also in robotic surgery, the latest novelty is called "robotic single port", a single access, made by the American company Intuitive and distributed in Italy by AB Medica (Lue et al., 2012; Capozzi et al., 2021).

The single port is a project that was developed in response to needs that, although they may seem minor, are full of amazing advancements. For surgeries on benign and malignant abdominal diseases, this single access is employed as an alternative to the customary 4-5 accesses needed for surgical intervention with the robot. Placed at the umbilical level, it makes it possible to work with the robot in a very short time, with minimal invasiveness. The hardware and software of the new instrument allow the surgeon to operate from the console just like in traditional robotic intervention. The true 3D vision (not virtual) offers a very clear field of action, and the two instruments are governed in a natural way. In order to give the operator enough room to maneuver, the operating and assistant instruments must be introduced through a single access that is only 2 cm in diameter at the umbilical level. To do this, the instruments must cross each other inside the abdominal cavity: in the operating field, the right hand controls the instrument that comes from the left, and the left hand controls the instrument that comes from the right, much like working in the m. The surgeon can use the operating instruments with complete ease and no need for dexterity effort thanks to the robot's specific program that automatically "straightens" the commands of the robotic arms. With a significantly better cosmetic outcome and less postoperative pain, this results in an outstanding improvement in the outcomes in terms of operating times and conversion rate in traditional laparoscopy or even open surgery (Autorino et al., 2013; Prodromidou et al., 2020).

CONCLUSION

Technology is the mantra of modern times. But is this always the case or do we sometimes risk falling in love with the new out of all proportion? The question is circulating among surgeons after the publication of the first randomized and controlled studies which compared classical "open" surgery and robotic: the data say that 3 months after the operation, the "flesh and blood" surgeon and the "robotic" one guarantee the same results. On closer inspection with the robot, adverse events during surgery and postoperative complications are reduced, but in terms of tumor removal, and urinary and sexual function, the two approaches would seem more similar than expected. The authors admit that it will take more time to understand if the results really overlap, but in the meantime, they encourage patients not to choose one surgical technique or another, but in general a good surgeon. A slowdown in enthusiasm not entirely shared in our country, where the experience with the robot-surgeon, the famous DaVinci, is among the greatest. The DaVinci robot, which is a sort of evolution of laparoscopic surgery, contains an HD 3D camera on one of its four "arms" that are introduced into the body through tiny holes. The other three "arms" are intended to hold surgical equipment and are "hands" that can move 360° while being controlled by a real surgeon seated at a control panel. If an experienced surgeon operates in the traditional way, the result is equally good, but on average, open surgery causes sequelae negatively affecting the patient's quality of life, such as the risk of sexual dysfunctions and urinary incontinence (Rodrigues et al., 2019; Alshowaikh et al., 2021).

With the robot, the probability drops because the instrument allows to magnify the image of the operating field, which is clear and clean, several times, and to make movements that are impossible for the human hand. The robotic intervention is quick and so precise that it greatly reduces the risk of relapses, because because of a magnified view of the working area, it is ensured that all the diseased tissue is removed. In oncological surgery, alongside the excellent results in terms of disease-free survival (DFS), the robotic technique, thanks to a greater ability to spare the nerve bundles that regulate bladder functions, avoids urinary incontinence, which is a problem for 5% of patients undergoing open surgery (Mettler & Meinhold-Heerlein, 2009; Kampers et al., 2021).

Like laparoscopic interventions, the robot guarantees a decrease in bleeding and postoperative discomfort, barely visible scars, and a short hospital stay; in addition, however, it allows a better technical execution, thanks to the 3D of the visual field and the robotic hand that can make movements wider and finer than the human one, and without any tremor. With the robot, therefore, it is possible to operate minimally, and with better results, patients who previously could not be treated in this way and is indicated above all in cases of high complexity (Chen & Falcone, 2009; Gitas et al., 2022).

The disadvantage of high cost remains. The robot is priced between 1.5 and 2.7 million € plus tens of thousands € for annual maintenance; for the materials used for each intervention, 4-5 thousand € are needed, to which hospital costs are added. Certainly not a little, but we must compare them to the cost of the entire patient path. For this reason, it makes sense that in each Region there are 2-3 reference centers with a large volume of interventions: this guarantees maximum experience, the best possible results and an optimal cost-benefit ratio. There are few doubts about the theory: the ideal is to have the right number of robots, well distributed in centers of excellence where many interventions can be carried out to amortize the cost of the machine and make high technology sustainable and efficient. The data from the Italian societies of surgery tell of another reality: in Italy there are over 100 robot-surgeons, which would be enough. The problems are the patchy distribution and the difference between the country's North and South. In the North there are 44 robots, in the Center 25 of which 11 in Tuscany, and in the South and in the islands 15, of which 9 between Campania and Puglia. In Lombardy there are 21, in Molise not even one, in Sicily one, and Calabria awaits its first shortly. A similar situation favors the health migration of patients for whom the robot is more suitable, with a considerable increase in costs. But the solution is not a robot in every department: the installation only makes sense if a minimum of 350 interventions per year can be performed. A redistribution, so that each Region has one or more centers of excellence, would be the best solution (van Dam et al., 2011; Knight & Escobar, 2014).

The robot-surgeons were born as "war surgeons", to operate in field hospitals during conflicts: in these situations, there is hardly the necessary expert, so it was thought of surgeons-robots that can be maneuvered remotely by the physician, far from the theater of war. These applications are still experimental, because the transmission of signals at a distance can be inaccurate and compromise the results, the costs are too high, and a real surgeon is still needed next to the operating table. In civil hospitals, the use of robots has been approved since 2000 and Italy has been in the forefront since the beginning: among the founding members of the international Clinical Robotic Surgery Association (CRSA), Italians are second only to Americans and the Italian Pier Cristoforo Giulianotti is considered one of the pioneers of world robotic surgery (Lauterbach, 2017).

Almost 12,000 studies, conducted all over the world on thousands of patients: this is the experience with the robot-surgeon, but in the majority of cases these are interventions in oncology, as it is on the removal of the tumor that the costs of robotics are more justified because the precision guarantees to save the nearby areas maximum effectiveness and minimum damage. Little by little, however, robots are entering the operating room also for benign pathologies: for now, it happens only in large centers, but it is not excluded that with the spread of technology it will not happen more often. In women, the robot may have further uses in the near future: beyond the ocean, the robot is often used for benign hysterectomies. What prevents from doing this also in Italy is the high cost but, when it will be comparable to that of laparoscopy, the new technique will be more used (Lim et al., 2020; Connor et al., 2021).

In conclusion, automated surgery today represents a qualitative leap for a healthcare facility. The evolution of nanotechnology and mechanics will ensure that current robotic surgery is only the first step towards ever less invasive and precise surgery. However, it must be borne in mind that the robot does not replace the surgeon (Berkowitz & Minkoff, 2016; Carbonnel et al., 2020).

REFERENCES

Aceto, P., Beretta, L., Cariello, C., Claroni, C., Esposito, C., Forastiere, E. M., Guarracino, F., Perucca, R., Romagnoli, S., Sollazzi, L., Cela, V., Ercoli, A., Scambia, G., Vizza, E., Ludovico, G. M., Sacco, E., Vespasiani, G., Scudeller, L., & Corcione, A. (2019). Joint consensus on anesthesia in urologic and gynecologic robotic surgery: Specific issues in management from a task force of the SIAARTI, SIGO, and SIU. *Minerva Anestesiologica*, *85*(8), 871–885. doi:10.23736/S0375-9393.19.13360-3 PMID:30938121

Advincula, A. P. (2006). Surgical techniques: Robot-assisted laparoscopic hysterectomy with the da Vinci surgical system. *International Journal of Medical Robotics and Computer Assisted Surgery*, *2*(4), 305–311. doi:10.1002/rcs.111 PMID:17520648

Advincula, A. P., & Song, A. (2007). The role of robotic surgery in gynecology. *Current Opinion in Obstetrics & Gynecology*, *19*(4), 331–336. doi:10.1097/GCO.0b013e328216f90b PMID:17625414

Advincula, A. P., & Wang, K. (2009). Evolving role and current state of robotics in minimally invasive gynecologic surgery. *Journal of Minimally Invasive Gynecology*, *16*(3), 291–301. doi:10.1016/j.jmig.2009.03.003 PMID:19423061

Agdi, M., & Tulandi, T. (2010). Minimally invasive approach for myomectomy. *Seminars in Reproductive Medicine*, *28*(03), 228–234. doi:10.1055-0030-1251479 PMID:20414845

Al-Akash, M., Boyle, E., & Tanner, W. A. (2009). N.O.T.E.S.: The progression of a novel and emerging technique. *Surgical Oncology*, *18*(2), 95–103. doi:10.1016/j.suronc.2008.11.003 PMID:19110418

Alkatout, I., Mettler, L., Maass, N., & Ackermann, J. (2016). Robotic surgery in gynecology. *Journal of the Turkish German Gynecological Association*, *17*(4), 224–232. doi:10.5152/jtgga.2016.16187 PMID:27990092

Alshowaikh, K., Karpinska-Leydier, K., Amirthalingam, J., Paidi, G., Iroshani Jayarathna, A. I., Salibindla, D. B. A. M. R., & Ergin, H. E. (2021). Surgical and Patient Outcomes of Robotic Versus Conventional Laparoscopic Hysterectomy: A Systematic Review. *Cureus*, *13*, e16828. doi:10.7759/cureus.16828 PMID:34367836

American College of Obstetricians and Gynecologists' Committee on Gynecologic Practice, The Society of Gynecologic Surgeons. (2020). Robot-Assisted Surgery for Noncancerous Gynecologic Conditions: ACOG Committee Opinion, Number 810. *Obstetrics and Gynecology*, *136*(3), e22–e30. doi:10.1097/AOG.0000000000004048 PMID:32826597

Andres, M. P., Souza, C., Villaescusa, M., Vieira, M., & Abrao, M. S. (2022). The current role of robotic surgery in endometriosis management. *Expert Review of Endocrinology & Metabolism*, *17*(1), 63–73. doi:10.1080/17446651.2022.2031976 PMID:35073819

Ardon, A., & Hernandez, N. (2022). The Use of Peripheral Nerve Blockade in Laparoscopic and Robotic Surgery: Is There a Benefit? *Current Pain and Headache Reports*, *26*(1), 25–31. doi:10.100711916-022-01002-4 PMID:35076876

Autorino, R., Kaouk, J. H., Stolzenburg, J. U., Gill, I. S., Mottrie, A., Tewari, A., & Cadeddu, J. A. (2013). Current status and future directions of robotic single-site surgery: A systematic review. *European Urology*, *63*(2), 266–280. doi:10.1016/j.eururo.2012.08.028 PMID:22940173

Ayala-Yáñez, R., Olaya-Guzmán, E. J., & Haghenbeck-Altamirano, J. (2013). Robotics in Gynecology: Why is this Technology Worth Pursuing? *Clinical Medicine Insights. Reproductive Health*, *7*, 71–77. doi:10.4137/CMRH.S10850 PMID:24453521

Azadi, S., Green, I. C., Arnold, A., Truong, M., Potts, J., & Martino, M. A. (2021). Robotic Surgery: The Impact of Simulation and Other Innovative Platforms on Performance and Training. *Journal of Minimally Invasive Gynecology*, *28*(3), 490–495. doi:10.1016/j.jmig.2020.12.001 PMID:33310145

Bafort, C., Beebeejaun, Y., Tomassetti, C., Bosteels, J., & Duffy, J. M. (2020). Laparoscopic surgery for endometriosis. *Cochrane Database of Systematic Reviews*, *10*, CD011031. PMID:33095458

Bailey, A. P., Jaslow, C. R., & Kutteh, W. H. (2015). Minimally invasive surgical options for congenital and acquired uterine factors associated with recurrent pregnancy loss. *Women's Health (London)*, *11*(2), 161–167. doi:10.2217/WHE.14.81 PMID:25776290

Behbehani, S., Suarez-Salvador, E., Buras, M., Magtibay, P., & Magrina, J. (2020). Mortality Rates in Benign Laparoscopic and Robotic Gynecologic Surgery: A Systematic Review and Meta-analysis. *Journal of Minimally Invasive Gynecology*, *27*(3), 603–612. doi:10.1016/j.jmig.2019.10.005 PMID:31627007

Berkowitz, R. L., & Minkoff, H. (2016). A Call for Change in a Changing World. *Obstetrics and Gynecology*, *127*(1), 153–156. doi:10.1097/AOG.0000000000001193 PMID:26646136

Blache, G., Jauffret, C., Mathis, J., Knight, S., Houvenaeghel, G., & Lambaudie, E. (2018). Can we use robotic surgery for the treatment of pelvic recurrence and locally advanced tumors in gynecological surgery? *Journal of Gynecology Obstetrics and Human Reproduction*, *47*(9), 431–435. doi:10.1016/j.jogoh.2018.08.008 PMID:30149209

Boggess, J. F. (2007). Robotic surgery in gynecologic oncology: Evolution of a new surgical paradigm. *Journal of Robotic Surgery*, *1*(1), 31–37. doi:10.100711701-007-0011-4 PMID:25484936

Bouquet de Joliniere, J., Librino, A., Dubuisson, J. B., Khomsi, F., Ben Ali, N., Fadhlaoui, A., Ayoubi, J. M., & Feki, A. (2016). Robotic Surgery in Gynecology. *Frontiers in Surgery*, *3*, 26. doi:10.3389/fsurg.2016.00026 PMID:27200358

Burke, W. M., Gossner, G., & Goldman, N. A. (2011). Robotic surgery in the obese gynecologic patient. *Clinical Obstetrics and Gynecology*, *54*(3), 420–430. doi:10.1097/GRF.0b013e31822b37b1 PMID:21857172

Čapek, K. (1920). R.U.R. Dover Publications.

Capozzi, V. A., Armano, G., Rosati, A., Tropea, A., & Biondi, A. (2021). The robotic single-port platform for gynecologic surgery: A systematic review of the literature and meta-analysis. *Updates in Surgery*, *73*(3), 1155–1167. doi:10.100713304-020-00812-8 PMID:32472402

Carbonnel, M., Dahm-Kähler, P., Revaux, A., Brännström, M., & Ayoubi, J. M. (2020). Adapting surgical skills from robotic-assisted radical hysterectomy in cervical cancer to uterine transplantation: A look to an optimistic future! *Journal of Robotic Surgery*, *14*(6), 841–847. doi:10.100711701-020-01058-7 PMID:32088836

Cardenas-Trowers, O., Kjellsson, K., & Hatch, K. (2018). Ergonomics: Making the OR a comfortable place. *International Urogynecology Journal and Pelvic Floor Dysfunction*, *29*(7), 1065–1066. doi:10.100700192-018-3674-7 PMID:30298292

Catanzarite, T., Tan-Kim, J., & Menefee, S. A. (2018). Ergonomics in gynecologic surgery. *Current Opinion in Obstetrics & Gynecology*, *30*(6), 432–440. doi:10.1097/GCO.0000000000000502 PMID:30299323

Chen, C. C., & Falcone, T. (2009). Robotic gynecologic surgery: Past, present, and future. *Clinical Obstetrics and Gynecology*, *52*(3), 335–343. doi:10.1097/GRF.0b013e3181b08adf PMID:19661749

Chesson, R., & Hallner, B. (2013). Why complex pelvic organ prolapse should be repaired vaginally. *Current Opinion in Urology*, *23*(4), 312–316. doi:10.1097/MOU.0b013e3283619e7f PMID:23882718

Cho, J. E., & Nezhat, F. R. (2009). Robotics and gynecologic oncology: Review of the literature. *Journal of Minimally Invasive Gynecology*, *16*(6), 669–681. doi:10.1016/j.jmig.2009.06.024 PMID:19896593

Cho, J. E., Shamshirsaz, A. H., Nezhat, C., Nezhat, C., & Nezhat, F. (2010). New technologies for reproductive medicine: Laparoscopy, endoscopy, robotic surgery and gynecology. A review of the literature. *Minerva Ginecologica*, *62*, 137–167. PMID:20502426

Cianci, S., Arcieri, M., Vizzielli, G., Martinelli, C., Granese, R., La Verde, M., Fagotti, A., Fanfani, F., Scambia, G., & Ercoli, A. (2021). Robotic Pelvic Exenteration for Gynecologic Malignancies, Anatomic Landmarks, and Surgical Steps: A Systematic Review. *Frontiers in Surgery*, *8*, 790152. doi:10.3389/fsurg.2021.790152 PMID:34917648

Clair, K. H., & Tewari, K. S. (2020). Robotic surgery for gynecologic cancers: Indications, techniques and controversies. *Journal of Obstetrics and Gynaecology Research*, *46*(6), 828–843. doi:10.1111/jog.14228 PMID:32410262

Closon, F., & Tulandi, T. (2016). Uterine myomata: Organ-preserving surgery. *Best Practice & Research. Clinical Obstetrics & Gynaecology*, *35*, 30–36. doi:10.1016/j.bpobgyn.2015.09.005 PMID:26542930

Cohen, S. L., & Einarsson, J. I. (2011). The role of mechanical bowel preparation in gynecologic laparoscopy. *Reviews in Obstetrics & Gynecology*, *4*, 28–31. PMID:21629496

Connor, A., Pasic, R., Quevedo, A., & Chamseddine, P. (2021). Robotic Applications for Benign Gynecologic Procedures. *Surgical Technology International*, *39*, 232–240. doi:10.52198/21.STI.39.GY1409 PMID:34181238

Courcier, H., Azaïs, H., Belghiti, J., Uzan, C., & Canlorbe, G. (2021). Minimally invasive surgery for early-stage cervical cancer: Rediscovering the Schautheim robot-assisted procedure. *Journal of Gynecology Obstetrics and Human Reproduction*, *50*(2), 101980. doi:10.1016/j.jogoh.2020.101980 PMID:33186773

Cuschieri, A. (1994). Therapeutic laparoscopy. *Endoscopy*, *26*(01), 134–147. doi:10.1055-2007-1008940 PMID:8205987

Das, D., Propst, K., Wechter, M. E., & Kho, R. M. (2019). Evaluation of Positioning Devices for Optimization of Outcomes in Laparoscopic and Robotic-Assisted Gynecologic Surgery. *Journal of Minimally Invasive Gynecology*, *26*(2), 244–252. doi:10.1016/j.jmig.2018.08.027 PMID:30176363

Desai, P. H., Lin, J. F., & Slomovitz, B. M. (2014). Milestones to optimal adoption of robotic technology in gynecology. *Obstetrics and Gynecology*, *123*(1), 13–20. doi:10.1097/AOG.0000000000000055 PMID:24463658

Dornhöfer, N., & Höckel, M. (2008). New developments in the surgical therapy of cervical carcinoma. *Annals of the New York Academy of Sciences*, *1138*(1), 233–252. doi:10.1196/annals.1414.029 PMID:18837903

Dubuisson, J., Veit-Rubin, N., Bouquet de Jolinière, J., & Dubuisson, J. B. (2016). Laparoscopic Lateral Suspension: Benefits of a Cross-shaped Mesh to Treat Difficult Vaginal Vault Prolapse. *Journal of Minimally Invasive Gynecology*, *23*(5), 672. doi:10.1016/j.jmig.2016.01.028 PMID:26867700

Elliott, D. S., Chow, G. K., & Gettman, M. (2006). Current status of robotics in female urology and gynecology. *World Journal of Urology*, *24*(2), 188–192. doi:10.100700345-006-0071-5 PMID:16557388

Eltabbakh, G. H., & Bower, N. A. (2008). Laparoscopic surgery in endometriosis. *Minerva Ginecologica*, *60*, 323–330. PMID:18560348

Falcone, T., & Goldberg, J. M. (2003). Robotics in gynecology. *The Surgical Clinics of North America*, *83*(6), 1483–1489. doi:10.1016/S0039-6109(03)00163-4 PMID:14712881

Falcone, T., & Parker, W. H. (2013). Surgical management of leiomyomas for fertility or uterine preservation. *Obstetrics and Gynecology*, *121*(4), 856–868. doi:10.1097/AOG.0b013e3182888478 PMID:23635687

Falcone, T., & Steiner, C. P. (2002). Robotically assisted gynaecological surgery. *Human Fertility (Cambridge, England)*, *5*(2), 72–74. doi:10.1080/1464727022000198962 PMID:12082211

Fanfani, F., Restaino, S., Ercoli, A., Chiantera, V., Fagotti, A., Gallotta, V., Monterossi, G., Cappuccio, S., & Scambia, G. (2016). Robotic versus laparoscopic surgery in gynecology: Which should we use? *Minerva Ginecologica*, *68*, 423–430. PMID:26633042

Fanning, J., Hojat, R., Johnson, J., & Fenton, B. (2009). Robotic radical hysterectomy. *Minerva Ginecologica*, *61*, 53–55. PMID:19204661

Ferrero, S., Stabilini, C., Barra, F., Clarizia, R., Roviglione, G., & Ceccaroni, M. (2021). Bowel resection for intestinal endometriosis. *Best Practice & Research. Clinical Obstetrics & Gynaecology*, *71*, 114–128. doi:10.1016/j.bpobgyn.2020.05.008 PMID:32665125

Fong, Y., Woo, Y., Hyung, W. J., Lau, C., & Strong, V. E. (2018). *The SAGES Atlas of Robotic Surgery*. Springer. doi:10.1007/978-3-319-91045-1

Foote, J. R., & Valea, F. A. (2016). Robotic surgical training: Where are we? *Gynecologic Oncology*, *143*(1), 179–183. doi:10.1016/j.ygyno.2016.05.031 PMID:27246302

Frick, A. C., & Falcone, T. (2009). Robotics in gynecologic surgery. *Minerva Ginecologica*, *61*, 187–199. PMID:19415063

Gadonneix, P., Ercoli, A., Scambia, G., & Villet, R. (2005). The use of laparoscopic sacrocolpopexy in the management of pelvic organ prolapse. *Current Opinion in Obstetrics & Gynecology*, *17*(4), 376–380. doi:10.1097/01.gco.0000175355.48802.7b PMID:15976543

Gala, R. B., Margulies, R., Steinberg, A., Murphy, M., Lukban, J., Jeppson, P., Aschkenazi, S., Olivera, C., South, M., Lowenstein, L., Schaffer, J., Balk, E. M., & Sung, V.Society of Gynecologic Surgeons Systematic Review Group. (2014). Systematic review of robotic surgery in gynecology: Robotic techniques compared with laparoscopy and laparotomy. *Journal of Minimally Invasive Gynecology*, *21*(3), 353–361. doi:10.1016/j.jmig.2013.11.010 PMID:24295923

Gastrich, M. D., Barone, J., Bachmann, G., Anderson, M., & Balica, A. (2011). Robotic surgery: Review of the latest advances, risks, and outcomes. *Journal of Robotic Surgery*, *5*(2), 79–97. doi:10.100711701-011-0246-y PMID:27637534

George, E. I., Brand, T. C., LaPorta, A., Marescaux, J., & Satava, R. M. (2018). Origins of Robotic Surgery: From Skepticism to Standard of Care. *Journal of the Society of Laparoscopic & Robotic Surgeons, 22*.

Gingold, J. A., Gueye, N. A., & Falcone, T. (2018). Minimally Invasive Approaches to Myoma Management. *Journal of Minimally Invasive Gynecology*, *25*(2), 237–250. doi:10.1016/j.jmig.2017.07.007 PMID:28734973

Giovannopoulou, E., Prodromidou, A., Blontzos, N., & Iavazzo, C. (2021). The Emerging Role of Robotic Single-site Approach for Myomectomy: A Systematic Review of the Literature. *Surgical Innovation*, *28*(3), 352–359. doi:10.1177/1553350620988227 PMID:33555235

Gitas, G., Hanker, L., Rody, A., Ackermann, J., & Alkatout, I. (2022). Robotic surgery in gynecology: Is the future already here? *Minimally Invasive Therapy & Allied Technologies*, *31*(6), 1–10. doi:10.1080/13645706.2021.2010763 PMID:34989636

Giulianotti, P. C. (2020). The future of robotic liver surgery. *Hepatobiliary Surgery and Nutrition*, *9*(4), 546–547. doi:10.21037/hbsn.2020.03.17 PMID:32832516

Guerin, S., Huaulmé, A., Lavoue, V., Jannin, P., & Timoh, K. N. (2022). Review of automated performance metrics to assess surgical technical skills in robot-assisted laparoscopy. *Surgical Endoscopy*, *36*(2), 853–870. doi:10.100700464-021-08792-5 PMID:34750700

Gültekin, İ. B., Karabük, E., Köse, M. F., Gültekin, İ. B., Karabük, E., & Köse, M. F. (2021). "Hey Siri! Perform a type 3 hysterectomy. Please watch out for the ureter!" What is autonomous surgery and what are the latest developments? *Journal of the Turkish German Gynecological Association*, *22*(1), 58–70. doi:10.4274/jtgga.galenos.2021.2020.0187 PMID:33624493

Hamet, P., & Tremblay, J. (2017). Artificial intelligence in medicine. *Metabolism: Clinical and Experimental, 69S*, S36–S40. doi:10.1016/j.metabol.2017.01.011 PMID:28126242

Han, E. S., & Advincula, A. P. (2019). Safety in Minimally Invasive Surgery. *Obstetrics and Gynecology Clinics of North America, 46*(2), 389–398. doi:10.1016/j.ogc.2019.01.013 PMID:31056139

Han, E. S., & Advincula, A. P. (2021). Robotic Surgery: Advancements and Inflection Points in the Field of Gynecology. *Obstetrics and Gynecology Clinics of North America, 48*(4), 759–776. doi:10.1016/j.ogc.2021.07.004 PMID:34756295

Harding, V., & Williams, J. (2016). Improving efficiency in robotic theatres in the Royal Wolverhampton NHS Trust: A gynaecology theatres perspective. *Journal of Perioperative Practice, 26*(3), 46–49. doi:10.1177/175045891602600304 PMID:27149833

Hiramatsu, Y. (2021). Nerve-Sparing Robotic Radical Hysterectomy for the Beginner in Robotic Surgery. *Surgery Journal (New York, N.Y.), 7*(S 02), S84–S96. doi:10.1055-0041-1739121 PMID:35111934

Hockstein, N. G., Gourin, C. G., Faust, R. A., & Terris, D. J. (2007). A history of robots: From science fiction to surgical robotics. *Journal of Robotic Surgery, 1*(2), 113–118. doi:10.100711701-007-0021-2 PMID:25484946

Holloway, R. W., Patel, S. D., & Ahmad, S. (2009). Robotic surgery in gynecology. *Scandinavian Journal of Surgery, 98*(2), 96–109. doi:10.1177/145749690909800205 PMID:19799047

Hottenrott, S., Schlesinger, T., Helmer, P., Meybohm, P., Alkatout, I., & Kranke, P. (2020). Do Small Incisions Need Only Minimal Anesthesia? – Anesthetic Management in Laparoscopic and Robotic Surgery. *Journal of Clinical Medicine, 9*(12), 4058. doi:10.3390/jcm9124058 PMID:33334057

Hur, C., & Falcone, T. (2021). Robotic treatment of bowel endometriosis. *Best Practice & Research. Clinical Obstetrics & Gynaecology, 71*, 129–143. doi:10.1016/j.bpobgyn.2020.05.012 PMID:32684433

Iavazzo, C., & Gkegkes, I. D. (2017). Cost-benefit analysis of robotic surgery in gynaecological oncology. *Best Practice & Research. Clinical Obstetrics & Gynaecology, 45*, 7–18. doi:10.1016/j.bpobgyn.2017.03.008 PMID:28526229

Iglesias, D. A., & Ramirez, P. T. (2011). Role of minimally invasive surgery in staging of ovarian cancer. *Current Treatment Options in Oncology, 12*(3), 217–229. doi:10.100711864-011-0155-3 PMID:21503633

Jacofsky, D. J., & Allen, M. (2016). Robotics in Arthroplasty: A Comprehensive Review. *The Journal of Arthroplasty, 31*(10), 2353–2363. doi:10.1016/j.arth.2016.05.026 PMID:27325369

Jung, Y. W., Kim, S. W., & Kim, Y. T. (2009). Recent advances of robotic surgery and single port laparoscopy in gynecologic oncology. *Journal of Gynecologic Oncology, 20*(3), 137–144. doi:10.3802/jgo.2009.20.3.137 PMID:19809546

Kampers, J., Gerhardt, E., Sibbertsen, P., Flock, T., Klapdor, R., Hertel, H., Jentschke, M., & Hillemanns, P. (2021). Protective operative techniques in radical hysterectomy in early cervical carcinoma and their influence on disease-free and overall survival: A systematic review and meta-analysis of risk groups. *Archives of Gynecology and Obstetrics, 304*(3), 577–587. doi:10.100700404-021-06082-y PMID:34021804

Kenngott, H. G., Fischer, L., Nickel, F., Rom, J., Rassweiler, J., & Müller-Stich, B. P. (2012). Status of robotic assistance – A less traumatic and more accurate minimally invasive surgery? *Langenbeck's Archives of Surgery, 397*(3), 333–341. doi:10.100700423-011-0859-7 PMID:22038293

Kho, R. M., Hilger, W. S., Hentz, J. G., Magtibay, P. M., & Magrina, J. F. (2007). Robotic hysterectomy: Technique and initial outcomes. *American Journal of Obstetrics and Gynecology, 197*(1), 113. doi:10.1016/j.ajog.2007.05.005 PMID:17618782

Kim, Y. T., Kim, S. W., & Jung, Y. W. (2008). Robotic surgery in gynecologic field. *Yonsei Medical Journal, 49*(6), 886–890. doi:10.3349/ymj.2008.49.6.886 PMID:19108009

Knight, J., & Escobar, P. F. (2014). Cost and robotic surgery in gynecology. *Journal of Obstetrics and Gynaecology Research, 40*(1), 12–17. doi:10.1111/jog.12197 PMID:24118557

Koo, Y. J. (2018). Recent advances in minimally invasive surgery for gynecologic indications. *Yeungnam University Journal of Medicine, 35*(2), 150–155. doi:10.12701/yujm.2018.35.2.150 PMID:31620587

Korsholm, M., Sørensen, J., Mogensen, O., Wu, C., Karlsen, K., & Jensen, P. T. (2018). A systematic review about costing methodology in robotic surgery: Evidence for low quality in most of the studies. *Health Economics Review, 8*(1), 21. doi:10.118613561-018-0207-5 PMID:30194567

La Russa, M., Liakou, C., & Burbos, N. (2021). Ultra-minimally invasive approaches for endometrial cancer treatment: Review of the literature. *Minerva Medica, 112*(1), 31–46. doi:10.23736/S0026-4806.20.07073-1 PMID:33205639

Lauterbach, R., Matanes, E., & Lowenstein, L. (2017). Review of Robotic Surgery in Gynecology – The Future Is Here. *Rambam Maimonides Medical Journal, 8*(2), e0019. doi:10.5041/RMMJ.10296 PMID:28467761

Lawrie, T. A., Liu, H., Lu, D., Dowswell, T., Song, H., Wang, L., & Shi, G. (2019). Robot-assisted surgery in gynaecology. *Cochrane Database of Systematic Reviews, 4*, CD011422. PMID:30985921

Lee, Y. J., van den Berg, N. S., Orosco, R. K., Rosenthal, E. L., & Sorger, J. M. (2021). A narrative review of fluorescence imaging in robotic-assisted surgery. *Laparoscopic Surgery, 5*, 31. doi:10.21037/ls-20-98 PMID:34549180

Lee, Y. L., Kilic, G. S., & Phelps, J. Y. (2011). Medicolegal review of liability risks for gynecologists stemming from lack of training in robot-assisted surgery. *Journal of Minimally Invasive Gynecology, 18*(4), 512–515. doi:10.1016/j.jmig.2011.04.002 PMID:21777841

Lee, Z., Kaplan, J., Giusto, L., & Eun, D. (2016). Prevention of iatrogenic ureteral injuries during robotic gynecologic surgery: A review. *American Journal of Obstetrics and Gynecology, 214*(5), 566–571. doi:10.1016/j.ajog.2015.10.150 PMID:26519785

Levinson, K. L., Auer, M., & Escobar, P. F. (2013). Evolving technologies in robotic surgery for minimally invasive treatment of gynecologic cancers. *Expert Review of Medical Devices, 10*(5), 603–610. doi:10.1586/17434440.2013.827509 PMID:24053253

Lim, C. S., Griffith, K. C., Travieso, J., & As-Sanie, S. (2020). To Robot or Not To Robot: The Use of Robotics in Benign Gynecologic Surgery. *Clinical Obstetrics and Gynecology*, *63*(2), 327–336. doi:10.1097/GRF.0000000000000491 PMID:31634158

Lim, P. C., & Kang, E. (2017). How to prepare the patient for robotic surgery: Before and during the operation. *Best Practice & Research. Clinical Obstetrics & Gynaecology*, *45*, 32–47. doi:10.1016/j.bpobgyn.2017.04.008 PMID:28579145

Lin, P. S., Wakabayashi, M. T., & Han, E. S. (2009). Role of robotic surgery in endometrial cancer. *Current Treatment Options in Oncology*, *10*(1-2), 33–43. doi:10.100711864-009-0086-4 PMID:19353274

Liu, H., Cao, Y., Li, L., Bai, Y., & Liu, J. (2021). Effectiveness of robotic surgery for endometrial cancer: A systematic review and meta-analysis. *Archives of Gynecology and Obstetrics*, *2021*(Sep), 8. PMID:34498124

Liu, H., Lawrie, T. A., Lu, D., Song, H., Wang, L., & Shi, G. (2014). Robot-assisted surgery in gynaecology. *Cochrane Database of Systematic Reviews*. PMID:25493418

Liu, H., Lu, D., Shi, G., Song, H., & Wang, L. (2014). Robotic surgery for benign gynaecological disease. *Cochrane Database of Systematic Reviews*. PMID:25502433

Liu, H., Lu, D., Wang, L., Shi, G., Song, H., & Clarke, J. (2012). Robotic surgery for benign gynaecological disease. *Cochrane Database of Systematic Reviews*, (2), CD008978. PMID:22336855

Lucidi, A., Chiantera, V., Gallotta, V., Ercoli, A., Scambia, G., & Fagotti, A. (2017). Role of robotic surgery in ovarian malignancy. *Best Practice & Research. Clinical Obstetrics & Gynaecology*, *45*, 74–82. doi:10.1016/j.bpobgyn.2017.05.005 PMID:28602522

Lue, J. R., Murray, B., & Bush, S. (2012). Single port robotic hysterectomy technique improving on multiport procedure. *Journal of Minimal Access Surgery*, *8*(4), 156–157. doi:10.4103/0972-9941.103130 PMID:23248446

Lungu, D. A., Foresi, E., Belardi, P., Nuti, S., Giannini, A., & Simoncini, T. (2021). The Impact of New Surgical Techniques on Geographical Unwarranted Variation: The Case of Benign Hysterectomy. *International Journal of Environmental Research and Public Health*, *18*(13), 6722. doi:10.3390/ijerph18136722 PMID:34206452

Mach, P., Kimmig, R., & Buderath, P. (2020). The role of sentinel-node biopsy in ovarian cancer. *Minerva Ginecologica*, *72*, 399–403. PMID:33140629

Magrina, J. F. (2007). Robotic surgery in gynecology. *European Journal of Gynaecological Oncology*, *28*, 77–82. PMID:17479665

Magrina, J. F., & Magtibay, P. M. (2020). Robotic Resection of Diaphragm Metastases in Ovarian Cancer: Technical Aspects. *Journal of Minimally Invasive Gynecology*, *27*(6), 1417–1422. doi:10.1016/j.jmig.2019.12.018 PMID:31917330

Marchand, G., Taher Masoud, A., Ware, K., Govindan, M., King, A., Ruther, S., Brazil, G., Calteux, N., Coriell, C., Ulibarri, H., Parise, J., Arroyo, A., Filippelli, C., Loli, H., & Sainz, K. (2021). Systematic review and meta-analysis of all randomized controlled trials comparing gynecologic laparoscopic procedures with and without robotic assistance. *European Journal of Obstetrics, Gynecology, and Reproductive Biology*, *265*, 30–38. doi:10.1016/j.ejogrb.2021.07.038 PMID:34418694

Marescaux, J., & Rubino, F. (2003). The ZEUS robotic system: Experimental and clinical applications. *The Surgical Clinics of North America*, *83*(6), 1305–1315. doi:10.1016/S0039-6109(03)00169-5 PMID:14712867

Marín-Buck, A., Karaman, E., Amer-Cuenca, J. J., Lisón, J. F., Török, P., Karaaslan, O., Valenti, G., Zito, G., Biondi, A., & Chiofalo, B. (2021). Minimally Invasive Myomectomy: An Overview on the Surgical Approaches and a Comparison with Mini-Laparotomy. *Journal of Investigative Surgery*, *34*(4), 443–450. doi:10.1080/08941939.2019.1642422 PMID:31322011

Marra, A. R., Puig-Asensio, M., Edmond, M. B., Schweizer, M. L., & Bender, D. (2019). Infectious complications of laparoscopic and robotic hysterectomy: A systematic literature review and meta-analysis. *International Journal of Gynecological Cancer*, *29*(3), 518–530. doi:10.1136/ijgc-2018-000098 PMID:30833440

Martinello, N., & Loshak, H. (2020). *Experiences with and Expectations of Robotic Surgical Systems: A Rapid Qualitative Review.* https://cadth.ca/experiences-and-expectations-robotic-surgical-systems-rapid-qualitative-review

Mattson, J. N., & Bender, D. P. (2020). Minimally Invasive Robotic Surgery for Gynecologic Cancers: A Review. *Clinical Obstetrics and Gynecology*, *63*(1), 24–29. doi:10.1097/GRF.0000000000000492 PMID:31850943

McDermott, C. D., & Hale, D. S. (2009). Abdominal, laparoscopic, and robotic surgery for pelvic organ prolapse. *Obstetrics and Gynecology Clinics of North America*, *36*(3), 585–614. doi:10.1016/j.ogc.2009.09.004 PMID:19932417

McLarney, J. T., & Rose, G. L. (2011). Anesthetic Implications of Robotic Gynecologic Surgery. *Journal of Gynecological Endoscopy and Surgery*, *2*(2), 75–78. doi:10.4103/0974-1216.114077 PMID:26085748

Medical Advisory Secretariat. (2010). Robotic-assisted minimally invasive surgery for gynecologic and urologic oncology: An evidence-based analysis. *Ontario Health Technology Assessment*, *10*, 1–118. PMID:23074405

Meinzer, A., Alkatout, I., Krebs, T. F., Baastrup, J., Reischig, K., Meiksans, R., & Bergholz, R. (2020, December 10). Advances and Trends in Pediatric Minimally Invasive Surgery. *Journal of Clinical Medicine*, *9*(12), 3999. doi:10.3390/jcm9123999

Menderes, G., Gysler, S. M., Vadivelu, N., & Silasi, D. A. (2019). Challenges of Robotic Gynecologic Surgery in Morbidly Obese Patients and How to Optimize Success. *Current Pain and Headache Reports*, *23*(7), 51. doi:10.100711916-019-0788-7 PMID:31263977

Mettler, L., & Meinhold-Heerlein, I. (2009). The value of laparoscopic surgery to stage gynecological cancers: Present and future. *Minerva Ginecologica*, *61*, 319–337. PMID:19745797

Mettler, L., Sammur, W., Alkatout, I., & Schollmeyer, T. (2011). Imaging in gynecologic surgery. *Women's Health (London)*, *7*(2), 239–248. doi:10.2217/WHE.11.6 PMID:21410349

Moawad, G., Tyan, P., & Louie, M. (2019). Artificial intelligence and augmented reality in gynecology. *Current Opinion in Obstetrics & Gynecology*, *31*(5), 345–348. doi:10.1097/GCO.0000000000000559 PMID:31259843

Moglia, A., Georgiou, K., Georgiou, E., Satava, R. M., & Cuschieri, A. (2021). A systematic review on artificial intelligence in robot-assisted surgery. *International Journal of Surgery*, *95*, 106151. doi:10.1016/j.ijsu.2021.106151 PMID:34695601

Moon, A. S., Garofalo, J., Koirala, P., Vu, M. T., & Chuang, L. (2020). Robotic Surgery in Gynecology. *The Surgical Clinics of North America*, *100*(2), 445–460. doi:10.1016/j.suc.2019.12.007 PMID:32169189

Morling, J. R., McAllister, D. A., Agur, W., Fischbacher, C. M., Glazener, C. M., Guerrero, K., Hopkins, L., & Wood, R. (2017). Adverse events after first, single, mesh and non-mesh surgical procedures for stress urinary incontinence and pelvic organ prolapse in Scotland, 1997-2016: A population-based cohort study. *Lancet*, *389*(10069), 629–640. doi:10.1016/S0140-6736(16)32572-7 PMID:28010993

Mourtzinos, A., & Raz, S. (2006). Repair of vaginal vault prolapse and pelvic floor relaxation using polypropylene mesh. *Current Opinion in Obstetrics & Gynecology*, *18*(5), 555–559. doi:10.1097/01.gco.0000242960.70985.c2 PMID:16932052

Mueller, E. R. (2013). Why complex pelvic organ prolapse should be approached abdominally. *Current Opinion in Urology*, *23*(4), 317–322. doi:10.1097/MOU.0b013e3283619f10 PMID:23882719

Nahas, S., Feigenberg, T., & Park, S. (2016). Feasibility and safety of same-day discharge after minimally invasive hysterectomy in gynecologic oncology: A systematic review of the literature. *Gynecologic Oncology*, *143*(2), 439–442. doi:10.1016/j.ygyno.2016.07.113 PMID:27475523

Neto, J. S., Siufi, D. F., & Magrina, J. F. (2015). Robotic nerve-sparing radical hysterectomy. *Minerva Ginecologica*, *67*, 281–287. PMID:25714873

Nezhat, C., Cho, J., King, L. P., Hajhosseini, B., & Nezhat, F. (2011). Laparoscopic management of adnexal masses. *Obstetrics and Gynecology Clinics of North America*, *38*(4), 663–676. doi:10.1016/j.ogc.2011.09.003 PMID:22134015

Nezhat, C., Falik, R., McKinney, S., & King, L. P. (2017). Pathophysiology and management of urinary tract endometriosis. *Nature Reviews. Urology*, *14*(6), 359–372. doi:10.1038/nrurol.2017.58 PMID:28467398

Nezhat, C., & Lakhi, N. (2016). Learning Experiences in Robotic-Assisted Laparoscopic Surgery. *Best Practice & Research. Clinical Obstetrics & Gynaecology*, *35*, 20–29. doi:10.1016/j.bpobgyn.2015.11.009 PMID:26707192

Nezhat, F. (2008). Minimally invasive surgery in gynecologic oncology: Laparoscopy versus robotics. *Gynecologic Oncology*, *111*(2), S29–S32. doi:10.1016/j.ygyno.2008.07.025 PMID:18762326

Ng, A. T., & Tam, P. C. (2014). Current status of robot-assisted surgery. *Hong Kong Medical Journal*, *20*, 241–250. PMID:24854139

Oehler, M. K. (2009). Robot-assisted surgery in gynaecology. *Australian and New Zealand Journal of Obstetrics and Gynaecology*, *49*(2), 124–129. doi:10.1111/j.1479-828X.2009.00950.x PMID:19441160

Palep, J. H. (2009). Robotic assisted minimally invasive surgery. *Journal of Minimal Access Surgery*, *5*(1), 1–7. doi:10.4103/0972-9941.51313 PMID:19547687

Patel, M. N., & Hemal, A. K. (2018). Does Advancing Technology Improve Outcomes? Comparison of the Da Vinci Standard/S/Si to the Xi Robotic Platforms During Robotic Nephroureterectomy. *Journal of Endourology*, *32*(2), 133–138. doi:10.1089/end.2017.0477 PMID:29205061

Pelizzo, G., Nakib, G., & Calcaterra, V. (2019). Pediatric and adolescent gynecology: Treatment perspectives in minimally invasive surgery. *Pediatric Reports*, *11*(4), 8029. doi:10.4081/pr.2019.8029 PMID:31871603

Picerno, T., Sloan, N. L., Escobar, P., & Ramirez, P. T. (2017). Bowel injury in robotic gynecologic surgery: Risk factors and management options. A systematic review. *American Journal of Obstetrics and Gynecology*, *216*(1), 10–26. doi:10.1016/j.ajog.2016.08.040 PMID:27640938

Poddar, P., & Maheshwari, A. (2021). Surgery for cervical cancer: Consensus & controversies. *The Indian Journal of Medical Research*, *2021*(Dec), 2. PMID:34854431

Prodromidou, A., Spartalis, E., Tsourouflis, G., Dimitroulis, D., & Nikiteas, N. (2020). Robotic versus laparoendoscopic single-site hysterectomy: A systematic review and meta-analysis. *Journal of Robotic Surgery*, *14*(5), 679–686. doi:10.100711701-020-01042-1 PMID:31956926

Pugin, F., Bucher, P., & Morel, P. (2011). History of robotic surgery: From AESOP® and ZEUS® to da Vinci®. *Journal of Visceral Surgery*, *148*(5), e3–e8. doi:10.1016/j.jviscsurg.2011.04.007 PMID:21974854

Pushkar, D. Y., Kasyan, G. R., & Popov, A. A. (2021). Robotic sacrocolpopexy in pelvic organ prolapse: A review of current literature. *Current Opinion in Urology*, *31*(6), 531–536. doi:10.1097/MOU.0000000000000932 PMID:34506336

Quemener, J., Boulanger, L., Rubod, C., Cosson, M., Vinatier, D., & Collinet, P. (2012). The place of robotics in gynecologic surgery. *Journal of Visceral Surgery*, *149*(5), e289–e301. doi:10.1016/j.jviscsurg.2012.07.002 PMID:22951086

Raju, R., & Linder, B. J. (2021). Evaluation and Management of Pelvic Organ Prolapse. *Mayo Clinic Proceedings*, *96*(12), 3122–3129. doi:10.1016/j.mayocp.2021.09.005 PMID:34863399

Ramirez, P. T., Soliman, P. T., Schmeler, K. M., dos Reis, R., & Frumovitz, M. (2008). Laparoscopic and robotic techniques for radical hysterectomy in patients with early-stage cervical cancer. *Gynecologic Oncology*, *110*(3), S21–S24. doi:10.1016/j.ygyno.2008.03.013 PMID:18486966

Rardin, C. R. (2011). Minimally invasive urogynecology. *Obstetrics and Gynecology Clinics of North America*, *38*(4), 639–649. doi:10.1016/j.ogc.2011.09.001 PMID:22134013

Rardin, C. R. (2014). The debate over robotics in benign gynecology. *American Journal of Obstetrics and Gynecology*, *210*(5), 418–422. doi:10.1016/j.ajog.2014.01.016 PMID:24418227

Ravivarapu, K. T., Omidele, O., Pfail, J., Tomer, N., Small, A. C., & Palese, M. A. (2021). Robotic-assisted simple prostatectomy versus open simple prostatectomy: A New York statewide analysis of early adoption and outcomes between 2009 and 2017. *Journal of Robotic Surgery*, *15*(4), 627–633. doi:10.100711701-020-01152-w PMID:33009988

Reza, M., Maeso, S., Blasco, J. A., & Andradas, E. (2010). Meta-analysis of observational studies on the safety and effectiveness of robotic gynaecological surgery. *British Journal of Surgery*, *97*(12), 1772–1783. doi:10.1002/bjs.7269 PMID:20949554

Rocha, A., Domínguez, A. M., Lécuru, F., & Bourdel, N. (2016). Indocyanine green and infrared fluorescence in detection of sentinel lymph nodes in endometrial and cervical cancer staging – A systematic review. *European Journal of Obstetrics, Gynecology, and Reproductive Biology*, *206*, 213–219. doi:10.1016/j.ejogrb.2016.09.027 PMID:27750179

Rodgers, A. K., Goldberg, J. M., Hammel, J. P., & Falcone, T. (2007). Tubal anastomosis by robotic compared with outpatient minilaparotomy. *Obstetrics and Gynecology*, *109*(6), 1375–1380. doi:10.1097/01.AOG.0000264591.43544.0f PMID:17540810

Rodrigues, R. C., Rodrigues, M. R. K., Freitas, N. O., Rudge, M. V. C., & Lima, S. A. M. (2019). Quality of life in patients who undergo conventional or robotic-assisted total laparoscopic hysterectomy: Protocol for a systematic review of randomized controlled trials. *Medicine*, *98*(23), e15974. doi:10.1097/MD.0000000000015974 PMID:31169730

Rosenblatt, P. L., McKinney, J., & Adams, S. R. (2013). Ergonomics in the operating room: Protecting the surgeon. *Journal of Minimally Invasive Gynecology*, *20*(6), 744. doi:10.1016/j.jmig.2013.07.006 PMID:23969139

Ross, J. W., & Preston, M. R. (2009). Update on laparoscopic, robotic, and minimally invasive vaginal surgery for pelvic floor repair. *Minerva Ginecologica*, *61*, 173–186. PMID:19415062

Rueda, M. A., Riga, C., & Hamady, M. S. (2018). Flexible robotics in pelvic disease: Does the catheter increase applicability of embolic therapy? *The Journal of Cardiovascular Surgery*, *59*(3), 322–327. doi:10.23736/S0021-9509.18.10422-8 PMID:29430891

Sanabria, D., Rodriguez, J., Pecci, P., Ardila, E., & Pareja, R. (2020). Same-Day Discharge in Minimally Invasive Surgery Performed by Gynecologic Oncologists: A Review of Patient Selection. *Journal of Minimally Invasive Gynecology*, *27*(4), 816–825. doi:10.1016/j.jmig.2019.10.023 PMID:31715304

Santoro, E., & Pansadoro, V. (2013). Robotic surgery in Italy national survey (2011). *Updates in Surgery*, *65*(1), 1–9. doi:10.100713304-012-0190-z PMID:23224637

Satava, R. M. (2003). Robotic surgery: From past to future – A personal journey. *The Surgical Clinics of North America*, *83*(6), 1491–1500. doi:10.1016/S0039-6109(03)00168-3 PMID:14712882

Schachar, J. S., & Matthews, C. A. (2020). Robotic-assisted repair of pelvic organ prolapse: A scoping review of the literature. *Translational Andrology and Urology*, *9*(2), 959–970. doi:10.21037/tau.2019.10.02 PMID:32420212

Scheib, S. A., Tanner, E. III, Green, I. C., & Fader, A. N. (2014). Laparoscopy in the morbidly obese: Physiologic considerations and surgical techniques to optimize success. *Journal of Minimally Invasive Gynecology*, *21*(2), 182–195. doi:10.1016/j.jmig.2013.09.009 PMID:24100146

Schreuder, H. W., Wolswijk, R., Zweemer, R. P., Schijven, M. P., & Verheijen, R. H. (2012). Training and learning robotic surgery, time for a more structured approach: A systematic review. *British Journal of Obstetrics and Gynaecology*, *119*(2), 137–149. doi:10.1111/j.1471-0528.2011.03139.x PMID:21981104

Shafer, A., & Boggess, J. F. (2008). Robotic-assisted endometrial cancer staging and radical hysterectomy with the da Vinci surgical system. *Gynecologic Oncology*, *111*(2), S18–S23. doi:10.1016/j.ygyno.2008.07.023 PMID:18771796

Shull, B. L., Bachofen, C., Coates, K. W., & Kuehl, T. J. (2000). A transvaginal approach to repair of apical and other associated sites of pelvic organ prolapse with uterosacral ligaments. *American Journal of Obstetrics and Gynecology*, *183*(6), 1365–1373. doi:10.1067/mob.2000.110910 PMID:11120498

Sinha, R., Sanjay, M., Rupa, B., & Kumari, S. (2015). Robotic surgery in gynecology. *Journal of Minimal Access Surgery*, *11*(1), 50–59. doi:10.4103/0972-9941.147690 PMID:25598600

Steil, J., Finas, D., Beck, S., Manzeschke, A., & Haux, R. (2019). Robotic Systems in Operating Theaters: New Forms of Team-Machine Interaction in Health Care. *Methods of Information in Medicine*, *58*(S 01), e14–e25. doi:10.1055-0039-1692465 PMID:31342471

Stein, K., & Ascher-Walsh, C. (2009). A comprehensive approach to the treatment of uterine leiomyomata. *The Mount Sinai Journal of Medicine, New York*, *76*(6), 546–556. doi:10.1002/msj.20145 PMID:20014416

Swan, K., & Advincula, A. P. (2011). Understanding the financial impact of robotics in gynecologic surgery. *Clinical Obstetrics and Gynecology*, *54*(3), 449–457. doi:10.1097/GRF.0b013e31822ce22c PMID:21857174

Theodore, N., Arnold, P. M., & Mehta, A. I. (2018). Introduction: The rise of the robots in spinal surgery. *Neurosurgical Focus*, *45*(videosuppl1), 45. doi:10.3171/2018.7.FocusVid.Intro PMID:29963916

Thomas, D., Medoff, B., Anger, J., & Chughtai, B. (2020). Direct-to-consumer advertising for robotic surgery. *Journal of Robotic Surgery*, *14*(1), 17–20. doi:10.100711701-019-00989-0 PMID:31243703

Thubert, T., Foulot, H., Vinchant, M., Santulli, P., Marzouk, P., Borghese, B., & Chapron, C. (2016). Surgical treatment: Myomectomy and hysterectomy; Endoscopy: A major advancement. *Best Practice & Research. Clinical Obstetrics & Gynaecology*, *34*, 104–121. doi:10.1016/j.bpobgyn.2015.11.021 PMID:27400649

Thurston, J., Murji, A., Scattolon, S., Wolfman, W., Kives, S., Sanders, A., & Leyland, N. (2019). No. 377 – Hysterectomy for Benign Gynaecologic Indications. *Journal of Obstetrics and Gynaecology Canada*, *41*(4), 543–557. doi:10.1016/j.jogc.2018.12.006 PMID:30879487

Tiwari, V. (2014). Calculating the true cost of robotic hysterectomy. *Healthcare Financial Management*, *68*, 80–85. PMID:25145035

Truong, M., Kim, J. H., Scheib, S., & Patzkowsky, K. (2016). Advantages of robotics in benign gynecologic surgery. *Current Opinion in Obstetrics & Gynecology*, *28*(4), 304–310. doi:10.1097/GCO.0000000000000293 PMID:27362711

Tsuda, S., Oleynikov, D., Gould, J., Azagury, D., Sandler, B., Hutter, M., Ross, S., Haas, E., Brody, F., & Satava, R. (2015). SAGES TAVAC safety and effectiveness analysis: Da Vinci® Surgical System (Intuitive Surgical, Sunnyvale, CA). *Surgical Endoscopy*, *29*(10), 2873–2884. doi:10.100700464-015-4428-y PMID:26205559

Ulmer, B. C. (2010). Best practices for minimally invasive procedures. *Association of periOperative Registered Nurses Journal, 91*, 558-572.

van Dam, P., Hauspy, J., Verkinderen, L., Trinh, X. B., van Dam, P. J., Van Looy, L., & Dirix, L. (2011). Are costs of robot-assisted surgery warranted for gynecological procedures? *Obstetrics and Gynecology International*, *2011*, 973830. doi:10.1155/2011/973830 PMID:21941556

van Seeters, J. A. H., Chua, S. J., Mol, B. W. J., & Koks, C. A. M. (2017). Tubal anastomosis after previous sterilization: A systematic review. *Human Reproduction Update*, *23*(3), 358–370. doi:10.1093/humupd/dmx003 PMID:28333337

Varghese, A., Doglioli, M., & Fader, A. N. (2019). Updates and Controversies of Robotic-Assisted Surgery in Gynecologic Surgery. *Clinical Obstetrics and Gynecology*, *62*(4), 733–748. doi:10.1097/GRF.0000000000000489 PMID:31524659

Visco, A. G., & Advincula, A. P. (2008). Robotic gynecologic surgery. *Obstetrics and Gynecology*, *112*(6), 1369–1384. doi:10.1097/AOG.0b013e31818f3c17 PMID:19037049

Weickert, U., Jakobs, R., & Riemann, J. F. (2005). Diagnostic laparoscopy. *Endoscopy*, *37*(1), 33–37. doi:10.1055-2004-826081 PMID:15657855

Weinberg, L., Rao, S., & Escobar, P. F. (2011). Robotic surgery in gynecology: An updated systematic review. *Obstetrics and Gynecology International*, *2011*, 852061. doi:10.1155/2011/852061 PMID:22190948

Wexner, S. D., Bergamaschi, R., Lacy, A., Udo, J., Brölmann, H., Kennedy, R. H., & John, H. (2009). The current status of robotic pelvic surgery: Results of a multinational interdisciplinary consensus conference. *Surgical Endoscopy*, *23*(2), 438–443. doi:10.100700464-008-0202-8 PMID:19037694

Whiteside, J. L., Kaeser, C. T., & Ridgeway, B. (2019). Achieving high value in the surgical approach to hysterectomy. *American Journal of Obstetrics and Gynecology*, *220*(3), 242–245. doi:10.1016/j.ajog.2018.11.124 PMID:30419200

Wimberger, P., & Schindelhauer, A. (2014). Robotic surgery in gynecology. *Archives of Gynecology and Obstetrics*, *289*(1), 5–6. doi:10.100700404-013-3066-7 PMID:24170162

Wisner, K. P., Gupta, S., Ahmad, S., & Holloway, R. W. (2015). Indications and techniques for robotic pelvic and para-aortic lymphadenectomy in gynecologic oncology. *Journal of Surgical Oncology*, *112*(7), 782–789. doi:10.1002/jso.24005 PMID:26288990

Wisner, K. P. A., Ahmad, S., & Holloway, R. W. (2017). Indications and techniques for robotic pelvic and para-aortic lymphadenectomy with sentinel lymph node mapping in gynecologic oncology. *Best Practice & Research. Clinical Obstetrics & Gynaecology*, *45*, 83–93. doi:10.1016/j.bpobgyn.2017.04.006 PMID:28533155

Xie, Y. (2015). Cost-effectiveness of robotic surgery in gynecologic oncology. *Current Opinion in Obstetrics & Gynecology*, *27*(1), 73–76. doi:10.1097/GCO.0000000000000134 PMID:25490379

Yurashevich, M., & Habib, A. S. (2021). Anesthetic management for robotic hysterectomy in obese women. *Current Opinion in Anaesthesiology*, *34*(3), 238–245. doi:10.1097/ACO.0000000000000987 PMID:33935171

Zakashansky, K., Bradley, W. H., & Nezhat, F. R. (2008). New techniques in radical hysterectomy. *Current Opinion in Obstetrics & Gynecology*, *20*(1), 14–19. doi:10.1097/GCO.0b013e3282f2288a PMID:18197000

Chapter 15
Kinematic Modeling and Real-Time Implementation of an Indigenous Control System for a Novel Three-Link Flexible Robot

Pratik Prakash Chothe
SVR InfoTech, India

Aniket Harish Bhelsaikar
SVR InfoTech, India

Shreyash Pandit Gajlekar
Tata Technologies Ltd., India

Vinod Dnyandeo Atpadkar
SVR InfoTech, India

Debanik Roy
Bhabha Atomic Research Centre (BARC), Mumbai, India

ABSTRACT

The domain of flexible robotics is gaining momentum due to its widespread advantages despite major technological barricades, such as in-situ vibration and joint instability. Flexible robotic system (FRS) pertains to unique designs of serial-chain assistive robots that exhibit the said inherent vibration impromptu. This chapter focuses on the control system algorithm of a prototype serial-chain FRS (PARv1.0), backed up by the indigenous firmware of the system controller and run-time software. The chapter also brings out the parametric relationship of the control system hardware over the electromechanical output of the FRS. The design function architecture of the prototype multi-degrees-of-freedom FRS, highlighted in this chapter, is a rare combination of three revolute and one prismatic (RRRP) joints intended for pick and place applications, having a 1500 mm radial span reach in the horizontal plane. A prototype miniature gripper actuated by a mini-servomotor is another novel development of this FRS.

DOI: 10.4018/978-1-6684-5381-0.ch015

INTRODUCTION

The prevalence of in-situ vibration of serial-chain Flexible Robotic System (FRS), manifested through trembling of its slender links and/or twisting of the joints, pose critical bottlenecks in real-time control of the FRS so far as attainment of accuracy of end-of-arm tooling is concerned. The vibration in the FRS in dynamic condition is self-propagating and perpetual in nature and it does not follow much of the rules of analytical modelling.

Quite naturally this in-situ trembling produces prolonged jitter, especially in case of multi-degrees-of-freedom FRS. Hence, we need to go for suitable vibration harnessing mechanism, ideally through the system electronics before deploying such FRS in application-field. Two approaches have been used hitherto to reduce the vibration challenge of FRS, namely: a] input command shaping and b] closed-loop feedback for vibration control. In the present research, the technique of input command shaping has been used in order to reduce run-time vibration in the prototype FRS. Chain of commands have been used judiciously in input command shaping to nullify the own vibration of the FRS. In order to attain the desired degree of vibration attenuation, our work has been focused on a bi-variate approach under input command shaping, viz. velocity and torque shaping of the prototype FRS.

By mirroring the fundamental design-layout of SCARA type static robots, the prototype FRS is conceptualized as a serial-chain system with sufficiently long horizontal reach. Our prototype flexible robot is having 3 degrees-of-freedom (d.o.f) in total, involving worm gear driven revolute joints (2 nos), one spur gear driven revolute joint, two metallic flexible shafts, three links, a recirculating ball screw, a miniature gripper instrumented with sensor, a tripod and electrical actuators, motion controllers & drivers. In order to implement input command shaping on the prototype FRS it is essential to know the nature of the built-in vibration. Vibrational analysis in prototype FRS has been carried out extensively using finite element analysis. The design paradigm of a typical multi-degrees-of-freedom FRS is essentially task-driven, wherein one end of the manipulator is connected to the supporting base while the other end is attached to the end-effector. The relative motions of the joints give the position of the FRS-links in desired orientations. The kinematics of the flexible robot arm deals with the motion of the robot with respect to the fixed reference co-ordinate system. As a matter of fact, it is quite interesting to know a-priori whether there are any obstacles present in the path, prescribed for the FRS.

Vibrational analysis becomes a decisive pre-requisite for real-time control of multi-d.o.f FRS because of the slender lengths of its links and incorporation of flexible shafts. We have duly observed through experimentation that the motor-power transmission in robot-links via flexible shafts is a distinct source of vibration. While the system is in motion, the vibration gets induced in the assembly that gets further intensified by the augmentation of flexible shafts. The ensemble cantilever structure of the planar FRS dictates the amount of such in-situ vibration as well as its modal frequencies. Thus, attenuation of vibration requires very robust control system & program level manipulations for vibration reduction and achieving the required precision at the end-effector.

The control problem of a flexible manipulator is ideally divided into two coherent sub-problems, namely: the trajectory planning and the motion control. Nonetheless, real-time control of inherent randomized vibration of a prototype FRS becomes a challenging problem indeed, if it has multiple degrees-of-freedom. The novelty of our control system architecture is rooted through joint navigation paradigm using trajectory-based control in real-time.

Background

Various techniques aiming at reduction of vibration at end-effector point in the prototype FRS have been examined by researchers in the past. With that perspective, significant research efforts have been streamlined over the past few decades to model and control various flexible robotic systems having varying degrees-of-freedom. Singer & Seering (1990) demonstrated the Preshaping of input for reduction of vibration in FRS. A new approach for dynamic modelling of multi- d.o.f FRS was reported by Chen Wen (2001), while the same was tackled through spring and rigid bodies by Zhang J., Tian Y., Zhang M. (2014). The issues of frequency damping and control of vibration in a single-link FRS thereof were studied through adaptive resonant control of the variables by Tjahyadi, H., & Sammut, K. (2006), which was further strengthened by using novel methods for algebraic identification by Trapero-Arenas, J.R. et al (2008). and integral resonant control by Pereirea, Aphale et al. (2011). Smith (1958) proposed input shaping as technique for generating non-oscillatory response from a lightly-damped system subject to a step input. The response is being achieved by generating two transient oscillations such that the end-result is cancellation of oscillations. In its earliest form, command Preshaping was used as motion templates for flexible manipulators as well as perturbation techniques by Fraser & Daniel (1991). Junkins & James (1986) and Chun et al. (1985) presented practical solutions of the optimal control formulation for flexible systems. Gupta (1980) presented frequency shaping terms in the optimal formulation of the feedback control equation of FRS, which was substantiated by the closed-form solution approach by Juang et al. (1985). Artificial intelligence model, using fuzzy logic for FRS control, was proposed by Moudgal et al. (1995). A treatise on various available techniques for the control of FRS can be referred for new research & application-based work done by Benosman & Vey (2004).

It may be commented that while real-time control of multi-d.o.f is a real challenge for the researchers, a pretty large number of papers on the control semantics of single-link FRS (with fixed and variable payload at end-effector) have been published till date. Kotnik et al (1988) presented the acceleration feedback control for a single-link FRS, based on experimental results. A considerable number of researches have been conducted on the modelling and control of 2 d.o.f flexible systems as well by Feliu et al (1999), Lin & Yih (1996), Moallem et al (1997), Yim & Singh (1996). In comparison to the research conducted with single and two-link FRS, 3-link FRS has its own set of challenges. Feliu et al (2003) explained the modelling of 3-d.o.f. planar FRS with feedback control and inverse kinematics in spherical co-ordinates. Somolinos et al (2002) reported how an FRS-arm can be modelled as a structure with only one oscillation mode in each of the three orthogonal directions. Design, development of prototype FRS along with finite element analysis has been reported by our group Warude et al (2019), Rauniyar et al (2018), Roy Debanik (2019).

The chapter has been organized in seven sections. *Ab initio* kinematic modeling and analysis of the prototype FRS are addressed in the next section, with an aim towards establishing connotation with the run-time analysis of the vibration. Section 3 details out the ensemble mathematical perspective of the real-time FRS, piggybacked by vibrational analysis that un-wraps its nature & characteristics. The schematic and syntax of the control system architecture and design of device controller of the prototype FRS are attributed in section 4. Section 5 addresses the real-time programming semantics of the prototype FRS, along with the Graphical User Interface (GUI) developed. An overview of the hardware architecture of the controller of the prototype FRS is provided in section 6 and finally section 7 concludes the chapter.

MODELING AND KINEMATIC ANALYSIS OF THE PROTOTYPE FLEXIBLE ROBOTIC SYSTEM

Highlights of the Design

The design function architecture of the prototype FRS having 3 d.o.f. is a rare combination of three revolute & one prismatic (RRRP) joints intended for pick and place applications, having 1500 mm radial span reach in horizontal plane. The prototype[1]*PARv1.0* has ball screw driven translational motion (with three linear bearing support in vertical plane Z) and spur gear driven rotary mechanism (for radial span reach in horizontal plane XY). PARv1.0 has three identical cross-sectioned links of unequal length, made up of carbon fiber reinforced plastic, that cover the 1500 reach span (in XY plane). The novel worm- worm gear pairs drive the miniature joints that reduce localized vibration, backlash and also harness the global randomized vibration to a significant extent. The servomotors for joint actuation are fitted in the robot top canopy and the rotational motion generated thereof is transferred to the joint-shaft(s) via flexible shafts, designed & manufactured indigenously. The incorporation of our novel cost-effective flexible shafts helps reducing the overhung weight on the FRS-links, which has been a crucial aspect in this research although. However, on the flip side, these flexible shafts, by nature, boost up the vibration in the system and cause loss in motion transfer to a certain extent. A prototype miniature gripper, actuated by a mini-servomotor, is another novel development of PARv1.0. The FRS prototype has a lightweight mini-gripper, which is a manufacturing marvel because of its tiny external envelope (50 mm x 40 mm x 15 mm). The design modelling of the prototype FRS is done by exploiting the principles of homogeneous transformation performed using Python scripting. Successful estimation of joint-angle for each link has been done through the forward kinematics scripting and end-effector pose is estimated through scripting using inverse kinematics. The validation of these two kinematic routines is done using *RoboAnalyzer*™ software. The novel indigenous control system of our FRS is validated through grasp experiments using a large variety of tiny objects. The deflection, modal and stress analysis of prototype FRS have been carried out with major priority on doing so on links and joints. The static analysis of the flexible robot links gives the glimpse of deflection contour and stress pattern for the load applied on the flexible arms. Notwithstanding the rheology of the FRS, the tip deflection due to payload is crucial in understanding the role of design modification to strengthen the structure. Interestingly, our research reveals that the fundamental frequency of vibration has the same deflection contour as that of static deflection. Thus, the amplitude of vibration can cause the prototype FRS to vibrate more and can cause severe problem in the ensemble design. It goes without amplifying that achieving the desired value of positional accuracy of the end-effector is the major challenge in FRS due to its inherent vibration, which was addressed and handled optimally in our hardware through different levels of software-control strategies.

The Kinematics of the Prototype Flexible Manipulator

A sound kinematic model and analysis thereof plays a vital role in designing the control system functionaries of a robot. For our prototype, *RoboAnalyzer*™ has been used for the purpose of overall modeling and study of the kinetics. The modeling and indigenous scripting under the platform of this software was used to simulate the links through forward and inverse kinematics as well. Presently, both forward kinematics & inverse kinematics are used in the study as the inputs to the controller semantics through customized programming. The code is written in-house in Python. The results of forward and inverse kinematics have been calculated using this novel Python code and the same was validated using *RoboAnalyzer*™ software. Figure 1 schematically illustrates the basic kinematic disposition of the prototype flexible manipulator that was used in the simulation. The conceptual paradigms of '*deflected postures of the links*' and '*trembling of the terminal link*' may be noted.

Figure 1. Kinematic Disposition of the Prototype Flexible Manipulator

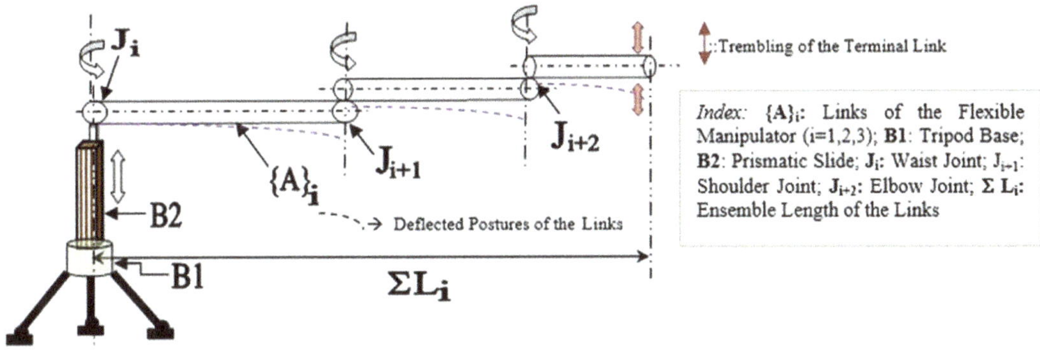

Based on the basic kinematic disposition of our prototype FRS (refer figure 1), we proceeded towards simulation & kinematic analysis of the system through *RoboAnalyzer*™. In order to have complete handshaking with the said software, the link length for our prototype FRS were selected suitably and fed to it along with the offsets. The three links are selected from the Graphical User Interface (GUI) of *RoboAnalyzer*™ and the relevant Denavit-Hartenberg (D-H) parameters are defined as inputs. As the operation of the prototype FRS is purely based on the rotation of its links, revolute joints are described between the two successive links. As per the semantics of the software, the initial and final values are utmost important to simulate the rotational degrees-of-freedom of the links, ranging from initial angular posture to a final position.

Forward Kinematics

The kinematic layout of the prototype FRS is modulated a bit in order to fit in the capsule of *RoboAnalyzer*™, which calls for an analytical shift from 'three-link flexible manipulator' to 'three-link planar arm'. Figure 2a shows the schematic of the said planar arm, with requisite co-ordinates frames for processing the D-H parameters. The kinematics of the same arm with the locations of end-effector as well as wrist is highlighted in Figure 2b.

It may be noted here that the legends a_i & θ_i (for i= 1,2,3) represent link lengths and joint angles of the planar arm respectively. For the planar reference Frame 1, (X1, Y1) is fixed at #0 as shown in figure 2a. The subsequent frames, viz. (X2, Y2) to (X4, Y4) are built at respective fixed locations (#1, #2 & #3). Likewise, we have got renamed locational frames for 'wrist' (Wx, Wy) and 'End-effector' (Px, Py), as clarified pictorially in Figure 2b. Various D-H parameters of the robotic arm are shown in Table 1 below, augmented by the final transformation (D-H) matrix [T] that represents the end- effector's pose with respect to its base frame, i.e., frame#1. The usual legends of [T], namely b_i, θ_i, a_i & α_i represent link offset, joint angles, link lengths and angle of twist respectively. Here twist angle α_i and link offset b_i are considered as zero.

Table 1. D-H Parameters of the Three-Link Arm and the Final D-H Matrix

Link	b_i	θ_i	a_i	α_i
1	0	θ_1	a1	0
2	0	θ_2	a2	0
3	0	θ_3	a3	0

$$T=\begin{bmatrix} C123 & -S123 & 0 & a1C1+a2C12+a3C123 \\ S123 & C123 & 0 & a1S1+a2S12+a3S123 \\ 0 & 0 & 1 & 0 \\ 0 & 0 & 0 & 0 \end{bmatrix}$$

['C123': Cos (θ1+θ2+θ3); 'S123': Sin (θ1+θ2+θ3)]

[As per the geometry of Figure 2b, we have: $a_1c_1+a_2c_{12}+a_3c_{123} = P_x$ and $a_1s_1+a_2s_{12}+a_3s_{123} = P_y$]

Figure 2. Schematic of the Three-Link Planar Arm as per the Kinematic Model: [a] only Manipulator; [b] with End-effector

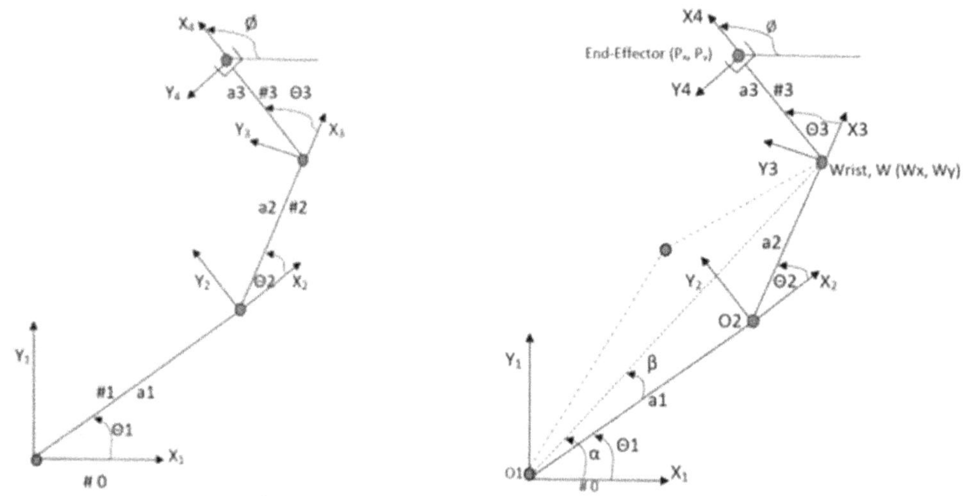

The screenshot of the arm model, done in *RoboAnalyzer*™ is shown in fig 3 below. The curved path traced by the manipulator in course of its movement from initial to final location is depicted by dark black line in Figure 3. It may be noted that *RoboAnalyzer*™ has default values in meters (m) and degrees (Deg).

During a pre-assigned movement from say point A to point B, the actual nature of the joint traverse with a change in the position of the end-effector can be noted. The forward kinematics solver of *RoboAnalyzer*™ finds out the position of the end-effector for the defined set of angles to each joint. The screenshots of figures 4 & 5 illustrate the said movement from 'A': (1.4, 0, 0.34) to 'B' (0.14, 0.84, 0.34).

Figure 3. The Screenshot of the Model and D-H Parameters of the Robotic Arm as per RoboAnalyzer™

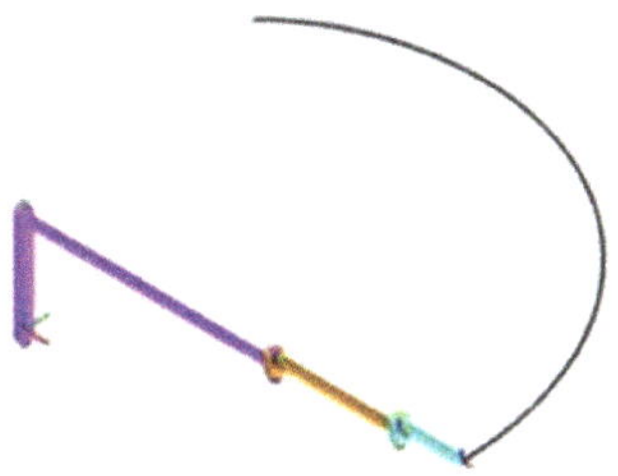

Joint No	Joint Type	Joint Offset (b) m	Joint Angle (theta) deg	Link Length (a) m	Twist Angle (alpha) deg	Initial Value (JV) deg or m	Final Value (JV) deg or m
1	Revolute	0.3	Variable	0.8	0	0	60
2	Revolute	0.02	Variable	0.4	0	0	90
3	Revolute	0.02	Variable	0.2	0	0	45

Figure 4. RoboAnalyzer™ Screenshot Showing the Initial Position of the Manipulator

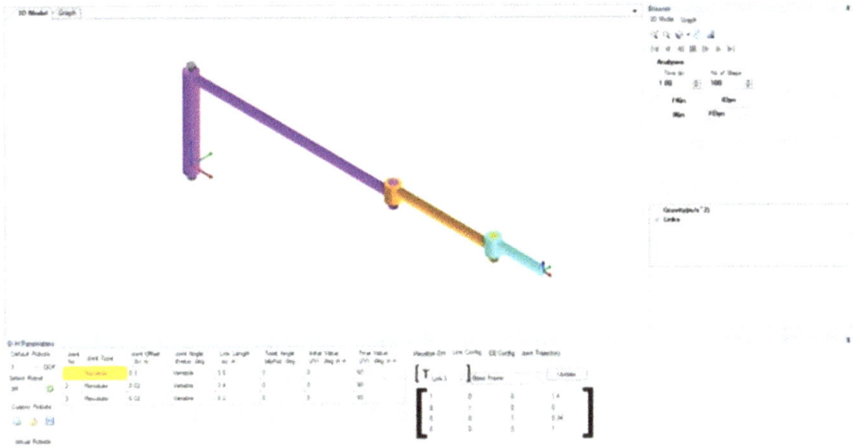

Based on the solution of forward kinematics module, the target position ('B') is found out to be 0.139595m in X direction and 0.841057m in Y direction. Z direction is the elevation of FRS and in most of the case Z is kept fixed and is equal to 0.34m. Figure 6 represents the transformation of rotation and position matrix from 'A' to 'B' with respect to the base frame, as obtained through the software.

The joint trajectories are crucial in determining the movement of the joints from initial value to the final value, while traversing between two points. Figure 7 illustrates the variation of joint trajectories for the three joint-angles of the planar flexible manipulator, wherein Y-axis represents the angle of each link in degrees and X-axis is pseudo time-scale from 0 to 1. As per the simulation, joint 1 will move from 0° to 90°, joint 2 will vary between 0° & 60° and joint 3 will have the rotary motion from 0° to 45°. Similarly, angular velocities and angular accelerations of the joints are plotted as shown in the graphs of Figures 8 & 9. Note that the joint velocities are in degrees/sec and joint accelerations are in degrees /sec^2.

The results of forward kinematics (vide plots of figs 7,8&9), although self-explanatory, need proper benchmarking. In order to attain that, we have used indigenously-developed scripting in *Python*. This code has been drafted in strict adherence to the kinematic propositions and lemmas of the prototype FRS.

Figure 5. RoboAnalyzer™ Screenshot Showing the Final Position of the Manipulator

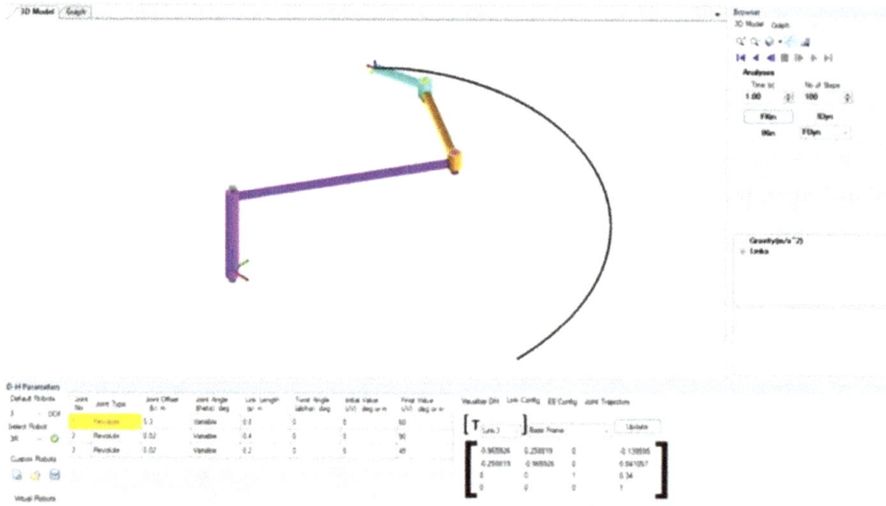

Figure 6. Screenshot of Transformation Matrices for a Representative Point-to-Point Movement of the Flexible Manipulator

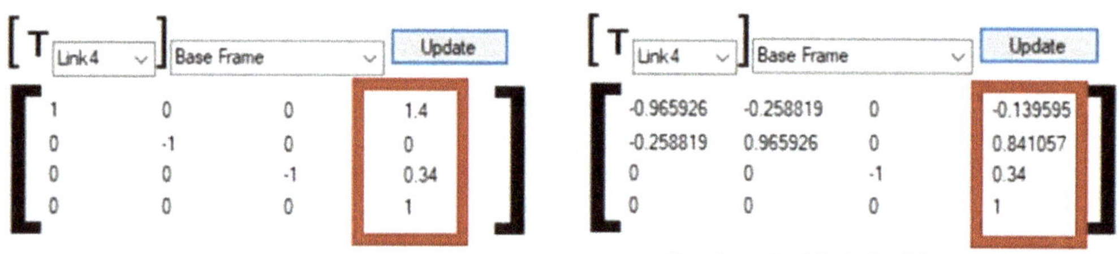

Transformation Matrix for 'A' Transformation Matrix for 'B'

Figure 7. Variation of the Joint Trajectories of the Planar Flexible Manipulator

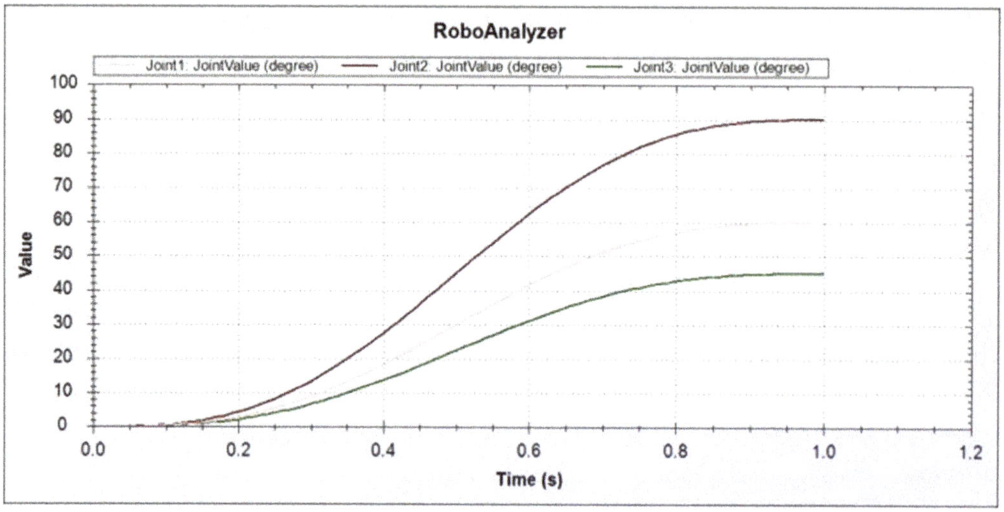

Figure 8. Variation of the Angular Velocities of the Joints of the Planar Flexible Manipulator

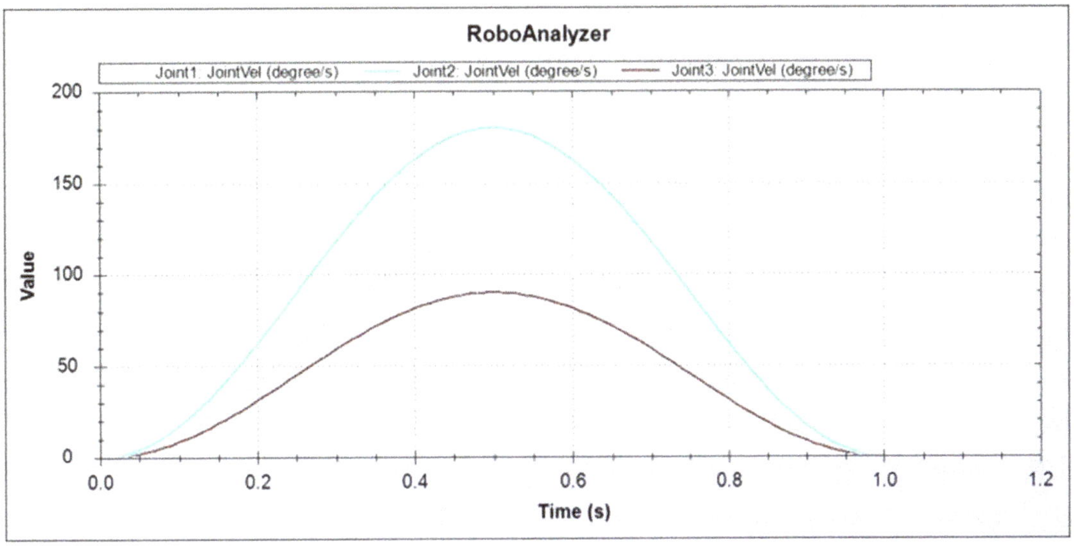

Figure 9. Variation of the Angular Accelerations of the Joints of the Planar Flexible Manipulator

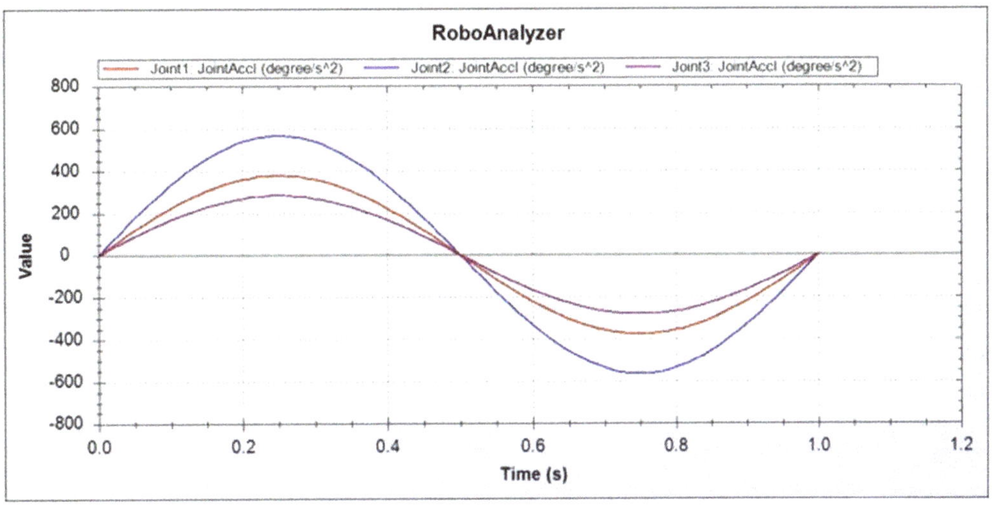

The said Python scripting for the forward kinematics is shown below:

import NumPy as np
T1= np. radians (0) # *theta1* value in radians
T2= np. radians (0) # *theta2* value in radians
T3= np. radians (0) # *theta3* value in radians
T4= np. radians (0) # *theta4* value in radians
a1 = 800 #*lenght of first link* in mm
a2 = 400 #*length of second link* in mm

```
a3 = 200 #length of third link in mm
a4 = 0 #length of link in mm (end-effector part)
S1=np.radians(0) #alpha1 in radians
S2=np.radians(0) #alpha2 in radians
S3=np.radians(180) #alpha3 in radians
S4=np.radians(0) #alpha4in radians
d1=0.3 #link offset in mm
d2=0.02 #link offset in mm
d3=0.02 #link offset in mm
d4=0 #link offset in mm
#A1=4x4Transformation Matrix
A1= [[np.cos(T1), -np.sin(T1) *np.cos(S1), np.sin(T1) *np.sin(S1), a1*np.cos(T1)],
[np.sin(T1), np.cos(T1) *np.cos(S1), -np.cos(T1) *np.sin(S1), a1*np.sin(T1)],
[0, np.sin(S1), np.cos(S1), d1],
[0,0,0,1]]
#A2= 4x4 Transformation Matrix
A2= [[np.cos(T2), -np.sin(T2) *np.cos(S2), np.sin(T2) *np.sin(S2), a2*np.cos(T2)],
[np.sin(T2), np.cos(T2) *np.cos(S2), -np.cos(T2) *np.sin(S2), a2*np.sin(T2)],
[0, np.sin(S2), np.cos(S2), d2],
[0,0,0,1]]
#A3= 4x4 Transformation Matrix
A3= [[np.cos(T3), -np.sin(T3) *np.cos(S3), np.sin(T3) *np.sin(S3), a3*np.cos(T3)],
[np.sin(T3), np.cos(T3) *np.cos(S3), -np.cos(T3) *np.sin(S3), a3*np.sin(T3)],
[0, np.sin(S3), np.cos(S3), d3],
[0,0,0,1]]
#A4= 4x4 Transformation Matrix
A4 = [[np.cos(T4), -np.sin(T4) *np.cos(S4), np.sin(T4) *np.sin(S4), a4*np.cos(T4)],
#[np.sin(T4), np.cos(T4) *np.cos(S4), -np.cos(T4) *np.sin(S4), a4*np.sin(T4)],
# [0, np.sin(S4), np.cos(S4), d4],
# [0,0,0,1]]
X= np.dot (A1, A2)
Y= np.dot (X, A3)
Z= np.dot (Y, A4)
print (np. matrix(Y))
```

The above scripting was tested for functionality for several combinations of joint-angles towards validation of the forward kinematics routine. The programming console of Python being open source, we have the liberty of stretching out the kinematic formulation to the extent that is needed via-a-vis stipulated simulation protocols of the commercial software, viz. *RoboAnalyzer*™. Table 2 presents the comparative results of the forward kinematics routine, as performed by our indigenous scripting and *RoboAnalyzer*™, which clearly indicate that our indigenous scripting attained perfect precision.

Table 2. Comparative results of forward kinematics using Python Script and RoboAnalyzer™

Sl. No.	Inputs (Joint-angles)			End-Effector Position [X, Y]	
	θ_1	θ_2	θ_3	Indigenous *Python* Scripting	*RoboAnalyzer™*
1	60°	90°	45°	-0.1395,0.8410	-0.1395, 0.8410
2	50°	70°	60°	0.1142,0.9592	0.1142,0.9592
3	60°	70°	40°	-0.0540, 1.033	-0.0540,1.033
4	50°	60°	70°	0.1774,0.9887	0.1774,0.9887

Inverse Kinematics

The Inverse Kinematics solver of *RoboAnalyzer™* (abbreviated as *Ikin*) was invoked once all the inputs on manipulator geometry as well as end-effector location were loaded. The typical layout of the inverse kinematics routine is highlighted via the screenshot of Figure 10.

Figure 10. RoboAnalyzer™ Screenshot for Inverse Kinematics with Modelling of Links and Allied D-H Parameters

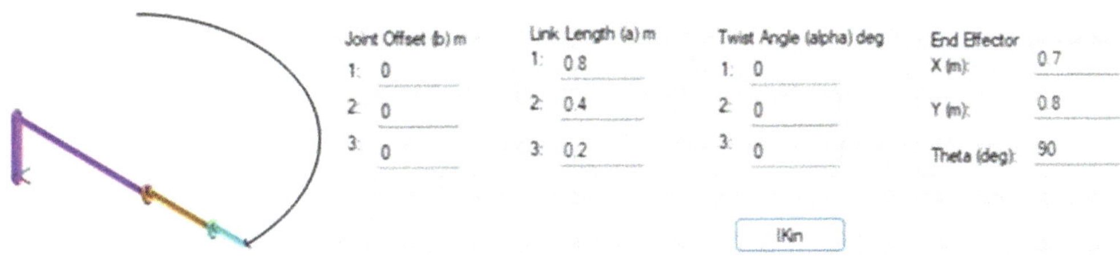

It may be noted that although defining only the (X, Y) co-ordinates of the end-effector is sufficient for the inverse kinematics routine for two-link planar manipulator, we need to have additional input of angle Ø (refer Figure 2) for higher-order manipulators. In other words, inverse kinematics cannot be solved without defining 'Ø' in manipulators beyond two links. Figure 11 illustrates the screenshot for the inverse kinematics of a typical problem-set, having two feasible solutions of joint-angle triads.

The solution sets of inverse kinematics routine (depicted via red-colored boxes in Figure 11) were tried in *RoboAnalyzer™* in order to get the positional layout of the 3R manipulator under investigation. Figures 12 & 13 illustrate the screenshots against these two solutions, which are essentially the computed forward kinematics for the manipulator. A line diagram-based nomenclature of the 3R manipulator (namely, ABCD) is presented as inset of Figure 12, which is not a part of the screenshot.

Figure 11. RoboAnalyzer™ Screenshot for Inverse Kinematics of a Typical Problem-set having Dual Solutions

Figure 12. Disposition of the 3R Manipulator against Solution 1 of Figure 11 via Screenshot of Robo-Analyzer™

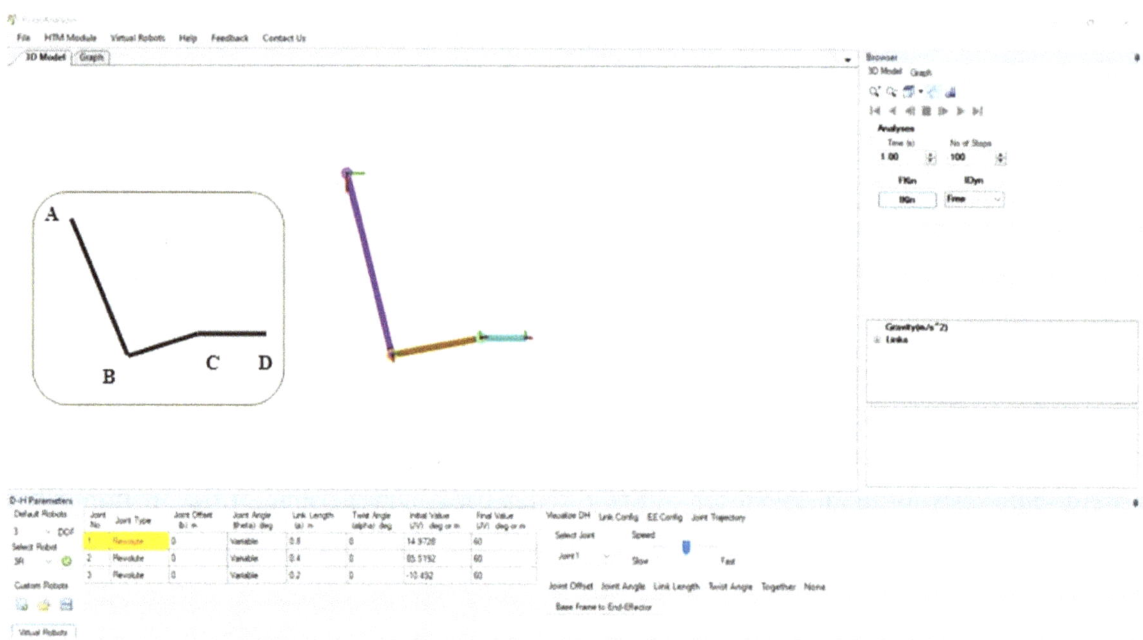

Figure 13. Screenshot of RoboAnalyzer™ Showing the Disposition of the 3R Manipulator against Solution2 of Figure 11

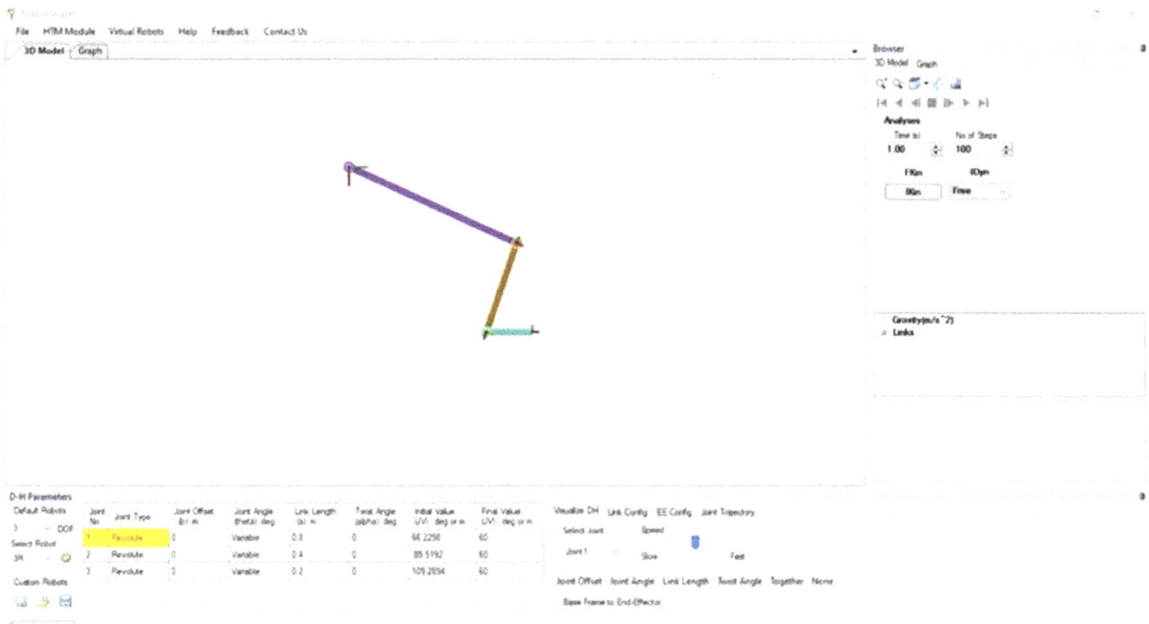

Like forward kinematics, we have used indigenous scripting for inverse kinematics as well using the open-source architecture of python. The said python code for computing the requisite inverse kinematics data is presented below:

```
Import NumPy as np
a1 = 0.8 # Link AB in meter
a2 = 0.4 # Link BC in meter
a3 = 0.2 # Link CD in meter
# Desired Position of End effector
X = 0.5
Y = 0.8
PhiDeg = 90 # Theta is in degrees
Phi = np. radians (PhiDeg)
# Coordinate at C i.e [w_x, w_y]
wx = X-a3*np.cos (Phi)
wy = Y-a3*np.sin (Phi)
W = (wy/wx)
# From forward kinematics, we have following equations
C2 = (wx**2+wy**2-a1**2-a2**2)/(2*a1*a2) # C2 is value of cos (Theta2)
if C2 <= 1:
s2_1 = np. sqrt(1-C2**2); # S2_1 is sin(theta2_1)
theta2_1 = np. arctan2(s2_1, C2); # It gives value of theta2_1 [values are verified and correct]
```

```
s2_2 = - (np. sqrt(1-C2**2)); # S2_2 is sin(theta2_2)
theta2_2 = np. arctan2(s2_2, C2); # It gives value of theta2_2 [values are verified and correct]
dm_1 = a1**2+a2**2+2*a1*a2*np.cos(theta2_1);
dm_2 = a1**2+a2**2+2*a1*a2*np.cos(theta2_2);
# Terminology for theta1_1 calculation
s1_1 = (wy*(a1+a2*np.cos(theta2_1))-a2*np.sin(theta2_1) *wx)/dm_1; # S1_1 is sin theta1_1
c1_1 = ((wx*(a1+a2*np.cos(theta2_1))) +a2*np.sin(theta2_1) *wy)/dm_1; # C1_1 is cos theta1_1
# Terminology for theta1_2 calculations
s1_2 = (wy*(a1+a2*np.cos(theta2_2))-a2*np.sin(theta2_2) *wx)/dm_2; # S1_2 is sin theta1_2
c1_2 = ((wx*(a1+a2*np.cos(theta2_2))) +a2*np.sin(theta2_2) *wy)/dm_1; # C1_2 is cos theta1_2
theta1_1 = np. arctan2(s1_1, c1_1);
theta1_2 = np. arctan2(s1_2, c1_2);
theta3_1 = Phi-theta1_1-theta2_1;
theta3_2 = Phi-theta1_2-theta2_2;
print (“””Solution 1”””)
print (np. degrees(theta1_1))
print (np. degrees(theta2_1))
print (np. degrees(theta3_1))

Solution 1
    20.915740850367705
    107.27000469497447
    -38.18574554534218
    print (“””Solution 2”””)
    print (np. degrees(theta1_2))
    print (np. degrees(theta2_2))
    print (np. degrees(theta3_2))
Solution 2
    79.47311696510191
    -107.27000469497447
    117.79688772987255
```

It is interesting to note that the result of the two methods of inverse kinematics routine, viz. through *RoboAnalyzer*™ and python scripting, almost matches for various angular positions the 3R manipulator. We observe slight variations in the results at 4$^{th.}$ or 5$^{th.}$ places of decimal, which can have alteration in vibration harnessing algorithm of the control system circuit for the manipulator. Table 3 presents a comparative analysis of the said inverse kinematics routine using *RoboAnalyzer*™ and our indigenous scripting in python. The deviations of joint-angle values at decimal level are indicated in red font Colour.

The indigenous scripting in python has become instrumental in quick assimilation of via-points so as to make the ensemble inverse kinematics routine more comprehensive. The so-called minor differences in joint-angle values are the indicators for better fine-tuning of the controller afterwards towards combating the in-situ oscillation of the FRS.

Table 3. Comparative results of inverse kinematics using RoboAnalyzer™ and Python Script

Sl. No.	Inputs			Methodology Used	Joint Angle values (in degree) of the Manipulator as obtained from Inverse Kinematics					
					Solution 1			*Solution 2*		
	X (m)	Y (m)	Ø°		θ_1	θ_2	θ_3	θ_1	θ_2	θ_3
1	0.5	0.7	90	*RoboAnalyzer™*	20.9158	107.27	-38.1857	79.473	-107.27	117.7969
2	0.7	0.9	90		22.1854	73.6652	-5.8506	67.8146	-73.6652	95.8506
3	0.5	0.6	90		8.9696	127.5443	-46.514	68.35	-127.5443	149.1944
4	0.4	0.7	90		21.65	127.5443	-59.1944	81.0304	-127.5443	136.514
1	0.5	0.7	90	Python Script	20.91574	107.27000	-38.18574	79.47311	-107.27000	117.79688
2	0.7	0.9	90		22.18537	73.66517	-5.85055	67.81462	-73.66517	95.85055
3	0.5	0.6	90		8.96964	127.54432	-46.51396	68.34997	-127.54432	149.19435
4	0.4	0.7	90		21.65002	127.54432	-59.19435	81.03035	-127.54432	136.51396

ANALYTICAL PERSPECTIVE OF THE REAL-TIME VIBRATION SCENARIO OF THE PROTOTYPE FLEXIBLE ROBOTIC SYSTEM

Origin of Vibration

The origin of the in-built system vibration of the prototype FRS is essentially intrinsic. In other words, very nature of the design of the prototype was responsible for the initial trembling, which gets transformed into jitter and finally, random vibration. The slenderness of the FRS-links coupled with low-density material of fabrication as well as long horizontal span of the FRS have catalytic effect on the spread of this vibration. As a matter of fact, this self-vibrating system will be more challenging in course of force closure of the grasp, with the help of a miniature gripper at its free end. Finite Element Analysis (FEA) was invoked for the rheological assessment of the prototype FRS (tare weight: ~ 18 kg.), especially for the run-time jitter of the mini-gripper with an effective (end-effector) reach of 1500 mm and 800 gm of maximum payload capacity.

Figure 14. Schematic of the Nature of Vibration in a typical Multi-Link Flexible Manipulator

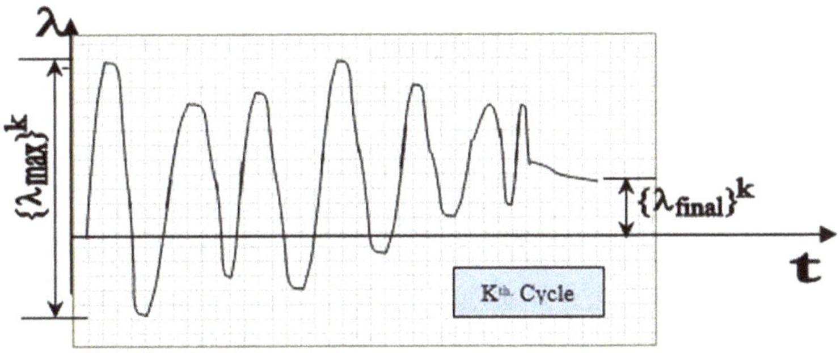

T: Time-period of oscillation; λ: Amplitude of oscillation

The propagation of the in-built system vibration is interesting, particularly between the links and link-joint interface zones. Figure 14 schematically illustrates the nature of this vibration, during $K^{th.}$ cycle.

Characterization of In-situ Vibration

In accordance with the typical nature of the in-situ vibration as depicted in Figure 14, evaluation was made for the first 10 modal frequencies of the vibration. As per FEA, we obtained the natural frequency of vibration at first mode as 3.3798 Hz. This base frequency gets graded increment at higher modes with successive modal frequencies. The second modal frequency became 3.7728 Hz, which was 11.63% higher than the natural frequency of vibration. In fact, the first two modal frequencies have many commonalities so far as the control system algorithm is concerned. The real vigor of this vibration is felt from third modal frequency onward, which shows a distinct quantum jump in comparison to the first two frequencies. The trend is same for the fourth mode of vibration as well; the frequencies of vibration for third & fourth mode are 20.063 Hz & 20.806 Hz. respectively. Severe jerking of the flexible manipulator system is observed during excitation through fifth mode of vibration. This modal frequency (36.24 Hz.) is 74% higher than its immediately preceding frequency and by virtue, it creates sufficiently large jitter in the system along with its immediate next modal frequency, i.e. sixth (39.85 Hz.). The buckling of the flexible links begins with severe trembling of the ensemble system during seventh & eighth modal frequencies. As a matter of fact, these modal frequencies are to be avoided during operation to the best possible extent in order to make the end-point control semantics smooth. Computational values of seventh & eighth modal frequencies for the prototype FRS were 59.796 Hz. & 60.016 Hz. respectively, which are nearly 51% higher than sixth modal frequency. The last two modes of vibration are extremely critical from the stand-point of control system and if invoked, those may lead to complete catastrophe in extreme moments. In our prototype, these two frequencies, namely the nineth & tenth, are 102.14 Hz. & 105.95 Hz respectively. It may be noted that the tenth modal frequency has the largest step-up in comparison to its previous tuple of fifth-sixth mode, which is nearly 76.5%. Hence, all these higher-order modes of vibration, especially fifth mode onward are crucial from the control system aspect that calls for dedicated algorithm to harness system vibration. All of these data on modal frequency become helpful in assessing the ensemble dynamic behaviour of the FRS at a certain frequency.

The in-situ vibration is characterized through measure / estimation of the curvature of the neutral axis of the flexible links (refer fig 1). This curvature can be of two types, depending on the severity of the vibration, especially the higher modes. The first type of curvature is essentially a single curve that typically corresponds to the first and second modal frequencies. During this phase, the bending of the neutral axis of the links is mild and the resultant deflection is manageable from the perspective of the control system architecture. On the contrary, the higher modes of vibration, namely third mode onward, do get 'double buldge' of the neutral axis during vibration-induced deflection. The phenomenon of double buldge is nothing but the combined effect of twisting & bending of the flexible link(s).

Figure 15 shows the prototype FRS having pattern of the deformation under first mode of vibration (modal frequency of 3.3798Hz), as envisaged in FEA. It may be noted that the ensemble deformation pattern under first mode of vibration, i.e., at the natural frequency follow non-linear deflection. Due to the payload at third link, the deflection is more at third link zone, however, it is considerable at second link too. The natural frequency of vibration can be controlled at second link, but comparison of the frequency with other natural frequency is also important.

Figure 15. First Mode of Vibration and Deformation Pattern of the Prototype FRS

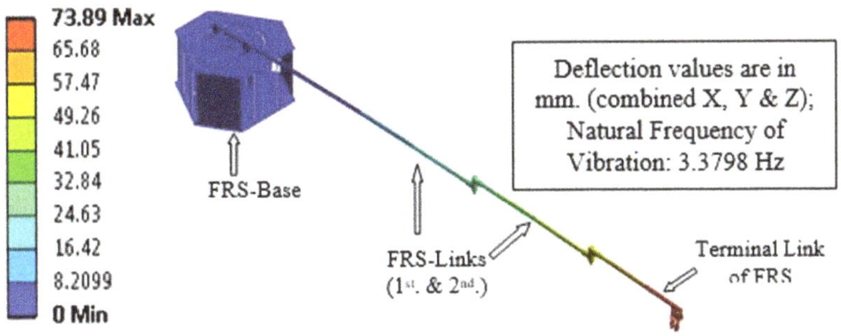

It is very important to note that we will always try to restrict the level of in-situ deflection (due to modal vibration) within third mode of vibration due to complications in control system paradigms. The ensemble system will be highly non-linear and quite unmanageable beyond third mode of vibration and the controller may not function properly. This sort of situation will be undesirable under any circumstances, because no effective utilization of the FRS can be made, e.g., successful grasp of objects by the miniature gripper (of PARv1.0). It is not necessary that maximum value of the in-situ deflection will increase with higher modes of vibration. For example, the maximum deflection of 73.89 mm. occurring at the terminal link in course of vibration at the natural frequency does not certify that higher value of deflection will take place at say second or third mode of vibration. The amount of in-situ deflection is coupled with non-linear dynamics of the system that in general, it is the 'spread' of the deflection that takes the pie at higher modes of vibration. Hence, the deformation pattern of the FRS gets altered to some extent under next higher modes of vibration, i.e., second & Third modes. While maximum deformation values get reduced, the effect of deformation becomes more spreading over the links during these modes of vibration. Figure 16 presents the FEA-screenshot of the pattern of the deflection in FRS under third mode of vibration. The screenshot clearly shows the said 'spread', wherein maximum value of deflection (~ 66.53 mm.) is occurring at two different locations.

Figure 16. Third Mode of Vibration and Corresponding Deformation Pattern of the Prototype FRS

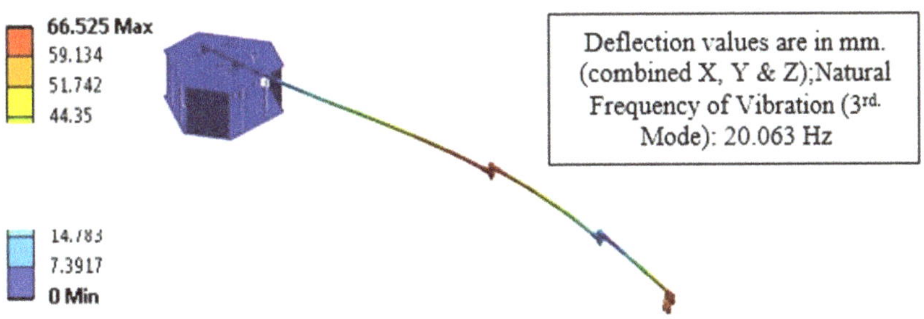

Control Semantics of In-situ Vibration

It is important to note that in-situ vibration in FRS ensemble manifest in several perpetual cycles of trembling~ from mild to severe. These cycles are not correlated to one another and the maximum amplitude of any cycle, say, $(\lambda_{max})^k$ (for $K^{th.}$ cycle: refer Figure 14) does vary between the cycles. Likewise, the final settling value of a particular pattern of vibration under a specific cycle, e.g. $(\lambda_{final})^k$ will not be the same amongst the cycles. The main facet of this in-situ vibration is its perpetuality that ensures continuity of successive cycles of small vibrations. Nevertheless, the final settling value of the vibration, i.e., the final deflection of the FRS-link will be the terminal settling value of the amplitude for the last cycle of vibration. The logic of the control system of FRS towards combating this perpetual vibration is based on attenuating the final settling value of the amplitude as well as reducing number of cycles.

It is to be noted here that deflection of the system is apparent even at a frequency lower than the natural frequency of vibration, e.g., 3 Hz. It is interesting to note that this sort of sub-modal frequency of vibration can cause some deflection in the system, although not very significant. Although deviation from natural frequency of vibration is very small [(3.3798-3.0)→ ~ 0.38 Hz.], yet it is always better to observe the deflection pattern of the FRS-links under such sub-modal frequency of vibration for fine-tuning the controller as a whole.

Figure 17 illustrates the FEA-plot of the graded deflection of the prototype FRS at 3 Hz. The accumulated 'displacements' at various zones of the FRS are Colour-coded with numerical average values shown in mm. The data of Figure 17 can act as an *upper bound* for the ensemble deflection of the FRS, which will be equally important for the evaluation of the gains of the amplifiers of the system controller optimally.

Figure 17. FEA-Screenshot Showing Upper Bound Deflection Values of the Prototype FRS at 3 Hz.

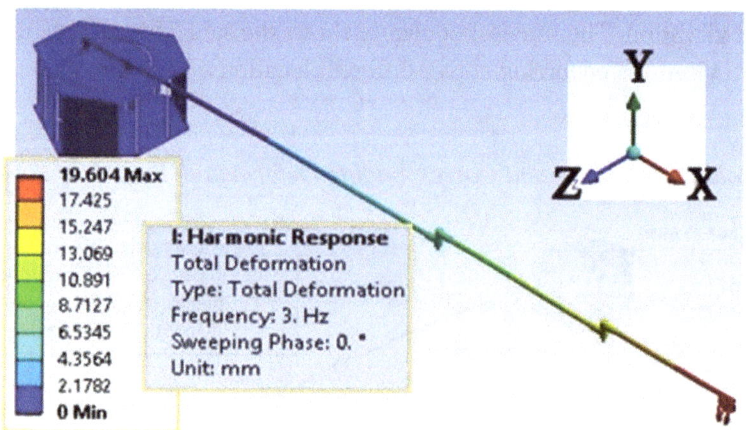

It may be stated here that the deformation in the ensemble FRS was obtained by FEA using *frequency response* technique. The modal analysis, carried out for our prototype, was a pre-requisite for the frequency response technique. The analysis was carried out considering actual load of 800 gram at the end of the third. link. Ideally, we need to consider the effect of damping in frequency response analysis, while performing FEA simulation in real-time. However, for upper bound analysis (refer Figure 17), the deflection characteristics were studied without considering the system damping. Nonetheless, it is important to accommodate damping in the system for unaccurate simulation. Accordingly, we have considered 2% damping in the FEA, in order to get more authentic data on the modal frequencies. In this context, we should also note that the predominant frequency of the system remained unaltered throughout the response analysis as there was no change in the phase angle.

Extending the concept of amplitude analysis as shown schematically in Figure 14, we have studied the variation of peak amplitude of the system with change in natural frequency of vibration. Figure 18 shows the said graph of amplitude vs. frequency, wherein the system was found to be moderately stable in the range of higher frequencies (approx. 90-120 Hz.). It may be noted that we have taken a safe upper threshold of 120 Hz for the maximum modal frequency, although 10th. modal frequency was found to be nearly 106 Hz. as per FEA simulation of the prototype FRS.

Figure 18. Amplitude vs. Frequency of Vibration of the Prototype FRS as per FEA Simulation

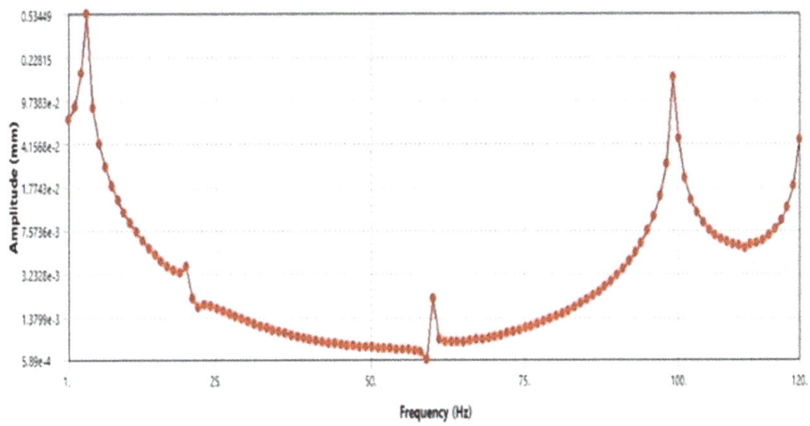

Figure 19. FEA Simulation-based Amplitude vs. Frequency of Vibration of the Prototype FRS at 2% Damping

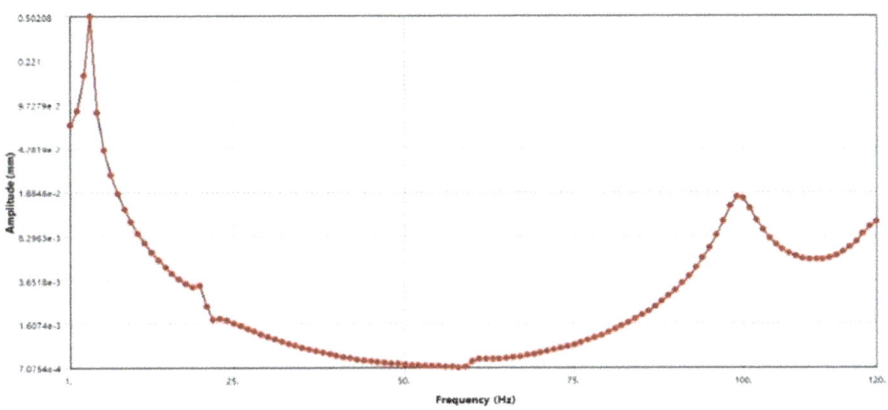

Figure 19 shows the similar plot of amplitude vs. frequency with 2% damping, which shows distinct improvement in the stability in the higher range of frequencies.

Variation in amplitude of vibration was studied under a frequency range of 1 to 120 Hz for both the plots at a spacing of 25 Hz, within an amplitude range of 5.89×10^{-4}mm to 0.53449 mm (for Figure 18) and 7.0754×10^{-4}mm to 0.50208 mm. (for Figure 19).

With this backdrop, control semantics of in-situ vibration of the prototype FRS was designed and developed in a complete indigenous manner, taking insight from FEA-based deformation analysis on modal frequencies.

CONTROL SYSTEM LOGIC AND DESIGN OF THE DEVICE CONTROLLER

The design and firmware of the device controller of the prototype FRS has been orchestrated by the joint tuple of kinematic modeling and customized programming of the AVR microcontroller. We shall now focus on the indigenously developed controller of the prototype FRS, by addressing the logistics of the said control system.

Incorporation of Kinematic Model for the Design Syntax

As elucidated earlier, we have used homogeneous transformation for defining the final kinematic model of our prototype planar flexible manipulator using D-H matrix. The said kinematic model was conceptualized as chain of semi-rigid bodies; namely, the semi-compliant links that are connected by revolute joints. The reference co-ordinate frame of the FRS-link(s) may revolve due to angular rotation and linear translation of the links, resulting in total spatial displacement of the end-effector. The systematic and generalized approach of D-H formulation helped us to find out an optimal measure for utilizing matrix algebra to describe the spatial geometry of the flexible links with respect to the fixed reference frame (located at the base of the prototype FRS). Accordingly, we have used a 4x4 homogeneous transformation matrix in order to describe the spatial relationship between two adjacent semi-rigid mechanical links that culminate in the desired forward kinematics of the prototype FRS. The evaluated D-H parameters will give the relative position and orientation of the links, considering Z-axis is in the direction of the joint axis and X-axis is parallel to the common normal. Figure 20 shows the kinematic model scheme of the prototype 3 d.o.f. flexible robot with its D-H matrix. Here a_i symbolizes the shortest distance between Z_{i-1}& Z_i axes and θ_i is defined as the angle from X_{i-1} to X_i along Z_{i-1}. The kinematic model also represents twist of the links, viz. angle (α_i) from Z_{i-1} to Z_i along $X_{i.}$ and link offset (d_i), viz. the distance from the intersection of Z_{i-1} with X_i to the origin of (i-1) $^{th.}$ co-ordinate system.

Once the position, velocity and acceleration for each joint are known, computation for the corresponding variables for the FRS-end-effecter are taken up in a given reference frame. By forward kinematic model, a function gets defined between the joint space J and work space W, such that, x= f(q) where x € J, q € W. Figure 21a presents the simulation results of the developed kinematic model of the FRS at 0° angle at all joints. Likewise, the simulation result on the developed kinematic model of the FRS at 90° is shown in Figure 21b.

Figure 20. Kinematic Model of the Prototype Flexible Robot with D-H Parameters

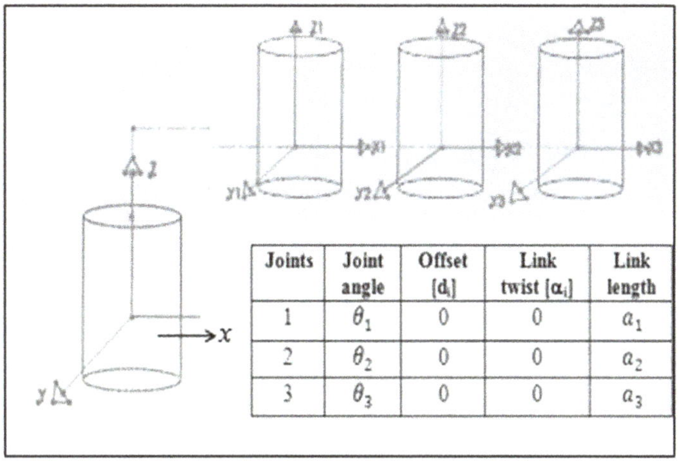

Joints	Joint angle	Offset [d_i]	Link twist [α_i]	Link length
1	θ_1	0	0	a_1
2	θ_2	0	0	a_2
3	θ_3	0	0	a_3

Figure 21. Simulation Results of the Kinematic Model of the Prototype FRS at: [a] 0° Angle & [b] 90° Angle for all Joints

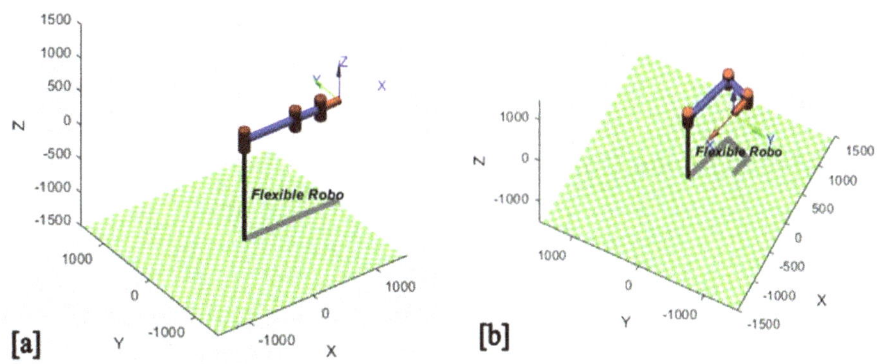

As part of the inverse kinematics protocol, we carried out computation of the relevant variables (position, velocity, acceleration) pertaining to each joint of the prototype FRS. This computation was essentially accomplished from the work space to the joint space and was tackled by the indigenously developed inverse kinematic model, viz:

g = $f^{1}$from J to W.

q = g(x) = f^{1} (x)

Figure 22 illustrates the simulation results of the inverse kinematics model of the developed FRS.

Figure 22. Simulation Results of the Inverse Kinematic Model of the Prototype FRS

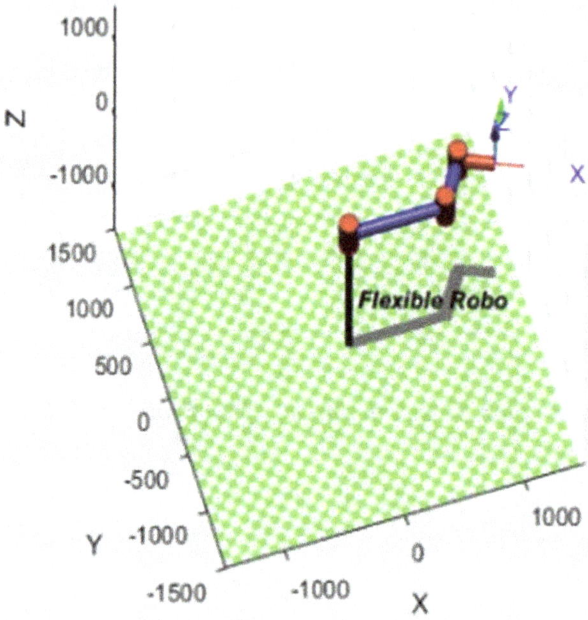

Programming of the Device Controller

Central Semantics

The methodology of '*Input Shaping*' has been invoked for designing the customized control system architecture of the prototype FRS. This is essentially a feed forward control technique used for harnessing the residual vibration of flexible systems, in general. Simulation on this basic control system layout of the prototype FRS was performed using MATLAB® software. The generalized block diagram of input shaping control configuration is shown in Figure 23, which has been primarily utilized in our controller.

Figure 23. Block Diagram of a Generic Control System Using Input Shaping Technique

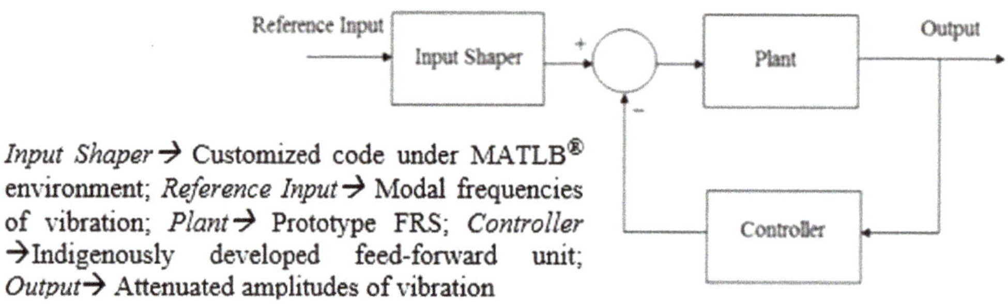

The paradigms of Input Shaping were utilized primarily to tune all the ten modal frequencies of vibration of the prototype FRS, as obtained via FEA. These are indexed as '*Reference Input*' in Figure 23. Indigenous and customized code was written in order to invoke input shaping over these modal frequencies. This code, running under MATLAB® environment, has been referred as '*Input Shaper*' in the block diagram. The '*Input Shaper*' module works hand-in-hand with '*Reference Input*' towards better authentication of the run-time vibration signature of the prototype FRS, which is indexed as '*Plant*' in Figure 23. The basic idea of introducing input shaping is to reduce the amplitude of these vibrations, without altering the phase angle. In other words, our customized code for the input shaping takes care of reducing the maximum amplitude under each modal frequency of vibration, under each command cycle for the activation of the joint-angle(s). Since the design of controller of the FRS is for feed-forward actuation exploiting both forward as well as inverse kinematics, it is of utmost importance that we take a robust stand for the positional accuracy of the co-ordinates of the joint(s). It is important to note that attenuated amplitudes of vibration (referred as '*Output*' in Figure 23) gets fed back again in the '*Controller*' as error quotient in each run-time so as to produce more refined and fine-tuned output after being processed through the '*Plant*'. This very conceptual layout is the novelty of our controller, wherein we have tackled both feed-forward motion of the prototype FRS as well as improvement over the in-situ vibration in the form of harnessing the modal frequencies via input shaping.

Naturally, our comprehensive program-code on input shaping has served the purpose of minimization of the error between the theoretically-computed co-ordinates (using D-H matrix) of the set-point and the run-time (experimental) co-ordinates of the same.

Programming Semantics

The indigenously designed & developed controller of the prototype FRS has been orchestrated by the customized programming of the AVR microcontroller. The design as well as computation for the FRS-control system, are vital to use two programming environments in conjunction, namely, MATLABR2018a® and Arduino™ IDE. Accordingly, the programming for the prototype FRS was done in two segments: the first section dealt with complex mathematical computations in the time and frequency domain for *input shaping* while the second part addressed the implementation of the coding algorithm to deploy the extracted parameters to the actuators for intricate movements. The first section was programmed using MATLAB R2018a®, which is a dynamic tool for computation. Figure 24 shows the flowchart of the program used for the controller of the prototype FRS. It may be stated here that the ensemble programming was attributed to two inter-linked modules of coding exercise, viz. a] kinematic code & b] input shaping code.

Mathematical Model of Input Shaping

Generally, a system with perpetual vibration of any order can be considered as a superposition of second order systems having transfer function, such as:

$$G(s)=\frac{\omega^2}{s^2+2\zeta\omega s+\omega^2} \tag{1}$$

Figure 24. Flowchart for the Programming of the Controller of the Prototype FRS

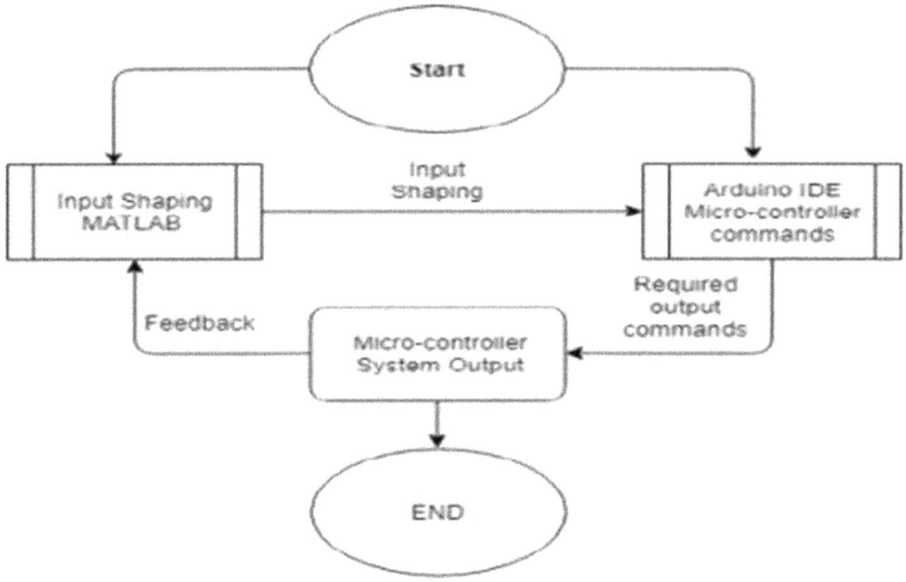

where, 'ω' is the natural frequency and 'ζ' is the damping ratio of the vibrating system. The response of the system in time domain can be represented as,

$$y(t)=\frac{A\omega}{\sqrt{1-\zeta^2}}exp^{-\zeta\omega(t-t_0)}\sin\left(\omega(t-t_0)\sqrt{1-\zeta^2}\right) \tag{2}$$

where, 'A' and 't_0' are amplitude and time location of impulse respectively. As response of sequence is nothing but the superposition of the impulse responses, for 'n' number of responses, with damped frequency of vibration, $\omega_d=\omega\left(\sqrt{1-\zeta^2}\right)$, impulse response is given by,

$$y(t) = M\sin(\omega_d t + \beta) \tag{3}$$

Considering the series of impulse (i=1, 2, ...n), we get,

$$M=\sqrt{\left\{\sum_{i=1}^{n}B_i\cos\varphi_i\right\}^2+\left\{\sum_{i=1}^{n}B_i\sin\varphi_i\right\}^2} \tag{4a}$$

$$B_i=\frac{\omega A_i}{\sqrt{1-\zeta^2}}exp^{-\zeta\omega(t-t_0)},\ \varphi_i=\omega_d t_i \tag{4b}$$

Equations 4a & 4b are instrumental in the evaluation of the frequency of vibration under the application of external forcing function, with specific importance on two parameters, namely, A_i and t_i, which are the amplitudes and time locations of the impulses respectively. Now, in order to obtain zero vibration at the end of the terminal impulse in an ideal real-time situation, the necessary analytical requirement is related to the two terms of eqn. 4a, viz. $\left\{\sum_{i=1}^{n} B_i \cos\varphi_i\right\}^2$ (say, V1) and $\left\{\sum_{i=1}^{n} B_i \sin\varphi_i\right\}^2$ (say, V2) that are independently be zero. In order to produce similar rigid body motion as unshaped reference command, sum of the amplitudes of the impulses is required to be unity, i.e.

$$\sum_{i=1}^{n} A_i = 1 \tag{5}$$

For improvement in the system response, the first impulse is selected at time t_1(-0) and last impulse is at t_n.

The robustness of the FRS-controller can be improved by equating derivatives of V1 and V2 to zero. The robustness of the errors in natural frequencies of the system can be refined further by enhancing the order of the derivative thereof. After solving the derivatives for three consecutive sequences of impulse, we are able to find the following results:

$$t_1 = 0,\ t_2 = \frac{\pi}{\omega_d},\ t_3 = \frac{2\pi}{\omega_d}$$

$$A_1 = \frac{1}{1+2K+K^2},\ A_2 = \frac{2K}{1+2K+K^2},\ A_3 = \frac{1}{1+2K+K^2}$$

where K is,

$$K = e^{-\zeta\pi/\sqrt{1-\zeta^2}} \tag{6}$$

Simulation Results

It may be stated that the numerical values of the natural frequencies of vibration of all the 10 modes were taken into consideration during the prototype FRS control system development. The impulse and step responses were obtained using MATLAB R2018a® simulation environment. Figures 25a, b shows the impulse response of the prototype FRS controller before and after *Input Shaping*, against input from ten modal frequencies.

The effectivity of invoking input shaping becomes prominent when we supply the input amplitude (of the modal vibration) in the form of a pre-assigned step response. The trembling of the FRS-link, from slow to high, is a near-perfect match for step response. Figures 26a, b illustrates step response of the prototype FRS controller before & after Input Shaping. We can very well observe that the usual upshoot of amplitude (between 40 to 45 secs) of the step response prior to invocation of input shaping gets successfully reversed to a significantly low amplitude, post input shaping. Hence, these plots not only render an idea of the vibrational amplitudes of the prototype FRS with respect to time but also certify the usefulness of our novel program for using the technique of input shaping.

Figure 25. Impulse Response of the Prototype FRS: [a] before and [b] after Input Shaping

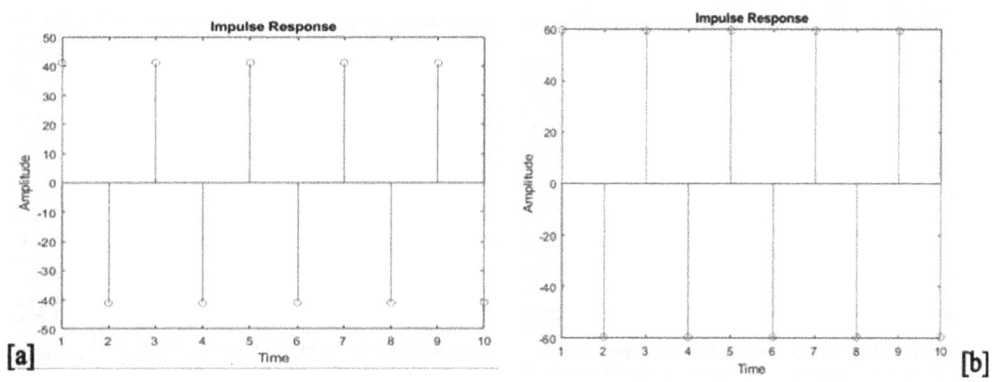

Figure 26. Step Response of the Prototype FRS: [a] before and [b] after Input Shaping

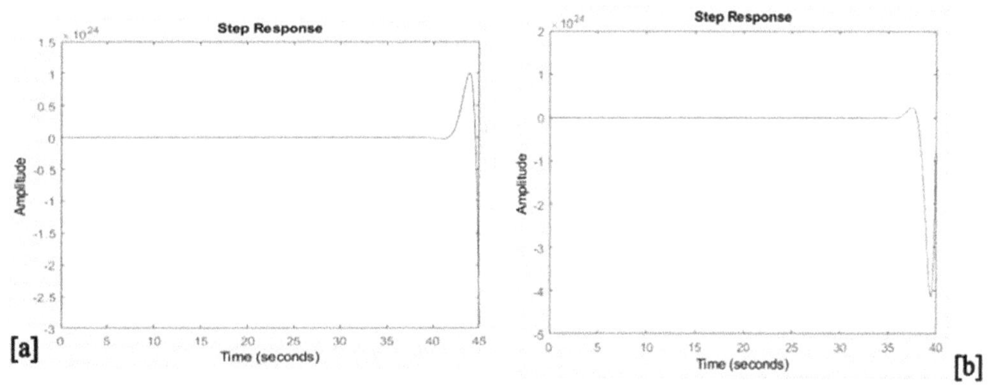

REAL-TIME PROGRAMMING AND DEVELOPMENT OF GRAPHICAL USER INTERFACE OF THE CONTROLLER

Real-Time Programming of the Device

Indigenous programming was invoked in order to harness the technology of input shaping towards mitigating run-time vibration of the prototype FRS to a significant extent. Although quantum of dampening of vibration may not be ascertained precisely, yet we can have moderately precise control over the force closure of the grasp by the end-effector of the FRS.

Figure 27 shows the screenshot of the relevant program-code in MATLAB® for invoking Input Shaping. The representative sample of the microcontroller program, coded through Arduino IDE™, has been illustrated in Figure 28. It may be noted that programming semantics of each peripheral of the overall control system logic are essentially divided segment-wise, of which a major contribution arises from the serial communication protocol in MATLAB®. In fact, serial communication is an essential part for the ensemble logic of the controller as it helps MATLAB® software and Arduino Mega™ communicate with each other. Serial communication in MATLAB® is established by using serial port interface feature of MATLAB®, which lets the user read a stream of ASCII terminated data from an ARDUINO. A software-check of the available serial ports on the computer a-priori gives a quick gateway for establishing the interface with the Arduino Mega™ by creating *serial port object* under a pre-assigned baud rate.

Figure 27. Screenshot of the MATLAB® Program Code for Input Shaping

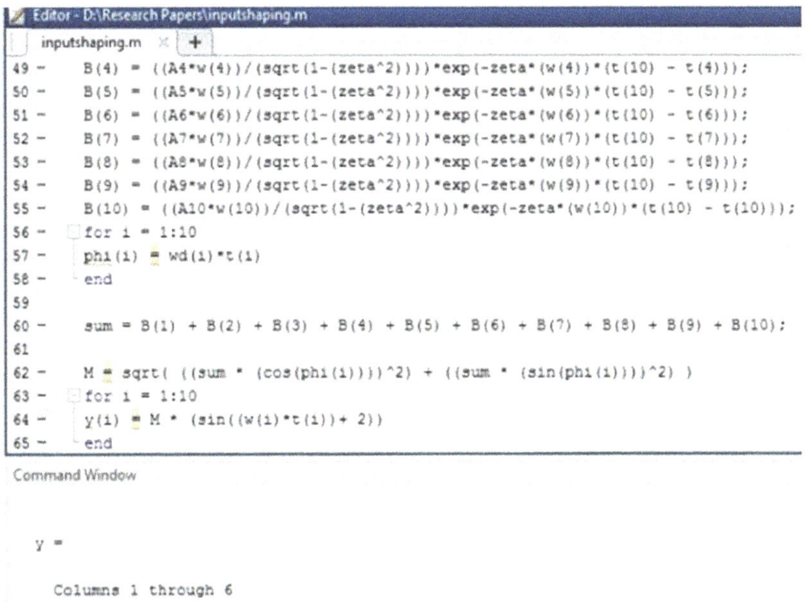

For setting up serial connection between MATLAB® and Arduino™, there are two methods, viz. 'Command window' and MATLAB GUI. For both the technique, same Arduino™ code can be used. However, in our case, MATLAB GUI has been utilized for the serial communication.

Development of the Graphical User Interface

A technology-enriched Graphical User Interface (GUI) was developed in order to provide ease of operation of the device controller for running the prototype FRS. We have introduced a customized GUI-based application for controlling the prototype FRS in real-time. In order to keep programming tools minimal and to provide dynamic control over prototype FRS, GUI was composed using MATLAB R2018a®. Figure 29 shows the block diagram of the GUI for our prototype FRS.

Figure 28. Screenshot of Microcontroller Program in Arduino™ IDE

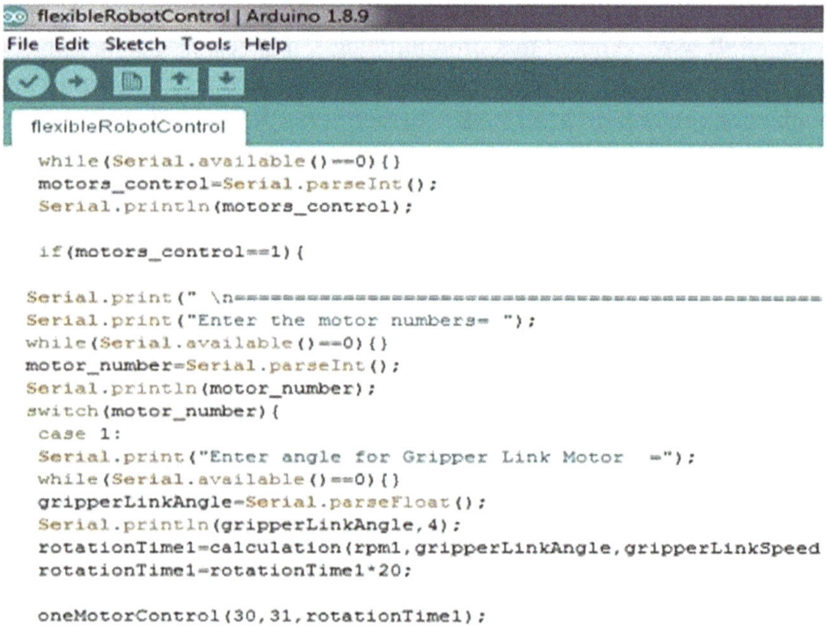

Figure 29. Block Diagram for Invoking the GUI of the Prototype FRS

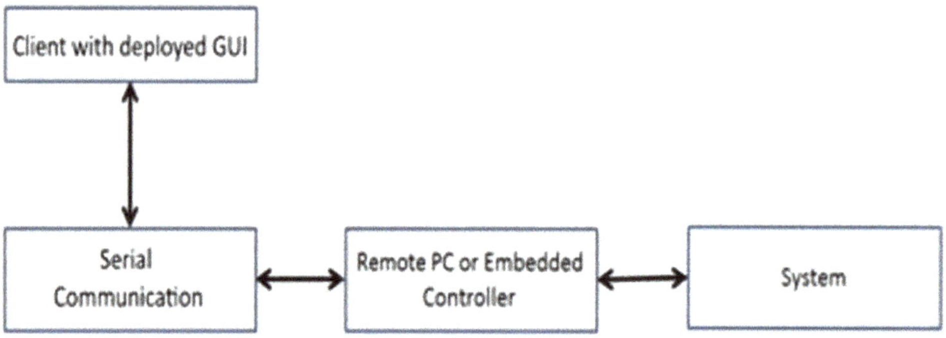

The GUI is the user interface via which the system receives input from the user in the form of X, Y, Z coordinates or joint angles. The GUI consists of two sections for preforming the forward kinematics and inverse kinematics. Both sections are capable of accepting inputs from user, which are processed in MATLAB R2018a® that provides output in the form of angles that can be passed to Arduino™ Idea's variables. Figure 30 shows the screenshot of developed GUI for the prototype FRS.

As reported earlier, the two components of the developed GUI are 'forward kinematics' and 'inverse kinematics'. Both components may accept user inputs, which are analyzed in MATLAB R2018a® to generate the joint angles required to achieve the desired position, which are further sent to the Arduino IDE as a *String*. In this situation, the serial communication protocol employed is UART (Universal Asynchronous Receiver-Transmitter).

Figure 30. Screenshot of the GUI Developed for the Prototype FRS

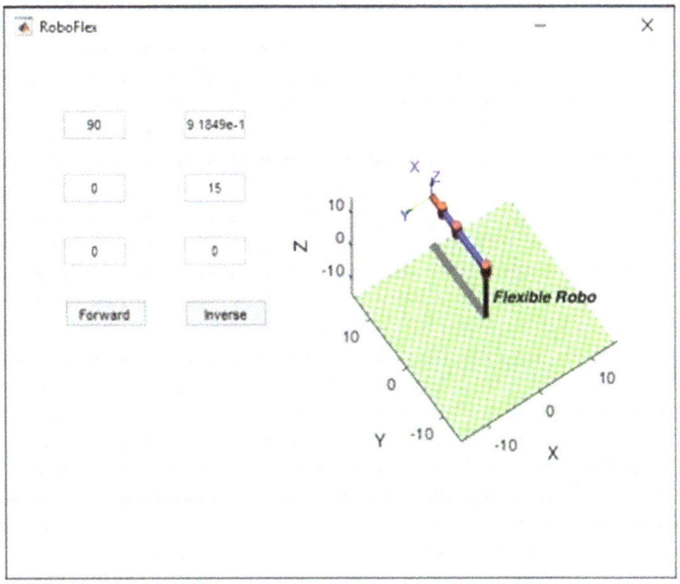

The MATLAB R2018a® software at the backend will process the inputs and run those through the inverse kinematics (*Input in the form of X, Y, Z*) or forward kinematics algorithm, depending on the inputs received (*Input in form of Joint angles*). These algorithms are resource heavy and need a lot of processing power, hence those can't be run on Arduino boards with little computing power. To meet this need, we used a computer to perform the algorithm and then send the output to the Arduino board through serial connection. The Arduino application is intended to recognize this information and issue commands to move the joints accordingly.

HARDWARE OF THE CONTROLLER OF THE PROTOTYPE FLEXIBLE ROBOT

Indigenous design of the ensemble control circuit, instrumentation and interfacing of modules are the key facets of the controller hardware of the prototype FRS. The circuit design was finalized taking into consideration of the functionality, operations, architecture and features of each device of the controller. Based on the syntax of the indigenous computer program, the hardware of the developed controller consists of serial communicator module and microcontroller module. The controller for the prototype FRS was built up using *National Instruments®* myRIO®, which provided dynamic interface with DC servomotors and motor drivers of the prototype FRS. The myRIO® operates on Linux®-based operating system that gives way for specific operations such as multi-threading and multi-tasking. Besides, it has built-in serial communication module, which reduces the number of modules in the system.

The testing hardware of the prototype FRS is designed using ATMega2560™ microcontroller and Arduino Mega built-in serial communication module. Arduino Mega™ is a programmable digital module, which incorporates the function of the central processing unit on a single Printed Circuit Board. ATMega2560™ has 54 digital input-output(I/O) pins of which 15 can produce high speed pulses for pulse width modulation (PWM). ATMega2560™ works with ~16MHz clock speed that allows high speed data acquisition. Figure31 illustrates the block diagram of the hardware architecture of the controller of the prototype FRS.

It may be noted that the actuators driving the links of the prototype FRS consist of DC servomotors. The motors were selected as per the torque and velocity requirement of the system. Notwithstanding the efficiency of the servomotors, the residual vibration in the system needs to be suppressed by the control system ensemble. Motor driver modules were selected optimally so as to provide a perfect interface between the motor and microcontroller unit. Power supply of the prototype FRS is a switched mode power supply with the voltage amplitude of 12V DC and maximum current of 10A. Figure 32 shows the photographic view of the control system implemented on real-time prototype of the planar three-link FRS.

Figure 31. Block Diagram of the Hardware Architecture of Prototype FRS

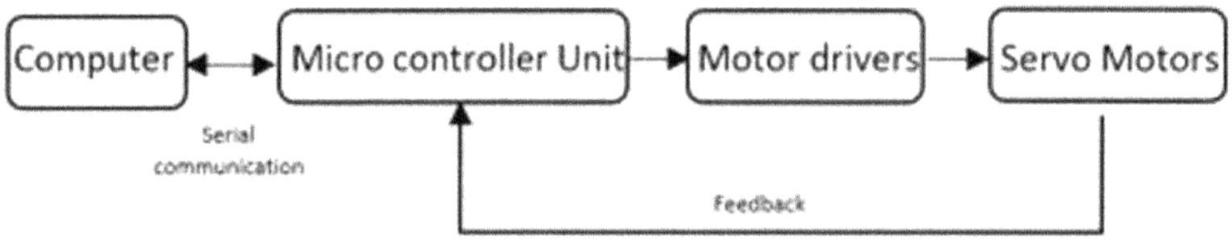

Figure 32. Real-time Implementation of Hardware Architecture of the Prototype FRS

Finite element analysis was used in conjunction with the CAD model to exhibit the real-time development of the test-beds and finally the prototype of the FRS. The induced vibration was experimented and validated using analytical solutions. The influence of the electronic control system hardware over electromechanical output of the FRS was studied in depth in real-time. The electronic sub-system of the prototype FRS was explored in detail along with implementation of hardware control through a series of codes. Motor driver modules were selected optimally so as to provide a perfect interface between the joint-specific DC servomotor and microcontroller unit. Indigenous design of the ensemble control circuit, instrumentation and interfacing of modules are the key facets of the controller hardware of the prototype FRS. The circuit design was finalized taking into consideration the functionality, operations, architecture and features of each device of the controller.

The design function architecture of the prototype multi-degrees-of-freedom FRS, highlighted herein, is a rare combination of three revolute & one prismatic (RRRP) joints intended for pick and place applications, having 1500 mm radial span reach in horizontal plane. The prototype *PARv1.0* has ball screw driven translational motion (with three linear bearing support in vertical plane Z) and spur gear driven rotary mechanism (for radial span reach in horizontal plane XY). The FRS has three identical cross-sectioned links of unequal length, made up of carbon fiber reinforced plastic, that cover the 1500 reach span (in XY plane). These links are driven by novel worm-worm gear-based miniature joints, which reduce the backlash as well as harness the randomized vibrations to a significant extent. The servomotors for joint actuation are fitted in the robot top canopy and the rotational motion generated thereof is transferred to the joint-shaft(s) via flexible shafts, designed & manufactured indigenously. The incorporation of our novel cost-effective flexible shafts helps reducing the overhung weight on the FRS-links, which has been a crucial aspect in this research although. However, on the flip side, these flexible shafts, by nature, boost up the vibration in the system and cause loss in motion transfer to a certain extent. A prototype miniature gripper, actuated by a mini-servomotor, is another novel development of this FRS, since the overall envelope of the mini gripper is 50 mm x 40 mm x 15 mm only.

The developed hardware test-bed setup of the planar three-link flexible robotic system (*PARv1.0*) is shown in Figure 33. The feed-forward command-centric motion sequences as well as grasp performance of the fully working prototype of PARv1.0 can be viewed at: https://www.youtube.com/watch?v=Nwbt94E8Tds&t=98s

Interestingly, our research reveals that the fundamental frequency of vibration has the same deflection contour as that of static deflection. Thus, the amplitude of vibration can cause the prototype FRS to vibrate more and can cause severe problem in the ensemble design. The inherent vibration occurring due to post-manufacturing compliance of FRS provides numerous challenges to achieve accuracy at the end-effector, which were addressed and handled optimally in our hardware through different levels of software-control strategies.

Figure 33. Developed Hardware Set-up of Complete Assembly of the Prototype FRS

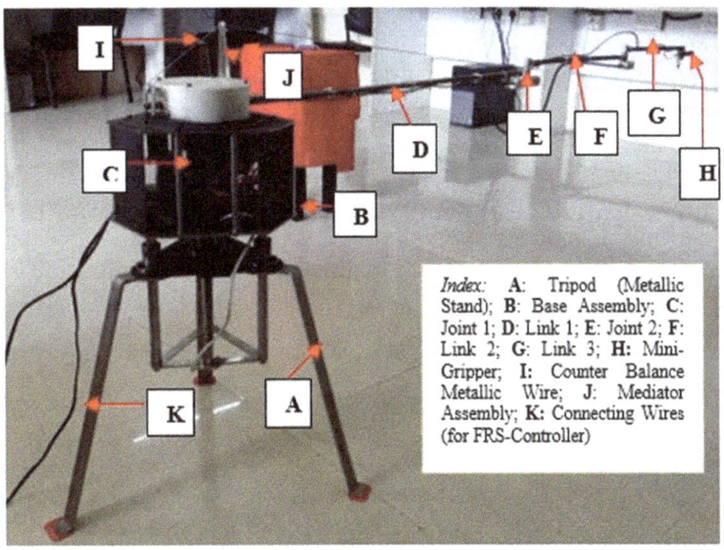

CONCLUSION

Design and prototype development of an indigenous robust control system for three-link serial flexible robotic system have been discussed in this chapter. The investigation of inherent vibration, its nature, natural frequencies and harnessing of amplitudes were studied under various modes of vibration. A novel design of control system was established in real-time including input shaper that was implemented to dampen the natural vibrational frequencies obtained using FE-simulation. We have investigated the design-basis of the FRS-controller for kinematics and implemented the firmware in an elegant way. Developed control system provides accuracy as well as stability to the prototype FRS. The research-quotient earned through modelling& hardware manifestation can be benchmarked. The next level of engineering sophistication of our indigenous controller will primarily include sensor feedback-based position control of the end-effector to compensate for the error and help to achieve the target position of the FRS so as to have a stable slip-free grasp. The software pathways as well as indigenous GUI have made the foundation for the firmware for the development of similar planar serial-chain flexible robots that can be deployed as assistive robot in near future.

ACKNOWLEDGMENT

We would like to express our heartfelt appreciation to Shri Shubham Chaudhari, Ex-CAE Support Engineer, SVR InfoTech, Pune, India for his valuable work on FE Analysis of the present research. Sincere appreciation is due to Shri Akshay Dalvi and Shri Amit Kumar Deka, Ex- Research Engineers, SVR Infotech for realizing some of the hardware issues of the developed controller and its software routines. We also thank Shri Pankaj Pandit, Shri Piyush Tailor, Shri Jaipal Solanki. Shri Kaifali Bhojani and Ms. Vibhavari Nirmale, Ex-Research Engineers, SVR Infotech for extending generous help in performing experiments with the FRS as well as run-time troubleshooting.

REFERENCES

Benosman, M., & Vey, G. (2004). Control of Flexible Manipulators: A Survey. *Robotica*, *22*(05), 533–545. doi:10.1017/S0263574703005642

Chun, H. M., Turner, J. D., & Juang, J.-N. (1985). Disturbance Accommodating Tracking Maneuvers of Flexible Spacecraft. *The Journal of the Astronautical Sciences*, *33*(2), 197–216.

Feliu, J. J., Feliu, V., & Cerrada, C. (1999). Load Adaptive Control of Single-Link Flexible Arms Based on a New Modeling Technique. *IEEE Transactions on Robotics and Automation*, *15*(5), 793–804. doi:10.1109/70.795785

Feliu, V., Somolinos, J. A., & García, A. (2003). Inverse Dynamics Based Control System for a Three-Degree-of-Freedom Flexible Arm. *IEEE Transactions on Robotics and Automation*, *19*(6), 1007–1014. doi:10.1109/TRA.2003.819596

Fraser, A. R., & Daniel, R. W. (1991). *Perturbation Techniques for Flexible Manipulator*. Kluwer. doi:10.1007/978-1-4615-3974-2

Gupta, N. K. (1980). Frequency-Shaped Cost Functionals: Extension of Linear-Quadratic-Gaussian Design Methods. *Journal of Guidance and Control*, *3*(6), 529–335. doi:10.2514/3.19722

Juang, J.-N., Turner, J. D., & Chun, H. M. (1985). Closed-Form Solutions for Feedback Control with Terminal Constraints. *Journal of Guidance, Control, and Dynamics*, *8*(1), 39–43. doi:10.2514/3.19932

Junkins, J. L., & Turner, J. D. (1986). *Optimal Spacecraft Rotational Maneuvers*. Elsevier Science Publishers.

Kotnik, P. T., Yurkovich, S., & Ozfuner, U. (1988). Acceleration Feedback for Control of a Flexible Manipulator Arm. *Journal of Robotic Systems*, *5*(3), 181–196. doi:10.1002/rob.4620050302

Lin, L. C., & Yih, T. W. (1996). Rigid Model-based Neural Network Control of Flexible-link Manipulators. *IEEE Transactions on Robotics and Automation*, *12*(4), 595–602. doi:10.1109/70.508442

Moallem, M., Patel, R., & Khorasani, K. (1997). An Inverse Dynamics Control Strategy for Tip Position Tracking of Flexible Multi-link Manipulators. *Journal of Robotic Systems*, *14*(9), 649–658. doi:10.1002/(SICI)1097-4563(199709)14:9<649::AID-ROB2>3.0.CO;2-L

Moudgal, V. G., Kwong, W. A., Passino, K. M., & Yurkovich, S. (1995). Fuzzy Learning Control for a Flexible-link Robot. *IEEE Transactions on Fuzzy Systems*, *3*(2), 199–210. doi:10.1109/91.388164

Pereirea, E., Aphale, S. S., Feliu, V., & Moheimani, S. O. R. (2011). Integral Resonant Control for Vibration Damping and Precise Tip-positioning of a Single-link Flexible Manipulator. IEEE/ASME Transactions on Mechatronics.

Rauniyar, A., Roy, D., Pandit, P., & Atpadkar, V. (2018). Design Model for the Drive & Actuator of the Test Set-up of a Novel Flexible Robotic System. *Proceedings of the 2018 IEEE International Conference on Computational Intelligence and Computing Research*. 10.1109/ICCIC.2018.8782388

Roy, D. (2019). Towards the Control of Inherent Vibration of Flexible Robotic Systems and Associated Dynamics: New Proposition and Model. *International Journal of Robotics Research, Applications and Automation*, *1*(1), 6–17.

Singer, N. C., & Seering, W. P. (1990). Preshaping Command Inputs to Reduce System Vibration. *Journal of Dynamic Systems, Measurement, and Control*, *112*(1), 76–82. doi:10.1115/1.2894142

Smith, O. J. M. (1958). *Feedback Control Systems*. McGraw-Hill.

Somolinos, J., Feliu, V., & Sánchez, L. (2002). Design, Dynamic Modeling and Experimental Validation of a New Three-Degree-of-Freedom Flexible Arm. *Mechatronics*, *12*(7), 919–948. doi:10.1016/S0957-4158(01)00033-2

Tjahyadi, H., & Sammut, K. (2006). Multi-mode Vibration Control of a Flexible Cantilever Beam using Adaptive Resonant Control. *Smart Materials and Structures*, *15*(2), 270–278. doi:10.1088/0964-1726/15/2/005

Trapero-Arenas, J. R., Mboup, M., Pereira-Gonzalez, E., & Feliu, V. (2008). Online Frequency and Damping Estimation in a Single-Link Flexible Manipulator based on Algebraic Identification. *Proceedings of the 16th. Mediterranean Conference on Control & Automation*.

Warude, P., Patel, M., Pandit, P., Patil, V., Pawar, H., Nate, C., Gajlekar, S., Atpadkar, V., & Roy, D. (2019). *On the Design and Vibration Analysis of a Three-Link Flexible Robot Interfaced with a Mini-Gripper. In Lecture Notes in Mechanical Engineering: Recent Trends in Wave Mechanics and Vibrations*. Springer.

Wen, C. (2001). Dynamic Modeling of Multi-link Flexible Robotic Manipulators. *Computers & Structures*, *79*(2), 183–195. doi:10.1016/S0045-7949(00)00129-2

Yim, W., & Singh, S. N. (1996). Predictive End-point Trajectory Control of Elastic Manipulators. *Journal of Robotic Systems*, *13*(9), 561–569. doi:10.1002/(SICI)1097-4563(199609)13:9<561::AID-ROB1>3.0.CO;2-K

Zhang, J., Tian, Y., & Zhang, M. (2014). Dynamic Model and Simulation of Flexible Manipulator based on Spring & Rigid Bodies (2460-2464). In *Proceedings of the 2014 IEEE International Conference on Robotics & Biomimetics*. 10.1109/ROBIO.2014.7090709

KEY TERMS AND DEFINITIONS

Flexible Robot: A robot that has flexibility in its operation and/or links Framed of with flexible material.

In-Situ Vibration: The vibration of the robotic system that is impromptu and instantaneous in nature.

Input Shaping: A control design technique for fine-tuning input signal/parameter of a controller.

Inverse Kinematics: Analytical procedure to compute joint co-ordinates from a pre-assigned end-point co-ordinates of a robotic system.

Modal Frequency: Any frequency of vibration that is higher than natural frequency of that system.

Rheology: Study on stress and strain of a mechanical/mechatronic system.

Rotary Joint: A joint of the robot that allows rotational movement around a specific axis.

Trajectories: The pre-computed paths of a robot, based on a known position of the target (e.g., gripper).

ENDNOTE

1 The prototype is christened as: Patient Assistant Robot version 1.0.

Compilation of References

Abe, Y., Fujimoto, T., Nakatani, M., Shigeta, M., & Tanaka, M. (2021). Study on Proper Welding Condition for Ultra-narrow Gap Submerged Arc Welding. *Welding International*, *35*(7-9), 369–381. doi:10.1080/09507116.2021.1980298

Abo-Hammour, Z. S., Alkayyali, Khasawneh, M. J., Hussam, I., Al Saadeh, Mohammed (2018). On the design of the integer and fractional PID controller using particle swarm optimization, In *Proceedings of International Conference on Fractional Differentiation and its Applications (ICFDA)*. 10.2139srn.3277839

Aceto, P., Beretta, L., Cariello, C., Claroni, C., Esposito, C., Forastiere, E. M., Guarracino, F., Perucca, R., Romagnoli, S., Sollazzi, L., Cela, V., Ercoli, A., Scambia, G., Vizza, E., Ludovico, G. M., Sacco, E., Vespasiani, G., Scudeller, L., & Corcione, A. (2019). Joint consensus on anesthesia in urologic and gynecologic robotic surgery: Specific issues in management from a task force of the SIAARTI, SIGO, and SIU. *Minerva Anestesiologica*, *85*(8), 871–885. doi:10.23736/S0375-9393.19.13360-3 PMID:30938121

Advincula, A. P. (2006). Surgical techniques: Robot-assisted laparoscopic hysterectomy with the da Vinci surgical system. *International Journal of Medical Robotics and Computer Assisted Surgery*, *2*(4), 305–311. doi:10.1002/rcs.111 PMID:17520648

Advincula, A. P., & Song, A. (2007). The role of robotic surgery in gynecology. *Current Opinion in Obstetrics & Gynecology*, *19*(4), 331–336. doi:10.1097/GCO.0b013e328216f90b PMID:17625414

Advincula, A. P., & Wang, K. (2009). Evolving role and current state of robotics in minimally invasive gynecologic surgery. *Journal of Minimally Invasive Gynecology*, *16*(3), 291–301. doi:10.1016/j.jmig.2009.03.003 PMID:19423061

Agdi, M., & Tulandi, T. (2010). Minimally invasive approach for myomectomy. *Seminars in Reproductive Medicine*, *28*(03), 228–234. doi:10.1055-0030-1251479 PMID:20414845

Agrawal, S. S., Prakash, N., & Jain, A. (2010). Transformation of emotion based on acoustic features of intonation patterns for Hindi speech. *African Journal of Mathematics and Computer Science Research*, *3*(10), 255–266.

Ahuja, A. S. (2019). The impact of artificial intelligence in medicine on the future role of the physician. *PeerJ*, *7*, e7702. doi:10.7717/peerj.7702 PMID:31592346

Al-Akash, M., Boyle, E., & Tanner, W. A. (2009). N.O.T.E.S.: The progression of a novel and emerging technique. *Surgical Oncology*, *18*(2), 95–103. doi:10.1016/j.suronc.2008.11.003 PMID:19110418

Alici, G. (2018). Softer is Harder: What Differentiates Soft Robotics from Hard Robotics? *MRS Advances*, *3*(28), 1557–1568. doi:10.1557/adv.2018.159

Aljalal, M., Ibrahim, S., Djemal, R., & Ko, W. (2020). Comprehensive review on brain-controlled mobile robots and robotic arms based on electroencephalography signals. *Intelligent Service Robotics*, *13*(4), 539–563. doi:10.100711370-020-00328-5

Alkan, R. M., Karaman, H., & Şahin, M. (2005). GPS, GALILEO and GLONASS satellite navigation systems & GPS modernization. *RAST 2005 - Proceedings of 2nd International Conference on Recent Advances in Space Technologies*, 390–394. 10.1109/RAST.2005.1512598

Alkatout, I., Mettler, L., Maass, N., & Ackermann, J. (2016). Robotic surgery in gynecology. *Journal of the Turkish German Gynecological Association*, *17*(4), 224–232. doi:10.5152/jtgga.2016.16187 PMID:27990092

Alshowaikh, K., Karpinska-Leydier, K., Amirthalingam, J., Paidi, G., Iroshani Jayarathna, A. I., Salibindla, D. B. A. M. R., & Ergin, H. E. (2021). Surgical and Patient Outcomes of Robotic Versus Conventional Laparoscopic Hysterectomy: A Systematic Review. *Cureus*, *13*, e16828. doi:10.7759/cureus.16828 PMID:34367836

American College of Obstetricians and Gynecologists' Committee on Gynecologic Practice, The Society of Gynecologic Surgeons. (2020). Robot-Assisted Surgery for Noncancerous Gynecologic Conditions: ACOG Committee Opinion, Number 810. *Obstetrics and Gynecology*, *136*(3), e22–e30. doi:10.1097/AOG.0000000000004048 PMID:32826597

Amisha, M., Malik, P., Pathania, M., & Rathaur, V. K. (2019). Overview of artificial intelligence in medicine. *Journal of Family Medicine and Primary Care*, *8*(7), 2328–2331. doi:10.4103/jfmpc.jfmpc_440_19 PMID:31463251

Amoozadeh, M., Deng, H., Chuah, C. N., Zhang, H. M., & Ghosal, D. (2015). Platoon management with cooperative adaptive cruise control enabled by VANET. *Vehicular Communications*, *2*(2), 110–123. doi:10.1016/j.vehcom.2015.03.004

Ananda, M. P., Bernstein, H., Cunningham, K., Feess, A., & Stroud, G. (1990). Global Positioning System (GPS). *IEEE Symposium on Position Location and Navigation. A Decade of Excellence in the Navigation Sciences.*, 497–508. 10.1109/PLANS.1990.66220

Anayor, C., Gao, W., & Odekunle, A. (2018). Cooperative Adaptive Cruise Control of A Mixture of Human-driven and Autonomous Vehicles. *Conference Proceedings - IEEE SOUTHEASTCON*, 1–3. 10.1109/SECON.2018.8479268

Anderson, T. L. (2017). *Fracture Mechanics—Fundamentals and Applications* (3rd ed.). CRC. doi:10.1201/9781315370293

Andreopoulos, A., & Tsotsos, J. K. (2013). 50 Years of object recognition: Directions forward. *Computer Vision and Image Understanding*, *117*(8), 827–891. doi:10.1016/j.cviu.2013.04.005

Andres, M. P., Souza, C., Villaescusa, M., Vieira, M., & Abrao, M. S. (2022). The current role of robotic surgery in endometriosis management. *Expert Review of Endocrinology & Metabolism*, *17*(1), 63–73. doi:10.1080/17446651.2022.2031976 PMID:35073819

Antonacci, F., Sarti, A., & Tubaro, S. (2010). Geometric reconstruction of the environment from its response to multiple acoustic emissions. *IEEE International Conference on Acoustics, Speech and Signal Processing*, 2822–2825. 10.1109/ICASSP.2010.5496186

Ardon, A., & Hernandez, N. (2022). The Use of Peripheral Nerve Blockade in Laparoscopic and Robotic Surgery: Is There a Benefit? *Current Pain and Headache Reports*, *26*(1), 25–31. doi:10.100711916-022-01002-4 PMID:35076876

Arturo, D., & Nombela, M. (2010). Sartre: Safe road trains for the environment. *Conference on Personal Rapid Transit PRT@LHR*, 2–3.

Arya, P. P., & Chakrabarty, S. (2020). A robust internal model based fractional order controller for fractional order plus time delay processes. *IEEE Control Systems Letters*, *4*(4), 862–867. doi:10.1109/LCSYS.2020.2994606

ASME. (2019). *Boiler and Pressure Vessel Code*. Welding General Requirements.

ASTM D412. (1981). Standard test methods for vulcanized rubber and thermoplastic elastomers–tension. *Annual Book of ASTM Standards*, Philadelphia.

Audette, M. A., Bordas, S., & Blatt, J. E. (2020). Robotically Steered Needles: A Survey of Neurosurgical Applications and Technical Innovations. *Robotic Surgery (Auckland)*, *7*, 1–23. doi:10.2147/RSRR.S224446 PMID:32258180

Autorino, R., Kaouk, J. H., Stolzenburg, J. U., Gill, I. S., Mottrie, A., Tewari, A., & Cadeddu, J. A. (2013). Current status and future directions of robotic single-site surgery: A systematic review. *European Urology*, *63*(2), 266–280. doi:10.1016/j.eururo.2012.08.028 PMID:22940173

Ayala-Yáñez, R., Olaya-Guzmán, E. J., & Haghenbeck-Altamirano, J. (2013). Robotics in Gynecology: Why is this Technology Worth Pursuing? *Clinical Medicine Insights. Reproductive Health*, *7*, 71–77. doi:10.4137/CMRH.S10850 PMID:24453521

Ay, M., Korkmaz, D., Ozmen Koca, G., Bal, C., Akpolat, Z. H., & Bingol, M. C. (2018). Mechatronic Design and Manufacturing of the Intelligent Robotic Fish for Bio-Inspired Swimming Modes. *Electronics (Basel)*, *7*(7), 118. doi:10.3390/electronics7070118

Azadi, S., Green, I. C., Arnold, A., Truong, M., Potts, J., & Martino, M. A. (2021). Robotic Surgery: The Impact of Simulation and Other Innovative Platforms on Performance and Training. *Journal of Minimally Invasive Gynecology*, *28*(3), 490–495. doi:10.1016/j.jmig.2020.12.001 PMID:33310145

Azar, A. T., & Vaidyanathan, S. (Eds.). (2014). *Handbook of Research on Advanced Intelligent Control Engineering and Automation*. IGI Global. doi:10.4018/978-1-4666-7248-2

Baba, Y. E., Walther, A., & Habets, E. A. P. (2017). 3D room geometry inference based on room impulse response stacks. *IEEE/ACM Transactions on Audio, Speech, and Language Processing*, *26*(5), 857–872. doi:10.1109/TASLP.2017.2784298

Bafort, C., Beebeejaun, Y., Tomassetti, C., Bosteels, J., & Duffy, J. M. (2020). Laparoscopic surgery for endometriosis. *Cochrane Database of Systematic Reviews*, *10*, CD011031. PMID:33095458

Bahram, M., Hubmann, C., Lawitzky, A., Aeberhard, M., & Wollherr, D. (2016). A Combined Model- and Learning-Based Framework for Interaction-Aware Maneuver Prediction. *IEEE Transactions on Intelligent Transportation Systems*, *17*(6), 1538–1550. doi:10.1109/TITS.2015.2506642

Baig, M. Z., Javed, E., Ayaz, Y., Afzal, W., Gillani, S. O., Naveed, M., & Jamil, M. (2015). Classification of left/right hand movement from EEG signal by intelligent algorithms. *ISCAIE 2014 - 2014 IEEE Symposium on Computer Applications and Industrial Electronics*, 163–168. 10.1109/ISCAIE.2014.7010230

Bailey, A. P., Jaslow, C. R., & Kutteh, W. H. (2015). Minimally invasive surgical options for congenital and acquired uterine factors associated with recurrent pregnancy loss. *Women's Health (London)*, *11*(2), 161–167. doi:10.2217/WHE.14.81 PMID:25776290

Bailey, T., & Durrant-Whyte, H. (2006). Simultaneous localization and mapping (SLAM): Part II. *IEEE Robotics & Automation Magazine*, *13*(3), 108–117. doi:10.1109/MRA.2006.1678144

Balcerzak, M., Pikunov, D., & Dabrowski, A. (2018). The fastest, simplified method of Lyapunov exponents spectrum estimation for continuous-time dynamical systems. *Nonlinear Dynamics*, *94*(4), 3053–3065. doi:10.100711071-018-4544-z

Ban, C., Cai, G., & Wei, W. (2021). Dynamic Response and Chaotic Behavior of a Controllable Flexible Robot. Research Square, 1-14. doi:10.21203/rs.3.rs-1116482/v1

Barman, T., & Deb, N. (2017). State of the art review of speech recognition using genetic algorithm. *IEEE International Conference on Power, Control, Signals and Instrumentation Engineering (ICPCSI)*, 2944–2946. 10.1109/ICPCSI.2017.8392264

Bartz, D. (2009, November 16). *Autonomous Cars Will Make Us Safer*. Retrieved from https://www.wired.com/2009/11/autonomous-cars/

Barua, R., Das, S., Datta, P., & RoyChowdhury, A. (2022). Computational FEM Application on Percutaneous Nephrolithotomy (PCNL) Minimum Invasive Surgery Through Needle Insertion Process. In P. Pain, S. Banerjee, & G. Bose (Ed.), *Advances in Computational Approaches in Biomechanics,* (pp. 210-222). IGI Global. doi:10.4018/978-1-7998-9078-2.ch013

Barua, R., & Das, S. (2022). Improvements of Virtual and Augmented Reality for Advanced Treatments in Urology. In *Emerging Advancements for Virtual and Augmented Reality in Healthcare,* (pp. 117–131). IGI Global. doi:10.4018/978-1-7998-8371-5.ch008

Barua, R., Das, S., Datta, S., Datta, P., & Chowdhury, A. R. (2022). Analysis of surgical needle insertion modeling and viscoelastic tissue material interaction for minimally invasive surgery (MIS). *Materials Today: Proceedings*, *57*, 259–264. doi:10.1016/j.matpr.2022.02.498

Barua, R., Das, S., Datta, S., Roy Chowdhury, A., & Datta, P. (2022). Experimental Study of the Robotically Controlled Surgical Needle Insertion for Analysis of the Minimum Invasive Process. In *Emergent Converging Technologies and Biomedical Systems,* (pp. 473–482). Springer. doi:10.1007/978-981-16-8774-7_38

Bathe, K. J. (2014). *Finite Element Procedures,* (2nd ed.). Prentice-Hall.

Beard, R., Kingston, D., Quigley, M., Snyder, D., Christiansen, R., Johnson, W., … Goodrich, M. A. (2005). Autonomous vehicle technologies for small fixed-wing UAVs. *Journal of Aerospace Computing, Information and Communication*, 92–108. doi:10.2514/1.8371

Becker, A. (2019). Artificial intelligence in medicine: What is it doing for us today? *Health Policy and Technology*, *8*(2), 198–205. doi:10.1016/j.hlpt.2019.03.004

Behbehani, S., Suarez-Salvador, E., Buras, M., Magtibay, P., & Magrina, J. (2020). Mortality Rates in Benign Laparoscopic and Robotic Gynecologic Surgery: A Systematic Review and Meta-analysis. *Journal of Minimally Invasive Gynecology*, *27*(3), 603–612. doi:10.1016/j.jmig.2019.10.005 PMID:31627007

Belcher, S., Merlis, E., McNew, J., & Wright, M. (2018). *Roadmap To Vehicle Connectivity*. Academic Press.

Benosman, M., & Le Vey, G. (2004). Control of flexible manipulators: A survey. *Robotica*, *22*(5), 533–545. doi:10.1017/S0263574703005642

Berkowitz, R. L., & Minkoff, H. (2016). A Call for Change in a Changing World. *Obstetrics and Gynecology*, *127*(1), 153–156. doi:10.1097/AOG.0000000000001193 PMID:26646136

Bernardo, R., Sousa, J. M. C., & Gonçalves, P. J. S. (2022). Survey on robotic systems for internal logistics. *Journal of Manufacturing Systems*, *65*, 339–350. doi:10.1016/j.jmsy.2022.09.014

Bertozzi, M., Bombini, L., Broggi, A., Buzzoni, M., Cardarelli, E., Cattani, S., ... Zani, P. (2010a). The VisLab Intercontinental Autonomous challenge: 13,000 km, 3 months,··· no driver. *17th ITS World Congress*, 1–11.

Bertozzi, M., Bombini, L., Broggi, A., Buzzoni, M., Cardarelli, E., Cattani, S., ... Zani, P. (2010b). The VisLab Intercontinental Autonomous challenge: 13,000 km. *17th ITS World Congress*, 1–11.

Bettayab, M., & Rachid, M. (2014). IMC PID fractional order filter controllers design for integer order systems. *ISA Transactions*, *53*(5), 1620–1628. doi:10.1016/j.isatra.2014.05.007 PMID:24957276

Bhatia, J., Jackson, H., Tian, Y., Xu, J., & Matusik, W. (2021). Evolution Gym: A Large-Scale Benchmark for Evolving Soft Robots. *Advances in Neural Information Processing Systems, 34*, 2201–2214. https://proceedings.neurips.cc/paper/2021/hash/118921efba23fc329e6560b27861f0c2-Abstract.html

Bienvenu, G. (2002). Signal sonar processing. *2002 14th International Conference on Digital Signal Processing Proceedings. DSP 2002 (Cat. No.02TH8628), 1*, 441–446. 10.1109/ICDSP.2002.1027914

Bijelic, M., Gruber, T., & Ritter, W. (2018). Benchmarking image sensors under adverse weather conditions for autonomous driving. *2018 IEEE Intelligent Vehicles Symposium (IV)*, 1773–1779. 10.1109/IVS.2018.8500659

Billard, A., & Kragic, D. (2019). Trends and challenges in robot manipulation. *Science*, *364*(6446), eaat8414. doi:10.1126cience.aat8414 PMID:31221831

Bin Ng, W., Saidatul, A., Chong, Y. F., & Ibrahim, Z. (2019). PSD-Based Features Extraction for EEG Signal during Typing Task. *IOP Conference Series. Materials Science and Engineering*, *557*(1), 012032. doi:10.1088/1757-899X/557/1/012032

Bishop, C. M., & Nasrabadi, N. M. (2006). Pattern recognition and machine learning, 4(4). Springer.

Bishop, R. (2005). *Intelligent Vehicle Technology and Trends*. Artech House.

Bizopoulos, P., & Koutsouris, D. (2019). Deep Learning in Cardiology. *IEEE Reviews in Biomedical Engineering*, *12*, 168–193. doi:10.1109/RBME.2018.2885714 PMID:30530339

Blache, G., Jauffret, C., Mathis, J., Knight, S., Houvenaeghel, G., & Lambaudie, E. (2018). Can we use robotic surgery for the treatment of pelvic recurrence and locally advanced tumors in gynecological surgery? *Journal of Gynecology Obstetrics and Human Reproduction*, *47*(9), 431–435. doi:10.1016/j.jogoh.2018.08.008 PMID:30149209

Blackburn, D., Singhose, W., Kitchen, J., Patrangenaru, V., Lawrence, J., Kamoi, T., & Taura, A. (2010). Command shaping for nonlinear crane dynamics. *Journal of Vibration and Control*, *16*(4), 477–501. doi:10.1177/1077546309106142

Blum, J., & Eskandarian, A. (2004). The threat of intelligent collisions. *IT Professional*, *6*(1), 24–29. doi:10.1109/MITP.2004.1265539

Bodily, D. M., Allen, T. F., & Killpack, M. D. (2017). Multi-objective design optimization of a soft, pneumatic robot. *IEEE International Conference on Robotics and Automation (ICRA)*, 1864–1871. 10.1109/ICRA.2017.7989218

Boggess, J. F. (2007). Robotic surgery in gynecologic oncology: Evolution of a new surgical paradigm. *Journal of Robotic Surgery*, *1*(1), 31–37. doi:10.100711701-007-0011-4 PMID:25484936

Bojarski, M., Del Testa, D., Dworakowski, D., Firner, B., Flepp, B., Goyal, P., . . . Zieba, K. (2016). *End to End Learning for Self-Driving Cars*. Retrieved from https://arxiv.org/abs/1604.07316

Bokobza, L. (2012). Multiwall carbon nanotube-filled natural rubber: Electrical and mechanical properties. *Express Polymer Letters*, *6*(3), 213–223. doi:10.3144/expresspolymlett.2012.24

Bonnefon, J.-F., Shariff, A., & Rahwan, I. (2016). The social dilemma of autonomous vehicles. *Science*, *352*(6293), 1573–1576. doi:10.1126cience.aaf2654 PMID:27339987

Borowski, D. W., Kanakala, V., Agarwal, A. K., Tabaqchali, M. A., Garg, D. K., & Gill, T. S. (2011). Single-port access laparoscopic reversal of Hartmann operation. *Diseases of the Colon and Rectum*, *54*(8), 1053–1056. doi:10.1097/DCR.0b013e318221aa13 PMID:21730797

Boudette, N. E. (2016). Autopilot cited in death of Chinese tesla driver. *The New York Times*. Retrieved from https://www.nytimes.com/2016/09/15/business/fatal-tesla-crash-in-china-involved-autopilot-government-tv-says.html

Boukens, M. (2017). *Contribution à la commande des robots manipulateurs mobiles* [*Contribution to the control of mobile manipulator robots.*]. University of Jijel.

Bouquet de Joliniere, J., Librino, A., Dubuisson, J. B., Khomsi, F., Ben Ali, N., Fadhlaoui, A., Ayoubi, J. M., & Feki, A. (2016). Robotic Surgery in Gynecology. *Frontiers in Surgery*, *3*, 26. doi:10.3389/fsurg.2016.00026 PMID:27200358

Bousseta, R., El Ouakouak, I., Gharbi, M., & Regragui, F. (2018). EEG Based Brain Computer Interface for Controlling a Robot Arm Movement Through Thought. *IRBM*, *39*(2), 129–135. doi:10.1016/j.irbm.2018.02.001

Braco, R., Prates, P., Costa, J. D. M., & Berto, F. (2018). New methodology of fatigue life evaluation for multiaxially loaded notched components based on two uniaxial strain-controlled tests. *International Journal of Fatigue*, *111*, 308–320. doi:10.1016/j.ijfatigue.2018.02.027

Brandt, A. C., Frame, E. A., Warden, R. W., & Antonio, S. (2010). *SAE J1321 testing using M1083A1 FMTVS interim report.* Force Projection Technologies.

Brandt, G., Zimolong, A., Carrat, L., Merloz, P., Staudte, H. W., Lavallée, S., Radermacher, K., & Rau, G. (1999). CRIGOS: a compact robot for image-guided orthopedic surgery. *IEEE transactions on information technology in biomedicine: a publication of the IEEE Engineering in Medicine and Biology Society, 3*(4), 252–260.

Broggi, A., Buzzoni, M., Debattisti, S., Grisleri, P., Laghi, M. C., Medici, P., & Versari, P. (2013). Extensive tests of autonomous driving technologies. *IEEE Transactions on Intelligent Transportation Systems*, *14*(3), 1403–1415. doi:10.1109/TITS.2013.2262331

Broggi, A., Medici, P., Zani, P., Coati, A., & Panciroli, M. (2012). Autonomous vehicles control in the VisLab Intercontinental Autonomous Challenge. *Annual Reviews in Control*, *36*(1), 161–171. doi:10.1016/j.arcontrol.2012.03.012

Brown, A. K., & Sturza, M. A. (1993). *Patent No. 5,225,842.* United STATES PATENT.

Bucher, P., Pugin, F., & Morel, P. (2008). Single port access laparoscopic right hemicolectomy. *International Journal of Colorectal Disease*, *23*(10), 1013–1016. doi:10.100700384-008-0519-8 PMID:18607608

Burdea, G. C. (1996). *Force and Touch Feedback for Virtual Reality*. John Wiley & Sons.

Burgard, W., Brock, O., & Stachniss, C. (2008). Map-based precision vehicle localization in urban environments. *Robotics: Science and Systems, 3*, 121–128. Retrieved from https://ieeexplore.ieee.org/document/6280129

Burgard, W., Fox, D., & Thrun, S. (1997). Active mobile robot localization. *IJCAI (United States)*, 1346–1352.

Burke, W. M., Gossner, G., & Goldman, N. A. (2011). Robotic surgery in the obese gynecologic patient. *Clinical Obstetrics and Gynecology*, *54*(3), 420–430. doi:10.1097/GRF.0b013e31822b37b1 PMID:21857172

Bylov, B. F., Vinograd, R. E., Grobman, D. M., & Nemytsky, V. V. (1966). *Theory of Lyapunov indexes and its application to stability problems*. Nauka.

Cantillo-Negrete, J., Carino-Escobar, R. I., Carrillo-Mora, P., Elias-Vinas, D., & Gutierrez-Martinez, J. (2018). Motor imagery-based brain-computer interface coupled to a robotic hand orthosis aimed for neurorehabilitation of stroke patients. *Journal of Healthcare Engineering*, *2018*, 1–10. doi:10.1155/2018/1624637 PMID:29849992

Cao, Y., & Liu, S. (2017). Visual servo control for wheeled robot platooning based on homography. *Proceedings of 2017 IEEE 6th Data Driven Control and Learning Systems Conference, DDCLS 2017*, 628–632. 10.1109/DDCLS.2017.8068145

Cao, Y., Liu, Y., Chen, Y., Zhu, L., Yan, Y., & Chen, X. (2017). A novel slithering locomotion mechanism for a snake-like soft robot. *Journal of the Mechanics and Physics of Solids*, *99*, 304–320. doi:10.1016/j.jmps.2016.11.019

Čapek, K. (1920). R.U.R. Dover Publications.

Capozzi, V. A., Armano, G., Rosati, A., Tropea, A., & Biondi, A. (2021). The robotic single-port platform for gynecologic surgery: A systematic review of the literature and meta-analysis. *Updates in Surgery*, *73*(3), 1155–1167. doi:10.100713304-020-00812-8 PMID:32472402

Carbonnel, M., Dahm-Kähler, P., Revaux, A., Brännström, M., & Ayoubi, J. M. (2020). Adapting surgical skills from robotic-assisted radical hysterectomy in cervical cancer to uterine transplantation: A look to an optimistic future! *Journal of Robotic Surgery*, *14*(6), 841–847. doi:10.100711701-020-01058-7 PMID:32088836

Cardenas-Trowers, O., Kjellsson, K., & Hatch, K. (2018). Ergonomics: Making the OR a comfortable place. *International Urogynecology Journal and Pelvic Floor Dysfunction*, *29*(7), 1065–1066. doi:10.100700192-018-3674-7 PMID:30298292

Casanova, F., Carney, P. R., & Sarntinoranont, M. (2014). Effect of needle insertion speed on tissue injury, stress, and backflow distribution for convection-enhanced delivery in the rat brain. *PLoS One*, *9*(4), e94919. doi:10.1371/journal.pone.0094919 PMID:24776986

Catanzarite, T., Tan-Kim, J., & Menefee, S. A. (2018). Ergonomics in gynecologic surgery. *Current Opinion in Obstetrics & Gynecology*, *30*(6), 432–440. doi:10.1097/GCO.0000000000000502 PMID:30299323

Cavusoglu, M. C., Villanueva, I., & Tendick, F. (2001, November). Workspace analysis of robotic manipulators for a teleoperated suturing task. In *Proceedings IEEE/RSJ International Conference on Intelligent Robots and Systems. Expanding the Societal Role of Robotics in the the Next Millennium (Cat. No. 01CH37180),* (Vol. 4, pp. 2234-2239). IEEE. 10.1109/IROS.2001.976402

Cerri, P., Soprani, G., Zani, P., Choi, J., Lee, J., Kim, D., ... Broggi, A. (2011). Computer vision at the Hyundai Autonomous Challenge. *IEEE Conference on Intelligent Transportation Systems, Proceedings, ITSC*, 777–783. 10.1109/ITSC.2011.6082859

Chai, L., Du, J., Liu, D. Y., Tu, Y. H., & Lee, C. H. (2021). Acoustic modeling for multi-array conversational speech recognition in the chime-6 challenge. *IEEE Spoken Language Technology Workshop (SLT)*, 912–918. 10.1109/SLT48900.2021.9383628

Chan, Y. H., Tse, Z., & Ren, H. (2017). Design evolution and pilot study for a kirigami-inspired flexible and soft anthropomorphic robotic hand. *18th International Conference on Advanced Robotics (ICAR)*, 432–437. 10.1109/ICAR.2017.8023645

Chanda, K., Ghosh, A., Dey, S., Bose, R., & Roy, S. (2022). Smart Self-Immolation Prediction Techniques: An Analytical Study for Predicting Suicidal Tendencies Using Machine Learning Algorithms. *EAI/Springer Innovations in Communication and Computing*, 69–91. doi:10.1007/978-3-030-71485-7_4

Chang, C., & Yuan, Z. (2018). Combined longitudinal and lateral control of vehicle platoons. *2017 International Conference on Computer Systems, Electronics and Control, ICCSEC 2017*, 848–852. 10.1109/ICCSEC.2017.8446824

Chang, H. J., Lee, C. S. G., Lu, Y. H., & Hu, Y. C. (2007). P-SLAM: Simultaneous localization and mapping with environmental-structure prediction. *IEEE Transactions on Robotics*, *23*(2), 281–293. doi:10.1109/TRO.2007.892230

Chang, K. S., Karl Hedrick, J., Zhang, W. B., Varaiya, P., Tomizuka, M., & Shladover, S. E. (1993). Automated highway system experiments in the path program. *Journal of Intelligent Transport Systems*, *1*(1), 63–87.

Chang, K. S., Li, W., Devlin, P., Shaikhbahai, A., Varaiya, P., & Hedrick, J. K. (1991). Experimentation with a vehicle platoon control system. *Vehicle Navigation and Information Systems Conference*, 1117–1124. 10.4271/912868

Chan, T., Godbole, K., & Hou, E. (2003). Optimal input shaper design for high-speed robotic workcells. *Journal of Vibration and Control*, *9*(12), 1359–1376. doi:10.1177/1077546304031165

Chaurasia, B. K., Verma, S., & Tomar, G. S. (2011). Attacks on anonymity in VANET. *Proceedings - 2011 International Conference on Computational Intelligence and Communication Systems, CICN 2011*, 217–221. 10.1109/CICN.2011.43

Chazan, S. E., Goldberger, J., & Gannot, S. (2018). DNN-based concurrent speakers detector and its application to speaker extraction with LCMV beamforming. *IEEE International Conference on Acoustics, Speech and Signal Processing (ICASSP)*, 6712–6716. 10.1109/ICASSP.2018.8462407

Chekari, T., Rachid, M., & Bettayab, M. (2019). Improved internal model control proportional integral derivative fractional order multiloop controller design for non-integer multivariable systems. *Journal of Dynamic Systems, Measurement, and Control, ASME, 141*(1), 011014. doi:10.1115/1.4041353

Chen, X., Kundu, K., Zhu, Y., Berneshawi, A., Ma, H., Fidler, S., & Urtasun, R. (2015). 3D object proposals for accurate object class detection. Advances in Neural Information Processing Systems, 424–432.

Chen, X., Ma, H., Wan, J., Li, B., & Xia, T. (2017). Multi-view 3d object detection network for autonomous driving. *Proceedings of the IEEE Conference on Computer Vision and Pattern Recognition*, 1907–1915.

Chen, C. C., & Falcone, T. (2009). Robotic gynecologic surgery: Past, present, and future. *Clinical Obstetrics and Gynecology, 52*(3), 335–343. doi:10.1097/GRF.0b013e3181b08adf PMID:19661749

Chen, C., Qin, C., Qiu, H., Tarroni, G., Duan, J., Bai, W., & Rueckert, D. (2020). Deep Learning for Cardiac Image Segmentation: A Review. *Frontiers in Cardiovascular Medicine, 7*, 25. doi:10.3389/fcvm.2020.00025 PMID:32195270

Chen, C., Seff, A., Kornhauser, A., & Xiao, J. (2015). Deepdriving: Learning affordance for direct perception in autonomous driving. *Proceedings of the IEEE International Conference on Computer Vision*, 2722–2730. 10.1109/ICCV.2015.312

Cheney, N., MacCurdy, R., Clune, J., & Lipson, H. (2014). Unshackling evolution: Evolving soft robots with multiple materials and a powerful generative encoding. *ACM SIGEVOlution, 7*(1), 11–23. doi:10.1145/2661735.2661737

Chen, F., & Wang, M. Y. (2020). Design Optimization of Soft Robots: A Review of the State of the Art. *IEEE Robotics & Automation Magazine, 27*(4), 27–43. doi:10.1109/MRA.2020.3024280

Chen, F., Xu, W., Zhang, H., Wang, Y., Cao, J., Wang, M. Y., Ren, H., Zhu, J., & Zhang, Y. F. (2018). Topology Optimized Design, Fabrication, and Characterization of a Soft Cable-Driven Gripper. *IEEE Robotics and Automation Letters, 3*(3), 2463–2470. doi:10.1109/LRA.2018.2800115

Cheng, J., Cheng, H., Meng, M. Q. H., & Zhang, H. (2019). Autonomous Navigation by Mobile Robots in Human Environments: A Survey. *2018 IEEE International Conference on Robotics and Biomimetics, ROBIO 2018*, 1981–1986. 10.1109/ROBIO.2018.8665075

Chen, X., Hu, W., Zhang, L., Shi, Z., & Li, M. (2018). Integration of low-cost GNSS and monocular cameras for simultaneous localization and mapping. *Sensors (Switzerland), 18*(7), 1–18. doi:10.339018072193 PMID:29986515

Chesson, R., & Hallner, B. (2013). Why complex pelvic organ prolapse should be repaired vaginally. *Current Opinion in Urology, 23*(4), 312–316. doi:10.1097/MOU.0b013e3283619e7f PMID:23882718

Chin, K., Hellebrekers, T., & Majidi, C. (2020). Machine Learning for Soft Robotic Sensing and Control. *Advanced Intelligent Systems, 2*(6), 1900171. doi:10.1002/aisy.201900171

Cho, J. E., & Nezhat, F. R. (2009). Robotics and gynecologic oncology: Review of the literature. *Journal of Minimally Invasive Gynecology, 16*(6), 669–681. doi:10.1016/j.jmig.2009.06.024 PMID:19896593

Cho, J. E., Shamshirsaz, A. H., Nezhat, C., Nezhat, C., & Nezhat, F. (2010). New technologies for reproductive medicine: Laparoscopy, endoscopy, robotic surgery and gynecology. A review of the literature. *Minerva Ginecologica, 62*, 137–167. PMID:20502426

Chun, H. M., Turner, J. D., & Juang, J.-N. (1985). Disturbance Accommodating Tracking Maneuvers of Flexible Spacecraft. *The Journal of the Astronautical Sciences, 33*(2), 197–216.

Cianci, S., Arcieri, M., Vizzielli, G., Martinelli, C., Granese, R., La Verde, M., Fagotti, A., Fanfani, F., Scambia, G., & Ercoli, A. (2021). Robotic Pelvic Exenteration for Gynecologic Malignancies, Anatomic Landmarks, and Surgical Steps: A Systematic Review. *Frontiers in Surgery*, *8*, 790152. doi:10.3389/fsurg.2021.790152 PMID:34917648

Ciui, B., Martin, A., Mishra, R. K., Nakagawa, T., Dawkins, T. J., Lyu, M., Cristea, C., & Wang, R. S. J. (2018). Chemical sensing at the robot fingertips: Toward automated taste discrimination in food samples. *ACS Sensors*, *3*(11), 2375–2384. doi:10.1021/acssensors.8b00778 PMID:30226368

Ciuti, G., Valdastri, P., Menciassi, A., & Dario, P. (2010). Robotic magnetic steering and locomotion of capsule endoscope for diagnostic and surgical endoluminal procedures. *Robotica*, *28*(2), 199–207. doi:10.1017/S0263574709990361

Clair, K. H., & Tewari, K. S. (2020). Robotic surgery for gynecologic cancers: Indications, techniques and controversies. *Journal of Obstetrics and Gynaecology Research*, *46*(6), 828–843. doi:10.1111/jog.14228 PMID:32410262

Claybrook, J., & Kildare, S. (2018). Autonomous vehicles: No driver… no regulation? *Science*, *361*(6397), 36–37. doi:10.1126cience.aau2715 PMID:29976817

Cleary, K., Melzer, A., Watson, V., Kronreif, G., & Stoianovici, D. (2006). Interventional robotic systems: Applications and technology state-of-the-art. *Minimally Invasive Therapy & Allied Technologies*, *15*(2), 101–113. doi:10.1080/13645700600674179 PMID:16754193

Closon, F., & Tulandi, T. (2016). Uterine myomata: Organ-preserving surgery. *Best Practice & Research. Clinical Obstetrics & Gynaecology*, *35*, 30–36. doi:10.1016/j.bpobgyn.2015.09.005 PMID:26542930

Cohen, S. L., & Einarsson, J. I. (2011). The role of mechanical bowel preparation in gynecologic laparoscopy. *Reviews in Obstetrics & Gynecology*, *4*, 28–31. PMID:21629496

Committee, S. (2016). *Sae j3061-cybersecurity guidebook for cyber-physical automotive systems.* SAE-Society of Automotive Engineers.

Comon, P., & Jutten, C. (2010). *Handbook of Blind Source Separation: Independent component analysis and applications*. Academic press.

Conker, C., Yavuz, H., & Bilgic, H. H. (2016). A review of command shaping techniques for elimination of residual vibrations in flexible-joint manipulators. *Journal of Vibroengineering*, *18*(5), 2947–2958. doi:10.21595/jve.2016.16725

Connor, A., Pasic, R., Quevedo, A., & Chamseddine, P. (2021). Robotic Applications for Benign Gynecologic Procedures. *Surgical Technology International*, *39*, 232–240. doi:10.52198/21.STI.39.GY1409 PMID:34181238

Cooper, J. B., & Taqueti, V. (2008). A brief history of the development of mannequin simulators for clinical education and training. *Postgraduate Medical Journal*, *84*(997), 563–570. doi:10.1136/qshc.2004.009886 PMID:19103813

Corti, A., Ongini, C., Tanelli, M., & Savaresi, S. M. (2013). Quantitative driving style estimation for energy-oriented applications in road vehicles. *Proceedings - 2013 IEEE International Conference on Systems, Man, and Cybernetics, SMC 2013*, 3710–3715. 10.1109/SMC.2013.632

Courcier, H., Azaïs, H., Belghiti, J., Uzan, C., & Canlorbe, G. (2021). Minimally invasive surgery for early-stage cervical cancer: Rediscovering the Schautheim robot-assisted procedure. *Journal of Gynecology Obstetrics and Human Reproduction*, *50*(2), 101980. doi:10.1016/j.jogoh.2020.101980 PMID:33186773

Cowden, S., & Singh, G. (2013). Sat-Nav education: A means to an end or an end to meaning. *Acts of Knowing: Critical Pedagogy in, against and beyond the University*, 41–59.

Cui, J., & Trinkle, J. (2021). Toward next-generation learned robot manipulation. *Science Robotics*, *6*(54), 9461. doi:10.1126cirobotics.abd9461 PMID:34043539

Cuschieri, A. (1994). Therapeutic laparoscopy. *Endoscopy*, *26*(01), 134–147. doi:10.1055-2007-1008940 PMID:8205987

Das, D., Propst, K., Wechter, M. E., & Kho, R. M. (2019). Evaluation of Positioning Devices for Optimization of Outcomes in Laparoscopic and Robotic-Assisted Gynecologic Surgery. *Journal of Minimally Invasive Gynecology*, *26*(2), 244–252. doi:10.1016/j.jmig.2018.08.027 PMID:30176363

Dasgupta, S., Raghuraman, V., Teja, T. N., & Dauwels, J. (2017). Merging and Splitting maneuver of platoons by means of a novel PID controller. *2017 IEEE Symposium Series on Computational Intelligence (SSCI)*, 1–8. Retrieved from https://ieeexplore.ieee.org/stamp/stamp.jsp?tp=&arnumber=8280871

Davies, A. (2016, February 23). Google's self-driving car caused its first crash. *Wired*. Retrieved from https://www.wired.com/2016/02/googles-self-driving-car-may-caused-first-crash/

Davies, B. (2000). A review of robotics in surgery. *Proceedings of the Institution of Mechanical Engineers. Part H, Journal of Engineering in Medicine*, *214*(1), 129–140. doi:10.1243/0954411001535309 PMID:10718057

Davies, B., Jakopec, M., Harris, S. J., Baena, F. R. Y., Barrett, A., Evangelidis, A., Gomes, P., Henckel, J., & Cobb, J. (2006). Active-constraint robotics for surgery. *Proceedings of the IEEE*, *94*(9), 1696–1704. doi:10.1109/JPROC.2006.880680

de Melo, C. M., Marsella, S., & Gratch, J. (2021). Risk of Injury in Moral Dilemmas With Autonomous Vehicles. *Frontiers in Robotics and AI*, *7*, 572529. Advance online publication. doi:10.3389/frobt.2020.572529 PMID:34212006

Deleforge, A. (2020). Drone audition for search and rescue: Datasets and challenges. *QUIET DRONES: International Symposium on UAV/UAS Noise*. https://hal.archives-ouvertes.fr/hal-03430293/

Della Santina, C., Duriez, C., & Rus, D. (2021). *Model Based Control of Soft Robots: A Survey of the State of the Art and Open Challenges*

Deloitte Global. (2019). *2019 Deloitte Global Automotive Consumer Study | Advanced vehicle technologies and multimodal transportation*. Retrieved from https://www2.deloitte.com/content/dam/Deloitte/us/Documents/manufacturing/us-global-automotive-consumer-study-2019.pdf

Demidovich, B. P. (1967). *Lektion on mathematical stability theory*. Nauka.

Demir, K. G., Zhang, Z., Yang, J., & Gu, G. X. (2020). Computational and Experimental Design Exploration of 3D-Printed Soft Pneumatic Actuators. *Advanced Intelligent Systems*, *2*(7), 2000013. doi:10.1002/aisy.202000013

Deng, C., Song, H., Zhang, Y., Sha, Y., & Li, X. (2020). *DNN-based Mask Estimation Integrating Spectral and Spatial Features for Robust Beamforming*, 4647–4651.

Deng, Q. (2016). A General Simulation Framework for Modeling and Analysis of Heavy-Duty Vehicle Platooning. *IEEE Transactions on Intelligent Transportation Systems*, *17*(11), 3252–3262. doi:10.1109/TITS.2016.2548502

Desai, P. H., Lin, J. F., & Slomovitz, B. M. (2014). Milestones to optimal adoption of robotic technology in gynecology. *Obstetrics and Gynecology*, *123*(1), 13–20. doi:10.1097/AOG.0000000000000055 PMID:24463658

Deurloo, E. E., Gilhuijs, K. G., Schultze Kool, L. J., & Muller, S. H. (2001). Displacement of breast tissue and needle deviations during stereotactic procedures. *Investigative Radiology*, *36*(6), 347–353. doi:10.1097/00004424-200106000-00008 PMID:11410756

Dey, D., Slomka, P. J., Leeson, P., Comaniciu, D., Shrestha, S., Sengupta, P. P., & Marwick, T. H. (2019). Artificial Intelligence in Cardiovascular Imaging: JACC State-of-the-Art Review. *Journal of the American College of Cardiology*, *73*(11), 1317–1335. doi:10.1016/j.jacc.2018.12.054 PMID:30898208

Dharne, A. G., & Jayasuriya, S. (2007). Robust adaptive control of residual vibration in point-to-point motion of flexible bodies. *Journal of Vibration and Control*, *13*(7), 951–968. doi:10.1177/1077546307078830

Dogangil, G., Ergeneman, O., Abbott, J. J., Pané, S., Hall, H., Muntwyler, S., & Nelson, B. J. (2008, September). Toward targeted retinal drug delivery with wireless magnetic microrobots. In *IEEE/RSJ International Conference on Intelligent Robots and Systems,* (pp. 1921-1926). IEEE. 10.1109/IROS.2008.4650768

Donatelli, C. M., Serlin, Z. T., Echols-Jones, P., Scibelli, A. E., Cohen, A., Musca, J.-M., Rozen-Levy, S., Buckingham, D., White, R., & Trimmer, B. A. (2017). Soft foam robot with caterpillar-inspired gait regimes for terrestrial locomotion. *2017 IEEE/RSJ International Conference on Intelligent Robots and Systems (IROS)*, 476–481. 10.1109/IROS.2017.8202196

Dorado-Díaz, P. I., Sampedro-Gómez, J., Vicente-Palacios, V., & Sánchez, P. L. (2019). Applications of artificial intelligence in cardiology. The future is already here. *Revista Espanola de Cardiologia*, *72*(12), 1065–1075. PMID:31611150

Dornhöfer, N., & Höckel, M. (2008). New developments in the surgical therapy of cervical carcinoma. *Annals of the New York Academy of Sciences*, *1138*(1), 233–252. doi:10.1196/annals.1414.029 PMID:18837903

Doshi, A., & Trivedi, M. M. (2010). Examining the impact of driving style on the predictability and responsiveness of the driver: Real-world and simulator analysis. *IEEE Intelligent Vehicles Symposium, Proceedings*, 232–237. 10.1109/IVS.2010.5547969

Dragon Skin. (2022). *Dragon Skin™ 20 Product Information*. Smooth-On, Inc. https://www.smooth-on.com/products/dragon-skin-20/

Dubuisson, J., Veit-Rubin, N., Bouquet de Jolinière, J., & Dubuisson, J. B. (2016). Laparoscopic Lateral Suspension: Benefits of a Cross-shaped Mesh to Treat Difficult Vaginal Vault Prolapse. *Journal of Minimally Invasive Gynecology*, *23*(5), 672. doi:10.1016/j.jmig.2016.01.028 PMID:26867700

Duchon, F., Babinec, A., Kajan, M., Beno, P., Florek, M., Fico, T., & Jurišica, L. (2014). Path planning with modified A star algorithm for a mobile robot. *Procedia Engineering*, *96*, 59–69. doi:10.1016/j.proeng.2014.12.098

Duga, J. J., Fisher, W. H., Buxaum, R. W., Rosenfield, A. R., Buhr, A. R., & Honton, E. J. (2008). Fatigue. In F. C. Campbell (Ed.), *Elements of Metallurgy and Engineering Alloys*. ASM International.

Dumlu, A., & Ayten, K. K. (2018). A combined control approach for industrial process systems using feed-forward and adaptive action based on second order sliding mode controller design, *Transactions of the Institute of Measurement and Control. Sage (Atlanta, Ga.)*, 1–12.

Duong, K., Vincent, E., & Gribonval, R. (2010). Under-determined reverberant audio source separation using a full-rank spatial covariance model. *IEEE Transactions on Audio, Speech, and Language Processing*, *18*(7), 1830–1840. doi:10.1109/TASL.2010.2050716

Du, T., Hughes, J., Wah, S., Matusik, W., & Rus, D. (2021). Underwater Soft Robot Modeling and Control With Differentiable Simulation. *IEEE Robotics and Automation Letters*, *6*(3), 4994–5001. doi:10.1109/LRA.2021.3070305

Dwivedy, S. K., & Eberhard, P. (2006). Dynamic analysis of flexible manipulators, a literature review. *Mechanism and Machine Theory*, *41*(7), 749–777. doi:10.1016/j.mechmachtheory.2006.01.014

Edla, D. R., Mangalorekar, K., Dhavalikar, G., & Dodia, S. (2018). Classification of EEG data for human mental state analysis using Random Forest Classifier. *Procedia Computer Science, 132*(Iccids), 1523–1532. doi:10.1016/j.procs.2018.05.116

Edwards, D. S., Allen, R., Papadopoulos, T., Rowan, D., Kim, S. Y., & Wilmot-Brown, L. (2009). Investigations of mammalian echolocation. *Annual International Conference of the IEEE Engineering in Medicine and Biology Society*, 7184–7187. 10.1109/IEMBS.2009.5335313

El Bana, H. G., & Abbas, A. (2020, February). A novel design of the utilisation of soft grippers in loading and unloading applications. *International Conference on Innovative Trends in Communication and Computer Engineering (ITCE)*, 158-163. 10.1109/ITCE48509.2020.9047779

El Sallab, A., Abdou, M., Perot, E., & Yogamani, S. (2017). Deep reinforcement learning framework for autonomous driving. *IS and T International Symposium on Electronic Imaging Science and Technology*, 70–76. 10.2352/ISSN.2470-1173.2017.19.AVM-023

Elango, N., & Faudzi, A. A. M. (2015). A review article: Investigations on soft materials for soft robot manipulations. *International Journal of Advanced Manufacturing Technology*, *80*(5), 1027–1037. doi:10.100700170-015-7085-3

El-Atab, N., Mishra, R. B., Al-Modaf, F., Joharji, L., Alsharif, A. A., Alamoudi, H., Diaz, M., Qaiser, N., & Hussain, M. M. (2020). Soft Actuators for Soft Robotic Applications: A Review. *Advanced Intelligent Systems*, *2*(10), 2000128. doi:10.1002/aisy.202000128

Eliakim, I., Cohen, Z., Kosa, G., & Yovel, Y. (2018). A fully autonomous terrestrial bat-like acoustic robot. *PLoS Computational Biology*, *14*(9), e1006406. doi:10.1371/journal.pcbi.1006406 PMID:30188901

Eliazar, A., & Parr, R. (2003). DP-SLAM: Fast, robust simultaneous localization and mapping without predetermined landmarks. *IJCAI International Joint Conference on Artificial Intelligence*, 1135–1142.

Elliott, D. S., Chow, G. K., & Gettman, M. (2006). Current status of robotics in female urology and gynecology. *World Journal of Urology*, *24*(2), 188–192. doi:10.100700345-006-0071-5 PMID:16557388

Elsayed, E. A. (2012). *Reliability Engineering*. John Wiley & Sons.

Elsayed, Y., Vincensi, A., Lekakou, C., Geng, T., Saaj, C. M., Ranzani, T., Cianchetti, M., & Menciassi, A. (2014). Finite Element Analysis and Design Optimization of a Pneumatically Actuating Silicone Module for Robotic Surgery Applications. *Soft Robotics*, *1*(4), 255–262. doi:10.1089oro.2014.0016

Eltabbakh, G. H., & Bower, N. A. (2008). Laparoscopic surgery in endometriosis. *Minerva Ginecologica*, *60*, 323–330. PMID:18560348

E-Network. (2013). *Programme for a European traffic system with highest efficiency and unprecedented safety*. Retrieved September 20, 2009, from https://www.eurekanetwork.org/project/id/45

Erić, M. M. (2011). Some research challenges of acoustic camera. *19th Telecommunications Forum, TELFOR Proceedings of Papers*, 1036–1039. 10.1109/TELFOR.2011.6143725

Eslami, S., Fischer, G. S., Song, S. E., Tokuda, J., Hata, N., Tempany, C. M., & Iordachita, I. (2013). Towards Clinically Optimized MRI-guided Surgical Manipulator for Minimally Invasive Prostate Percutaneous Interventions: Constructive Design. *International Conference on Robotics and Automation*, *20132*, 1228–1233. PMID:24683502

Fabricio, B., & Liang, Z. (2013). Fuzzy community structure detection by particle competition and cooperation. *Soft Computing*, *17*(4), 659–673. doi:10.100700500-012-0924-3

Falcone, T., & Goldberg, J. M. (2003). Robotics in gynecology. *The Surgical Clinics of North America*, *83*(6), 1483–1489. doi:10.1016/S0039-6109(03)00163-4 PMID:14712881

Falcone, T., & Parker, W. H. (2013). Surgical management of leiomyomas for fertility or uterine preservation. *Obstetrics and Gynecology*, *121*(4), 856–868. doi:10.1097/AOG.0b013e3182888478 PMID:23635687

Falcone, T., & Steiner, C. P. (2002). Robotically assisted gynaecological surgery. *Human Fertility (Cambridge, England)*, *5*(2), 72–74. doi:10.1080/1464727022000198962 PMID:12082211

Fanfani, F., Restaino, S., Ercoli, A., Chiantera, V., Fagotti, A., Gallotta, V., Monterossi, G., Cappuccio, S., & Scambia, G. (2016). Robotic versus laparoscopic surgery in gynecology: Which should we use? *Minerva Ginecologica*, *68*, 423–430. PMID:26633042

Fanning, J., Hojat, R., Johnson, J., & Fenton, B. (2009). Robotic radical hysterectomy. *Minerva Ginecologica*, *61*, 53–55. PMID:19204661

Fathy, A., Yousri, D., Rezk, H., Thanikanti, S., & Hasaniern, H. M. (2022). A robust fractional order PID controller based load frequency control using modified hunger games optimizer. *Energies*, *15*(361), 361. doi:10.3390/en15010361

Feliu, J. J., Feliu, V., & Cerrada, C. (1999). Load Adaptive Control of Single-Link Flexible Arms Based on a New Modeling Technique. *IEEE Transactions on Robotics and Automation*, *15*(5), 793–804. doi:10.1109/70.795785

Feliu, V., Somolinos, J. A., & García, A. (2003). Inverse Dynamics Based Control System for a Three-Degree-of-Freedom Flexible Arm. *IEEE Transactions on Robotics and Automation*, *19*(6), 1007–1014. doi:10.1109/TRA.2003.819596

Ferrero, S., Stabilini, C., Barra, F., Clarizia, R., Roviglione, G., & Ceccaroni, M. (2021). Bowel resection for intestinal endometriosis. *Best Practice & Research. Clinical Obstetrics & Gynaecology*, *71*, 114–128. doi:10.1016/j.bpobgyn.2020.05.008 PMID:32665125

Filos, J., Habets, E. A. P., & Naylor, P. A. (n.d.). *A two-step approach to blindly infer room geometries.*

Fitzgerald, S. G., Delaney, G. W., Howard, D., & Maire, F. (2021). Evolving soft robotic jamming grippers. *Proceedings of the Genetic and Evolutionary Computation Conference*, 102–110. 10.1145/3449639.3459331

Fong, Y., Woo, Y., Hyung, W. J., Lau, C., & Strong, V. E. (2018). *The SAGES Atlas of Robotic Surgery*. Springer. doi:10.1007/978-3-319-91045-1

Foote, J. R., & Valea, F. A. (2016). Robotic surgical training: Where are we? *Gynecologic Oncology*, *143*(1), 179–183. doi:10.1016/j.ygyno.2016.05.031 PMID:27246302

Fortin, F. A., De Rainville, F. M., Gardner, M. A., Parizeau, M., & Gagńe, C. (2012). DEAP: Evolutionary algorithms made easy. *Journal of Machine Learning Research*, *13*, 2171–2175.

Fraser, A. R., & Daniel, R. W. (1991). *Perturbation Techniques for Flexible Manipulator*. Kluwer. doi:10.1007/978-1-4615-3974-2

Freddi, A., Longhi, S., & Monteriù, A. (2015). Nonlinear decentralized model predictive control for unmanned vehicles moving in formation. *Information Technology and Control*, *44*(1), 89–97. doi:10.5755/j01.itc.44.1.7219

Frick, A. C., & Falcone, T. (2009). Robotics in gynecologic surgery. *Minerva Ginecologica*, *61*, 187–199. PMID:19415063

Fung, T. C. (1997). Unconditionally stable higher-order Newmark methods by sub-stepping procedure. *Computer Methods in Applied Mechanics and Engineering*, *147*(1-2), 61–84. doi:10.1016/S0045-7825(96)01243-1

Gadonneix, P., Ercoli, A., Scambia, G., & Villet, R. (2005). The use of laparoscopic sacrocolpopexy in the management of pelvic organ prolapse. *Current Opinion in Obstetrics & Gynecology*, *17*(4), 376–380. doi:10.1097/01.gco.0000175355.48802.7b PMID:15976543

Gal, Y. (2016). *Uncertainty in Deep Learning*. University of Cambridge. Retrieved from http://mlg.eng.cam.ac.uk/yarin/blog_2248.html

Gala, R. B., Margulies, R., Steinberg, A., Murphy, M., Lukban, J., Jeppson, P., Aschkenazi, S., Olivera, C., South, M., Lowenstein, L., Schaffer, J., Balk, E. M., & Sung, V.Society of Gynecologic Surgeons Systematic Review Group. (2014). Systematic review of robotic surgery in gynecology: Robotic techniques compared with laparoscopy and laparotomy. *Journal of Minimally Invasive Gynecology*, *21*(3), 353–361. doi:10.1016/j.jmig.2013.11.010 PMID:24295923

Gallego, G., Delbruck, T., Orchard, G., Bartolozzi, C., Taba, B., Censi, A., . . . Scaramuzza, D. (2019). *Event-based Vision: A Survey*. https://arxiv.org/abs/1904.08405

Gandhi, S., Mosleh, W., Shen, J., & Chow, C.-M. (2018). Automation, machine learning, and artificial intelligence in echocardiography: A brave new world. *Echocardiography (Mount Kisco, N.Y.)*, *35*(9), 1402–1418. doi:10.1111/echo.14086 PMID:29974498

Gandhi, V., Prasad, G., Coyle, D., Behera, L., & McGinnity, T. M. (2014). EEG-Based mobile robot control through an adaptive brain-robot interface. *IEEE Transactions on Systems, Man, and Cybernetics. Systems*, *44*(9), 1278–1285. doi:10.1109/TSMC.2014.2313317

García, E., Paredes, J. A., Álvarez, F. J., Pérez, M. C., & García, J. J. (2015). Spreading sequences in active sensing: A review. *Signal Processing*, *106*, 88–105. doi:10.1016/j.sigpro.2014.07.002

Gaspar, J., Winters, N., & Santos-Victor, J. (2000). Vision-based navigation and environmental representations with an omnidirectional camera. *IEEE Transactions on Robotics and Automation*, *16*(6), 890–898. doi:10.1109/70.897802

Gastrich, M. D., Barone, J., Bachmann, G., Anderson, M., & Balica, A. (2011). Robotic surgery: Review of the latest advances, risks, and outcomes. *Journal of Robotic Surgery*, *5*(2), 79–97. doi:10.100711701-011-0246-y PMID:27637534

Gaudenzi, P., Carbonaro, R., & Benzi, E. (2000). Control of beam vibrations by means of piezoelectric devices: Theory and experiments. *Composite Structures*, *50*(4), 373–379. doi:10.1016/S0263-8223(00)00114-8

Gencay, R., & Dechert, W. D. (1992). An algorithm for the n Lyapunov exponents of an n-dimensional unknown dynamical system. *Physica D. Nonlinear Phenomena*, *59*(1-3), 142–157. doi:10.1016/0167-2789(92)90210-E

George Thuruthel, T., Ansari, Y., Falotico, E., & Laschi, C. (2018). Control Strategies for Soft Robotic Manipulators: A Survey. *Soft Robotics*, *5*(2), 149–163. doi:10.1089oro.2017.0007 PMID:29297756

George, E. I., Brand, T. C., LaPorta, A., Marescaux, J., & Satava, R. M. (2018). Origins of Robotic Surgery: From Skepticism to Standard of Care. *Journal of the Society of Laparoscopic & Robotic Surgeons, 22*.

Ghamari, S. M., Narm, H. G., & Mollaee, H. (2022). Fractional order fuzzy PID controller design on buck converter with antlion optimization algorithm. *IET Control Theory & Applications*, *16*(3), 340–352. doi:10.1049/cth2.12230

Ghorbani, A., Ouyang, D., & Abid, A. (2020). Deep learning interpretation of echocardiograms. npj. *Digital Medicine*, *3*, 10. PMID:31993508

Ghosh, A., & Saha, S. (2020). *Interactive Game-Based Motor Rehabilitation Using Hybrid Sensor Architecture*, 312–337. doi:10.4018/978-1-5225-9643-1.ch015

Ghosh, A., & Dey, S. (2021). *"Sensing the Mind": An Exploratory Study About Sensors Used in E-Health and M-Health Applications for Diagnosis of Mental Health Condition*. Springer.

Ghosh, A., Saha, S., & Konar, A. (2020). Fuzzy Posture Matching for Pain Recovery Using Yoga. *Advances in Intelligent Systems and Computing*, *999*(January), 957–967. doi:10.1007/978-981-13-9042-5_82

Gibbs, S. (2017). Google sibling waymo launches fully autonomous ride-hailing service. *The Guardian, 7*.

Gingold, J. A., Gueye, N. A., & Falcone, T. (2018). Minimally Invasive Approaches to Myoma Management. *Journal of Minimally Invasive Gynecology*, *25*(2), 237–250. doi:10.1016/j.jmig.2017.07.007 PMID:28734973

Giovanni, B., & Olivier, L. M. (2020). Artificial Intelligence in Medicine: Today and Tomorrow. *Frontiers in Medicine*, *7*(27), 1–6. PMID:32118012

Giovannopoulou, E., Prodromidou, A., Blontzos, N., & Iavazzo, C. (2021). The Emerging Role of Robotic Single-site Approach for Myomectomy: A Systematic Review of the Literature. *Surgical Innovation*, *28*(3), 352–359. doi:10.1177/1553350620988227 PMID:33555235

Gitas, G., Hanker, L., Rody, A., Ackermann, J., & Alkatout, I. (2022). Robotic surgery in gynecology: Is the future already here? *Minimally Invasive Therapy & Allied Technologies*, *31*(6), 1–10. doi:10.1080/13645706.2021.2010763 PMID:34989636

Giulianotti, P. C. (2020). The future of robotic liver surgery. *Hepatobiliary Surgery and Nutrition*, *9*(4), 546–547. doi:10.21037/hbsn.2020.03.17 PMID:32832516

Glick, P., Suresh, S. A., Ruffatto, D., Cutkosky, M., Tolley, M. T., & Parness, A. (2018). A Soft Robotic Gripper With Gecko-Inspired Adhesive. *IEEE Robotics and Automation Letters*, *3*(2), 903–910. doi:10.1109/LRA.2018.2792688

Glyzin, D. S., Glyzin, S. D., Kolesov, A. Y., & Rozov, N. K. (2005). The Dynamic Renormalization Method for Finding the Maximum Lyapunov Exponent of a Chaotic Attractor. *Differential Equations*, *41*(2), 284–289. doi:10.100710625-005-0159-5

Golshahr, A., Natarajan, E., Santhosh, M. S., Sasikumar, R., Ramesh, S., & Durairaj, R. (2018). Multi wall carbon nanotube reinforced silicone for aerospace applications. *International Journal of Mechanical and Production Engineering Research Development*, *8*(4), 743–752. doi:10.24247/ijmperdaug201880

Goury, O., & Duriez, C. (2018). Fast, Generic, and Reliable Control and Simulation of Soft Robots Using Model Order Reduction. *IEEE Transactions on Robotics*, *34*(6), 1565–1576. doi:10.1109/TRO.2018.2861900

Gouta, H., Said, S. H., Turki, A., & Sahli, F. (2019). Experimental sensor less control for a coupled tank system using high gain adaptive observer and nonlinear generalized predictive strategy. *ISA Transactions*, *87*, 187–199. doi:10.1016/j.isatra.2018.11.046 PMID:30545770

Grove, A. (1967). Physics and Technology of Semiconductor Device (1st ed.). Wiley International Edition.

Guerin, S., Huaulmé, A., Lavoue, V., Jannin, P., & Timoh, K. N. (2022). Review of automated performance metrics to assess surgical technical skills in robot-assisted laparoscopy. *Surgical Endoscopy*, *36*(2), 853–870. doi:10.100700464-021-08792-5 PMID:34750700

Guesbaya, T. (n.d.). *Identification et commande de robot manipulateur rigide et flexible en utilisant les réseaux de neurones et la logique floue* [*Identification and control of rigid and flexible manipulator robots using neural networks and fuzzy logic.*]. University of Biskra.

Gu, G., Wang, D., Ge, L., & Zhu, X. (2021). Analytical Modeling and Design of Generalized Pneu-Net Soft Actuators with Three-Dimensional Deformations. *Soft Robotics*, *8*(4), 462–477. doi:10.1089oro.2020.0039 PMID:32822253

Guizzo, E. (2011). How Google's self-driving car works. *IEEE Spectrum Online*, *18*(7), 1132–1141.

Gul, J. Z., Sajid, M., Rehman, M. M., Siddiqui, G. U., Shah, I., Kim, K.-H., Lee, J.-W., & Choi, K. H. (2018). 3D printing for soft robotics – a review. *Science and Technology of Advanced Materials*, *19*(1), 243–262. doi:10.1080/14686996.2018.1431862 PMID:29707065

Gültekin, İ. B., Karabük, E., Köse, M. F., Gültekin, İ. B., Karabük, E., & Köse, M. F. (2021). "Hey Siri! Perform a type 3 hysterectomy. Please watch out for the ureter!" What is autonomous surgery and what are the latest developments? *Journal of the Turkish German Gynecological Association*, *22*(1), 58–70. doi:10.4274/jtgga.galenos.2021.2020.0187 PMID:33624493

Gunel, B. (2002). Room shape and size estimation using directional impulse response measurements. *Proc. Forum Acusticum Sevilla.*

Guo, G., & Wen, S. (2016). Communication Scheduling and Control of a Platoon of Vehicles in VANETs. *IEEE Transactions on Intelligent Transportation Systems*, *17*(6), 1551–1563. doi:10.1109/TITS.2015.2505407

Guo, N., Wang, M., Guo, W., Yu, J., & Feng, J. (2014). Study on forming mechanism of appearance defect in rotating arc narrow gap horizontal GMAW. *International Journal of Advanced Manufacturing Technology*, *2014*(75), 15–20. doi:10.100700170-014-6127-6

Gupta, K. C. (1997). Solution manual for mechanics and control of robots. In Mechanical Engineering Series. doi:10.1007/978-1-4612-1840-1

Gupta, N. K. (1980). Frequency-Shaped Cost Functionals: Extension of Linear-Quadratic-Gaussian Design Methods. *Journal of Guidance and Control*, *3*(6), 529–335. doi:10.2514/3.19722

Gürleyük, S. S., & Cinal, Ş. (2007). Robust three-impulse sequence input shaper design. *Journal of Vibration and Control*, *13*(12), 1807–1818. doi:10.1177/1077546307080012

Gutmann, J.-S., Burgard, W., Fox, D., & Konolige, K. (n.d.). An experimental comparison of localization methods. *Proceedings IEEE/RSJ International Conference on Intelligent Robots and Systems. Innovations in Theory, Practice and Applications (Cat. No.98CH36190)*, *2*(1), 736–743. 10.1109/IROS.1998.727280

Guzik, M., Kowalczyk, K., Woźniak, S., Fraś, M., Juros, K., Kaczor, D., & Walas, P. (2019). Acoustic source localization using drone-embedded microphone array. *The Journal of the Acoustical Society of America*, *146*(4), 3058–3059. doi:10.1121/1.5137614

Hadded, M., Lasgouttes, J. M., Nashashibi, F., & Xydias, I. (2018). Platoon Route Optimization for Picking up Automated Vehicles in an Urban Network. *IEEE Conference on Intelligent Transportation Systems, Proceedings, ITSC*, 1987–1992. 10.1109/ITSC.2018.8569809

Hahn, G. J., & Meeker, W. Q. (2004). *How to Plan an Accelerated Life Test (E-Book).* ASQ Quality Press.

Haido, J. H., Bakar, B. A., Abdul-Razzak, A. A., & Jayaprakash, J. (2010). Dynamic response simulation for reinforced concrete slabs. *Simulation Modelling Practice and Theory*, *18*(6), 696–711. doi:10.1016/j.simpat.2010.01.011

Haitao, G., Qingbao, Z., & Shoujiang, X. (2007). Rapid-exploring random tree algorithm for path planning of robot based on grid method. *Journal of Nanjing Normal University*, *2*(14).

Hallé, S., Laumonier, J., & Chaib-draa, B. (2004). A decentralized approach to collaborative driving coordination. *IEEE Conference on Intelligent Transportation Systems, Proceedings, ITSC*, 453–458. 10.1109/ITSC.2004.1398942

Hamet, P., & Tremblay, J. (2017). Artificial intelligence in medicine. *Metabolism: Clinical and Experimental*, *69S*, S36–S40. doi:10.1016/j.metabol.2017.01.011 PMID:28126242

Han, E. S., & Advincula, A. P. (2019). Safety in Minimally Invasive Surgery. *Obstetrics and Gynecology Clinics of North America*, *46*(2), 389–398. doi:10.1016/j.ogc.2019.01.013 PMID:31056139

Han, E. S., & Advincula, A. P. (2021). Robotic Surgery: Advancements and Inflection Points in the Field of Gynecology. *Obstetrics and Gynecology Clinics of North America*, *48*(4), 759–776. doi:10.1016/j.ogc.2021.07.004 PMID:34756295

Hanif, W. W., Risby, M. S., & Noor, M. M. (2015). Influence of carbon nanotube inclusion on the fracture toughness and ballistic resistance of twaron/epoxy composite panels. *Procedia Engineering*, *114*, 118–123. doi:10.1016/j.proeng.2015.08.049

Harding, V., & Williams, J. (2016). Improving efficiency in robotic theatres in the Royal Wolverhampton NHS Trust: A gynaecology theatres perspective. *Journal of Perioperative Practice*, *26*(3), 46–49. doi:10.1177/175045891602600304 PMID:27149833

Harfouch, Y. A., Yuan, S., & Baldi, S. (2017). Adaptive control of interconnected networked systems with application to heterogeneous platooning. *IEEE International Conference on Control and Automation, ICCA*, 212–217. 10.1109/ICCA.2017.8003062

Hassan, M., Dubay, R., Li, C., & Wang, R. (2007). Active vibration control of a flexible one-link manipulator using a multivariable predictive controller. *Mechatronics*, *17*(6), 311–323. doi:10.1016/j.mechatronics.2007.02.004

Hata, A., & Wolf, D. (2014). Road marking detection using LIDAR reflective intensity data and its application to vehicle localization. *2014 17th IEEE International Conference on Intelligent Transportation Systems, ITSC 2014*, 584–589. 10.1109/ITSC.2014.6957753

Hedrick, J., Chen, Y., & Mahal, S. (2001). *Optimized vehicle control/communication interaction in an automated highway system*. Retrieved from https://escholarship.org/uc/item/8p23v0q4

He, H., Li, J., & Stoica, P. (2012). *Waveform design for active sensing systems: a computational approach*. Cambridge University Press. doi:10.1017/CBO9781139095174

Held, D., Thrun, S., & Savarese, S. (2016). Learning to track at 100 FPS with deep regression networks. Lecture Notes in Computer Science, 9905, 749–765. doi:10.1007/978-3-319-46448-0_45

Heli, H., & Abutalebi, H. R. (2011). Localization of multiple simultaneous sound sources in reverberant conditions using blind source separation methods. *International Symposium on Artificial Intelligence and Signal Processing (AISP)*, 1–5. 10.1109/AISP.2011.5960978

He, M., Takeuchi, E., Ninomiya, Y., & Kato, S. (2016). Precise and efficient model-based vehicle tracking method using Rao-Blackwellized and scaling series particle filters. *IEEE International Conference on Intelligent Robots and Systems*, 117–124. 10.1109/IROS.2016.7759043

Herbordt, W., & Kellermann, W. (2003). Adaptive beamforming for audio signal acquisition. In *Adaptive Signal Processing*, (pp. 155–194). Springer. doi:10.1007/978-3-662-11028-7_6

Herman, A. (2012). *Freedom's forge: how American business produced victory in World War II*. Random House.

Herman, P., Prasad, G., McGinnity, T. M., & Coyle, D. (2008). Comparative analysis of spectral approaches to feature extraction for EEG-based motor imagery classification. *IEEE Transactions on Neural Systems and Rehabilitation Engineering*, *16*(4), 317–326. doi:10.1109/TNSRE.2008.926694 PMID:18701380

Hiller, J., & Lipson, H. (2012). Automatic Design and Manufacture of Soft Robots. *IEEE Transactions on Robotics*, *28*(2), 457–466. doi:10.1109/TRO.2011.2172702

Hiller, J., & Lipson, H.HillerJonathan. (2014). Dynamic Simulation of Soft Multimaterial 3D-Printed Objects. *Soft Robotics*, *1*(1), 88–101. doi:10.1089oro.2013.0010

Hinton, G. E. (2009). Deep belief networks. *Scholarpedia.*, *4*(5), 5947. doi:10.4249cholarpedia.5947

Hiramatsu, Y. (2021). Nerve-Sparing Robotic Radical Hysterectomy for the Beginner in Robotic Surgery. *Surgery Journal (New York, N.Y.)*, *7*(S 02), S84–S96. doi:10.1055-0041-1739121 PMID:35111934

Hochreiter, S., & Schmidhuber, J. (1997). Long short-term memory. *Neural Computation*, *9*(8), 1735–1780. doi:10.1162/neco.1997.9.8.1735 PMID:9377276

Hockstein, N. G., Gourin, C. G., Faust, R. A., & Terris, D. J. (2007). A history of robots: From science fiction to surgical robotics. *Journal of Robotic Surgery*, *1*(2), 113–118. doi:10.100711701-007-0021-2 PMID:25484946

Hoeger, R., Amditis, A., Kunert, M., Hoess, A., Flemish, F., Krueger, H.-P., ... Beutner, A. (2008). Highly automated vehicles for intelligent transport: Haveit approach. *ITS World Congress.*

Holloway, R. W., Patel, S. D., & Ahmad, S. (2009). Robotic surgery in gynecology. *Scandinavian Journal of Surgery*, *98*(2), 96–109. doi:10.1177/145749690909800205 PMID:19799047

Holzinger, A., Langs, G., Denk, H., Zatloukal, K., & Müller, H. (2019). Causability and explainability of artificial intelligence in medicine. *WIREs Data Mining Knowl Discov.*, *9*(4), e1312. doi:10.1002/widm.1312 PMID:32089788

Hooper, J. (2004). *From Darpa Grand Challenge 2004 DARPA's Debacle in the Desert.* Retrieved September 20, 2009, from Popular Science website: https://www.popsci.com/scitech/article/2004-06/darpa-grand-challenge-2004darpas-debacle-desert/

Hossain, A. B. M. A., Rahman, M. W., & Riheen, M. A. (2015). Left and right hand movements EEG signals classification using wavelet transform and probabilistic neural network. *Iranian Journal of Electrical and Computer Engineering*, *5*(1), 92–101. doi:10.11591/ijece.v5i1.pp92-101

Hottenrott, S., Schlesinger, T., Helmer, P., Meybohm, P., Alkatout, I., & Kranke, P. (2020). Do Small Incisions Need Only Minimal Anesthesia? – Anesthetic Management in Laparoscopic and Robotic Surgery. *Journal of Clinical Medicine*, *9*(12), 4058. doi:10.3390/jcm9124058 PMID:33334057

Hsiao, L. Y., Jing, L., Li, K., Yang, H., Li, Y., & Chen, P. Y. (2020). Carbon nanotube-integrated conductive hydrogels as multifunctional robotic skin. *Carbon*, *161*, 784–793. doi:10.1016/j.carbon.2020.01.109

Huang, R. H., Chang, B. J., Tsai, Y. L., & Liang, Y. H. (2018). Mobile Edge Computing-Based Vehicular Cloud of Cooperative Adaptive Driving for Platooning Autonomous Self Driving. *Proceedings - 2017 IEEE 7th International Symposium on Cloud and Service Computing, SC2 2017,* 32–39. 10.1109/SC2.2017.13

Huang, H., Lin, J., Wu, L., Fang, B., Wen, Z., & Sun, F. (2020). Machine learning-based multi-modal information perception for soft robotic hands. *Tsinghua Science and Technology*, *25*(2), 255–269. doi:10.26599/TST.2019.9010009

Huang, J., Chen, H., He, J., Yu, S., Pan, W., & Fan, D. (2019). Narrow gap applications of swing TIG-MIG hybrid weldings. *Journal of Materials Processing Technology*, *271*, 609–614. doi:10.1016/j.jmatprotec.2019.04.043

Huang, Z., Chu, D., Wu, C., & He, Y. (2019). Path Planning and Cooperative Control for Automated Vehicle Platoon Using Hybrid Automata. *IEEE Transactions on Intelligent Transportation Systems*, *20*(3), 959–974. doi:10.1109/TITS.2018.2841967

Hur, C., & Falcone, T. (2021). Robotic treatment of bowel endometriosis. *Best Practice & Research. Clinical Obstetrics & Gynaecology*, *71*, 129–143. doi:10.1016/j.bpobgyn.2020.05.012 PMID:32684433

Hu, Y., Liu, J., Spielberg, A., Tenenbaum, J. B., Freeman, W. T., Wu, J., Rus, D., & Matusik, W. (2019). ChainQueen: A Real-Time Differentiable Physical Simulator for Soft Robotics. *2019 International Conference on Robotics and Automation (ICRA)*, 6265–6271. 10.1109/ICRA.2019.8794333

Hwang, S., Kim, N., Choi, Y., Lee, S., & Kweon, I. S. (2016). Fast multiple objects detection and tracking fusing color camera and 3D LIDAR for intelligent vehicles. *2016 13th International Conference on Ubiquitous Robots and Ambient Intelligence, URAI 2016*, 234–239. 10.1109/URAI.2016.7625744

Iavazzo, C., & Gkegkes, I. D. (2017). Cost-benefit analysis of robotic surgery in gynaecological oncology. *Best Practice & Research. Clinical Obstetrics & Gynaecology*, *45*, 7–18. doi:10.1016/j.bpobgyn.2017.03.008 PMID:28526229

IEEE Standard Glossary of Software Engineering Terminology. (2002). *IEEE STD 610.12-1990. Standards Coordinating Committee of the Computer Society of IEEE.* Available online: https://ieeexplore.ieee.org/document/159342

Iglesias, D. A., & Ramirez, P. T. (2011). Role of minimally invasive surgery in staging of ovarian cancer. *Current Treatment Options in Oncology*, *12*(3), 217–229. doi:10.100711864-011-0155-3 PMID:21503633

İlman, M. M., Yavuz, Ş., & Taser, P. Y. (2022). Generalized Input Preshaping Vibration Control Approach for Multi-Link Flexible Manipulators using Machine Intelligence. *Mechatronics*, *82*, 102735. doi:10.1016/j.mechatronics.2021.102735

Imran, S. M., Talukdar, M. T. F., Sakib, S. K., Pathan, N. S., & Fattah, S. A. (2014). Motor imagery EEG signal classification scheme based on wavelet domain statistical features. *1st International Conference on Electrical Engineering and Information and Communication Technology, ICEEICT*. 10.1109/ICEEICT.2014.6919172

Ince, G., Nakadai, K., Rodemann, T., Hasegawa, Y., Tsujino, H., & Imura, J. I. (2009). *Ego noise suppression of a robot using template subtraction*. 199–204.

Ingle, V. K., & Proakis, J. G. (2008). Digital Signal Processing Using MATLAB, (3rd Edition). In Electronics {&} Communications Engineering Journal, 5(1).

Insausti, X., Hogstad, B. O., & Pätzold, M. (2020). Modelling and Simulation of Ego-Noise of Unmanned Aerial Vehicles. *IEEE 91st Vehicular Technology Conference (VTC2020-Spring)*, 1–5.

Institute, I. I. (2018, July 30). *Background on: Self-driving cars and insurance*. Retrieved October 29, 2019, from Insurance Information Institute website: https://www.iii.org/article/background-on-self-driving-cars-and-insurance

International, S. A. E. (2016). *Taxonomy and definitions for terms related to driving automation systems for on-road motor vehicles*. SAE International (J3016).

International, S. A. E. (2021, May). *SAE levels of Driving Automation™ refined for clarity and international audience.* SAE International. Retrieved from https://www.sae.org/blog/sae-j3016-update

Isermann, R. (2005). Mechatronic systems: Fundamentals. In *Mechatronic Systems*. Fundamentals. doi:10.1007/1-84628-259-4/COVER

ISO. (2016). *BS EN ISO 148-1: 2016- Metallic Materials*. Charpy Pendulum Impact Test. Test Method.

Ito, A., Kanayama, T., Suzuki, M., & Makino, S. (2005). Internal noise suppression for speech recognition by small robots. *Ninth European Conference on Speech Communication and Technology*. 10.21437/Interspeech.2005-260

Iwata, S., Murayama, M., & Kojima, Y. (2009). Application of narrow gap welding process with high speed rotating arc to box column joints of heavy thick plates. *Jef Technical Report*, *14*(14), 16–21.

Izobov, N. A. (2006). *Introduction in theory of Lyapunov indexes*. Minsk BGU.

Jacofsky, D. J., & Allen, M. (2016). Robotics in Arthroplasty: A Comprehensive Review. *The Journal of Arthroplasty, 31*(10), 2353–2363. doi:10.1016/j.arth.2016.05.026 PMID:27325369

Jain, A. K., Mao, J., & Mohiuddin, K. M. (1996). Artificial neural networks: A tutorial. *Computer, 29*(3), 31–44. https://doi.org/10.1109/2.485891

Jain, M., & Saxena, R. (2017). Overview of VANET: Requirements and its routing protocols. *2017 International Conference on Communication and Signal Processing (ICCSP)*, 1957–1961.

Jain, S., Hote, Y. V., Dehuri, P., Mittal, D., & Siddhartha, V. (2020). Experimental validation of fractional order internal model control design on buck boost converter. *Measurement and Control*, 1–19.

Jansen, W., Laurijssen, D., Kerstens, R., Daems, W., & Steckel, J. (2020). *In-Air Imaging Sonar Sensor Network with Real-Time Processing Using GPUs*. Springer. doi:10.1007/978-3-030-33509-0_67

Janssen, R., Zwijnenberg, H., Blankers, I., & de Kruijff, J. (2015). *Truck platooning. Driving The Future of Transportation*. TNO.

Jenita Amali Rani, B., & Umamakeswari, A. (2015). Electroencephalogram-based Brain Controlled Robotic Wheelchair. *Indian Journal of Science and Technology, 8*(S9), 188. doi:10.17485/ijst/2015/v8iS9/65580

Jensen, P. S., Grace, K. W., Attariwala, R., Colgate, J. E., & Glucksberg, M. R. (1997). Toward robot-assisted vascular microsurgery in the retina. *Graefes Archive for Clinical and Experimental Ophthalmology, 235*(11), 696–701. doi:10.1007/BF01880668 PMID:9407227

Jeyasingh, P., & Mohamed Ismail, M. (2018). Real-time Multi Source Speech Enhancement based on Sound Source Separation using Microphone Array. *Conference on Emerging Devices and Smart Systems (ICEDSS)*, 183–187. 10.1109/ICEDSS.2018.8544373

Jia, C., Yan, Q., Wei, B., & Wu, C. (2018). Rotating-Tungsten Narrow-Groove GTAW for Thick Plates. *Welding Journal*.

Jia, D., Lu, K., Wang, J., Zhang, X., & Shen, X. (2015). A survey on platoon-based vehicular cyber-physical systems. *IEEE Communications Surveys & Tutorials, 18*(1), 263–284.

Jiang, H., Wang, Z., Liu, X., Chen, X., Jin, Y., You, X., & Chen, X. (2017). A two-level approach for solving the inverse kinematics of an extensible soft arm considering viscoelastic behavior. *2017 IEEE International Conference on Robotics and Automation (ICRA)*, 6127–6133. 10.1109/ICRA.2017.7989727

Jian, X., & Wu, H. (2020). Influence of the longitudinal magnetic field on the formation of the bead in narrow gap gas tungsten arc welding. *Metals, 10*(10), 1351. doi:10.3390/met10101351

Jie, Z., & Gunes, H. (2020). Investigating Taste-liking with a Humanoid Robot Facilitator. *29th IEEE International Conference on Robot and Human Interactive Communication (RO-MAN)*, 131–136.

Joachimczak, M., & Wróbel, B. (2012). Co-evolution of morphology and control of soft-bodied multicellular animats. *Proceedings of the 14th Annual Conference on Genetic and Evolutionary Computation*, 561–568. 10.1145/2330163.2330243

Jolad, B., & Khanai, R. (2019). An Art of Speech Recognition: A Review. *2nd International Conference on Signal Processing and Communication (ICSPC)*, 31–35.

Jones, G. (2005). Echolocation. *Current Biology, 15*(13), R484–R488. doi:10.1016/j.cub.2005.06.051 PMID:16005275

Juang, J.-N., Turner, J. D., & Chun, H. M. (1985). Closed-Form Solutions for Feedback Control with Terminal Constraints. *Journal of Guidance, Control, and Dynamics, 8*(1), 39–43. doi:10.2514/3.19932

Jung, Y. W., Kim, S. W., & Kim, Y. T. (2009). Recent advances of robotic surgery and single port laparoscopy in gynecologic oncology. *Journal of Gynecologic Oncology*, *20*(3), 137–144. doi:10.3802/jgo.2009.20.3.137 PMID:19809546

Junkins, J. L., & Turner, J. D. (1986). *Optimal Spacecraft Rotational Maneuvers*. Elsevier Science Publishers.

Jurgen, R. K. (2006). *Adaptive cruise control*. Academic Press.

Jurgen, R. (2011). Design and Analysis of an Adaptive Real-Time Advisory System for Improving Real World Fuel Economy in a Hybrid Electric Vehicle (2010-01-0835). In *Electric and Hybrid-Electric Vehicles* (pp. 35–45). Overviews and Viewpoints. Retrieved from https://ieeexplore.ieee.org/document/8522869

Kakkavas, A., Casta, M. H., Stirling-gallacher, R. A., & Nossek, J. A. (2018). Multi-Array 5G V2V Relative Positioning: Performance Bounds. *2018 IEEE Global Communications Conference (GLOBECOM)*, 206–212. Retrieved from https://arxiv.org/abs/1808.07373

Kampers, J., Gerhardt, E., Sibbertsen, P., Flock, T., Klapdor, R., Hertel, H., Jentschke, M., & Hillemanns, P. (2021). Protective operative techniques in radical hysterectomy in early cervical carcinoma and their influence on disease-free and overall survival: A systematic review and meta-analysis of risk groups. *Archives of Gynecology and Obstetrics*, *304*(3), 577–587. doi:10.100700404-021-06082-y PMID:34021804

Kanade, T., Thorpe, C., & Whittaker, W. (1986). Autonomous land vehicle project at CMU. *CSC '86 Proceedings of the 1986 ACM Fourteenth Annual Conference on Computer Science*, 71–80. Retrieved from https://dl.acm.org/citation.cfm?id=325197

Kant, P., Hazarika, J., & Laskar, S. H. (2019). Wavelet transform based approach for EEG feature selection of motor imagery data for braincomputer interfaces. *Proceedings of the 3rd International Conference on Inventive Systems and Control, ICISC, Icisc*, 101–105. 10.1109/ICISC44355.2019.9036445

Kao, W. W. (1991). Integration of GPS and dead-reckoning navigation systems. *Proceedings - Society of Automotive Engineers, 2*(2), 635–643. doi:10.1109/vnis.1991.205808

Karabutov, N. (2015). Structural methods of estimation Lyapunov exponents linear dynamic system. *International Journal of Intelligent Systems and Applications*, *7*(10), 1-11. . doi:10.5815/ijisa.2015.10.01

Karabutov, N. (2018). About Lyapunov exponents identification for systems with periodic coefficients. *International Journal of Intelligent Systems and Applications*, *10*(11), 1-10. . doi:10.5815/ijisa.2018.11.01

Karabutov, N. N. (2009). *Structural Identification of Systems. Analysis of Information Structures*. URSS/ Librokom.

Karabutov, N. N. (2018). *Frameworks in identification problems: Construction and analysis*. Lenand.

Karginova, N., Byttner, S., & Svensson, M. (2012). *Data-driven methods for classification of driving styles in buses*. SAE Technical Papers. doi:10.4271/2012-01-0744

Karlsson, K., & Larsson, M. (2015). On the effect of vertical spatial diversity on V2V communication for three different platooning scenarios. *2015 9th European Conference on Antennas and Propagation (EuCAP), 5*(2), 1–5.

Karnadi, F. K., Mo, Z. H., & Lan, K. C. (2007). Rapid generation of realistic mobility models for VANET. *IEEE Wireless Communications and Networking Conference, WCNC*, 2508–2513. doi:10.1109/WCNC.2007.467

Karnopp, D. C., Margolis, D. L., & Rosenberg, R. C. (2012). *System Dynamics: Modeling, Simulation, and Control of Mechatronic Systems* (6th ed.). John Wiley & Sons. doi:10.1002/9781118152812

Katzschmann, R. K., Marchese, A. D., & Rus, D. (2016). Hydraulic Autonomous Soft Robotic Fish for 3D Swimming. In M. A. Hsieh, O. Khatib, & V. Kumar (Eds.), *Experimental Robotics: The 14th International Symposium on Experimental Robotics,* (pp. 405–420). Springer International Publishing. 10.1007/978-3-319-23778-7_27

Kaul, V., Enslin, S., & Gross, S. A. (2020). History of artificial intelligence in medicine. *Gastrointestinal Endoscopy, 92*(4), 807–812. doi:10.1016/j.gie.2020.06.040 PMID:32565184

Kaur, H., & Meenakshi. (2018). Analysis of VANET geographic routing protocols on real city map. *RTEICT 2017 - 2nd IEEE International Conference on Recent Trends in Electronics, Information and Communication Technology Proceedings*, 895–899. doi:10.1109/RTEICT.2017.8256727

Kavathekar, P., & Chen, Y. (2011). Vehicle platooning: A brief survey and categorization. *International Design Engineering Technical Conferences and Computers and Information in Engineering Conference, 54808*, 829–845.

Kay, S. M. (1993). *Fundamentals of statistical signal processing*. Prentice Hall PTR.

Kendall, A., Hawke, J., Janz, D., Mazur, P., Reda, D., & Allen, J. M. … Shah, A. (2019). Learning to drive in a day. *Proceedings - IEEE International Conference on Robotics and Automation,* 8248–8254. doi:10.1109/ICRA.2019.8793742

Kennedy, J., & Eberhart, R. (1995). Particle swarm optimization, *Proceedings of the IEEE International Conference on Neural Networks, 4*, 1942-1948. 10.1109/ICNN.1995.488968

Kenngott, H. G., Fischer, L., Nickel, F., Rom, J., Rassweiler, J., & Müller-Stich, B. P. (2012). Status of robotic assistance – A less traumatic and more accurate minimally invasive surgery? *Langenbeck's Archives of Surgery, 397*(3), 333–341. doi:10.100700423-011-0859-7 PMID:22038293

Kerstens, R., Jansen, W., & Steckel, J. (2022). *Radial Velocity Estimation for Multiple Objects using In-Air Sonar with MIMO Virtual Arrays*. IEEE. doi:10.1109/ACCESS.2022.3154395

Kerstens, R., Laurijssen, D., & Steckel, J. (2017). Low-cost One-bit MEMS Microphone Arrays for In-air Acoustic Imaging Using FPGA ' s. *IEEE SENSORS, 2017-. Decem, 1–3*, 1–3. doi:10.1109/ICSENS.2017.8234087

Kerstens, R., Laurijssen, D., & Steckel, J. (2019a). An Optimized Planar MIMO Array Approach to In-Air Synthetic Aperture Sonar. *IEEE Sensors Letters, 3*(11), 1–4. doi:10.1109/LSENS.2019.2951061

Kerstens, R., Laurijssen, D., & Steckel, J. (2019b). eRTIS: A Fully Embedded Real Time 3D Imaging Sonar Sensor for Robotic Applications. *International Conference on Robotics and Automation (ICRA)*, 1438–1443. 10.1109/ICRA.2019.8794419

Kerstens, R., Verellen, T., & Steckel, J. (2020). Comparing Adaptations of MUSIC beamforming for 3D In-Air Sonar Applications with Coherent Source-spaces. *Berlin Beamforming Conference (BeBeC)*, 1–12.

Kessler, A. M. (2015). *Elon musk says self-driving tesla cars will be in the us by summer. The New York Times*. Retrieved from https://www.nytimes.com/2015/03/20/business/elon-musk-says-self-driving-tesla-cars-will-be-in-the-us-by-summer.html?_r=0

Khandoker, A. H., Palaniswami, M., & Karmakar, C. K. (2009). Support vector machines for automated recognition of obstructive sleep apnea syndrome from ECG recordings. *IEEE Transactions on Information Technology in Biomedicine, 13*(1), 37–48. doi:10.1109/TITB.2008.2004495 PMID:19129022

Kho, R. M., Hilger, W. S., Hentz, J. G., Magtibay, P. M., & Magrina, J. F. (2007). Robotic hysterectomy: Technique and initial outcomes. *American Journal of Obstetrics and Gynecology, 197*(1), 113. doi:10.1016/j.ajog.2007.05.005 PMID:17618782

Kim, D., & Singhose, W. (2010). Performance studies of human operators driving double-pendulum bridge cranes. *Control Engineering Practice*, *18*(6), 567–576. doi:10.1016/j.conengprac.2010.01.011

Kimura, R., Matsunaga, N., Okajima, H., & Kotaki, G. (2017). Design of virtual platoon control system using augmented reality to assist welfare vehicle users. *International Conference on Control, Automation and Systems,* 330–335. doi:10.23919/ICCAS.2017.8204460

Kim, Y. T., Kim, S. W., & Jung, Y. W. (2008). Robotic surgery in gynecologic field. *Yonsei Medical Journal*, *49*(6), 886–890. doi:10.3349/ymj.2008.49.6.886 PMID:19108009

King, H. H., Tadano, K., Donlin, R., Friedman, D., Lum, M. J., Asch, V., Wang, C., Kawashima, K., & Hannaford, B. (2009, June). Preliminary protocol for interoperable telesurgery. In *International Conference on Advanced Robotics,* (pp. 1-6). IEEE.

Knapp, C., & Carter, G. (1976). The generalized correlation method for estimation of time delay. *IEEE Transactions on Acoustics, Speech, and Signal Processing*, *24*(4), 320–327. doi:10.1109/TASSP.1976.1162830

Knight, J., & Escobar, P. F. (2014). Cost and robotic surgery in gynecology. *Journal of Obstetrics and Gynaecology Research*, *40*(1), 12–17. doi:10.1111/jog.12197 PMID:24118557

Kojima, H., & Hiruma, T. (2006). Evolutionary Learning Acquisition of Optimal Joint Angle Trajectories of Flexible Robot Arm. J. *Robotics Mechatronics*, *18*(1), 103–110. doi:10.20965/jrm.2006.p0103

Koo, Y. J. (2018). Recent advances in minimally invasive surgery for gynecologic indications. *Yeungnam University Journal of Medicine*, *35*(2), 150–155. doi:10.12701/yujm.2018.35.2.150 PMID:31620587

Korosec, K. (2016, June 7). Autonomous Car Sales Will Hit 21 Million by 2035, IHS Says. *Fortune*. Retrieved from https://fortune.com/2016/06/07/autonomous-car-sales-ihs/

Korovesis, N., Kandris, D., Koulouras, G., & Alexandridis, A. (2019). Robot motion control via an eeg-based brain–computer interface by using neural networks and alpha brainwaves. *Electronics (Switzerland)*, *8*(12), 1–15. doi:10.3390/electronics8121387

Korsholm, M., Sørensen, J., Mogensen, O., Wu, C., Karlsen, K., & Jensen, P. T. (2018). A systematic review about costing methodology in robotic surgery: Evidence for low quality in most of the studies. *Health Economics Review*, *8*(1), 21. doi:10.118613561-018-0207-5 PMID:30194567

Kotnik, P. T., Yurkovich, S., & Ozfuner, U. (1988). Acceleration Feedback for Control of a Flexible Manipulator Arm. *Journal of Robotic Systems*, *5*(3), 181–196. doi:10.1002/rob.4620050302

Koutník, J., Cuccu, G., Schmidhuber, J., & Gomez, F. (2013). Evolving large-scale neural networks for vision-based reinforcement learning. *GECCO 2013 - Proceedings of the 2013 Genetic and Evolutionary Computation Conference*, 1061–1068. doi:10.1145/2463372.2463509

Kowadlo, G., & Russell, R. A. (2008). Robot odor localization: A taxonomy and survey. *The International Journal of Robotics Research*, *27*(8), 869–894. doi:10.1177/0278364908095118

Kreković, M., Dokmanić, I., & Vetterli, M. (2016). *EchoSLAM: Simultaneous localization and mapping with acoustic echoes*, 11–15.

Kreyszig, E. (2011). *Advanced Engineering Mathematics* (10th ed.). John Wiley and Son.

Krim, H., & Viberg, M. (1996). Two decades of array signal processing research: The parametric approach. *IEEE Signal Processing Magazine*, *13*(4), 67–94. doi:10.1109/79.526899

Kumarakulasingam, P., & Agah, A. (2008). Neural network-based single sensor sound localization using a mobile robot. *Intelligent Automation \&. Soft Computing*, *14*(1), 89–103.

Kusunose, K., Haga, A., Inoue, M., Fukuda, D., Yamada, H., & Sata, M. (2020). Clinically Feasible and Accurate View Classification of Echocardiographic Images Using Deep Learning. *Biomolecules*, *10*(5), 665. doi:10.3390/biom10050665 PMID:32344829

Kutila, M., Pyykonen, P., Holzhuter, H., Colomb, M., & Duthon, P. (2018). Automotive LiDAR performance verification in fog and rain. *IEEE Conference on Intelligent Transportation Systems, Proceedings, ITSC*, 1695–1701. 10.1109/ITSC.2018.8569624

Kuutti, S., Fallah, S., Katsaros, K., Dianati, M., Mccullough, F., & Mouzakitis, A. (2018). A Survey of the State-of-the-Art Localization Techniques and Their Potentials for Autonomous Vehicle Applications. *IEEE Internet of Things Journal*, *5*(2), 829–846. doi:10.1109/JIOT.2018.2812300

Kwakernaak, H., & Sivan, R. (1972). Linear optimal control systems. Wilev-Interscience.

La Russa, M., Liakou, C., & Burbos, N. (2021). Ultra-minimally invasive approaches for endometrial cancer treatment: Review of the literature. *Minerva Medica*, *112*(1), 31–46. doi:10.23736/S0026-4806.20.07073-1 PMID:33205639

Lakshmanprabu, S. K., & Banu, U. S. (2019). Adaptive multiloop IMC based PID controller tuning using bat optimization algorithm for two interacting conical tank process. *International Journal of Advanced Intelligence Paradigm*, *13*(3).

Laukkonen, J. (2019, November 17). Are Self-Driving Cars Legal in Your State? *LifeWire*. Retrieved from https://www.lifewire.com/are-self-driving-cars-legal-4587765

Lauterbach, R., Matanes, E., & Lowenstein, L. (2017). Review of Robotic Surgery in Gynecology – The Future Is Here. *Rambam Maimonides Medical Journal*, *8*(2), e0019. doi:10.5041/RMMJ.10296 PMID:28467761

LaValle, S. M., & Kuffner, J. J. (2001). Randomized kinodynamic planning. *The International Journal of Robotics Research*, *20*(5), 378–400. https://doi.org/10.1177/02783640122067453

Lawrie, T. A., Liu, H., Lu, D., Dowswell, T., Song, H., Wang, L., & Shi, G. (2019). Robot-assisted surgery in gynaecology. *Cochrane Database of Systematic Reviews*, *4*, CD011422. PMID:30985921

Lazarou, I., Nikolopoulos, S., Petrantonakis, P. C., Kompatsiaris, I., & Tsolaki, M. (2018). EEG-based brain–computer interfaces for communication and rehabilitation of people with motor impairment: A novel approach of the 21st century. *Frontiers in Human Neuroscience*, *12*(January), 1–18. doi:10.3389/fnhum.2018.00014 PMID:29472849

LeCun, Y., Muller, U., Ben, J., Cosatto, E., & Flepp, B. (2005). Off-road obstacle avoidance through end-to-end learning. *Advances in Neural Information Processing Systems*, 739–746.

Lee, C. H., Soong, F. K., & Juang, B. H. (1988). *A segment model based approach to speech recognition*, 501–502.

Lee, C., Kim, M., Kim, Y. J., Hong, N., Ryu, S., Kim, H. J., & Kim, S. (2017). Soft robot review. *International Journal of Control, Automation, and Systems*, *15*(1), 3–15. doi:10.100712555-016-0462-3

Lee, J. Y., Mucksavage, P., Sundaram, C. P., & McDougall, E. M. (2011). Best practices for robotic surgery training and credentialing. *The Journal of Urology*, *185*(4), 1191–1197. doi:10.1016/j.juro.2010.11.067 PMID:21334030

Lee, K.-H., Fu, D. K. C., Leong, M. C. W., Chow, M., Fu, H.-C., Althoefer, K., Sze, K. Y., Yeung, C.-K., & Kwok, K.-W. (2017). Nonparametric Online Learning Control for Soft Continuum Robot: An Enabling Technique for Effective Endoscopic Navigation. *Soft Robotics*, *4*(4), 324–337. doi:10.1089oro.2016.0065 PMID:29251567

Lee, Y. J., van den Berg, N. S., Orosco, R. K., Rosenthal, E. L., & Sorger, J. M. (2021). A narrative review of fluorescence imaging in robotic-assisted surgery. *Laparoscopic Surgery*, *5*, 31. doi:10.21037/ls-20-98 PMID:34549180

Lee, Y. L., Kilic, G. S., & Phelps, J. Y. (2011). Medicolegal review of liability risks for gynecologists stemming from lack of training in robot-assisted surgery. *Journal of Minimally Invasive Gynecology*, *18*(4), 512–515. doi:10.1016/j.jmig.2011.04.002 PMID:21777841

Lee, Z., Kaplan, J., Giusto, L., & Eun, D. (2016). Prevention of iatrogenic ureteral injuries during robotic gynecologic surgery: A review. *American Journal of Obstetrics and Gynecology*, *214*(5), 566–571. doi:10.1016/j.ajog.2015.10.150 PMID:26519785

Leibinger, A., Oldfield, M. J., Rodriguez, Y., & Baena, F. (2016). Minimally disruptive needle insertion: A biologically inspired solution. *Interface Focus*, *6*(3), 20150107. doi:10.1098/rsfs.2015.0107 PMID:27274797

Lepora, N. F., Evans, M., Fox, C. W., Diamond, M. E., Gurney, K., & Prescott, T. J. (2010). Naive Bayes texture classification applied to whisker data from a moving robot. *The 2010 International Joint Conference on Neural Networks (IJCNN)*, 1–8. 10.1109/IJCNN.2010.5596360

Levanon, N. (1988). *Radar principles*. Retrieved from https://ui.adsabs.harvard.edu/abs/1988wi...book.....L

Levanon, N., & Mozeson, E. (2004). *Radar Signals*. John Wiley & Sons. doi:10.1002/0471663085

Levinson, J., Askeland, J., Becker, J., Dolson, J., Held, D., Kammel, S., … Thrun, S. (2011). Towards fully autonomous driving: Systems and algorithms. *IEEE Intelligent Vehicles Symposium Proceedings*, (4), 163–168. doi:10.1109/IVS.2011.5940562

Levinson, K. L., Auer, M., & Escobar, P. F. (2013). Evolving technologies in robotic surgery for minimally invasive treatment of gynecologic cancers. *Expert Review of Medical Devices*, *10*(5), 603–610. doi:10.1586/17434440.2013.827509 PMID:24053253

Li, P., & Zhu, G. (2019). IMC based PID control of servo motors with extended state observer. Mechtronics, Elsevier, 62(2019).

Lien, G. S., Liu, C. W., Teng, M. T., & Huang, Y. M. (2009, May). Integration of two optical image modules and locomotion functions in capsule endoscope applications. In *IEEE 13th International Symposium on Consumer Electronics,* (pp. 828-829). IEEE.

Lim, C. S., Griffith, K. C., Travieso, J., & As-Sanie, S. (2020). To Robot or Not To Robot: The Use of Robotics in Benign Gynecologic Surgery. *Clinical Obstetrics and Gynecology*, *63*(2), 327–336. doi:10.1097/GRF.0000000000000491 PMID:31634158

Lim, J. H., & Leonard, J. J. (2000). Mobile robot relocation from echolocation constraints. *IEEE Transactions on Pattern Analysis and Machine Intelligence*, *22*(9), 1035–1041. doi:10.1109/34.877524

Lim, P. C., & Kang, E. (2017). How to prepare the patient for robotic surgery: Before and during the operation. *Best Practice & Research. Clinical Obstetrics & Gynaecology*, *45*, 32–47. doi:10.1016/j.bpobgyn.2017.04.008 PMID:28579145

Lin, L. C., & Yih, T. W. (1996). Rigid Model-based Neural Network Control of Flexible-link Manipulators. *IEEE Transactions on Robotics and Automation*, *12*(4), 595–602. doi:10.1109/70.508442

Lin, P. S., Wakabayashi, M. T., & Han, E. S. (2009). Role of robotic surgery in endometrial cancer. *Current Treatment Options in Oncology*, *10*(1-2), 33–43. doi:10.100711864-009-0086-4 PMID:19353274

Litjens, G., Ciompi, F., Wolterink, J. M., de Vos, B. D., Leiner, T., Teuwen, J., & Išgum, I. (2019). State-of-the-Art Deep Learning in Cardiovascular Image Analysis. *JACC: Cardiovascular Imaging*, *12*(8 Pt 1), 1549–1565. doi:10.1016/j.jcmg.2019.06.009 PMID:31395244

Litjens, G., Kooi, T., Bejnordi, B. E., Setio, A. A. A., Ciompi, F., Ghafoorian, M., van der Laak, J. A. W. M., van Ginneken, B., & Sánchez, C. I. (2017). A survey on deep learning in medical image analysis. *Medical Image Analysis*, *42*, 60–88. doi:10.1016/j.media.2017.07.005 PMID:28778026

Litvinov, S. (2020). Over-the-horizon detection radar ionosphere support system. *Journal of Physics: Conference Series*, *1632*(1), 012021. doi:10.1088/1742-6596/1632/1/012021

Liu, L., Ouyang, W., Wang, X., Fieguth, P., Chen, J., Liu, X., & Pietikäinen, M. (2019). Deep Learning for Generic Object Detection: A Survey. *International Journal of Computer Vision*. doi:10.1007/s11263-019-01247-4

Liu, A., Tendick, F., Cleary, K., & Kaufmann, C. (2003). A survey of surgical simulation: Applications, technology, and education. *Presence (Cambridge, Mass.)*, *12*(6), 599–614. doi:10.1162/105474603322955905

Liu, B., Jia, D., Lu, K., Ngoduy, D., Wang, J., & Wu, L. (2017). A joint control-communication design for reliable vehicle platooning in hybrid traffic. *IEEE Transactions on Vehicular Technology*, *66*(10), 9394–9409. https://doi.org/10.1109/TVT.2017.2702650

Liu, C.-H., Chen, T.-L., Chiu, C.-H., Hsu, M.-C., Chen, Y., Pai, T.-Y., Peng, W.-G., & Chiang, Y.-P. (2018). Optimal Design of a Soft Robotic Gripper for Grasping Unknown Objects. *Soft Robotics*, *5*(4), 452–465. doi:10.1089oro.2017.0121 PMID:29741987

Liu, H. H., Li, L. J., Shi, B., Xu, C. W., & Luo, E. (2017). Robotic surgical systems in maxillofacial surgery: A review. *International Journal of Oral Science*, *9*(2), 63–73. doi:10.1038/ijos.2017.24 PMID:28660906

Liu, H., Cao, Y., Li, L., Bai, Y., & Liu, J. (2021). Effectiveness of robotic surgery for endometrial cancer: A systematic review and meta-analysis. *Archives of Gynecology and Obstetrics*, *2021*(Sep), 8. PMID:34498124

Liu, H., Lu, D., Shi, G., Song, H., & Wang, L. (2014). Robotic surgery for benign gynaecological disease. *Cochrane Database of Systematic Reviews*. PMID:25502433

Li, Y. L., & Bing, J. Q. (2019). Analytical optimization of IMC-PID design based on performance/robustness trade-off tuning strategy for the modified Smith structure, *Journal of Low Frequency Noise, Vibration, and Active Control. Sage (Atlanta, Ga.)*, 1–16.

Li, Y., Zhou, G., Graham, D., & Holtzhauer, A. (2016). Towards an EEG-based brain-computer interface for online robot control. *Multimedia Tools and Applications*, *75*(13), 7999–8017. doi:10.100711042-015-2717-z

Loutfi, A., & Coradeschi, S. (2006). Smell, think and act: A cognitive robot discriminating odours. *Autonomous Robots*, *20*(3), 239–249. doi:10.100710514-006-7098-8

Lucas, M. A., Phys, F., & Krstic, A. R. (2005). Unmanned tele-operated robots as medical support on the battlefield. *ADF Health, 6*, 34-8.

Lucidi, A., Chiantera, V., Gallotta, V., Ercoli, A., Scambia, G., & Fagotti, A. (2017). Role of robotic surgery in ovarian malignancy. *Best Practice & Research. Clinical Obstetrics & Gynaecology*, *45*, 74–82. doi:10.1016/j.bpobgyn.2017.05.005 PMID:28602522

Lue, J. R., Murray, B., & Bush, S. (2012). Single port robotic hysterectomy technique improving on multiport procedure. *Journal of Minimal Access Surgery*, *8*(4), 156–157. doi:10.4103/0972-9941.103130 PMID:23248446

Luesutthiviboon, S., Malgoezar, A. M. N., Merino-Martinez, R., Snellen, M., Sijtsma, P., & Simons, D. G. (2019). Enhanced HR-CLEAN-SC for resolving multiple closely spaced sound sources. *International Journal of Aeroacoustics, 18*(4–5), 392–413. doi:10.1177/1475472X19852938

Lungu, D. A., Foresi, E., Belardi, P., Nuti, S., Giannini, A., & Simoncini, T. (2021). The Impact of New Surgical Techniques on Geographical Unwarranted Variation: The Case of Benign Hysterectomy. *International Journal of Environmental Research and Public Health, 18*(13), 6722. doi:10.3390/ijerph18136722 PMID:34206452

Lyapunov, A. M. (1992). *General Problem of the Stability of Motion*. Taylor & Francis. doi:10.1080/00207179208934253

M, H., & M.N, S. (2015). A Review on Evaluation Metrics for Data Classification Evaluations. *International Journal of Data Mining & Knowledge Management Process, 5*(2), 01–11. doi:10.5121/ijdkp.2015.5201

Mach, P., Kimmig, R., & Buderath, P. (2020). The role of sentinel-node biopsy in ovarian cancer. *Minerva Ginecologica, 72*, 399–403. PMID:33140629

Mack, M. J. (2001). Minimally invasive and robotic surgery. *Journal of the American Medical Association, 285*(5), 568–572. doi:10.1001/jama.285.5.568 PMID:11176860

Madan, A., & Gupta, D. (2014). Speech feature extraction and classification: A comparative review. *International Journal of Computers and Applications, 90*(9).

Madani, A., Arnaout, R., & Mofrad, M. (2018). Fast and accurate view classification of echocardiograms using deep learning. npj. *Digital Medicine, 1*, 6. PMID:30828647

Maddern, W., Pascoe, G., Linegar, C., & Newman, P. (2017). 1 year, 1000 km: The Oxford RobotCar dataset. *The International Journal of Robotics Research, 36*(1), 3–15.

Magaziner, I. C., & Patinkin, M. (1989). Cold competition: GE wages the refrigerator war. *Harvard Business Review, 89*, 114–124.

Magnusson, M. (2009). *The Three-Dimensional Normal-Distributions Transform – an Efficient Representation for Registration, Surface Analysis, and Loop Detection*. Örebro University. Retrieved from http://www.aass.oru.se/Research/Learning/publications/2009/Magnusson_2009-Doctoral_Thesis-3D_NDT.pdf

Magrina, J. F. (2007). Robotic surgery in gynecology. *European Journal of Gynaecological Oncology, 28*, 77–82. PMID:17479665

Magrina, J. F., & Magtibay, P. M. (2020). Robotic Resection of Diaphragm Metastases in Ovarian Cancer: Technical Aspects. *Journal of Minimally Invasive Gynecology, 27*(6), 1417–1422. doi:10.1016/j.jmig.2019.12.018 PMID:31917330

Mahapatro, S. R., Subudhi, B., & Ghosh, S. (2019). Design and experimental realization of robust decentralized PI controller for a coupled tank system. *ISA Transactions, 89*, 158–168. doi:10.1016/j.isatra.2018.12.003 PMID:30594337

Mamone, V., Condino, S., Cutolo, F., Tamadon, I., Menciassi, A., Murzi, M., Ferrari, M., & Ferrari, V. (2019). Low-Computational Cost Stitching Method in a Three-Eyed Endoscope. *Journal of Healthcare Engineering, 2019*, 5613931. doi:10.1155/2019/5613931 PMID:31316742

Mandel, M. I., Weiss, R. J., & Ellis, D. P. W. (2009). Model-based expectation-maximization source separation and localization. *IEEE Transactions on Audio, Speech, and Language Processing, 18*(2), 382–394. doi:10.1109/TASL.2009.2029711

Mann-Wellenhof, B., Lichtenegger, H., & Wasle, E. (2007). *GNSS-global navigation satellite systems: GPS, GLONASS, Galileo, and more*. Springer Science and Business Media.

Mansour, M., Smith, B. V., & Noaks, D. R. (1988). Active sonar signal simulation. *IEE Colloquium on Simulation Techniques Applied to Sonar*, 2/1-2/4.

Manzano, S. A., Xu, P., Ly, K., Shepherd, R., & Correll, N. (2021). High-Bandwidth Nonlinear Control for Soft Actuators with Recursive Network Models. In B. Siciliano, C. Laschi, & O. Khatib (Eds.), *Experimental Robotics,* (pp. 589–599). Springer International Publishing., doi:10.1007/978-3-030-71151-1_52

Manzoni, V., Corti, A., De Luca, P., & Savaresi, S. M. (2010). Driving style estimation via inertial measurements. *IEEE Conference on Intelligent Transportation Systems Proceedings,* 777–782. doi:10.1109/ITSC.2010.5625113

Ma, P., Xia, R., Wang, X., Zhang, X., Królczyk, G., Gardoni, P., & Li, Z. (2022). An active control method for vibration reduction of a single-link flexible manipulator. *Journal of Low Frequency Noise, Vibration and Active Control, 41*(4), 14613484221094982. doi:10.1177/14613484221094982

Marchand, G., Taher Masoud, A., Ware, K., Govindan, M., King, A., Ruther, S., Brazil, G., Calteux, N., Coriell, C., Ulibarri, H., Parise, J., Arroyo, A., Filippelli, C., Loli, H., & Sainz, K. (2021). Systematic review and meta-analysis of all randomized controlled trials comparing gynecologic laparoscopic procedures with and without robotic assistance. *European Journal of Obstetrics, Gynecology, and Reproductive Biology*, *265*, 30–38. doi:10.1016/j.ejogrb.2021.07.038 PMID:34418694

Marchese, A. D., Onal, C. D., & Rus, D. (2014). Autonomous Soft Robotic Fish Capable of Escape Maneuvers Using Fluidic Elastomer Actuators. *Soft Robotics*, *1*(1), 75–87. doi:10.1089oro.2013.0009 PMID:27625912

Marechal, L., Balland, P., Lindenroth, L., Petrou, F., Kontovounisios, C., & Bello, F. (2021). Toward a Common Framework and Database of Materials for Soft Robotics. *Soft Robotics*, *8*(3), 284–297. doi:10.1089oro.2019.0115 PMID:32589507

Marescaux, J., & Rubino, F. (2003). The ZEUS robotic system: Experimental and clinical applications. *The Surgical Clinics of North America*, *83*(6), 1305–1315. doi:10.1016/S0039-6109(03)00169-5 PMID:14712867

Marina Martinez, C., Heucke, M., Wang, F. Y., Gao, B., & Cao, D. (2018). Driving Style Recognition for Intelligent Vehicle Control and Advanced Driver Assistance: A Survey. *IEEE Transactions on Intelligent Transportation Systems*, *19*(3), 666–676. https://doi.org/10.1109/TITS.2017.2706978

Marín-Buck, A., Karaman, E., Amer-Cuenca, J. J., Lisón, J. F., Török, P., Karaaslan, O., Valenti, G., Zito, G., Biondi, A., & Chiofalo, B. (2021). Minimally Invasive Myomectomy: An Overview on the Surgical Approaches and a Comparison with Mini-Laparotomy. *Journal of Investigative Surgery*, *34*(4), 443–450. doi:10.1080/08941939.2019.1642422 PMID:31322011

Marino, A., Parker, L. E., Antonelli, G., Caccavale, F., & Chiaverini, S. (2009). A fault-tolerant modular control approach to multi-robot perimeter patrol. *2009 IEEE International Conference on Robotics and Biomimetics, ROBIO 2009*, 735–740. doi:10.1109/ROBIO.2009.5420581

Markoff, J. (2007). *Crashes and traffic jams in military test of robotic vehicles*. Retrieved from https://www.nytimes.com/2007/11/05/technology/05robot.html

Marques, L., Casimiro, A., & Calha, M. (2009). Design and development of a proof-of-concept platooning application using the HIDENETS architecture. *Proceedings of the International Conference on Dependable Systems and Networks*, 223–228. doi:10.1109/DSN.2009.5270334

Marra, A. R., Puig-Asensio, M., Edmond, M. B., Schweizer, M. L., & Bender, D. (2019). Infectious complications of laparoscopic and robotic hysterectomy: A systematic literature review and meta-analysis. *International Journal of Gynecological Cancer*, *29*(3), 518–530. doi:10.1136/ijgc-2018-000098 PMID:30833440

Martinello, N., & Loshak, H. (2020). *Experiences with and Expectations of Robotic Surgical Systems: A Rapid Qualitative Review*. https://cadth.ca/experiences-and-expectations-robotic-surgical-systems-rapid-qualitative-review

Mattson, J. N., & Bender, D. P. (2020). Minimally Invasive Robotic Surgery for Gynecologic Cancers: A Review. *Clinical Obstetrics and Gynecology*, *63*(1), 24–29. doi:10.1097/GRF.0000000000000492 PMID:31850943

Maurin, B., Gangloff, J., Bayle, B., De Mathelin, M., Piccin, O., Zanne, P., Doignon, C., Soler, L., & Gangi, A. (2004, September). A parallel robotic system with force sensors for percutaneous procedures under CT-guidance. In *International Conference on Medical Image Computing and Computer-Assisted Intervention*, (pp. 176-183). Springer, Berlin, Heidelberg. 10.1007/978-3-540-30136-3_23

McAllister, R., Gal, Y., Kendall, A., Van Der Wilk, M., Shah, A., Cipolla, R., & Weller, A. (2017). Concrete problems for autonomous vehicle safety: Advantages of Bayesian deep learning. *International Joint Conference on Artificial Intelligence*, 4745–4753. doi:10.24963/ijcai.2017/661

McBeth, P. B., Louw, D. F., Rizun, P. R., & Sutherland, G. R. (2004). Robotics in neurosurgery. *American Journal of Surgery*, *188*(4), 68–75. doi:10.1016/j.amjsurg.2004.08.004 PMID:15476655

McDermott, C. D., & Hale, D. S. (2009). Abdominal, laparoscopic, and robotic surgery for pelvic organ prolapse. *Obstetrics and Gynecology Clinics of North America*, *36*(3), 585–614. doi:10.1016/j.ogc.2009.09.004 PMID:19932417

McLarney, J. T., & Rose, G. L. (2011). Anesthetic Implications of Robotic Gynecologic Surgery. *Journal of Gynecological Endoscopy and Surgery*, *2*(2), 75–78. doi:10.4103/0974-1216.114077 PMID:26085748

McMillan, S. C. (1982). *The Economic Effects of Fracture in the United States*. Final Report.

McPherson, J. (1989). Accelerated testing. In Electronic Materials Handbook Volume 1: Packaging. ASM International Publishing.

McPherson, J. (2010). *Reliability Physics and Engineering: Time-to-Failure Modeling*. Springer. doi:10.1007/978-1-4419-6348-2

Medical Advisory Secretariat. (2010). Robotic-assisted minimally invasive surgery for gynecologic and urologic oncology: An evidence-based analysis. *Ontario Health Technology Assessment*, *10*, 1–118. PMID:23074405

Mei, J., Zheng, K., Zhao, L., Lei, L., & Wang, X. (2018). Joint radio resource allocation and control for vehicle platooning in lte-v2v network. *IEEE Transactions on Vehicular Technology*, *67*(12), 12218–12230. https://doi.org/10.1109/TVT.2018.2874722

Meinzer, A., Alkatout, I., Krebs, T. F., Baastrup, J., Reischig, K., Meiksans, R., & Bergholz, R. (2020, December 10). Advances and Trends in Pediatric Minimally Invasive Surgery. *Journal of Clinical Medicine*, *9*(12), 3999. doi:10.3390/jcm9123999

Mellal, M. A., & Williams, E. J. (2017). The cuckoo optimization algorithm and its applications. In *Handbook of Neural Computation*, (pp. 269–277). Elsevier. doi:10.1016/B978-0-12-811318-9.00014-4

Mellal, M. A., & Williams, E. J. (2018). A survey on ant colony optimization, particle swarm optimization, and cuckoo algorithms. In *Handbook of Research on Emergent Applications of Optimization Algorithms*. IGI Global. doi:10.4018/978-1-5225-2990-3.ch002

Mello, R. C., Ribeiro, M. R. N., & Frizera-Neto, A. (2023). Implementing cloud robotics for practical applications. In Springer Tracts in Advanced Robotics. doi:10.1007/978-3-031-16908-3

Melvin, W. L., & Scheer, J. A. (2013). Principles of modern radar: Advanced techniques. In Principles of Modern Radar: Advanced Techniques. Institution of Engineering and Technology. doi:10.1049/SBRA020E

Mena-Oreja, J., & Gozalvez, J. (2018). PERMIT - A SUMO Simulator for Platooning Maneuvers in Mixed Traffic Scenarios. *IEEE Conference on Intelligent Transportation Systems Proceedings*, 3445–3450. doi:10.1109/ITSC.2018.8569496

Mena-Oreja, J., Gozalvez, J., & Sepulcre, M. (2019). Effect of the Configuration of Platooning Maneuvers on the Traffic Flow under Mixed Traffic Scenarios. *IEEE Vehicular Networking Conference*. doi:10.1109/VNC.2018.8628381

Menderes, G., Gysler, S. M., Vadivelu, N., & Silasi, D. A. (2019). Challenges of Robotic Gynecologic Surgery in Morbidly Obese Patients and How to Optimize Success. *Current Pain and Headache Reports*, *23*(7), 51. doi:10.100711916-019-0788-7 PMID:31263977

Merlet, J. P. (2002, May). Optimal design for the micro parallel robot MIPS. In *Proceedings IEEE International Conference on Robotics and Automation (Cat. No. 02CH37292),* 2, pp. 1149-1154. IEEE. 10.1109/ROBOT.2002.1014698

Methenitis, G., Hennes, D., Izzo, D., & Visser, A. (2015). Novelty Search for Soft Robotic Space Exploration. *Proceedings of the 2015 Annual Conference on Genetic and Evolutionary Computation*, 193–200. 10.1145/2739480.2754731

Mettler, L., & Meinhold-Heerlein, I. (2009). The value of laparoscopic surgery to stage gynecological cancers: Present and future. *Minerva Ginecologica*, *61*, 319–337. PMID:19745797

Mettler, L., Sammur, W., Alkatout, I., & Schollmeyer, T. (2011). Imaging in gynecologic surgery. *Women's Health (London)*, *7*(2), 239–248. doi:10.2217/WHE.11.6 PMID:21410349

Meyer, J., & Elko, G. (2002). A highly scalable spherical microphone array based on an orthonormal decomposition of the soundfield. *IEEE International Conference on Acoustics, Speech, and Signal Processing, 2*, II--1781. 10.1109/ICASSP.2002.1006109

Meyer, J., & Elko, G. W. (2004). Spherical microphone arrays for 3D sound recording. In *Audio signal processing for next-generation multimedia communication systems,* (pp. 67–89). Springer. doi:10.1007/1-4020-7769-6_3

Miller, B. (2015, October 30). First Driverless Car Crash Study: Autonomous Vehicles Crash More, but Injuries Are Less Serious. *Government Technology*. Retrieved from https://www.govtech.com/fs/First-Driverless-Car-Crash-Study-Autonomous-Vehicles-Crash-More-Injuries-Are-Less-Serious.html

Mirjalili, S., & Lewis, A. (2016). The whale optimization algorithm. *Advances in Engineering Software, Elsevier*, *95*, 51–67. doi:10.1016/j.advengsoft.2016.01.008

Mirjalili, S., Mirjalili, S. M., & Lewis, A. (2014). Grey Wolf Optimizer. *Advances in Engineering Software*, *69*, 46–61. doi:10.1016/j.advengsoft.2013.12.007

Moallem, M., Patel, R., & Khorasani, K. (1997). An Inverse Dynamics Control Strategy for Tip Position Tracking of Flexible Multi-link Manipulators. *Journal of Robotic Systems*, *14*(9), 649–658. doi:10.1002/(SICI)1097-4563(199709)14:9<649::AID-ROB2>3.0.CO;2-L

Moawad, G., Tyan, P., & Louie, M. (2019). Artificial intelligence and augmented reality in gynecology. *Current Opinion in Obstetrics & Gynecology*, *31*(5), 345–348. doi:10.1097/GCO.0000000000000559 PMID:31259843

Modarres, M., Kaminskiy, M., & Krivtsov, V. (2016). *Reliability Engineering and Risk Analysis: A Practical Guide* (3rd ed.). CRC Press. doi:10.1201/9781315382425

Moglia, A., Georgiou, K., Georgiou, E., Satava, R. M., & Cuschieri, A. (2021). A systematic review on artificial intelligence in robot-assisted surgery. *International Journal of Surgery*, *95*, 106151. doi:10.1016/j.ijsu.2021.106151 PMID:34695601

Moit, H., Dwyer, A., De Sutter, M., Heinzel, S., & Crawford, D. (2019). A Standardized Robotic Training Curriculum in a General Surgery Program. *JSLS: Journal of the Society of Laparoendoscopic Surgeons, 23*(4), e2019.00045. doi:10.4293/JSLS.2019.00045

Montazeri, A., Montazeri, N., Pourshamsian, K., & Tcharkhtchi, A. (2011). The effect of sonication time and dispersing medium on the mechanical properties of multiwalled carbon nanotube (MWCNT)/epoxy composite. *International Journal of Polymer Analysis and Characterization, 16*(7), 465–476. doi:10.1080/1023666X.2011.600517

Moon, A. S., Garofalo, J., Koirala, P., Vu, M. T., & Chuang, L. (2020). Robotic Surgery in Gynecology. *The Surgical Clinics of North America, 100*(2), 445–460. doi:10.1016/j.suc.2019.12.007 PMID:32169189

Moore, A. H., Brookes, M., & Naylor, P. A. (2013). *Room geometry estimation from a single channel acoustic impulse response.*

Moreira, T., Lima, J., Costa, P., & Cunha, M. (2019). Low-cost Sonar based on the Echolocation. *ICINCO*, (1), 818–825.

Morikawa, D. (2014). Effect of Interaural Difference for Localization of Spatially Segregated Sound. *Tenth International Conference on Intelligent Information Hiding and Multimedia Signal Processing*, 602–605. 10.1109/IIH-MSP.2014.156

Morling, J. R., McAllister, D. A., Agur, W., Fischbacher, C. M., Glazener, C. M., Guerrero, K., Hopkins, L., & Wood, R. (2017). Adverse events after first, single, mesh and non-mesh surgical procedures for stress urinary incontinence and pelvic organ prolapse in Scotland, 1997-2016: A population-based cohort study. *Lancet, 389*(10069), 629–640. doi:10.1016/S0140-6736(16)32572-7 PMID:28010993

Moschioni, G. (2002). A new method for measurement of early sound reflections in theaters and halls. *IMTC/2002. Proceedings of the 19th IEEE Instrumentation and Measurement Technology Conference, 1*, 425–430 vol.1.

Moudgal, V. G., Kwong, W. A., Passino, K. M., & Yurkovich, S. (1995). Fuzzy Learning Control for a Flexible-link Robot. *IEEE Transactions on Fuzzy Systems, 3*(2), 199–210. doi:10.1109/91.388164

Moulines, E., Duhamel, P., Cardoso, J.-F., & Mayrargue, S. (1995). Subspace methods for the blind identification of multichannel FIR filters. *IEEE Transactions on Signal Processing, 43*(2), 516–525. doi:10.1109/78.348133

Mourtzinos, A., & Raz, S. (2006). Repair of vaginal vault prolapse and pelvic floor relaxation using polypropylene mesh. *Current Opinion in Obstetrics & Gynecology, 18*(5), 555–559. doi:10.1097/01.gco.0000242960.70985.c2 PMID:16932052

Mueller, E. R. (2013). Why complex pelvic organ prolapse should be approached abdominally. *Current Opinion in Urology, 23*(4), 317–322. doi:10.1097/MOU.0b013e3283619f10 PMID:23882719

Murayama, M., Oazamoto, D., & Ooe, K. (2015). Narrow gap gas metal arc (GMA) welding technologies. *JFE Technol. Rep, 20*, 147–153.

Muresan, C. I., Birs, I., & De Keyser, R. (2021). An alternative design approach for fractional order internal model controllers for time delay systems. *Journal of Advanced Research, 31*, 177–189. doi:10.1016/j.jare.2021.01.004 PMID:34194841

Murphy, K. P. (2012). *Machine learning: a probabilistic perspective*. MIT press.

Murphy, R. R. (2012). A decade of rescue robots. *IEEE International Conference on Intelligent Robots and Systems*, 5448–5449. 10.1109/IROS.2012.6386301

Murray, J. C., Erwin, H. R., & Wermter, S. (2009). Robotic sound-source localisation architecture using cross-correlation and recurrent neural networks. *Neural Networks, 22*(2), 173–189. doi:10.1016/j.neunet.2009.01.013 PMID:19233613

Murray, J. C., Wermter, S., & Erwin, H. R. (2006). Bioinspired auditory sound localisation for improving the signal to noise ratio of socially interactive robots. *IEEE/RSJ International Conference on Intelligent Robots and Systems*, 1206–1211. 10.1109/IROS.2006.281855

Naclerio, N. D., Karsai, A., Murray-Cooper, M., Ozkan-Aydin, Y., Aydin, E., Goldman, D. I., & Hawkes, E. W. (2021). Controlling subterranean forces enables a fast, steerable, burrowing soft robot. *Science Robotics*, *6*(55), eabe2922. doi:10.1126cirobotics.abe2922 PMID:34135117

Nahas, S., Feigenberg, T., & Park, S. (2016). Feasibility and safety of same-day discharge after minimally invasive hysterectomy in gynecologic oncology: A systematic review of the literature. *Gynecologic Oncology*, *143*(2), 439–442. doi:10.1016/j.ygyno.2016.07.113 PMID:27475523

Nakadai, K., Lourens, T., Okuno, H. G., & Kitano, H. (2000). Active audition for humanoid. *AAAI/IAAI*, 832–839.

Nasr, S., Abadi, A., Bouallegue, K., & Mekki, H. (2018). Chaos Engineering and Control in Mobile Robotics Applications. *Proceedings of the 15th International Conference on Informatics in Control, Automation and Robotics (ICINCO 2018)*, *2*, 364-371. 10.5220/0006867103640371

Natarajan, E., Hassan, C. S., Kit, A. C., Santhosh, M. S., Ramesh, S., & Sasikumar, R. (2021). Modeling of multiwall carbon nanotubes reinforced natural rubber for soft robotic applications–A comprehensive presentation. *Materials Today: Proceedings*, *46*, 3251–3258. doi:10.1016/j.matpr.2020.11.293

Nath, U. M., Dey, C., & Mudi, R. K. (2019). *Designing of IMC-PID controller for higher-order process based on model reduction method and fractional coefficient filter with real time verification*. Chemical Product and Process Modelling. doi:10.1515/cppm-2019-0089

Naus, G., Vugts, R., Ploeg, J., Van De Molengraft, R., & Steinbuch, M. (2010). Cooperative adaptive cruise control, design and experiments. *Proceedings of the 2010 American Control Conference*, (1), 6145–6150. doi:10.1109/acc.2010.5531596

NBS Special Publication 647-2. (n.d.). Battelle Laboratories.

Nemeth, B., Csikos, A., Varga, I., & Gaspar, P. (2012). Road inclinations and emissions in platoon control via multi-criteria optimization. *2012 20th Mediterranean Conference on Control and Automation, MED 2012 - Conference Proceedings*, 1524–1529. doi:10.1109/MED.2012.6265855

Neto, J. S., Siufi, D. F., & Magrina, J. F. (2015). Robotic nerve-sparing radical hysterectomy. *Minerva Ginecologica*, *67*, 281–287. PMID:25714873

News, B. (2019, November 6). *Uber in fatal crash had safety flaws say US investigators*. Retrieved from https://www.bbc.com/news/business-50312340

Nezhat, C., Cho, J., King, L. P., Hajhosseini, B., & Nezhat, F. (2011). Laparoscopic management of adnexal masses. *Obstetrics and Gynecology Clinics of North America*, *38*(4), 663–676. doi:10.1016/j.ogc.2011.09.003 PMID:22134015

Nezhat, C., Falik, R., McKinney, S., & King, L. P. (2017). Pathophysiology and management of urinary tract endometriosis. *Nature Reviews. Urology*, *14*(6), 359–372. doi:10.1038/nrurol.2017.58 PMID:28467398

Nezhat, C., & Lakhi, N. (2016). Learning Experiences in Robotic-Assisted Laparoscopic Surgery. *Best Practice & Research. Clinical Obstetrics & Gynaecology*, *35*, 20–29. doi:10.1016/j.bpobgyn.2015.11.009 PMID:26707192

Nezhat, F. (2008). Minimally invasive surgery in gynecologic oncology: Laparoscopy versus robotics. *Gynecologic Oncology*, *111*(2), S29–S32. doi:10.1016/j.ygyno.2008.07.025 PMID:18762326

Ng, A. T., & Tam, P. C. (2014). Current status of robot-assisted surgery. *Hong Kong Medical Journal*, *20*, 241–250. PMID:24854139

Nguyen, L., Miro, J. V, & Qiu, X. (2019). *Can a Robot Hear the Shape and Dimensions of a Room?* 5346–5351.

Norman, L. J., Dodsworth, C. D., Foresteire, D., & Thaler, L. (2021). Human click-based echolocation: Effects of blindness and age, and real-life implications in a 10-week training program. *PLoS One*, *16*(6), e0252330. doi:10.1371/journal.pone.0252330 PMID:34077457

Oehler, M. K. (2009). Robot-assisted surgery in gynaecology. *Australian and New Zealand Journal of Obstetrics and Gynaecology*, *49*(2), 124–129. doi:10.1111/j.1479-828X.2009.00950.x PMID:19441160

Ofner, P., Schwarz, A., Pereira, J., Wyss, D., Wildburger, R., & Müller-Putz, G. R. (2019). Attempted Arm and Hand Movements can be Decoded from Low-Frequency EEG from Persons with Spinal Cord Injury. *Scientific Reports*, *9*(1), 1–15. doi:10.103841598-019-43594-9 PMID:31073142

Oguamanam, D., Bosnjak, S., & Zrnic, N. (2006). On the dynamic modelling of flexible manipulators. *FME Trans.*, *34*, 231–237.

Okamura, A. M. (2004). Methods for haptic feedback in teleoperated robot-assisted surgery. *Industrial Robot. International Journal (Toronto, Ont.)*.

Okuno, H., & Nakadai, K. (2015). *Robot audition: Its rise and perspectives,* 5610–5614. doi:10.1109/ICASSP.2015.7179045

Omar, H. M., Yanzhong, Z., Bo, Z., & Gul, H. U. (2016). Integration of GPS and dead reckoning navigation system using moving horizon estimation. In Proceedings of 2016 IEEE Information Technology, Networking, Electronic and Automation Control Conference (pp. 553–556). ITNEC. https://doi.org/10.1109/ITNEC.2016.7560421.

Öncü, S. (2014). *String stability of interconnected vehicles : network- aware modelling, analysis and experiments.* doi:10.6100/IR762788

Ono, N., Saito, A., & Ando, S. (2003). Design and experiments of bio-mimicry sound source localization sensor with gimbal-supported circular diaphragm. *TRANSDUCERS '03. 12th International Conference on Solid-State Sensors, Actuators and Microsystems. Digest of Technical Papers, 1*, 935–938.

Ort, T., Paull, L., & Rus, D. (2018). Autonomous Vehicle Navigation in Rural Environments Without Detailed Prior Maps. *Proceedings - IEEE International Conference on Robotics and Automation*, 2040–2047. doi:10.1109/ICRA.2018.8460519

Otsuka, M. (2008). Double effect, triple effect and the trolley problem: Squaring the circle in looping cases. *Utilitas*, *20*(1), 92–110.

Owren, B., & Simonsen, H. H. (1995). Alternative integration methods for problems in structural dynamics. *Computer Methods in Applied Mechanics and Engineering*, *122*(1-2), 1–10. doi:10.1016/0045-7825(94)00717-2

Pagoli, A., Chapelle, F., Corrales-Ramon, J.-A., Mezouar, Y., & Lapusta, Y. (2020). Design and Optimization of a Dextrous Robotic Finger: Incorporating a Sliding, Rotating, and Soft-Bending Mechanism While Maximizing Dexterity and Minimizing Dimensions. *IEEE Robotics & Automation Magazine*, *27*(4), 56–64. doi:10.1109/MRA.2020.3024283

Palep, J. H. (2009). Robotic assisted minimally invasive surgery. *Journal of Minimal Access Surgery*, *5*(1), 1–7. doi:10.4103/0972-9941.51313 PMID:19547687

Pal, M. (2005). Random forest classifier for remote sensing classification. *International Journal of Remote Sensing*, *26*(1), 217–222. doi:10.1080/01431160412331269698

Park, H. S., Kim, T. J., Song, T., Kim, M. K., Lee, Y. Y., Choi, C. H., Lee, J.-W., Kim, B.-G., & Bae, D. S. (2011). Single-port access (SPA) laparoscopic surgery in gynecology: A surgeon's experience with an initial 200 cases. *European Journal of Obstetrics, Gynecology, and Reproductive Biology*, *154*(1), 81–84. doi:10.1016/j.ejogrb.2010.09.004 PMID:21056527

Patel, J., Khaked, A. A., Patel, J., & Patel, J. (2021). Heart Disease Prediction Using Machine Learning. *Lecture Notes in Networks and Systems, 203 LNNS*(6), 653–665. doi:10.1007/978-981-16-0733-2_46

Patel, H. R., & Shah, V. A. (2018). A framework for fault tolerant control for an interacting and non-interacting level control system using AI, In *Proceedings of the ICINCO*, (180-190).

Patel, M. N., & Hemal, A. K. (2018). Does Advancing Technology Improve Outcomes? Comparison of the Da Vinci Standard/S/Si to the Xi Robotic Platforms During Robotic Nephroureterectomy. *Journal of Endourology*, *32*(2), 133–138. doi:10.1089/end.2017.0477 PMID:29205061

Pati, A., & Negi, R. (2019). An optimized 2-DOF IMC-PID-based control scheme for real time magnetic levitation system. *International Journal of Automation and Control*, *13*(4), 413–439. doi:10.1504/IJAAC.2019.100468

Patronik, N. A., Zenati, M. A., & Riviere, C. N. (2011). A study ex vivo of the effect of epicardial fat on the heartlander robotic crawler. In *5th European Conference of the International Federation for Medical and Biological Engineering,* (pp. 227-230). Springer, Berlin, Heidelberg. 10.1007/978-3-642-23508-5_60

Pawlowski, B., Sun, J., Xu, J., Liu, Y., & Zhao, J. (2019). Modeling of Soft Robots Actuated by Twisted-and-Coiled Actuators. *IEEE/ASME Transactions on Mechatronics*, *24*(1), 5–15. doi:10.1109/TMECH.2018.2873014

Payandeh, S., & Stanisic, Z. (2002). On application of virtual fixtures as an aid for telemanipulation and training. *Proceedings 10th Symposium on Haptic Interfaces for Virtual Environment and Teleoperator Systems. HAPTICS*, 18-23. 10.1109/HAPTIC.2002.998936

Pelizzo, G., Nakib, G., & Calcaterra, V. (2019). Pediatric and adolescent gynecology: Treatment perspectives in minimally invasive surgery. *Pediatric Reports*, *11*(4), 8029. doi:10.4081/pr.2019.8029 PMID:31871603

Peng, H., Ziran, W., Kanok, B., & Matthew, B. (2017). Intra-platoon Vehicle Sequence Optimization for Eco-Cooperative Adaptive Cruise Control. *2017 IEEE 20th International Conference on Intelligent Transportation Systems (ITSC)*, 1–6.

Pereirea, E., Aphale, S. S., Feliu, V., & Moheimani, S. O. R. (2011). Integral Resonant Control for Vibration Damping and Precise Tip-positioning of a Single-link Flexible Manipulator. IEEE/ASME Transactions on Mechatronics.

Pew Research Center. (2017). *Many Americans would be hesitant to use various automation technologies*. Retrieved from https://www.pewresearch.org/internet/2017/10/04/automation-in-everyday-life/pi_2017-10-04_automation_0-02/

Pfeifenberger, L., Zöhrer, M., & Pernkopf, F. (2017). DNN-based speech mask estimation for eigenvector beamforming. *IEEE International Conference on Acoustics, Speech and Signal Processing (ICASSP)*, 66–70. 10.1109/ICASSP.2017.7952119

Picerno, T., Sloan, N. L., Escobar, P., & Ramirez, P. T. (2017). Bowel injury in robotic gynecologic surgery: Risk factors and management options. A systematic review. *American Journal of Obstetrics and Gynecology*, *216*(1), 10–26. doi:10.1016/j.ajog.2016.08.040 PMID:27640938

Pico, A., Schillaci, G., Hafner, V. V., & Lara, B. (2016). How do I sound like? forward models for robot ego-noise prediction. *Joint IEEE Int. Conf. on Development and Learning and Epigenetic Robotics*, 246–251. 10.1109/DEVLRN.2016.7846826

Pierce, A. D., & Beyer, R. T. (2005). Acoustics: An Introduction to Its Physical Principles and Applications. In *The Journal of the Acoustical Society of America*, *87*(4). Acoustical Society of America. https://asa.scitation.org/doi/10.1121/1.399390

Pierson, H. A., & Gashler, M. S. (2017). Deep learning in robotics: A review of recent research. *Advanced Robotics, 31*(16), 821–835. doi:10.1080/01691864.2017.1365009

Pimpale, M. R., Therese, S., & Shinde, V. (2016). A survey on: Sound source separation methods. *International Journal (Toronto, Ont.), 3*(11), 580–584.

Poddar, P., & Maheshwari, A. (2021). Surgery for cervical cancer: Consensus & controversies. *The Indian Journal of Medical Research, 2021*(Dec), 2. PMID:34854431

Podlubny, I. (1994). *Fractional order systems and fractional controllers, Slovak Academy of Sciences, Institute of Experimental Physics,* (1st ed.). Springer.

Polygerinos, P., Galloway, K. C., Sanan, S., Herman, M., & Walsh, C. J. (2015). EMG controlled soft robotic glove for assistance during activities of daily living. *IEEE International Conference on Rehabilitation Robotics (ICORR)*, 55–60. 10.1109/ICORR.2015.7281175

Pomerleau, D. a. (1989). ALVINN: An autonomous land vehicle in a neural network (Technical Report CMU-CS-89-107). *Advances in Neural Information Processing Systems, 1*, 305–313. Retrieved from https://dl.acm.org/citation.cfm?id=89851.89891

Presti, D. L., Massaroni, C., D'Abbraccio, J., Massari, L., Caponero, M., Longo, U. G., Formica, D., Oddo, C., & Schena, E. (2019). Wearable system based on flexible FBG for respiratory and cardiac monitoring. *Sensors Journal, 19*(17), 7391–7398. doi:10.1109/JSEN.2019.2916320

Prodromidou, A., Spartalis, E., Tsourouflis, G., Dimitroulis, D., & Nikiteas, N. (2020). Robotic versus laparoendoscopic single-site hysterectomy: A systematic review and meta-analysis. *Journal of Robotic Surgery, 14*(5), 679–686. doi:10.100711701-020-01042-1 PMID:31956926

PTV Group. (2015). *PTV Vissim.* Retrieved from http://vision-traffic.ptvgroup.com/en-uk/training-support/support/ptv-vissim/

Pugin, F., Bucher, P., & Morel, P. (2011). History of robotic surgery: From AESOP® and ZEUS® to da Vinci®. *Journal of Visceral Surgery, 148*(5), e3–e8. doi:10.1016/j.jviscsurg.2011.04.007 PMID:21974854

Pushkar, D. Y., Kasyan, G. R., & Popov, A. A. (2021). Robotic sacrocolpopexy in pelvic organ prolapse: A review of current literature. *Current Opinion in Urology, 31*(6), 531–536. doi:10.1097/MOU.0000000000000932 PMID:34506336

Puzyrov, V., Acho, L., Pujol, G., & Rodellar, J. (2019). Fault detection based on Lyapunov exponents estimation for stabilized mechanical systems. *Journal of Theoretical and Applied Mechanics, 57*(2), 519–531. doi:10.15632/jtam-pl/105471

Qi, Y., & Mac, H. T. (2019). QoS and Coverage Aware Dynamic High Density Vehicle Platooning (HDVP). *IEEE Vehicular Technology Conference,* 1–5. doi:10.1109/VTCFall.2018.8690882

Qin, J. (2013). *Commande hybride position/force robuste d'un robot manipulateur utilisé en usinageet/ou en soudage.* [*Robust position/force hybrid control of a manipulator robot used in machining and/or welding.*] Ecole nationale supérieure d'arts et métiers - ENSAM.

Qiu, Z. C., Li, C., & Zhang, X. M. (2019). Experimental study on active vibration control for a kind of two-link flexible manipulator. *Mechanical Systems and Signal Processing, 118*, 623–644. doi:10.1016/j.ymssp.2018.09.001

Quazi, A. (1981). An overview on the time delay estimate in active and passive systems for target localization. *IEEE Transactions on Acoustics, Speech, and Signal Processing, 29*(3), 527–533. doi:10.1109/TASSP.1981.1163618

Quemener, J., Boulanger, L., Rubod, C., Cosson, M., Vinatier, D., & Collinet, P. (2012). The place of robotics in gynecologic surgery. *Journal of Visceral Surgery*, *149*(5), e289–e301. doi:10.1016/j.jviscsurg.2012.07.002 PMID:22951086

Quirini, M., Webster, R. J., Menciassi, A., & Dario, P. (2007, April). Design of a pill-sized 12-legged endoscopic capsule robot. In *Proceedings IEEE International Conference on Robotics and Automation,* (pp. 1856-1862). IEEE. 10.1109/ROBOT.2007.363592

Raghavan, A., & Cesnik, C. E. (2007). Review of guided-wave structural health monitoring. *Shock & Vibration Digest*, *39*(2), 91–114. doi:10.1177/0583102406075428

Rahmadayana, F., & Sibaroni, Y. (2021). Sentiment Analysis of Work from Home Activity using SVM with Randomized Search Optimization. [Rekayasa Sistem Dan Teknologi Informasi]. *Jurnal RESTI*, *5*(5), 936–942. doi:10.29207/resti.v5i5.3457

Raju, R., & Linder, B. J. (2021). Evaluation and Management of Pelvic Organ Prolapse. *Mayo Clinic Proceedings*, *96*(12), 3122–3129. doi:10.1016/j.mayocp.2021.09.005 PMID:34863399

Ramezanpour, P., & Mosavi, M. R. (2020). DNN-based interference mitigation beamformer. *IET Radar, Sonar \& Navigation, 14*(11), 1788–1794.

Ramirez, P. T., Soliman, P. T., Schmeler, K. M., dos Reis, R., & Frumovitz, M. (2008). Laparoscopic and robotic techniques for radical hysterectomy in patients with early-stage cervical cancer. *Gynecologic Oncology*, *110*(3), S21–S24. doi:10.1016/j.ygyno.2008.03.013 PMID:18486966

Rannenberg, K. (2016). *Autonomous Driving: Technical, Legal and Social Aspects*. doi:10.1007/978-3-662-48847-8_24

Ranzato, M., Boureau, Y., & LeCun, Y. (2008). Sparse feature learning for deep belief networks. Proceedings of Advances in Neural Information Processing Systems, 1185-1192.

Rao, S. (2011). *Mechanical Vibrations,* (5th ed.). Prantice-Hall.

Rardin, C. R. (2011). Minimally invasive urogynecology. *Obstetrics and Gynecology Clinics of North America*, *38*(4), 639–649. doi:10.1016/j.ogc.2011.09.001 PMID:22134013

Rardin, C. R. (2014). The debate over robotics in benign gynecology. *American Journal of Obstetrics and Gynecology*, *210*(5), 418–422. doi:10.1016/j.ajog.2014.01.016 PMID:24418227

Rauniyar, A., Roy, D., Pandit, P., & Atpadkar, V. (2018). Design Model for the Drive & Actuator of the Test Set-up of a Novel Flexible Robotic System. *Proceedings of the 2018 IEEE International Conference on Computational Intelligence and Computing Research*. 10.1109/ICCIC.2018.8782388

Ravivarapu, K. T., Omidele, O., Pfail, J., Tomer, N., Small, A. C., & Palese, M. A. (2021). Robotic-assisted simple prostatectomy versus open simple prostatectomy: A New York statewide analysis of early adoption and outcomes between 2009 and 2017. *Journal of Robotic Surgery*, *15*(4), 627–633. doi:10.100711701-020-01152-w PMID:33009988

Rayalla, R., Ambati, S. R., & Uday Bhaskar Babu, U. (2019). *Improved fractional filter IMC-PID controller design for enhanced performance of integrating plus time delay processes, Indian Chemical Engineer*. Taylor & Francis.

Reddy, B. S., & Ghosal, A. (2016). Asymptotic Stability and Chaotic Motions in Trajectory Following Feedback Controlled Robots. *Journal of Computational and Nonlinear Dynamics*, *11*(5), 051012. doi:10.1115/1.4032389

Reddy, J. N. (2021). *An Introduction to Nonlinear Finite Element Method with Applications to Heat Transfer, Fluid Mechanics, and Solid Mechanics* (2nd ed.). Oxford Press.

Redmon, J., Divvala, S., Girshick, R., & Farhadi, A. (2016). You only look once: Unified, real-time object detection. *Proceedings of the IEEE Conference on Computer Vision and Pattern Recognition*, 779–788.

Reutebuch, S. E., Andersen, H.-E., & McGaughey, R. J. (2005). Light detection and ranging (LIDAR): An emerging tool for multiple resource inventory. *Journal of Forestry*, *103*(6), 286–292.

Reza, M., Maeso, S., Blasco, J. A., & Andradas, E. (2010). Meta-analysis of observational studies on the safety and effectiveness of robotic gynaecological surgery. *British Journal of Surgery*, *97*(12), 1772–1783. doi:10.1002/bjs.7269 PMID:20949554

Richards, M. A., Scheer, J. A., & Holm, W. A. (2010). Principles of modern radar: Basic principles. In Principles of Modern Radar: Basic Principles. Institution of Engineering and Technology. doi:10.1049/SBRA021E

Rieffel, J., Knox, D., Smith, S., & Trimmer, B. (2014). Growing and Evolving Soft Robots. *Artificial Life*, *20*(1), 143–162. doi:10.1162/ARTL_a_00101 PMID:23373976

Riopelle, N., Caspers, P., & Sofge, D. (2018). Terrain classification for autonomous vehicles using bat-inspired echolocation. *International Joint Conference on Neural Networks (IJCNN)*, 1–6. 10.1109/IJCNN.2018.8489046

Robinson, T., Chan, E., & Coelingh, E. (2010). An Introduction To The SARTRE Platooning Programme. *17th World Congress on Intelligent Transport Systems*, 1–11.

Rocha, A., Domínguez, A. M., Lécuru, F., & Bourdel, N. (2016). Indocyanine green and infrared fluorescence in detection of sentinel lymph nodes in endometrial and cervical cancer staging – A systematic review. *European Journal of Obstetrics, Gynecology, and Reproductive Biology*, *206*, 213–219. doi:10.1016/j.ejogrb.2016.09.027 PMID:27750179

Rodgers, A. K., Goldberg, J. M., Hammel, J. P., & Falcone, T. (2007). Tubal anastomosis by robotic compared with outpatient minilaparotomy. *Obstetrics and Gynecology*, *109*(6), 1375–1380. doi:10.1097/01.AOG.0000264591.43544.0f PMID:17540810

Rodrigues, R. C., Rodrigues, M. R. K., Freitas, N. O., Rudge, M. V. C., & Lima, S. A. M. (2019). Quality of life in patients who undergo conventional or robotic-assisted total laparoscopic hysterectomy: Protocol for a systematic review of randomized controlled trials. *Medicine*, *98*(23), e15974. doi:10.1097/MD.0000000000015974 PMID:31169730

Rodrigues, T., Muhammad, A. Q., Zahoor, M., Kamga, C., Kwok, E., & Lee, M. (2019). Analytical design of constraint handling optimal two parameter internal model control for dead-time processes. *Korean Journal of Chemical Engineering*, *36*(3), 356–367. doi:10.100711814-018-0215-5

Rolf, M., Neumann, K., Queißer, J. F., Reinhart, R. F., Nordmann, A., & Steil, J. J. (2015). A multi-level control architecture for the bionic handling assistant. *Advanced Robotics*, *29*(13), 847–859. doi:10.1080/01691864.2015.1037793

Rosenblatt, P. L., McKinney, J., & Adams, S. R. (2013). Ergonomics in the operating room: Protecting the surgeon. *Journal of Minimally Invasive Gynecology*, *20*(6), 744. doi:10.1016/j.jmig.2013.07.006 PMID:23969139

Rosenstein, M. T., Collins, J. J., & De Luca, C. J. (1993). A practical method for calculating largest Lyapunov exponents from small data sets. *Physica D. Nonlinear Phenomena*, *65*(1–2), 117–134. doi:10.1016/0167-2789(93)90009-P

Ross, J. W., & Preston, M. R. (2009). Update on laparoscopic, robotic, and minimally invasive vaginal surgery for pelvic floor repair. *Minerva Ginecologica*, *61*, 173–186. PMID:19415062

Ross, P. E. (2019). Taxis without drivers-or steering wheels. *IEEE Spectrum*, *56*(1), 33–34. https://doi.org/10.1109/MSPEC.2019.8594791

Rost, A., & Schädle, S. (2013). The SLS-Generated Soft Robotic Hand—An Integrated Approach Using Additive Manufacturing and Reinforcement Learning. *12th International Conference on Machine Learning and Applications, 1*, 215–220. 10.1109/ICMLA.2013.44

Roy, D. (2019). Towards the Control of Inherent Vibration of Flexible Robotic Systems and Associated Dynamics: New Proposition and Model. *International Journal of Robotics Research, Applications and Automation*, *1*(1), 6–17.

Rueda, M. A., Riga, C., & Hamady, M. S. (2018). Flexible robotics in pelvic disease: Does the catheter increase applicability of embolic therapy? *The Journal of Cardiovascular Surgery*, *59*(3), 322–327. doi:10.23736/S0021-9509.18.10422-8 PMID:29430891

Runge, G., Peters, J., & Raatz, A. (2017). Design optimization of soft pneumatic actuators using genetic algorithms. *IEEE International Conference on Robotics and Biomimetics (ROBIO)*, 393–400. 10.1109/ROBIO.2017.8324449

Rus, D., & Tolley, M. T. (2015). Design, fabrication and control of soft robots. *Nature*, *521*(7553), 467–475. doi:10.1038/nature14543 PMID:26017446

Russo, M., Stella, M., Sikora, M., & Pekić, V. (2019). Robust cochlear-model-based speech recognition. *Computers*, *8*(1), 5. doi:10.3390/computers8010005

Sadati, S. M. H., Naghibi, S., Cruz, L., & Bergeles, C. (2020). *Reduced-Order Modeling and Model Order Reduction for Soft Robots*. doi:10.13140/RG.2.2.24115.86568

SAE. (2013). Definitions for Terms Related to On-Road Motor Vehicle Automated Driving Systems-J3016. Society of Automotive Engineers: On-Road Automated Vehicle Standards Committee; SAE Pub. Inc.

Saha, S., & Ghosh, A. (2019). Rehabilitation using neighbor-cluster based matching inducing artificial bee colony optimization. *IEEE 16th India Council International Conference, INDICON Symposium Proceedings*, 5–8. 10.1109/INDICON47234.2019.9028975

Salehan, A., & Deldari, A. (2022). Corona virus optimization (CVO): A novel optimization algorithm inspired from the Corona virus pandemic. *The Journal of Supercomputing*, *78*(4), 5712–5743. doi:10.100711227-021-04100-z PMID:34629744

Salicrú, S., Gil-Moreno, A., Montero, A., Roure, M., Pérez-Benavente, A., & Xercavins, J. (2011). Laparoscopic radical hysterectomy with pelvic lymphadenectomy in early invasive cervical cancer. *Journal of Minimally Invasive Gynecology*, *18*(5), 555–568. doi:10.1016/j.jmig.2011.05.003 PMID:21783432

Salzman, O. (2019). Sampling-based robot motion planning. *Communications of the ACM*, *62*(10), 54–63.

Sanabria, D., Rodriguez, J., Pecci, P., Ardila, E., & Pareja, R. (2020). Same-Day Discharge in Minimally Invasive Surgery Performed by Gynecologic Oncologists: A Review of Patient Selection. *Journal of Minimally Invasive Gynecology*, *27*(4), 816–825. doi:10.1016/j.jmig.2019.10.023 PMID:31715304

Sano, M., & Sawada, Y. (1985). Measurement of the Lyapunov spectrum from a chaotic time series. *Physical Review Letters*, *55*(10), 1082–1085. doi:10.1103/PhysRevLett.55.1082 PMID:10031723

Santoro, E., & Pansadoro, V. (2013). Robotic surgery in Italy national survey (2011). *Updates in Surgery*, *65*(1), 1–9. doi:10.100713304-012-0190-z PMID:23224637

Saqib, U., Gannot, S., & Jensen, J. R. (2020). Estimation of acoustic echoes using expectation-maximization methods. *EURASIP Journal on Audio, Speech, and Music Processing*, *2020*(1), 1–15. doi:10.118613636-020-00179-z

SARTRE. (2019). *Project SARTRE (Safe Road Trains for the Environment).* Road Traffic Technology. Retrieved from https://www.roadtraffic-technology.com/projects/the-sartre-project/

Sasaki, Y., Tamai, Y., Kagami, S., & Mizoguchi, H. (2005). 2D sound source localization on a mobile robot with a concentric microphone array. *IEEE International Conference on Systems, Man and Cybernetics*, *4*, 3528-3533. 10.1109/ICSMC.2005.1571694

Satava, R. M. (2003). Robotic surgery: From past to future – A personal journey. *The Surgical Clinics of North America*, *83*(6), 1491–1500. doi:10.1016/S0039-6109(03)00168-3 PMID:14712882

Sathish Kumar, K., Kirubakaran, V., Radhakrishnan, T. K. (2017). Real time modeling and control of three tank hybrid system. *Chemical Product and Process Modelling*, 13(1).

Saunders, F., Golden, E., White, R. D., & Rife, J. (2011). Experimental verification of soft-robot gaits evolved using a lumped dynamic model. *Robotica*, *29*(6), 823–830. doi:10.1017/S0263574711000014

Sawade, O., & Radusch, I. (2015). Survey and classification of cooperative automated driver assistance systems. *2015 IEEE 82nd Vehicular Technology Conference (VTC2015-Fall)*, 1–5.

Saxena, S., & Birada, S. (2022). Fractional-Order IMC controller for high-order system using reduced-order modelling via Big-bang Big-crunch algorithm. *International Journal of Systems Science*, *53*(1), 168–181. doi:10.1080/00207721.2021.1942587

Sayers, C. (1999). *Remote Control Robotics*. Remote Control Robotics. doi:10.1007/978-1-4612-1430-4

Schachar, J. S., & Matthews, C. A. (2020). Robotic-assisted repair of pelvic organ prolapse: A scoping review of the literature. *Translational Andrology and Urology*, *9*(2), 959–970. doi:10.21037/tau.2019.10.02 PMID:32420212

Scheib, S. A., Tanner, E. III, Green, I. C., & Fader, A. N. (2014). Laparoscopy in the morbidly obese: Physiologic considerations and surgical techniques to optimize success. *Journal of Minimally Invasive Gynecology*, *21*(2), 182–195. doi:10.1016/j.jmig.2013.09.009 PMID:24100146

Schillebeeckx, F., De Mey, F., Vanderelst, D., & Peremans, H. (2011). Biomimetic Sonar: Binaural 3D Localization using Artificial Bat Pinnae. *The International Journal of Robotics Research*, *30*(8), 975–987. doi:10.1177/0278364910380474

Schlagenhauf, C., Bauer, D., Chang, K.-H., King, J. P., Moro, D., Coros, S., & Pollard, N. (2018). Control of Tendon-Driven Soft Foam Robot Hands. *IEEE-RAS 18th International Conference on Humanoid Robots (Humanoids)*, 1–7. 10.1109/HUMANOIDS.2018.8624937

Schmidt, A., Deleforge, A., & Kellermann, W. (2016). *Ego-noise reduction using a motor data-guided multichannel dictionary*. 1281–1286.

Schmidt, A., W., HEinrich, W. L., & Kellermann, W. (2018). *A Novel Ego-Noise Suppression Algorithm for Acoustic Signal Enhancement in Autonomous Systems*. 6583–6587.

Schmidt, T., Babic, B., Bruns, C. J., & Fuchs, H. F. (2021). ChirurgischeTherapie des Ösophaguskarzinoms – neueTechnologien, moderneKonzepte [Surgical treatment of esophageal cancer-New technologies, modern concepts]. *Der Chirurg; Zeitschrift fur alleGebiete der operativenMedizen, 92*(12), 1100–1106.

Schmidt, A., Löllmann, H. W., & Kellermann, W. (2020). Acoustic self-awareness of autonomous systems in a world of sounds. *Proceedings of the IEEE*, *108*(7), 1127–1149. doi:10.1109/JPROC.2020.2977372

Schramm, L., Sintov, A., & Boularias, A. (2020). *Learning to Transfer Dynamic Models of Underactuated Soft Robotic Hands*. doi:10.1109/ICRA40945.2020.9197300

Schreuder, H. W., Wolswijk, R., Zweemer, R. P., Schijven, M. P., & Verheijen, R. H. (2012). Training and learning robotic surgery, time for a more structured approach: A systematic review. *British Journal of Obstetrics and Gynaecology, 119*(2), 137–149. doi:10.1111/j.1471-0528.2011.03139.x PMID:21981104

Schultheiss, P. M., & Wagner, K. (1989). Active and passive localization: Similarities and differences. In *Underwater Acoustic Data Processing,* (pp. 215–232). Springer. doi:10.1007/978-94-009-2289-1_23

Sebastián-Romagosa, M., Udina, E., Ortner, R., Dinarès-Ferran, J., Cho, W., Murovec, N., Matencio-Peralba, C., Sieghartsleitner, S., Allison, B. Z., & Guger, C. (2020). EEG Biomarkers Related With the Functional State of Stroke Patients. *Frontiers in Neuroscience, 14*(July), 1–16. doi:10.3389/fnins.2020.00582 PMID:32733182

Secinaro, S., Calandra, D., Secinaro, A., Muthurangu, V., & Biancone, P. (2021). The role of artificial intelligence in healthcare: A structured literature review. *BMC Medical Informatics and Decision Making, 21*(1), 125. doi:10.118612911-021-01488-9 PMID:33836752

Seki, H., Yamamoto, K., & Nakagawa, S. (2017). *A deep neural network integrated with filterbank learning for speech recognition.* 5480–5484.

Shabani, S., & Norouzi, Y. (2016). Speech recognition using Principal Components Analysis and Neural Networks. *IEEE 8th International Conference on Intelligent Systems (IS)*, 90–95.

Shafer, A., & Boggess, J. F. (2008). Robotic-assisted endometrial cancer staging and radical hysterectomy with the da Vinci surgical system. *Gynecologic Oncology, 111*(2), S18–S23. doi:10.1016/j.ygyno.2008.07.023 PMID:18771796

Shang, D., Li, X., Yin, M., & Li, F. (2022). Vibration Suppression Method for Dual-flexible Manipulator Considering Bearing Friction Based on Adaptive PI Control Strategy with Nonlinear Disturbance Observer.

Shannon, C. E. (2001). A mathematical theory of communication. *Mobile Computing and Communications Review, 5*(1), 3–55. doi:10.1145/584091.584093

Shao, L., Zhang, L., Belkacem, A. N., Zhang, Y., Chen, X., Li, J., Liu, H., & Minati, L. (2020). EEG-Controlled Wall-Crawling Cleaning Robot Using SSVEP-Based Brain-Computer Interface. *Journal of Healthcare Engineering, 2020,* 1–11. doi:10.1155/2020/6968713 PMID:32399166

Shapiro, D. (2017). *Self-Driving Car Demo Like No Other: The Road to an Unforgettable Cup of Coffee.* Retrieved October 14, 2019, from Nvidia Blog website: https://blogs.nvidia.com/blog/2017/01/04/bb8-ces/

Sharma, P., Chattopadhyaya, S., & Singh, N. K. (2019). A review on magnetically supported gas metal arc welding process for magnesium alloys. *Materials Research Express, 6*(8), 082002. doi:10.1088/2053-1591/ab1e67

Shedeed, H. A., Issa, M. F., & El-Sayed, S. M. (2013). Brain EEG signal processing for controlling a robotic arm. *Proceedings, 8th International Conference on Computer Engineering and Systems, ICCES*, 152–157. 10.1109/ICCES.2013.6707191

Shekar, B. H., & Dagnew, G. (2019). Grid search-based hyperparameter tuning and classification of microarray cancer data. *2nd International Conference on Advanced Computational and Communication Paradigms, ICACCP,* 1–8. 10.1109/ICACCP.2019.8882943

Shen, Z., Zhang, X., Yang, D., Ensor, I. I. S., & Etwork, S. H. N. (2018). Performance Analysis of Extended Sensor Sharing in Vehicular Ad Hoc Networks. *2018 International Symposium on Antennas and Propagation (ISAP)*, 1–2.

Shimazu, H., Kobayashi, K., Hashimoto, A., & Kameoka, T. (2007). Tasting robot with an optical tongue: Real time examining and advice giving on food and drink. *Symposium on Human Interface and the Management of Information,* 950–957. 10.1007/978-3-540-73345-4_107

Shipp, S. (2007). Structure and function of the cerebral cortex. *Current Biology*, *17*(12), R443–R449. doi:10.1016/j.cub.2007.03.044 PMID:17580069

Shladover, S. E., Desoer, C. A., Hedrick, J. K., Tomizuka, M., Walrand, J., Zhang, W., Bin, … McKeown, N. (1991). Automatic Vehicle Control Developments in the PATH Program. *IEEE Transactions on Vehicular Technology, 40*(1), 114–130. doi:10.1109/25.69979

Shladover, S. E., Nowakowski, C., Lu, X. Y., & Ferlis, R. (2015). Cooperative adaptive cruise control: Definitions and operating concepts. *Transportation Research Record, 2489*(November), 145–152. doi:10.3141/2489-17

Shoham, M., Burman, M., Zehavi, E., Joskowicz, L., Batkilin, E., & Kunicher, Y. (2003). Bone-mounted miniature robot for surgical procedures: Concept and clinical applications. *IEEE Transactions on Robotics and Automation*, *19*(5), 893–901. doi:10.1109/TRA.2003.817075

Shull, B. L., Bachofen, C., Coates, K. W., & Kuehl, T. J. (2000). A transvaginal approach to repair of apical and other associated sites of pelvic organ prolapse with uterosacral ligaments. *American Journal of Obstetrics and Gynecology*, *183*(6), 1365–1373. doi:10.1067/mob.2000.110910 PMID:11120498

Shuttleworth, J. (2019). *SAE Standards News: J3016 automated-driving graphic update*. SAE International.

Siciliano, B., & Khatib, O. (2008). Springer handbook of robotics. In Springer Handbook of Robotics. doi:10.1007/978-3-540-30301-5

Simi, M., Valdastri, P., Quaglia, C., Menciassi, A., & Dario, P. (2010). Design, fabrication, and testing of a capsule with hybrid locomotion for gastrointestinal tract exploration. *IEEE/ASME Transactions on Mechatronics*, *15*(2), 170–180. doi:10.1109/TMECH.2010.2041244

Simmons, J. A., & Stein, R. A. (1980). Acoustic imaging in bat sonar: Echolocation signals and the evolution of echolocation. *Journal of Comparative Physiology*, *135*(1), 61–84. doi:10.1007/BF00660182

Singer, N. C., & Seering, W. P. (1990). Preshaping Command Inputs to Reduce System Vibration. *Journal of Dynamic Systems, Measurement, and Control*, *112*(1), 76–82. doi:10.1115/1.2894142

Singh, S., & Agrawal, S. (2014). VANET routing protocols: Issues and challenges. *2014 Recent Advances in Engineering and Computational Sciences (RAECS)*, 1–5.

Singhose, W., & Pao, L. (1997). A comparison of input shaping and time-optimal flexible-body control. *Control Engineering Practice*, *5*(4), 459–467. doi:10.1016/S0967-0661(97)00025-7

Singhose, W., Porter, L., Kenison, M., & Kriikku, E. (2000). Effects of hoisting on the input shaping control of gantry cranes. *Control Engineering Practice*, *8*(10), 1159–1165. doi:10.1016/S0967-0661(00)00054-X

Singla, R., Agarwal, A., Kumar, V., Pachauri, N., & Verma, O. P. (2019). Industrial simulation of PID and modified M-PID controller for coupled tank system. *Advances in Signal Processing and Communications, Springer*, *526*, 421–428. doi:10.1007/978-981-13-2553-3_41

Sinha, R., Sanjay, M., Rupa, B., & Kumari, S. (2015). Robotic surgery in gynecology. *Journal of Minimal Access Surgery*, *11*(1), 50–59. doi:10.4103/0972-9941.147690 PMID:25598600

Siqueira, A. A. G., Terra, M. H., & Bergerman, M. (2011). Robust control of robots. In Robust Control of Robots. doi:10.1007/978-0-85729-898-0

Sivaji, V. V., & Sailaja, M. (2013). Adaptive cruise control systems for vehicle modeling using stop and go manoeuvres. *International Journal of Engineering Research and Applications, 3*(4), 2453–2456. https://citeseerx.ist.psu.edu/viewdoc/download?doi=10.1.1.433.9908&rep=rep1&type=pdf

Skokos, C. (2010). *The Lyapunov Characteristic Exponents and Their Computation*. Lecture Notes in Physics, doi:10.1007/978-3-642-04458-8_2

Smirnov, V. I. (1968). A Course of Higher Mathematics: Vol. 5. *Integration And Functional Analysis*. Nauka.

Smith, D. E., & Starkey, J. M. (1995). Effects of Model Complexity on the Performance of Automated Vehicle Steering Controllers: Model Development, Validation and Comparison. *Vehicle System Dynamics, 24*(2), 163–181. https://doi.org/10.1080/00423119508969086

Smith, O. J. M. (1958). *Feedback Control Systems*. McGraw-Hill.

Sohl-Dickstein, J., Teng, S., Gaub, B. M., Rodgers, C. C., Li, C., DeWeese, M. R., & Harper, N. S. (2015). A Device for Human Ultrasonic Echolocation. *IEEE Transactions on Biomedical Engineering, 62*(6), 1526–1534. doi:10.1109/TBME.2015.2393371 PMID:25608301

Soloviev, V., Bielinskyi, A., Serdyuk, O., Solovieva, V., & Semerikov, S. (2020). *Lyapunov exponents as indicators of the stock market crashes*. http://ds.knu.edu.ua/jspui/handle/123456789/3080

Somolinos, J., Feliu, V., & Sánchez, L. (2002). Design, Dynamic Modeling and Experimental Validation of a New Three-Degree-of-Freedom Flexible Arm. *Mechatronics, 12*(7), 919–948. doi:10.1016/S0957-4158(01)00033-2

Song, S., & Xiao, J. (2014). Sliding shapes for 3d object detection in depth images. *European Conference on Computer Vision*, 634–651.

Song, Y., Cai, S., Yang, L., Li, G., Wu, W., & Xie, L. (2020). A Practical EEG-Based Human-Machine Interface to Online Control an Upper-Limb Assist Robot. *Frontiers in Neurorobotics, 14*(July), 1–13. doi:10.3389/fnbot.2020.00032 PMID:32754025

Son, Y. J., Kim, H. G., Kim, E. H., Choi, S., & Lee, S. K. (2010). Application of support vector machine for prediction of medication adherence in heart failure patients. *Healthcare Informatics Research, 16*(4), 253–259. doi:10.4258/hir.2010.16.4.253 PMID:21818444

Spanias, A. S., & Wu, F. H. (1991). Speech coding and speech recognition technologies: a review. IEEE International Sympoisum on Circuits and Systems, 572–577.

Steckel, J., & Peremans, H. (2012). Biomimetic sonar for biomimetic SLAM. *Sensors (Basel)*, 1–4.

Steckel, J., & Peremans, H. (2013). BatSLAM: Simultaneous Localization and Mapping Using Biomimetic Sonar. *PLoS One, 8*(1), e54076. doi:10.1371/journal.pone.0054076 PMID:23365647

Steil, J., Finas, D., Beck, S., Manzeschke, A., & Haux, R. (2019). Robotic Systems in Operating Theaters: New Forms of Team-Machine Interaction in Health Care. *Methods of Information in Medicine, 58*(S 01), e14–e25. doi:10.1055-0039-1692465 PMID:31342471

Stein, K., & Ascher-Walsh, C. (2009). A comprehensive approach to the treatment of uterine leiomyomata. *The Mount Sinai Journal of Medicine, New York, 76*(6), 546–556. doi:10.1002/msj.20145 PMID:20014416

Stevens, S. S., Volkmann, J., & Newman, E. B. (1937). A scale for the measurement of the psychological magnitude pitch. *The Journal of the Acoustical Society of America, 8*(3), 185–190. doi:10.1121/1.1915893

Stewart, J. (2018a, March 30). Tesla's Autopilot Was Involved in Another Deadly Car Crash. *Transportation, Wired.* Retrieved from https://www.wired.com/story/tesla-autopilot-self-driving-crash-california/

Stewart, J. (2018b, October 10). Why people keep rear-ending self-driving cars. *Wired.* Retrieved from https://www.wired.com/story/self-driving-car-crashes-rear-endings-why-charts-statistics/

Stilwell, D. J., & Bishop, B. E. (2001). *Decentralized Guidance, Navigation, and Control for Platoons of Cooperating UUVs*. Academic Press.

Su, D., & Ahn, S. (2016). Autonomous platoon formation for VANET-enabled vehicles. *2016 International Conference on Information and Communication Technology Convergence, ICTC 2016*, 247–250. doi:10.1109/ICTC.2016.7763478

Sugano, T., Okajima, H., & Matsunaga, N. (2015). Robust and precise platoon driving control of welfare vehicles along wheel track by using model error compensator. *IECON 2015 - 41st Annual Conference of the IEEE Industrial Electronics Society*, 256–261. doi:10.1109/IECON.2015.7392108

Sugano, T., Okajima, H., Samura, S., & Matsunaga, N. (2016). Narrow space driving of welfare vehicles using robust platoon control with adaptive way point tracking. *2016 55th Annual Conference of the Society of Instrument and Control Engineers of Japan, SICE 2016*, 1436–1441. doi:10.1109/SICE.2016.7749175

Sumi, H., Kataoka, T., & Kitani, Y. (2015). Application of narrow gap welding process with "J-STARTM welding" to shipbuilding and construction. *JFE Tech. Rep*, *20*, 112–117.

Suresh, A., Laha, D., Gaba, D., & Bhambri, S. (2019). The classification of EEG signal using different machine learning techniques for BCI application. In Robot Intelligence Technology and Applications, 5(January). Springer Singapore. doi:10.1007/978-981-13-7780-8

Swan, K., & Advincula, A. P. (2011). Understanding the financial impact of robotics in gynecologic surgery. *Clinical Obstetrics and Gynecology*, *54*(3), 449–457. doi:10.1097/GRF.0b013e31822ce22c PMID:21857174

Swaroop, D. (1997). *String Stability Of Interconnected Systems: An Application To Platooning In Automated Highway Systems*. UC Berkeley. Retrieved from https://escholarship.org/uc/item/86z6h1b1

Swaroop, D. (1996). String Stability of Interconnected Systems - Automatic Control, IEEE Transactions on. *IEEE Transactions on Automatic Control*, *41*(3), 349–357.

Szalay, Z., Tettamanti, T., Esztergár-Kiss, D., Varga, I., & Bartolini, C. (2018). Development of a test track for driverless cars: Vehicle design, track configuration, and liability considerations. *Periodica Polytechnica Transportation Engineering, 46*(1), 29–35. doi:10.3311/PPtr.10753

Szaniewski, M., Dąbrowski, A., Balcerzak, M., & Zawiasa, P. (2015). Application of the Lyapunov Exponent Based on Current Vibration Control Parameter (CVC) in Control of an Industrial Robot. *Mechanics and Mechanical Engineering*, *19*(1), 51–61.

Takens, F. (1981). *Detecting strange attractors in turbulence*. Springer. doi:10.1007/BFb0091924

Takeuchi, E., & Tsubouchi, T. (2006). A 3-D scan matching using improved 3-D normal distributions transform for mobile robotic mapping. *IEEE International Conference on Intelligent Robots and Systems*, 3068–3073. doi:10.1109/IROS.2006.282246

Tang, D., Qin, B., & Liu, T. (2015). Document modeling with gated recurrent neural network for sentiment classification. *Proceedings of the Conference on Empirical Methods in Natural Language Processing*, 1422–1432. 10.18653/v1/D15-1167

Tang, L. C. (2003). Multiple-steps Step-stress Accelerated Life Test. In *Handbook of Reliability Engineering* (pp. 441–455). Springer. doi:10.1007/1-85233-841-5_24

Tang, L., Gouttefarde, M., Sun, H., Yin, L., & Zhou, C. (2021). Dynamic modelling and vibration suppression of a single-link flexible manipulator with two cables. *Mechanism and Machine Theory*, *162*, 104347. doi:10.1016/j.mechmachtheory.2021.104347

Tanikawa, T., & Arai, T. (1999). Development of a micro-manipulation system having a two-fingered micro-hand. *IEEE Transactions on Robotics and Automation*, *15*(1), 152–162. doi:10.1109/70.744610

Tasoujian, S., Salavati, S., Franchek, M., & Grigoriadis, K. (2019). Robust IMC-PID and parameter varying control strategies for automated blood pressure regulation. *International Journal of Control and Automation, and Systems, Springer*, *17*(7), 1803–1813. doi:10.100712555-018-0631-7

Tavakoli, S., Sadeghi, J., Griffin, I., & Fleming, P. J. (2018). PI controller tuning for load disturbance rejection using constrained optimization. *International Journal of Dynamics and Control, Springer*, *6*(1), 188–199. doi:10.100740435-016-0290-6

Team, A. A. I. (2019). *ARGO AI*. Retrieved from https://www.argo.ai/

Team-Waymo. (2019). *WAYMO*. Retrieved from https://waymo.com/mission/

Tepljakov, A., Petlenkov, E., & Belikov, J. (2011). FOMCON: A MATLAB toolbox for fractional order system identification and control. *IJAMCS*, *2*(2), 51–62.

Terms, F. (1961). The Ten Twenty Electrode System: International Federation of Societies for Electroencephalography and Clinical Neurophysiology. *The American Journal of EEG Technology*, *1*(1), 13–19. doi:10.1080/00029238.1961.11080571

Tervo, S., & Tossavainen, T. (2012). *3D room geometry estimation from measured impulse responses*, 513–516.

Theodore, N., Arnold, P. M., & Mehta, A. I. (2018). Introduction: The rise of the robots in spinal surgery. *Neurosurgical Focus*, *45*(videosuppl1), 45. doi:10.3171/2018.7.FocusVid.Intro PMID:29963916

Thomas, D., Medoff, B., Anger, J., & Chughtai, B. (2020). Direct-to-consumer advertising for robotic surgery. *Journal of Robotic Surgery*, *14*(1), 17–20. doi:10.100711701-019-00989-0 PMID:31243703

Thomson, J. J. (1976). Killing, letting die, and the trolley problem. *The Monist*, *59*(2), 204–217.

Thrun, S. (2010). *What we're driving at*. Retrieved October 11, 2019, from The Official Google Blog 26 website: https://googleblog.blogspot.com/2010/10/what-were-driving-at.html

Thubert, T., Foulot, H., Vinchant, M., Santulli, P., Marzouk, P., Borghese, B., & Chapron, C. (2016). Surgical treatment: Myomectomy and hysterectomy; Endoscopy: A major advancement. *Best Practice & Research. Clinical Obstetrics & Gynaecology*, *34*, 104–121. doi:10.1016/j.bpobgyn.2015.11.021 PMID:27400649

Thurston, J., Murji, A., Scattolon, S., Wolfman, W., Kives, S., Sanders, A., & Leyland, N. (2019). No. 377 – Hysterectomy for Benign Gynaecologic Indications. *Journal of Obstetrics and Gynaecology Canada*, *41*(4), 543–557. doi:10.1016/j.jogc.2018.12.006 PMID:30879487

Thuruthel, T. G., Falotico, E., Renda, F., & Laschi, C. (2017). Learning dynamic models for open loop predictive control of soft robotic manipulators. *Bioinspiration &amp$\mathsemicolon$ Biomimetics, 12*(6), 066003. doi:10.1088/1748-3190/aa839f

Thuruthel, T. G., Falotico, E., Cianchetti, M., Renda, F., & Laschi, C. (2016). Learning Global Inverse Statics Solution for a Redundant Soft Robot. *Proceedings of the 13th International Conference on Informatics in Control, Automation and Robotics*, *2*, 303–310. 10.5220/0005979403030310

Tiwari, V. (2014). Calculating the true cost of robotic hysterectomy. *Healthcare Financial Management*, *68*, 80–85. PMID:25145035

Tjahyadi, H., & Sammut, K. (2006). Multi-mode Vibration Control of a Flexible Cantilever Beam using Adaptive Resonant Control. *Smart Materials and Structures*, *15*(2), 270–278. doi:10.1088/0964-1726/15/2/005

Tonguz, O., Wisitpongphan, N., Bai, F., Mudalige, P., & Sadekar, V. (2007). Broadcasting in VANET. *2007 Mobile Networking for Vehicular Environments, MOVE*, 7–12. doi:10.1109/MOVE.2007.4300825

Tong, W., Hussain, A., Bo, W. X., & Maharjan, S. (2019). Artificial intelligence for vehicle-to-everything: A survey. *IEEE Access: Practical Innovations, Open Solutions*, *7*, 10823–10843.

Tran, P. V. (2017). *A fully convolutional neural network for cardiac segmentation in short-axis MRI.* arXiv:1604.00494.

Traoré, Y., Hosseinzadeh, F., & Bouchard, P. J. (2014). Plasticity in the contour method of residual stress measurement. *Advanced Materials Research*, *996*, 337–342. doi:10.4028/www.scientific.net/AMR.996.337

Trapero-Arenas, J. R., Mboup, M., Pereira-Gonzalez, E., & Feliu, V. (2008). Online Frequency and Damping Estimation in a Single-Link Flexible Manipulator based on Algebraic Identification. *Proceedings of the 16th. Mediterranean Conference on Control & Automation.*

Trivedi, D., Rahn, C. D., Kier, W. M., & Walker, I. D. (2008). Soft robotics: Biological inspiration, state of the art, and future research. *Applied Bionics and Biomechanics*, *5*(3), 99–117. doi:10.1155/2008/520417

Truong, M., Kim, J. H., Scheib, S., & Patzkowsky, K. (2016). Advantages of robotics in benign gynecologic surgery. *Current Opinion in Obstetrics & Gynecology*, *28*(4), 304–310. doi:10.1097/GCO.0000000000000293 PMID:27362711

Tsai, T. C., & Hsu, Y. L. (2007). Development of a parallel surgical robot with automatic bone drilling carriage for stereotactic neurosurgery. *Biomedical Engineering: Applications, Basis and Communications*, *19*(04), 269–277.

Tsuda, S., Oleynikov, D., Gould, J., Azagury, D., Sandler, B., Hutter, M., Ross, S., Haas, E., Brody, F., & Satava, R. (2015). SAGES TAVAC safety and effectiveness analysis: Da Vinci® Surgical System (Intuitive Surgical, Sunnyvale, CA). *Surgical Endoscopy*, *29*(10), 2873–2884. doi:10.100700464-015-4428-y PMID:26205559

Tsugawa, S., Kato, S., Tokuda, K., Matsui, T., & Fujii, H. (2001). A cooperative driving system with automated vehicles and inter-vehicle communications in Demo 2000. *IEEE Conference on Intelligent Transportation Systems, Proceedings, ITSC*, 918–923. doi:10.1109/itsc.2001.948783

U. Saqib, A. D., & Jensen, J. R. (2021). *Detecting Acoustic Reflectors using a Robot's Ego-Noise*, 1–5.

Uchino, I., Uemura, T., & Gotoh, K. (2020). A study on adopting λ-shape groove for laser-arc hybrid welding to construct thick plate butt welded joints. *Welding International*, *34*(7-9), 357–371. doi:10.1080/09507116.2021.1936931

Ulmer, B. C. (2010). Best practices for minimally invasive procedures. *Association of periOperative Registered Nurses Journal, 91*, 558-572.

Üneri, A., Balicki, M. A., Handa, J., Gehlbach, P., Taylor, R. H., & Iordachita, I. (2010, September). New steady-hand eye robot with micro-force sensing for vitreoretinal surgery. In *3rd IEEE RAS & EMBS International Conference on Biomedical Robotics and Biomechatronics,* (pp. 814-819). IEEE. 10.1109/BIOROB.2010.5625991

United Nations. (2017). *World Population Prospects: The 2017 Revision, Key Findings and Advance Tables"*. The Department of Economic and Social Affairs: UN.

Ureña, J., Hernández, Á., García, J. J., Villadangos, J. M., Pérez, M. C., Gualda, D., Álvarez, F. J., & Aguilera, T. (2018). Acoustic Local Positioning with Encoded Emission Beacons. In *Proceedings of the IEEE,* 106(6), pp. 1042–1062. Institute of Electrical and Electronics Engineers Inc. 10.1109/JPROC.2018.2819938

Urmson, C., Anhalt, J., Bagnell, D., Baker, C., Bittner, R. C., Dolan, M., Duggins, J., Galatali, & Geyer, T. (2008). Autonomous driving in urban environments: Boss and the urban challenge. *Journal of Field Robotics, 25*, 425–466. doi:10.1002/rob

Vadamalu, R. S., & Beidl, C. (2019). Adaptive internal model based harmonic control for torsional vibration reduction. *IEEE Transactions on Industrial Electronics*, *67*(4), 3024–3032. doi:10.1109/TIE.2019.2908579

Vahidi, A., & Eskandarian, A. (2003). Research advances in intelligent collision avoidance and adaptive cruise control. *IEEE Transactions on Intelligent Transportation Systems*, *4*(3), 132–153. https://doi.org/10.1109/TITS.2003.821294

Vaisto, T. T. (2019). Thesis submitted to the University of Manchester for degree of Doctor of Philosophy in the Faculty of Science and Engineering.

Vakil, M., Fotouhi, R., Nikiforuk, P. N., & Heidari, F. (2011). A study of the free vibration of flexible-link flexible-joint manipulators. *Proceedings of the Institution of Mechanical Engineers. Part C, Journal of Mechanical Engineering Science*, *225*(6), 1361–1371. doi:10.1177/0954406211399517

Valdastri, P., Webster, R. J., Quaglia, C., Quirini, M., Menciassi, A., & Dario, P. (2009). A new mechanism for mesoscale legged locomotion in compliant tubular environments. *IEEE Transactions on Robotics*, *25*(5), 1047–1057. doi:10.1109/TRO.2009.2014127

Valdês, F., Espinosa, F., Awawdeh, A. B. M. H., & Pastor, J. (2006). Routes for splitting and merging maneuvers of platoon followers in Urban environments. *IEEE International Conference on Emerging Technologies and Factory Automation, ETFA*, 1153–1156. doi:10.1109/ETFA.2006.355236

Van Arem, B., Van Driel, C. J. G., & Visser, R. (2006). The impact of cooperative adaptive cruise control on traffic-flow characteristics. *IEEE Transactions on Intelligent Transportation Systems*, *7*(4), 429–436. https://doi.org/10.1109/TITS.2006.884615

Van Compernolle, D. (1990). Switching adaptive filters for enhancing noisy and reverberant speech from microphone array recordings. *International Conference on Acoustics, Speech, and Signal Processing*, 833–836. 10.1109/ICASSP.1990.115966

van Dam, P., Hauspy, J., Verkinderen, L., Trinh, X. B., van Dam, P. J., Van Looy, L., & Dirix, L. (2011). Are costs of robot-assisted surgery warranted for gynecological procedures? *Obstetrics and Gynecology International*, *2011*, 973830. doi:10.1155/2011/973830 PMID:21941556

Van De Hoef, S., Johansson, K. H., & Dimarogonas, D. V. (2018). Fuel-Efficient en Route Formation of Truck Platoons. *IEEE Transactions on Intelligent Transportation Systems*, *19*(1), 102–112. https://doi.org/10.1109/TITS.2017.2700021

van Seeters, J. A. H., Chua, S. J., Mol, B. W. J., & Koks, C. A. M. (2017). Tubal anastomosis after previous sterilization: A systematic review. *Human Reproduction Update*, *23*(3), 358–370. doi:10.1093/humupd/dmx003 PMID:28333337

Varghese, A., Doglioli, M., & Fader, A. N. (2019). Updates and Controversies of Robotic-Assisted Surgery in Gynecologic Surgery. *Clinical Obstetrics and Gynecology*, *62*(4), 733–748. doi:10.1097/GRF.0000000000000489 PMID:31524659

Vaseghi, S. V. (2008). Advanced Digital Signal Processing and Noise Reduction. In Advanced Digital Signal Processing and Noise Reduction: Fourth Edition. Wiley. doi:10.1002/9780470740156

Vavilala, S. K., & Vinopraba, T. (2018). Design of fractional order controllers for conical tank process, In *Proceedings of the IEEE International Conference ICETIETR*, Ernakulam, India. 10.1109/ICETIETR.2018.8529047

Veenstra, F., Jørgensen, J., & Risi, S. (2018). Evolution of fin undulation on a physical knifefish-inspired soft robot. *Proceedings of the Genetic and Evolutionary Computation Conference*, 157–164. 10.1145/3205455.3205583

Veluvolu, K., & Soh, Y. (2011). Fault reconstruction and state estimation with sliding mode observers for lipschitz non-linear systems. *IET Control Theory & Applications*, *5*(11), 1255–1263. doi:10.1049/iet-cta.2010.0171

Verellen, T., Kerstens, R., & Steckel, J. (2020). High-Resolution Ultrasound Sensing for Robotics Using Dense Microphone Arrays. *IEEE Access: Practical Innovations, Open Solutions*, *8*, 190083–190093. doi:10.1109/ACCESS.2020.3032177

Vermesan, O., & Bacquet, J. (2017). *Cognitive Hyperconnected Digital Transformation*. doi:10.13052/rp-9788793609105

Vinopraba, T., Sivakumaran, N., Narayanan, S., & Radhakrishnan, T. K. (2012). Design of internal model control based fractional order PID controller. *Journal of Control Theory and Applications*, *10*(3), 297–302. doi:10.100711768-012-1044-4

Viscelli, S. (2018). *Autonomous Trucks and the Future of the American Trucker*. Retrieved from https://laborcenter.berkeley.edu/driverless/

Visco, A. G., & Advincula, A. P. (2008). Robotic gynecologic surgery. *Obstetrics and Gynecology*, *112*(6), 1369–1384. doi:10.1097/AOG.0b013e31818f3c17 PMID:19037049

Visnic, B. (2018, May 10). SAE brings autonomy to the public in new demonstration program. *SAE International*. Retrieved from https://www.sae.org/news/2018/05/sae-demo-days-in-florida

Wagner, I. (2019a). *Autonomous vehicle technology - Statistics and Facts*. Statista. Retrieved from https://www.statista.com/topics/3573/autonomous-vehicle-technology/

Wagner, I. (2019b). *Projected size of the global autonomous car market in 2018 and 2030 (in billion U.S. dollars)*. Retrieved October 15, 2019, from Statista website: https://www.statista.com/statistics/428692/projected-size-of-global-autonomous-vehicle-market-by-vehicle-type/

Wagner, C. R., Lederman, S. J., & Howe, R. D. (2002). A tactile shape display using RC servomotors. In *Proceedings 10th Symposium on Haptic Interfaces for Virtual Environment and Teleoperator Systems. HAPTICS,* (pp. 354-355). IEEE. 10.1109/HAPTIC.2002.998981

Wagner, C. R., Stylopoulos, N., & Howe, R. D. (2002). The role of force feedback in surgery: analysis of blunt dissection. *Proceedings 10th Symposium on Haptic Interfaces for Virtual Environment and Teleoperator Systems. HAPTICS*, 68-74. 10.1109/HAPTIC.2002.998943

Wahlang, I., Maji, A. K., Saha, G., Chakrabarti, P., Jasinski, M., Leonowicz, Z., & Jasinska, E. (2021). Deep Learning Methods for Classification of Certain Abnormalities in Echocardiography. *Electronics (Basel)*, *10*(4), 495. doi:10.3390/electronics10040495

Walambe, R., Marathe, A., Kotecha, K., & Ghinea, G. (2021). Lightweight object detection ensemble framework for autonomous vehicles in challenging weather conditions. *Computational Intelligence and Neuroscience*.

Walker, J., Zidek, T., Harbel, C., Yoon, S., Strickland, F. S., Kumar, S., & Shin, M. (2020). Soft Robotics: A Review of Recent Developments of Pneumatic Soft Actuators. *Actuators*, *9*(1), 3. doi:10.3390/act9010003

Wang, Z., Wu, G., Member, S. S., Barth, M. J., Wu, G., Member, S. S., & Barth, M. J. (2018). *A Review on Cooperative Adaptive Cruise Control (CACC) Systems*. Retrieved from https://arxiv.org/ftp/arxiv/papers/1809/1809.02867.pdf

Wang, J., Zhu, J., Zhang, C., Xu, G., & Li, W. (2016). Effect of arc swing parameters on narrow gap vertical GMA weld formation. *ISIJ International*, *56*(5), 844–850. doi:10.2355/isijinternational.ISIJINT-2015-660

Wang, L., & Cavallaro, A. (2017). Microphone-array ego-noise reduction algorithms for auditory micro aerial vehicles. *IEEE Sensors Journal*, *17*(8), 2447–2455. doi:10.1109/JSEN.2017.2669262

Wang, M., Tamimi, H., & Zell, A. (2005). Robot navigation using biosonar for natural landmark tracking. *International Symposium on Computational Intelligence in Robotics and Automation*, 3–7. 10.1109/CIRA.2005.1554246

Wang, T., & Choy, Y. (2015). An approach for sound sources localization and characterization using array of microphones. *International Conference on Noise and Fluctuations (ICNF)*, 1–4. 10.1109/ICNF.2015.7288571

Wang, T., Wang, H., Hu, H., Qing, J., & Wang, C. (2022). Lion swarm based optimization based tuning method for generalized predictive fractional-order PI to control the speed of brushless direct current motor. *IET Electric Power Applications*, *16*(8), 879–895. doi:10.1049/elp2.12201

Wang, X., & Meng, M. Q. (2011). Perspective of active capsule endoscope: Actuation and localisation. *International Journal of Mechatronics and Automation*, *1*(1), 38–45. doi:10.1504/IJMA.2011.039154

Warude, P., Patel, M., Pandit, P., Patil, V., Pawar, H., Nate, C., Gajlekar, S., Atpadkar, V., & Roy, D. (2019). *On the Design and Vibration Analysis of a Three-Link Flexible Robot Interfaced with a Mini-Gripper. In Lecture Notes in Mechanical Engineering: Recent Trends in Wave Mechanics and Vibrations*. Springer.

Wasserman, G. (2003). *Reliability Verification, Testing, and Analysis in Engineering Design*. Marcel Dekker.

Wayve, T. (2019). *Learning to Drive like a Human.* Retrieved October 12, 2019, from https://wayve.ai/blog/driving-like-human

Wei, B., Jia, C., Wu, W., Fang, C., & Wu, C. (2020). Stirring effect of the rotating arc on the molten pool during non-axisymmetric tungsten NG-GTAW. *Journal of Materials Processing Technology*, *285*, 116769. doi:10.1016/j.jmatprotec.2020.116769

Weickert, U., Jakobs, R., & Riemann, J. F. (2005). Diagnostic laparoscopy. *Endoscopy*, *37*(1), 33–37. doi:10.1055-2004-826081 PMID:15657855

Weinberg, L., Rao, S., & Escobar, P. F. (2011). Robotic surgery in gynecology: An updated systematic review. *Obstetrics and Gynecology International*, *2011*, 852061. doi:10.1155/2011/852061 PMID:22190948

Weingart, R.G., & Stephen, P. (2007, Aug.). Timoshenko: Father of Engineering Mechanics in the U.S. *Structure Magazine*.

Wen, C. (2001). Dynamic Modeling of Multi-link Flexible Robotic Manipulators. *Computers & Structures*, *79*(2), 183–195. doi:10.1016/S0045-7949(00)00129-2

Wen, S., Hu, X., Zhang, D., Zhang, B., Sheng, M., Lam, H. K., & Zhao, Y. (2020). Fractional order internal model control algorithm based on the force/position control structure of redundant actuation parallel robot. *International Journal of Advanced Robotic Systems*, *17*(1), 1–13. doi:10.1177/1729881419892143

Wexner, S. D., Bergamaschi, R., Lacy, A., Udo, J., Brölmann, H., Kennedy, R. H., & John, H. (2009). The current status of robotic pelvic surgery: Results of a multinational interdisciplinary consensus conference. *Surgical Endoscopy*, *23*(2), 438–443. doi:10.100700464-008-0202-8 PMID:19037694

White, J. (2019, July 24). *GM Cruise to delay commercial launch of self-driving cars beyond 2019*. Retrieved from https://www.reuters.com/article/us-gm-cruise/gm-cruise-to-delay-commercial-launch-of-self-driving-cars-to-beyond-2019-idUSKCN1UJ1NA

Whiteside, J. L., Kaeser, C. T., & Ridgeway, B. (2019). Achieving high value in the surgical approach to hysterectomy. *American Journal of Obstetrics and Gynecology*, *220*(3), 242–245. doi:10.1016/j.ajog.2018.11.124 PMID:30419200

Wimberger, P., & Schindelhauer, A. (2014). Robotic surgery in gynecology. *Archives of Gynecology and Obstetrics*, *289*(1), 5–6. doi:10.100700404-013-3066-7 PMID:24170162

Wischhof, L., Ebner, A., & Rohling, H. (2005). Information dissemination in multihop inter-vehicle networks. *IEEE Conference on Intelligent Transportation Systems Proceedings*, *6*(1), 90–101. https://doi.org/10.1109/ITSC.2002.1041301

Wisner, K. P. A., Ahmad, S., & Holloway, R. W. (2017). Indications and techniques for robotic pelvic and para-aortic lymphadenectomy with sentinel lymph node mapping in gynecologic oncology. *Best Practice & Research. Clinical Obstetrics & Gynaecology*, *45*, 83–93. doi:10.1016/j.bpobgyn.2017.04.006 PMID:28533155

Wisner, K. P., Gupta, S., Ahmad, S., & Holloway, R. W. (2015). Indications and techniques for robotic pelvic and para-aortic lymphadenectomy in gynecologic oncology. *Journal of Surgical Oncology*, *112*(7), 782–789. doi:10.1002/jso.24005 PMID:26288990

Wolf, A., Swift, J. B., Swinney, H. L., & Vastano, J. A. (1985). Determining Lyapunov exponents from a time series. *Physica D. Nonlinear Phenomena*, *16*(3), 285–317. doi:10.1016/0167-2789(85)90011-9

Woo, S., & O'Neal, D. (2021). Reliability Design of Mechanical Systems Such as Compressor Subjected to Repetitive Stresses. *Metals*, *11*(8), 1261. doi:10.3390/met11081261

Woo, S., Pecht, M., & O'Neal, D. (2019). Reliability design and case study of the domestic compressor subjected to repetitive internal stresses. *Reliability Engineering & System Safety*, *193*, 106604. doi:10.1016/j.ress.2019.106604

Worldometer. (2019). *Annual Global Car Production*. Retrieved from https://www.worldometers.info/cars/

Worley, R., Yu, Y., & Anderson, S. (2020). Acoustic Echo-Localization for Pipe Inspection Robots. *IEEE International Conference on Multisensor Fusion and Integration for Intelligent Systems (MFI)*, 160–165. 10.1109/MFI49285.2020.9235225

Wu, Z., Yuan, J., Chen, Y., Li, D., Chen, Y. Q. (2019). Fractional order [PI] controller and Smith like predictor design for a class of high order systems, *Systems and Control*.

Wu, X., Abrahantes, M., & Edgington, M. (2016). MUSSE: A designed multi-ultrasonic-sensor system for echolocation on multiple robots. *Asia-Pacific Conference on Intelligent Robot Systems (ACIRS)*, 79–83. 10.1109/ACIRS.2016.7556192

Wu, X., D'Orazio, P., Edgington, M., & Abrahantes, M. (2015). Robotics echolocation test platform. *IEEE International Conference on Electro/Information Technology (EIT)*, 558–562. 10.1109/EIT.2015.7293397

Xianmin, Z., Changjian, S., & Erdman, A. G. (2002). Active vibration controller design and comparison study of flexible linkage mechanism systems. *Mechanism and Machine Theory*, *37*(9), 985–997. doi:10.1016/S0094-114X(02)00025-3

Xie, Y. (2015). Cost-effectiveness of robotic surgery in gynecologic oncology. *Current Opinion in Obstetrics & Gynecology*, *27*(1), 73–76. doi:10.1097/GCO.0000000000000134 PMID:25490379

Yetkin, S., & Koca, G. O. (2021). Esnek Robotların Tasarım, Kontrol ve İmalat Çalışmaları. *International Journal of Engineering Research and Development*, *13*(1), 74–86. doi:10.29137/umagd.706041

Yim, W., & Singh, S. N. (1996). Predictive End-point Trajectory Control of Elastic Manipulators. *Journal of Robotic Systems*, *13*(9), 561–569. doi:10.1002/(SICI)1097-4563(199609)13:9<561::AID-ROB1>3.0.CO;2-K

Yiwere, M., & Rhee, E. J. (2020). Sound Source Distance Estimation Using Deep Learning: An Image Classification Approach. *Sensors (Basel)*, *20*(1), 172. doi:10.339020010172 PMID:31892213

Yumuk, E., Guzelkaya, M., & Eksin, I. (2019). Analytical fractional PID controller design based on Bode's ideal transfer function plus time delay. *ISA Transactions, 91*, 196–206. doi:10.1016/j.isatra.2019.01.034 PMID:30738584

Yunping, L., Lipeng, W., Ping, M., & Kai H. (2013). Stability Analysis of Bipedal Robots Using the Concept of Lyapunov Exponents. In *Robust Control with Engineering Applications*. doi:10.1155/2013/546520

Yurashevich, M., & Habib, A. S. (2021). Anesthetic management for robotic hysterectomy in obese women. *Current Opinion in Anaesthesiology, 34*(3), 238–245. doi:10.1097/ACO.0000000000000987 PMID:33935171

Zaher, M. E., Gechter, F., Hajjar, M., & Gruer, P. (2016). An interaction model for a local approach to vehicle platoons. *International Journal of Vehicle Autonomous Systems, 13*(2), 91–113. Retrieved from https://www.inderscienceonline.com/doi/pdf/10.1504/IJVAS.2016.078760

Zakashansky, K., Bradley, W. H., & Nezhat, F. R. (2008). New techniques in radical hysterectomy. *Current Opinion in Obstetrics & Gynecology, 20*(1), 14–19. doi:10.1097/GCO.0b013e3282f2288a PMID:18197000

Zaltieri, M., Massaroni, C., Presti, D. L., Bravi, M., Sabbadini, R., Miccinilli, S., Sterzi, S., Formica, D., & Schena, E. (2020). A wearable device based on a fiber bragg grating sensor for low back movements monitoring. *Sensors (Basel), 20*(14), 3825. doi:10.339020143825 PMID:32659958

Zang, S., Ding, M., Smith, D., Tyler, P., Rakotoarivelo, T., & Kaafar, M. A. (2019). The impact of adverse weather conditions on autonomous vehicles: How rain, snow, fog, and hail affect the performance of a self-driving car. *IEEE Vehicular Technology Magazine, 14*(2), 103–111.

Zhang, C., Zang, Y., Calvo, J. A. L., & Mathar, R. (2017). A novel V2V assisted platooning system: Control scheme and MAC layer designs. *2017 IEEE 28th Annual International Symposium on Personal, Indoor, and Mobile Radio Communications (PIMRC)*, 1–7.

Zhang, M., Geng, X., Bruce, J., Caluwaerts, K., & Vespignani, M., SunSpiral, V., Abbeel, P., & Levine, S. (2017). Deep reinforcement learning for tensegrity robot locomotion. *2017 IEEE International Conference on Robotics and Automation (ICRA)*, 634–641. 10.1109/ICRA.2017.7989079

Zhang, X. L. (2021). Deep ad-hoc beamforming. *Computer Speech \& Language, 68*, 101201.

Zhang, D., Liu, J., Zhang, L., & Yang, G. Z. (2020). Hamlyn CRM: A compact master manipulator for surgical robot remote control. *International Journal of Computer Assisted Radiology and Surgery, 15*(3), 503–514. doi:10.100711548-019-02112-y PMID:31956954

Zhang, G., Davoodnia, V., Sepas-Moghaddam, A., Zhang, Y., & Etemad, A. (2020). Classification of Hand Movements from EEG Using a Deep Attention-Based LSTM Network. *IEEE Sensors Journal, 20*(6), 3113–3122. doi:10.1109/JSEN.2019.2956998

Zhang, H., Cao, R., Zilberstein, S., Wu, F., & Chen, X. (2017). Toward Effective Soft Robot Control via Reinforcement Learning. In Y. Huang, H. Wu, H. Liu, & Z. Yin (Eds.), *Intelligent Robotics and Applications,* (pp. 173–184). Springer International Publishing. doi:10.1007/978-3-319-65289-4_17

Zhang, H., Kumar, A. S., Chen, F., Fuh, J. Y. H., & Wang, M. Y. (2019). Topology Optimized Multimaterial Soft Fingers for Applications on Grippers, Rehabilitation, and Artificial Hands. *IEEE/ASME Transactions on Mechatronics, 24*(1), 120–131. doi:10.1109/TMECH.2018.2874067

Zhang, H., Kumar, A. S., Fuh, J. Y. H., & Wang, M. Y. (2018). Design and Development of a Topology-Optimized Three-Dimensional Printed Soft Gripper. *Soft Robotics, 5*(5), 650–661. doi:10.1089oro.2017.0058 PMID:29985781

Zhang, J., Tian, Y., & Zhang, M. (2014). Dynamic Model and Simulation of Flexible Manipulator based on Spring & Rigid Bodies (2460-2464). In *Proceedings of the 2014 IEEE International Conference on Robotics & Biomimetics*. 10.1109/ROBIO.2014.7090709

Zhang, L., Zu, J. W., & Zheng, Z. (1999). The stochastic Newmark algorithm for random analysis of multi-degree-of-freedom nonlinear systems. *Computers & Structures*, *70*(5), 557–568. doi:10.1016/S0045-7949(98)00203-X

Zhang, Z., & Zhao, Z. (2014). A multiple mobile robots path planning algorithm based on a-star and dijkstra algorithm. *International Journal of Smart Home*, *8*(3), 75–86. https://doi.org/10.14257/ijsh.2014.8.3.07

Zhao, S., Zhang, T., Wu, N., Ogai, H., & Tateno, S. (2015). Vehicle to vehicle communication and platooning for EV with wireless sensor network. *2015 54th Annual Conference of the Society of Instrument and Control Engineers of Japan, SICE 2015*, 1435–1440. doi:10.1109/SICE.2015.7285493

Zhao, Y., & Ogai, H. (2011). Development of a platooning control algorithm based on RoboCar. *SICE Annual Conference 2011*, 352–355.

Zheng, R., Nakano, K., Kato, S., Ogitsu, T., Yamabe, S., Aoki, K., & Suda, Y. (2013). Human-machine interface system for simulation-based automatic platooning of trucks. *16th International IEEE Conference on Intelligent Transportation Systems (ITSC 2013)*, 535–539.

Zheng, R., Nakano, K., Yamabe, S., Aki, M., Nakamura, H., & Suda, Y. (2014). Study on emergency-avoidance braking for the automatic platooning of trucks. *IEEE Transactions on Intelligent Transportation Systems*, *15*(4), 1748–1757. https://doi.org/10.1109/TITS.2014.2307160

Zhou, L., Li, M., Chen, H., Ma, G., & Wang, W. (2016). Research on ultrasonic inspection of the narrow gap welds of the main coolant line (MCL) in Chinese evolutionary pressurized reactor (CEPR) nuclear power plant. In *19th World Conference on Non-Destructive Testing* (pp. 13-17). Academic Press.

Zhou, B., Elbadry, M., Gao, R., & Ye, F. (2017). BatMapper: Acoustic sensing based indoor floor plan construction using smartphones. *Proceedings of the 15th Annual International Conference on Mobile Systems, Applications, and Services*, 42–55. 10.1145/3081333.3081363

Zhu, M., Wang, Z., Hirai, S., & Kawamura, S. (2017). Design and fabrication of a soft-bodied gripper with integrated curvature sensors. *24th International Conference on Mechatronics and Machine Vision in Practice (M2VIP)*, 1–6. 10.1109/M2VIP.2017.8211521

Ziegler, J., Bender, P., Schreiber, M., Lategahn, H., Strauss, T., Stiller, C., … Zeeb, E. (2014). Making bertha drive-an autonomous journey on a historic route. *IEEE Intelligent Transportation Systems Magazine*, *6*(2), 8–20. doi:10.1109/MITS.2014.2306552

Zou, C., & Li, H. (2019). Event-Driven Connected Vehicular Platoon Control with Mixed Time-Varying Delay. *IEEE Access*. doi:10.1109/ACCESS.2019.2927522

Zuo, Z., Shuai, B., Wang, G., Liu, X., Wang, X., Wang, B., & Chen, Y. (2015). Convolutional recurrent neural networks: Learning spatial dependencies for image representation. *Proceedings of the IEEE Conference on Computer Vision and Pattern Recognition Workshops*, 18–26. 10.1109/CVPRW.2015.7301268

About the Contributors

Mohamed Arezki Mellal has a Ph.D. in Mechatronics. He is a Professor at the Department of Mechanical Engineering, Faculty of Technology, M'Hamed Bougara University, Algeria. He was a research scholar at various universities. He has edited seven books. He has authored and coauthored ten book chapters. He has written over fifty-four journal and conference papers. He was also a committee member of over sixty international conferences. He serves as a regular reviewer for several SCIE journals. His research interests include mechatronics and soft computing methods for engineering problems.

* * *

Doğukan Akgöl received Bs.C. degree in Mechanical Engineering from Dokuz Eylul University, Izmir, TR, in 2020. Since 2021, he has been Research Assistant at Mechanical Engineering Department, Dokuz Eylul University. His current research interests include vibration control, unmanned aerial vehicles, robotics, and robot programming.

Vinod Atpadkar is a CEO at SVR InfoTech, with expertise in Robotics and automation. He has been instrumental in development of Novel project to society like, Robotics Farm Worker, Mango Plucking Robot, miniature electrical gripper. He has led and executed robotics and automation projects at various commercial as well as an educational organizations. He has published his research work at national and international conferences.

Aniket Bhelsaikar has technical expertise in Robotics including core competency in mechanical engineering. he has more than 6 years of experience in the field of Finite element analysis. He has completed his MTECH in Machine design. he has published his research work at national and international conferences.

Gökçe Naz Biricik is a BSc. student in Industrial Engineering at Dokuz Eylul University, Izmir, TR. Her current research interests include vibration control, robotics and robot programming.

Ana Maria Castillo Rivera completed her Medical Degree at the University of Granada (Spain). She obtained her Master of Advanced Studies in Internal Medicine at the University of Granada (Spain). She was awarded a PhD in Medicine at the University of Granada (Spain).

Salvador Castillo Rivera completed his BSc in Physics and majoring in Fundamental Physics at the University of Granada (Spain). He obtained his Master of Advanced Studies in Physics and Mathematics at the University of Granada (Spain). He was awarded a PhD in Aeronautical Engineering at the City University London (United Kingdom).

Pratik Chothe has technical expertise in Robotics includes core competency in mechanical engineering and working knowledge of electronics and computer systems. Pratik has 4 years of industry experience in robotics. It includes Mechatronics/ Robotics system design, Product development, Robotics development in nuclear applications, CAE, End to End project execution, project management. Pratik has done B. Tech in automobile engineering. He has published his robotics research work in national and international publications including ARME, IEEE (Digital library), Springer Book, etc. Recently he was awarded the Best research paper award at an international conference organised by IEEE Bangalore.

Shreyash Gajlekar has technical expertise in Robotics including core competency in Electronics and software domain. he has more than 4 years of experience in Robotics and Automation. He has completed BTech in Electronics and telecommunication. he has published his research work at various national and international conferences.

Ahona Ghosh is a B.Tech., M.Tech. in Computer Science and Engineering and presently is an AICTE Doctoral Fellow in the Department of Computer Science and Engineering, Maulana Abul Kalam Azad University of Technology, West Bengal (In house). Before joining her Ph.D., she was an Assistant Professor in the Department of Computational Science at Brainware University, Barasat, Kolkata. She has one patent and published more than 30 papers in international conference proceedings, peer-reviewed international journals, and books with IEEE, Springer, Elsevier, IGI Global, CRC Press, etc. She is also a reviewer of different international journals. Her research interests include Machine Learning and the Internet of Things.

Alex Gunagwera received his M.S. degree in Computer Engineering in 2017, and his Ph.D. degree in Computer Engineering in 2022 from Istanbul S. Zaim University, Istanbul, Turkey.

Yimer Mohammed Hassen has a BSc and MSc in Manufacturing Technology, and he is a Ph.D candidate in Material Science and Engineering at University of Science and Technology Beijing. He has expertise in material science, reliability design of mechanical components, and failure Analysis of environmental components in marketplace. In 2005–2017 he worked in polytechnic college, Addis Ababa, Ethiopia, where he is a senior instructor and department head in charge of manufacturing Technology department. He is working in faculty of mechanical Technology and is a Dean of the faculty, Addis Ababa Ethiopian Technical University, Addis Ababa, Ethiopia. He has 5 published papers.

Mehmet Mert Ilman is a research assistant in the Department of Mechanical Engineering and the program of Machine Theory & Dynamics (MTD) at the M. Celal Bayar University since 2017. He finished his Ph.D. in the MTD program at Dokuz Eylül University. His main research interests include industrial flexible or soft robotic systems, active vibration control or control via machine intelligence, modeling, and simulation. He is also a reviewer for scientific journals and a member of the Turkish Machine Theory Association.

Robin Kerstens received his BSc. degree in Applied Engineering Electronics & ICT from the University of Antwerp (Antwerp, Belgium) in 2014 for a bachelor thesis focussed on RADAR signal processing. This was followed by a MSc. Eng. in Electronics and ICT Engineering Technology, also at the University of Antwerp but this time with a thesis focussed on sonar beamforming. After a short spell in the automotive industry designing pressure sensors, dr. Kerstens completed a Ph.D. degree on Advanced Array Signal Processing for In-Air sonar. His research interests include in-air sonar, signal processing, radar, wave propagation, acoustics, machine learning and algorithm development.

Gezae Mebrahtu is a Lecturer at Federal TVET Institute.

Sivakumaran N. received Ph.D. from NIT Tiruchirapalli, Trichy, India. He is working as Professor in the Department of Instrumentation and Control Engineering, NIT Tiruchirapalli, Trichy, India. His research interests include process control, biomedical instrumentation, physiological control systems and intelligent systems.

Dennis L. O'Neal has been the Dean of Engineering and Computer Science at Baylor University since 2012. Prior to joining Baylor University, he served as the Associate Dean for Research for the College of Engineering and as the Head of the Department of Mechanical Engineering at Texas A&M University. He primary research areas are in refrigeration, heat pumps, and air conditioning systems. He has over 190 refereed publications and two book chapters. His papers have won several awards as best technical papers in ASHRAE.

Debanik Roy is a Senior Scientist at the Division of remote handling, BARC Mumbai. Dr. Roy has authored five book chapters, published by InTech® & others, in the domain of robotic gripper and sensor fusion and a book on applications of robotics (published by Lambert®, Germany). Also, he has published his research work at national and international conferences. In October 2014, Dr. Roy received the DAE Award for Excellence in Science & Technology (as part of the Group Achievement Award) for his contribution made in DAE programmes till date.

Sriparna Saha received her B.Tech. degree from RCC Institute of Information Technology, Kolkata, India and her M.E. and Ph.D. degrees from Electronics and Tele-Communication Engineering department of Jadavpur University, Kolkata, India. She is currently an Assistant Professor in the Department of Computer Science and Engineering of Maulana Abul Kalam Azad University of Technology, West Bengal, India. Prior to that, she was associated as a faculty with Jadavpur University and two other institutions. She has more than 8 years of experience in teaching and research. Her research area includes artificial intelligence, image processing, machine learning and recently she is giving more focus on deep learning. She has over 70 publications in international journals and conference proceedings including IEEE, Elsevier, Springer, etc. She is the author of a book on Gesture Recognition published by Studies in Computational Intelligence, Springer, currently available in many online book retailers. She is also the reviewer for many international journals. Her major research proposal is accepted for Start Up Grant under UGC Basic Scientific Research Grant. Not only that, she also got University seed money for setting up her laboratory.

Usama Saqib received his B.Sc. degree in Electrical Engineering from American University of Sharjah, U.A.E. in 2010, and a M.Sc. degree in Embedded Systems Engineering from University of Bedfordshire, U.K., in 2015. He has experience working with several consumer electronics and robotic startups. Currently, he is working towards his Ph.D. degree from Aalborg University, Denmark. Along with his Ph.D. studies, he is also working at GN Audio (Jabra) within R&D department. His research interest includes combining audio signal processing techniques with robotics, biomimicry and beamforming techniques.

Radhakrishnan T. K. received Ph. D. from Bharathidasan University Trichy, India in 2001. He is working as a Professor (HAG) in the Department of Chemical Engineering, NIT Tiruchirapalli, Trichy, India. His research interests include process modelling and simulation, model-based control, control relevant identification and process systems engineering.

Hamza Taş is a Research Assistant Dr. in the Department of Mechanical Engineering at Manisa Celal Bayar University. Hamza received his PhD degree from the Manisa Celal Bayar University where he studied as a Research Assistant. He studied the fluid body armors in his PhD thesis. Hamza's research mainly focuses on the mechanics of nano-composite materials and ballistics of body armors.

Vinopraba Thirumavalavan received Ph.D. from the Department of Instrumentation and Control NIT Tiruchirapalli, Trichy, India in 2013. She is currently working as an Assistant Professor in the Department of Electrical and Electronics Engineering, NIT Puducherry, Karaikal, India. Her research interests include fractional order PID controller, and Instrumentation system design.

Sateesh K. Vavilala received PhD from the Department of Electrical and Electronics Engineering, NIT Puducherry, Karaikal, India. He is currently working as an Associate Professor in the Department of Electrical and Electronics Engineering, VLITS, Guntur, India. His research interests include fractional control systems, nonlinear systems, and adaptive control.

Seongwoo Woo has a BS and MS in Mechanical Engineering, and he has obtained PhD in Mechanical Engineering from Texas A&M. He majors in energy system such as HVAC and its heat transfer, optimal design and control of refrigerator, reliability design of mechanical components, and failure Analysis of thermal components in marketplace using the Non-destructive such as SEM & XRAY. Especially, he developed parametric accelerated life testing (ALT) as new reliability methodology. If there is design fault in the mechanical system that is subjected to repetitive stress, it will fail in its lifetime. Engineer should find the design faults by parametric ALT before product launches. In 1992–1997 he worked in Agency for Defense Development, Chinhae, South Korea, where he has researcher in charge of Development of Naval weapon System. In 2000-2010 he had been working as a Senior Reliability Engineer in Side-by-Side Refrigerator Division, Digital Appliance, SAMSUNG Electronics, where he focused on enhancing the life of refrigerator as using parametric the accelerating life testing. He is working in faculty of mechanical engineering, Ethiopian Technical University, Addis Ababa, Ethiopia. He has published 50 research articles in the national and international journals.

Şahin Yavuz received the Ph.D. degree in Mechanical Engineering from Dokuz Eylul University, Izmir, TR, in 2016. From 2010 to 2016, he was a Research Assistant and from 2017 to 2018, he was an Assistant Professor at Mechanical Engineering Department, Dokuz Eylul University. Since 2018, he works as an Associated Professor. His current research interests include vibration control, finite element method, flexible systems, unmanned aerial vehicles, robotics, and robot programming.

Pelin Yildirim Taşer is the Assistant Professor Doctor in Computer Engineering Department of İzmir Bakırçay University. She received her B.Sc., M.Sc., and Ph.D. degrees in Computer Engineering from Dokuz Eylül University in 2013, 2015, and 2019 respectively. After her B.Sc. graduation, she worked as a project assistant at the Computer Engineering Department of Dokuz Eylül University for six months. Between 2014 – 2019, she worked as a research assistant at the Software Engineering Department of Manisa Celal Bayar University, Turkey. Her research interests include data mining, machine learning, intelligent systems, and deep learning. She is the author of several book chapters and more than 20 publications. She is also the advisory board member of some journals and a reviewer for major scientific journals. She is an editor of IGI Global Reference Book. She also served on the organization committee of some conferences and scientific activities. She referees for the assessment of the industrial R&D project proposals. Dr. Yildirim Taser has graduation awards, including certificates of high achievement from her M.S. and Ph.D. educations and certificate of publication honor from her Ph.D.

Aydin Tarik Zengin received his electrical and electronics engineering B.S. degree from Ege University, Turkey, in 2007. Later on, received M.E. and Ph.D. degrees from the Department of Computer Science and Electrical Engineering, Kumamoto University, Japan, in 2010 and 2013, respectively. Currently, he is an Asst. Prof. at Istanbul Technical University. His research interests include Autonomous systems and control theory.

Index

K

L

M

N

O

P

R

S

T

U

V

W

www.ingramcontent.com/pod-product-compliance
Lightning Source LLC
LaVergne TN
LVHW081253011026
849162LV00007BA/1605